Tamás Matolcsi

Spacetime Without Reference Frames

New Publication

MINKOWSKI
Institute Press

Tamás Matolcsi
Eötvös Loránd University
Budapest, Hungary

Cover: Figure on page 150

ISBN: 978-1-927763-94-0 (softcover)
ISBN: 978-1-927763-95-7 (ebook)

Minkowski Institute Press
Montreal, Quebec, Canada
http://minkowskiinstitute.org/mip/

For information on all Minkowski Institute Press publications
visit our website at http://minkowskiinstitute.org/mip/books/

Publisher's Preface

This is a new publication of the second revised edition of *Spacetime Without Reference Frames* by Professor Tamás Matolcsi of Eötvös Loránd University, Budapest, Hungary. We thank the *Society for the Unity of Science and Technology*, Budapest for the permission to republish Professor Matolcsi's valuable book.

The first edition was published in 1993 by Akadémiai Kiadó, Budapest.

The second revised edition was published in 2018 by the *Society for the Unity of Science and Technology*, Budapest and was edited by Tamás Fülöp and Áron Szabó.

Minkowski Institute Press

I express my gratitude to Tamás Fülöp and Áron Szabó for their hard work in helping me to accomplish this revised version of my book and to the Society for the Unity of Science and Technology for realizing the edition.

CONTENTS

PART TWO

MATHEMATICAL TOOLS

The star (*) indicates paragraphs or sections which can be skipped at first reading without loss in understanding the essential meaning of the book.

PREFACE

1. Mathematics reached a crisis at the end of the last century when a number of paradoxes came to light. Mathematicians surmounted the difficulties by revealing the origin of the troubles: the obscure notions, the inexact definitions; then the modern mathematical exactness was created and all the earlier notions and results were reappraised. After this great work nowadays mathematics is firmly based upon its exactness.

Theoretical physics—in quantum field theory—reached its own crisis in the last decades. The reason of the troubles is the same. Earlier physics treated common, visible and palpable phenomena, everything used to be obvious. On the other hand, modern physics deals with phenomena of the microworld where nothing is common, nothing is visible, nothing is obvious. Most of the notions applied to the description of phenomena of the microworld are the old ones and in the new framework they are necessarily confused.

It is quite evident that we have to follow a way similar to that followed by mathematicians to create a firm theory based on mathematical exactness; having mathematical exactness as a guiding principle, we must reappraise physics, its most common, most visible and most palpable notions as well. Doing so we can hope we shall be able to overcome the difficulties.

2. According to a new concept, *mathematical physics* should be a *mathematical theory* of the *whole physics,* a mathematical theory based on mathematical exactness, a mathematical theory in which only *mathematically defined notions* appear and in which *all the notions used in physics* are defined in a mathematically exact way.

Since physics is a natural science, its criterion of truth is experiment. As a straightforward consequence, theoretical physics has become a mixture of mathematics and

– **intuitive notions** which are supposed 'obviously known' for everybody, not requiring a deep explanation and therefore do not have a precise definition,

– **tacit assumptions** i.e. seemingly suitable 'natural properties' of the intuitive notions.

Since the education of physicists starts with the classical theories which are left more or less as they were at the beginning of the last century, the acquired style of thinking is the mixture mentioned above and this is applied further on to describe phenomena in regions where nothing is obvious, resulting in confusion, leading to contradictions and apparent paradoxes.

Mathematical exactness means that we formulate all the 'intuitive motions' and 'tacit assumptions' in the language of mathematics starting at the very beginning, with the most natural, most palpable notions. Following this method, we have a good chance of making an important step forward in modern theoretical physics.

At first sight this seems to a physicist like creating unnecessary confusion around obvious things. Such a feeling is quite natural; if one has never driven a car before, the first few occasions are terrible. But after a while it becomes easy and comfortable and much faster than walking on foot; it is worth spending a part of our valuable time on learning to drive.

3. To build up such a form of mathematical physics, we must start with the simplest, most common notions of physics; we cannot start with quantum field theory but we hope that we can end up with it.

The fundamental notion of mathematical physics is that of *models.* Our aim is to construct mathematical models for physical phenomena. The modelling procedure has two sides of equal importance: the mathematical model and the modelled part of physical reality. We shall sharply distinguish between these two sides. Physical reality is *independent* of our mind, it is as it is. A mathematical model *depends* on our mind, it is as it is constructed by us. The confusion of physical reality and its models has led to grave misunderstandings in connection with quantum mechanics.

A mathematical model is constructed as a result of experiments and theoretical considerations; conclusions based on the model are controlled by experiments. The mathematical model is a *mathematical structure* which is expected to reflect some properties of the modelled part of reality. It lies outside the model to answer what and how it reflects and to decide in what sense it is good or bad. To answer these questions, we have to go *beyond the exact framework* of the model.

4. The whole world is an indivisible unity. However, to treat physics, we are forced by our limited biological, mental etc. capacity to divide it into parts in theory.

Today's physics suggests the arrangement of physical phenomena in three groups; the corresponding three entities can be called Spacetime, Matter and Field.

The phenomena of these three entities interact and determine each other mutually. At present it is impossible to give a good description of the complex situation in which everything interacts with everything, which can be illustrated as follows:

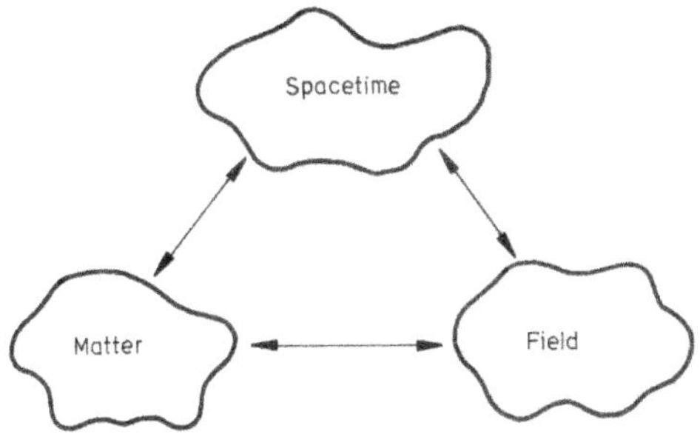

5. Fortunately, a great number of phenomena allows us to neglect some aspects of the interactions. More precisely, we can construct a good theory if we can *replace interaction by action,* i.e. we can work as if the phenomena of two of the entities above were given, fixed, 'stiff' and only the phenomena of the third one were 'flexible', unknown and looked for. The stiff phenomena of the two entities are supposed to act upon and even determine the phenomena of the third one which do not react. We obtain different theories according to the entities considered to be fixed.

Mechanics (classical and quantal), if spacetime and field phenomena are given to determine phenomena of matter, can be depicted as:

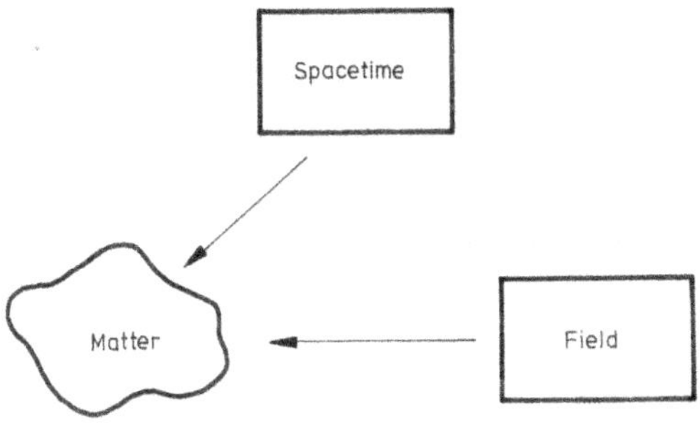

In some sense *continuum physics* and *thermodynamics,* too, are such theories.

Field theory (classical, i.e. electrodynamics), if spacetime and matter phenomena are given to determine phenomena of field, is:

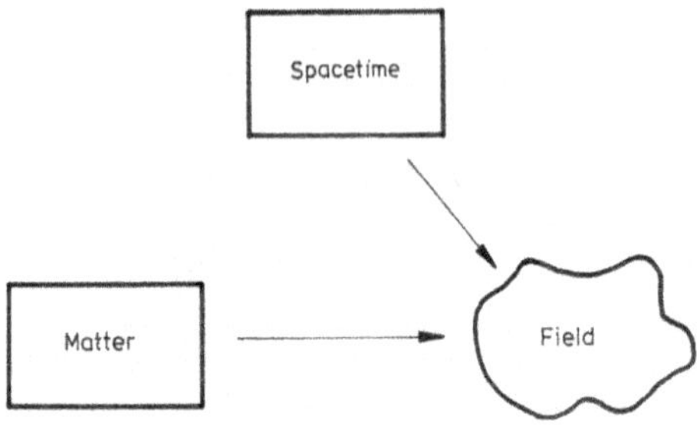

Gravitation theory, if matter and field are given to determine spacetime, is:

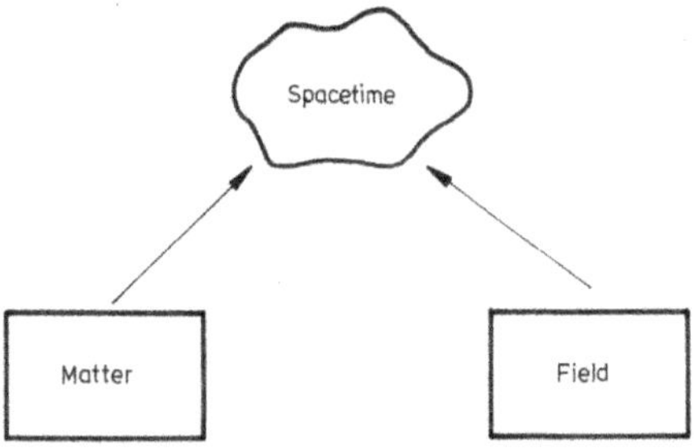

These theories in usual formulation are relatively simple and well applicable to describe a number of phenomena: it is clear, however, that they draw *roughly simplified* pictures of the actually existing physical world.

6. Difficulties arise when we want to describe complicated situations in which only one of the three entities can be regarded as known and interactions occur among the phenomena of the other two entities. The following graphically delineated possibilities exist:

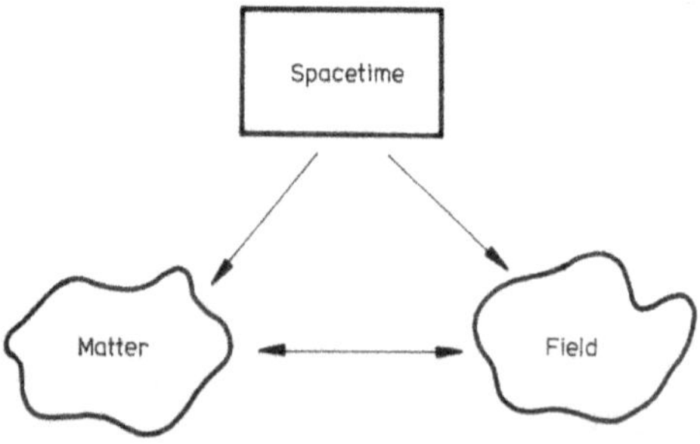

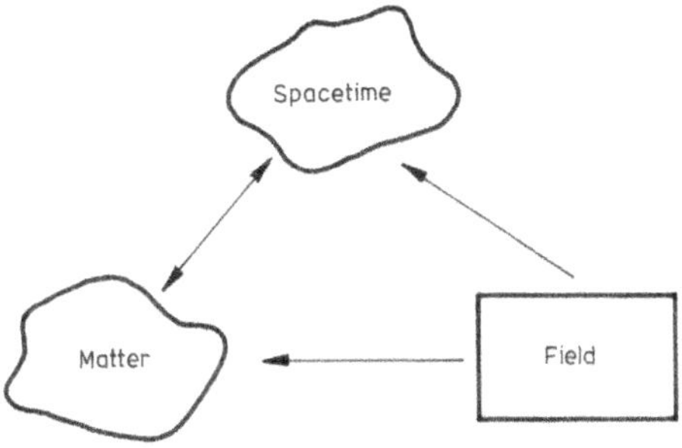

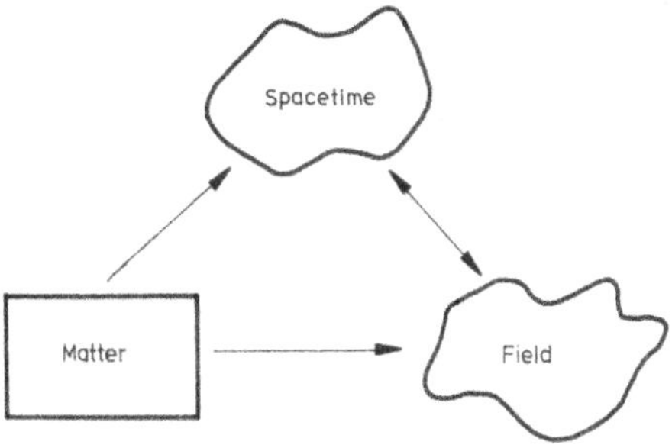

The third one is of no physical interest, so far. However, the other two are very important and we are forced to deal with them. They represent qualitatively new problems and they cannot be reduced to the previous well-known theories, except some special cases treated in the next item.

Electromagnetic radiation of microparticles is, for instance, a phenomenon, which needs such a theory. Usual quantum electrodynamics serves as a theory for its description, and in general usual quantum field theory is destined to describe the interaction of field and matter in a given spacetime.

As it is well known, usual quantum field theory has failed to be completely correct and satisfactory. One might suspect the reason of the failure is that usual quantum field theory was created in such a way that the notions and formulae of *mechanics* were mixed with those of *field theory*. This way leads to nowhere: in mechanics the field phenomena are fixed, in field theory the matter phenomena are fixed; the corresponding notions 'stiff' on one side cannot be fused correctly to produce notions 'flexible' on both sides.

The complicated mathematics of quantum field theory does not allow us to present a simple example to illustrate what has been said, whereas classical electrodynamics offers an excellent example. The electromagnetic field of a point charge moving on a *prescribed* path is obtained by the Liénard–Wiechert potential which allows us to calculate the force due to electromagnetic radiation acting upon the charge. Then the Newtonian equation is supplemented with this radiation reacting force—which is deduced for a point charge *moving on a given path*—to get the so-called Lorentz–Dirac equation for *giving the motion* of a point charge in an electromagnetic field. No wonder, the result is the nonsense of 'runaway solutions'.

Electromagnetic radiation is an irreversible process; in fact every process in Nature is irreversible. *The description of interactions must reflect irreversibility.* Mechanics (Newtonian equation, Schrödinger equation) and electrodynamics (Maxwell equations) i.e. the theories dealing with action instead of interaction do not know irreversibility. Evidently, no amalgamation of these theories can describe interaction and irreversibility.

7. There is a special case in which interaction can be reduced to some combination of actions yielding a good approximation. Assume that matter phenomena can be divided into two parts, a 'big' one and a 'small' one. The big one and field (or spacetime) are considered to be given and supposed to produce spacetime (or field) which in turn acts on the small matter phenomena to determine them. The situations can be illustrated graphically as follows:

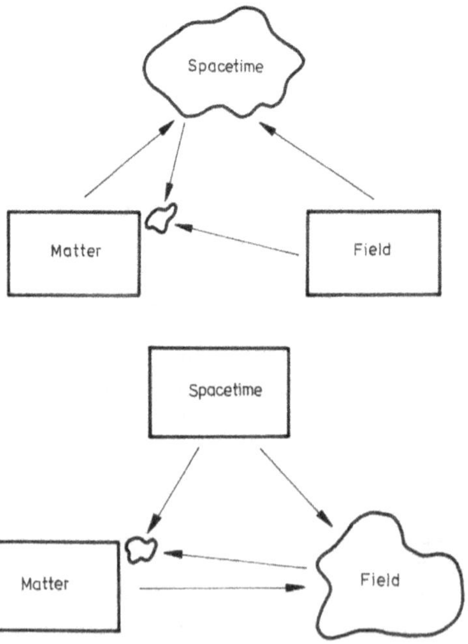

An example for the application of this trick is the description of planetary motion in general relativity, more closely, the advance of the perihelion of Mercury. The field is supposed to be absent, the big Sun produces spacetime and this spacetime determines the motion of the small Mercury. Doing so we neglect that spacetime is influenced by Mercury and the motion of the Sun is influenced by spacetime as well, i.e. we neglect interaction.

The second example is similar. Suppose we want to determine how a light charged particle moves in the field generated by a heavy charged particle. We assume that spacetime and the heavy point charge are given and that they produce an electromagnetic field and that this electromagnetic field determines the motion of a light point charge. Doing so we neglect that the electromagnetic field is influenced by the light point charge and the motion of the heavy point charge is influenced by the electromagnetic field, i.e. we neglect interaction.

PART ONE

SPACETIME MODELS

INTRODUCTION

1. The principles of covariance and of relativity

1.1. Today the guiding principle for finding appropriate laws of Nature is the principle of general relativity: any kind of coordinates (reference frame) should finally conclude the same laws of Nature; the laws are independent of the way we look at them. The usual mathematical method of applying this principle is the following: in a certain coordinate system (reference frame) we have an equation that, as we suspect, expresses some law independent of the coordinates (reference frame). The way to check this is referred to as the principle of covariance: transfer the equation into another coordinate system (reference frame) with an appropriate transformation (Galilean, Lorentzian, or a general coordinate transformation), and if the form of the equation remains the same after this procedure, then it can be a law of some phenomenon. This method can be illustrated in the following way:

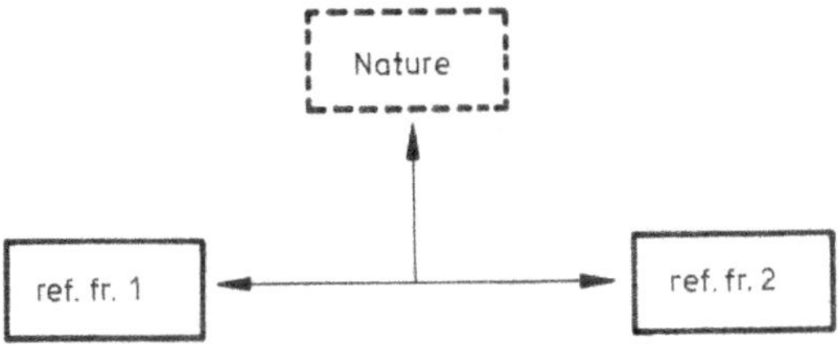

It seems quite natural to organize the procedure in such a way; this is how Galileo and Newton started it and this is how Einstein finally concluded the principle of general relativity. What could be the next step? Very simple: since the laws of Nature are the same for all coordinate systems (reference frames), the theoretical description does not need coordinates (reference frame) any longer; there should exist a way of describing Nature without coordinate systems (reference frames). In fact, at that time Einstein said this in another way: "the description of Nature should be coordinate-free".

This was some 100 years ago but if we take a glance at some books on theoretical physics today, we stumble upon an enormous amount of indices; thinking starts with coordinate systems (reference frames) and remains there; the program of coordinate-free description has not yet been accomplished.

1.2 Getting rid of coordinate systems (reference frames) in establishing mathematical models in physics, we can reorganize the method of description using *absolute objects*. Of course, if we wish to test our theory by experiments, we have to convert absolute quantities into relative ones corresponding to reference frames and then to turn them into numbers by introducing coordinates. Compared with the previous situation, this can be illustrated as follows:

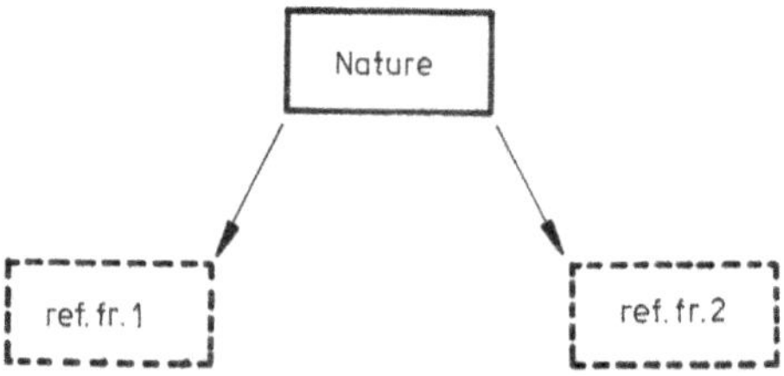

1.3. The most important result of the present book is this reorganizing of the whole method of theoretical description. In this framework the principle of covariance and the principle of relativity sound very simple (encouraging us that this might be the right way).

Principle of covariance: *according to our present knowledge, the description of Nature should be done by first choosing one of the nonrelativistic*, special relativistic and general relativistic spacetime models and then using the tools of the chosen model.*

Principle of relativity: *there must be a rule in the spacetime models that says how an arbitrary reference frame derives its own quantities from the absolute ones describing the phenomena.*

2. Measure lines

2.1. In practice, the magnitudes of a physical quantity (observable) are usually related to some unit of measurement, i.e., to a chosen and fixed value.

---

* *also called Galilean*

We decide, for instance, which distance is called *meter* and then we express all distances as non-negative multiples of meter.

In general the following can be said. Let A be the set of the magnitudes of an observable. Taking an arbitrary element a of A and a non-negative real number α, we can establish which element of A is α times a, denoted by αa. In other words, we give a mapping, called *multiplication by non-negative numbers,*

$$\mathbb{R}_0^+ \times \mathrm{A} \to \mathrm{A}, \quad (\alpha, a) \to \alpha a$$

with the following properties: for all $a \in \mathrm{A}$

(i) $0a$ is the same element, called the *zero* of A and is denoted by 0 as well;

(ii) $1a = a$

(iii) $\beta(\alpha a) = (\beta\alpha)a$ for all $\alpha, \beta \in \mathbb{R}_0^+$ and $a \in \mathrm{A}$;

(iv) if $a \neq 0$ then $J_a : \mathbb{R}_0^+ \to \mathrm{A}, \ \alpha \mapsto \alpha a$ is bijective.

In customary language we can say that A is a one-dimensional cone.

An addition can be defined on this one-dimensional cone. It is easy to see that the mapping, called *addition,*

$$\mathrm{A} \times \mathrm{A} \to \mathrm{A}, \quad (b, c) \mapsto J_a\left({J_a}^{-1}(b) + {J_a}^{-1}(c)\right) =: b + c$$

is independent of a.

Let us introduce the notations

$$-\mathrm{A} := \{\, (-1, a) \mid 0 \neq a \in \mathrm{A}\}, \qquad \mathbb{A} := (-\mathrm{A}) \cup \mathrm{A}.$$

Then we can give a multiplication by real numbers

$$\mathbb{R} \times \mathbb{A} \to \mathbb{A}, \qquad (\alpha, \boldsymbol{a}) \mapsto \alpha\boldsymbol{a}$$

and an addition

$$\mathbb{A} \times \mathbb{A} \to \mathbb{A}, \qquad (\boldsymbol{b}, \boldsymbol{c}) \mapsto \boldsymbol{b} + \boldsymbol{c}$$

that are trivial extensions of the operations given on A, so that $\mathbb{A}$ becomes a one-dimensional real vector space. For instance,

$$\begin{aligned} \alpha a &:= -|\alpha| a \quad \text{for} \quad \alpha < 0, \quad a \in \mathrm{A}, \\ \alpha(-a) &:= -\alpha a \quad \text{for} \quad \alpha > 0, \quad a \in \mathrm{A}, \\ \alpha(-a) &:= |\alpha| a \quad \text{for} \quad \alpha < 0, \quad a \in \mathrm{A}. \end{aligned}$$

Furthermore, the two 'halves' of this vector space have different importance: the original cone contains the physically meaningful elements. We express this fact mathematically by orienting $\mathbb{A}$ with the elements of A (see IV.5).

The preceding construction works e.g. for distance, mass, force magnitude, etc. In some cases—e.g. for electric charge—we are given originally a one-dimensional real vector space of observable values.

Thus we accept that the magnitudes of observables are represented by elements of oriented one-dimensional real vector spaces called *measure lines.* Choosing a unit of measurement means that we pick a positive element of the measure line.

2.2. In practice some units of measurement are deduced from other ones by multiplication and division; for instance, if $\boldsymbol{kg}$, $\boldsymbol{m}$ and $\boldsymbol{s}$ are units of mass, distance and time period, respectively, then $\frac{\boldsymbol{kg}\,\boldsymbol{m}}{\boldsymbol{s}^2}$ is the unit of force. The question arises immediately: how can we give a mathematically exact meaning to such a symbol? According to what has been said, $\boldsymbol{kg}$, $\boldsymbol{m}$ and $\boldsymbol{s}$ are elements of one-dimensional vector spaces; how can we take their product and quotient? To give an answer let us list the rules associated usually with these operations; for instance,

$$(\alpha\boldsymbol{kg})(\beta\boldsymbol{m}) = (\alpha\beta)(\boldsymbol{kg}\,\boldsymbol{m}) \qquad (\alpha,\beta \in \mathbb{R}_0^+),$$
$$\frac{\alpha\boldsymbol{m}}{\beta\boldsymbol{s}} = \frac{\alpha}{\beta}\frac{\boldsymbol{m}}{\boldsymbol{s}} \qquad (\alpha \in \mathbb{R}_0^+, \beta \in \mathbb{R}^+).$$

Extending these rules to negative numbers, too, we see that the usual multiplication is a bilinear map on the measure lines and the usual division is a linear-quotient map, with the additional property that the product and quotient of nonzero elements are not zero.

Consequently, we can state that the product and quotient of units of measurements are to be defined by their tensor product and tensor quotient, respectively (see IV.3 and IV.4).

Thus if $\mathbb{L}$, $\mathbb{T}$ and $\mathbb{M}$ denote the measure line of distance, time period and mass, respectively, $\boldsymbol{m} \in \mathbb{L}$, $\boldsymbol{s} \in \mathbb{T}$, $\boldsymbol{kg} \in \mathbb{M}$, then $\frac{\boldsymbol{kg}\,\boldsymbol{m}}{\boldsymbol{s}^2} := \frac{\boldsymbol{kg}\otimes\boldsymbol{m}}{\boldsymbol{s}\otimes\boldsymbol{s}} \in \frac{\mathbb{M}\otimes\mathbb{L}}{\mathbb{T}\otimes\mathbb{T}}$.

3. Spacetime heuristics

3.1. Space

3.1.1. Sitting in a room, we conceive that a corner of the room, a spot on the carpet are points and the table is a part of our space. Looking through the window we see trees, chimneys, hills that form other parts of our space. A car travelling on the road is not a part of this space.

On the other hand, the seats, the dashboard, etc. constitute a space for someone sitting in the car. Looking out he sees that the trees, the houses, the hills are running, they are not parts of the space corresponding to the car.

Consequently, the space for us in the room and the space for the one in the car are different. We have ascertained that space itself does not exist, i.e. *there is no absolute space*, there are only spaces relative to material objects. *A space is constituted by material objects.*

3.1.2. Let us call the room (more widely, the Earth) and the car *observers*. Then we can say that every observer has its own space, different observers have different spaces. The space points of observers are material points.

We warn the reader that the term observer appears in the literature in different senses. Frequently one considers a single point as an observer. Making experiments, however, a single point is not sufficient; for instance, a cloud chamber, showing the path of an elementary particle in no manner can be considered as a single point. So we emphasize that in our treatment an observer is a collection of material points.

We find the following properties of our space:

(**S1**) There are *straight lines* in it and an oriented straight line segment, a *vector* can be drawn between any two of our space points; the vectors obey the well known rules of addition and multiplication by real numbers.

(**S2**) Our space is *three-dimensional*: there are three essentially different directions – right-left, forward-backward, up-down – from which every other direction can be 'combined'.

(**S3**) It is not trivial – but experiments show (the asymmetry of K mezon decay) – that our space is *oriented*: the order right-forward-up and the order left-backward-down are not equivalent.

(**S4**) The *distance* between two space points – in other words, the magnitude of a vector – is meaningful and the distances obey the triangle inequality; moreover, we experience an *angle* between two vectors.

The listed properties can be summarized by stating that *our space is a three dimensional oriented Euclidean affine space*. Of course, this is valid for the space of a 'good observer', e.g. for an *inertial observer* whose space points are inertial i.e. free of any action and do not move with respect to each other (the space of a room having rubber walls and shaken by an earthquake is not such).

3.2. Time

3.2.1. Time is much more sophisticated than space. Processes indicate that time progresses: the Sun proceeds in the sky, someone is speaking, a clock is ticking etc. *Time, too, is constituted by material objects.*

In everyday conversations and in the usual terminology of physics, too, unfortunately, the word time can refer to

– a *time period*: 'long time ago', 'it will take time';

– a *time point (instant)*: 'what is the time?' 'at the same time'.

It is extremely important to distinguish between these two essentially different notions. Note that a clock *measures time periods* by the number of its tickings and does not measure but *indicates time points*.

3.2.2. What is a time period exactly? I experience the time period between my breakfast and my dinner, an elementary particle experiences its own life time period between its birth and decay etc. We can accept as a fundamental fact that time progresses for every material point. Visualizing this, we imagine that a tiny *chronometer* (a quartz crystal) is tied to any material point and the ticks (oscillations) measure the time period between arbitrary two occurrences of the material point. *Time progresses for every material point individually*, i.e. every material point has its *proper time.*

It may happen – why not? – that different proper time periods pass for two material points between two of their meetings. Indeed, experiments with elementary particles prove that this is the situation.

We find that time progresses in one direction only and the notion earlier-later is meaningful for the occurrences of a material point: past (my breakfast) and future (my dinner) cannot be interchanged. Summarizing, we find:

(**T**) Proper time of any material point is a physical reality, its progress has a one-dimensional character and it is oriented.

3.2.3. Now let us examine what a time point is. It can be best grasped by considering the notion 'at the same time', i.e. simultaneity. How can we determine that an explosion in London and an other one in Paris occurred at the same time, i.e. simultaneously? Saying that both occurred at two o'clock puts off the answer because the question remains: how is two o'clock is established in London and in Paris?

Of course, there are some conventions to establish simultaneity in different space points on the Earth (using the position of the Sun or using radio signals). A *synchronization* settles simultaneity again and again continuously and it creates *time points* as simultaneous occurrences. A synchronization must have the fundamental property that different occurrences of any material points cannot be simultaneous.

It is evident that *a synchronization is a human invention, not a physical fact.*

Let us consider two space points L and P of an observer (e.g. London and Paris) and a time point t of a synchronization. Then an occurrence l_t in L and an occurrence p_t in P are simultaneous with respect to the synchronization in question. Take another time point s of the synchronization, too.

As said, time progresses for every material point, in particular, for every space point of an observer. It may happen – why not? – that the time passed in L between l_t and l_s is not equal to the time passed in P between p_t and p_s.

3.3. Motions

3.3.1. We experience material objects moving; in general, a material point can move with respect to observers. The same material object moves differently with respect to different observers.

Velocity is an everyday notion connected with a motion. It makes sense, however, only with the use of a synchronization. Indeed, a motion is described by a function that assigns space points to time points and velocity is derived from this function.

Now we shall state some fundamental facts concerning motions without using synchronizations.

The *path of a motion in the space of an observer* is the collection of space points that the material point meets with.

Our first experience is:

(**M1**) Every straight line in an observer space can be the path of a motion.

As said, a synchronization is necessary to determine how fast a motion is. However, we can declare the following important statement: *it makes sense without synchronization which one is faster of two motions having the same path relative to an observer, and meeting in some space point.*

Indeed, let us consider a foot race between a valley and a hill. The racers leave the start together (their departure is the same occurrence at the start) but they arrive at the goal separately (their arrivals are different occurrences at the goal). The notion earlier-later is a physical fact at a space point, in particular at the goal. The racer who arrives earlier is faster, who arrives later is slower. Thus, we know which of the racers is faster or slower without knowing how fast or slow they are (what their speeds are) according to a synchronization.

We emphasize that such a faster-slower relation is independent of synchronizations but does depend on observers. A (long-long) car travelling along the racing track observes that the winner runs backward slower than the others.

Then we can formulate our experience:

(**M2**) For every motion there is a faster motion on the same path.

(**M3**) For every motion all the possible slower ones can be realized on the same path.

The first assertion is evident. The second means that given an arbitrary (long) time interval $\boldsymbol{t}$, then there is a racer (e.g. a snail) which arrives at the goal $\boldsymbol{t}$ time later than the winner.

3.3.2. Newton's first law concerns uniform motions on straight lines. It can be well formulated as follows: every inertial observer can choose at least one synchronization in such a way that the motion of an arbitrary inertial material point has constant velocity (which implies that the path of the motion is a straight line).

Note that this formulation does not involve the explicit use of a synchronization; it merely declares the possibility of some synchronizations. Nevertheless, we look for another statement.

Namely, let the running time be measured by the material point and the covered distance by the observer. This can be done as follows: after each determined amount of ticks of its chronometer, the material point marks the space point where it is and the observer measures the distance between these marked space points. Then we can state that "an inertial material point moves (if does not stand) in the space of any inertial observer in such a way that its path is a straight line and it covers equal distances during equal proper time periods".

From that we conclude:

(**U**) Inertial proper times and inertial space distances are uniformly related to each other.

3.3.3. There is a further, extremely important notion in connection with motions.

Let us consider a race where the start and the goal coincide. Then the time period between the starting and the arrival of a racer is measured by the chronometer of the start-goal, hence we can say without any synchronization how fast a racer is.

In general, *the round-way speed* is meaningful, without any synchronization, for a material point moving on a closed path in an observer's space.

3.4. Spacetime

3.4.1. We will construct models for spacetime on the base of our experience concerning physical facts. Observers, space points, space vectors as well as proper time periods are physical entities but synchronizations and time points are not. Therefore synchronizations and the usual notion of velocity have no place in our following arguments.

The model must be formulated in terms of *absolute objects.* The physical facts listed above, in general, are not absolute, they are related to material objects, differently to different material objects.

Of course, all of our experience is relative, too. We have to find the absolute objects behind them.

A child claps in the car; this occurs sometime in some space point of the car as well as sometime in some space point of the Earth. The clapping is a physical fact characterized by 'here and now' or 'there and then' according to different points of view. Similarly, an explosion, a flash of a lamp, a collision of two vehicles happen in reality; such possible (pointlike) *occurrences* are conceived to be spacetime points.

One often says event instead of occurrence which causes a number of misunderstandings. Namely, the notion of an event is well defined in probability theory and in physics as well. An event always happens to the object in question: the clap is an event of the child, the explosion is an event of the bomb etc. they are not events of spacetime. That is why we adhere to say occurrence.

Properties (**S1**)-(**S3**), (**T**) and (**U**) suggest us first of all that *spacetime is a four-dimensional oriented affine space.*

3.4.2. Of course, we have to pose the uneasy questions: have we reasoned properly? have we not made some mistakes? have we not left anything out of consideration?

There is a serious objection to our reasoning: *we have extrapolated our experience gained in human size to much larger and much smaller size, too.*

The affine structure includes a concept of continuity. According to our common experience, i.e. from human point of view, water is a continuous material. However, we already know that water is in fact rather coarse: a microbe does not perceive it to be continuous at all. Are perhaps space and time coarse as well? At present no experimental fact supports the coarse nature of space and time but we cannot exclude it in good faith.

Let us accept the continuity of space and time. Our conviction that a vector can be associated with two space points is based on the fact that e.g. we can span a thread between the corner of the room and a spot on the carpet, or we can produce a light beam between them. But how can we determine the vector between two points whose distance is much smaller than the diameter of the thread or the light beam? If we can define vectors for such near points, too, do they obey the customary rules of addition and multiplication by real numbers?

We meet a similar problem if we want to give sense of vectors corresponding to points very far from each other. A thread cannot but a light beam can draw a straight line between Earth and Moon; however, it is not evident at all that addition and multiplication by real numbers of such huge vectors make sense with the customary properties.

Indeed, some experiments show that in astronomical size the vectorial operations cannot be defined for segments defined by light beams. At present we have no similar knowledge regarding minute size.

Evidently, the same problems arise for small and large time periods.

3.4.3. The objections above do not matter much. By accepting our experience regarding human size as *global*, i.e. extrapolating it to very small and large size, too, we have made some abstractions to create mathematical models. Such a model is not reality itself; it is an image—a necessarily simplified and distorted one—of reality. *Reality and model should not be confused!*

If we admit that our experience is only *local,* i.e. considering it approximately true even in human size, we have to give up the affine structure and we make models in which *spacetime is a four-dimensional manifold.*

This book deals with models in which spacetime is accepted to has an affine structure: the nonrelativistic model and the special relativistic model.

General relativistic spacetime models are based on differential manifolds.

The nonrelativistic spacetime model is suitable for the description of 'sluggish' mechanical phenomena—when bodies move relative to each other with velocities much smaller than light speed—and of static electromagnetic phenomena.

The special relativistic spacetime model is suitable for the description of all mechanical and electromagnetic phenomena, but it has a more complicated structure than the nonrelativistic one, therefore it is suitable for 'brisk' mechanical phenomena and nonstatic electromagnetic phenomena.

To describe cosmic phenomena we have to adopt general relativistic spacetime models.

3.4.4. Up to now we employed statments (**S1**)-(**S3**), (**T**) and (**U**) to accept the affine structure of spacetime. It is noteworthy that (**U**) indicates also that *the proper time progress and the Euclidean structure of inertial observers are connected* somehow. Further properties of spacetime models will be obtained later from (**M1**)-(**M3**) and from further special assumptions.

3.4.5. Of course, having a spacetime model as a mathematical structure, all the intuitive notions appeared in our heuristics such as observers, proper times etc. must be exactly defined in that mathematical framework. This holds for synchronizations, too, which are important from a technical (not fundamental) point of view.

A synchronization and an observer together will be called a *reference frame.*

I. NONRELATIVISTIC SPACETIME MODEL

1. Fundamentals

1.1. Absolute time progress

1.1.1. As mentioned earlier, experiments show that different proper time periods can pass for elementary particles between their two meetings. On the other hand, according to our (superficial) everyday experience, both the clock at home and my watch have ticked just as many times between my departure and return. Now we construct a spacetime model based on this assumption:

(**A1**) Proper time progress is absolute, i.e. the same time period elapses for any two material points between two meetings.

According to another (superficial) everyday experience there is no upper limit for the velocity of material objects; this can be rightly formulated as follows:

(**A2**) In the space of any inertial observer arbitrary round-way speed is possible.

1.1.2 According to what has been said in the Introduction, spacetime will be modelled by an affine space.

Then it can be shown – the proof lies outside the purpose of this book – that (**M1**)–(**M3**) and (**A1**)–(**A2**) imply that the absolute proper time progress is described by a nonzero linear map $\boldsymbol{\tau}$ defined on the underlying vector space of spacetime.

Moreover, (**U**) and (**S4**) result in that there is a Euclidean structure $\boldsymbol{h}$ on the kernel of $\boldsymbol{\tau}$.

1.2. The spacetime model

1.2.1. Now we are ready to formulate a correct definition.

Definition. A *nonrelativistic spacetime model* is $(\mathrm{M}, \mathbb{T}, \mathbb{L}, \boldsymbol{\tau}, \boldsymbol{h})$ where

— M is spacetime, an oriented four-dimensional real affine space (over the vector space $\mathbf{M}$),
— $\mathbb{T}$ is the measure line of time periods,
— $\mathbb{L}$ is the measure line of distances,
— $\boldsymbol{\tau}: \mathbf{M} \to \mathbb{T}$ describes the absolute proper time progress, a linear surjection,
— $\boldsymbol{h}: \mathbf{S} \times \mathbf{S} \to \mathbb{L} \otimes \mathbb{L}$ is the absolute Euclidean structure, a positive definite symmetric bilinear map where $\mathbf{S} := \operatorname{Ker} \boldsymbol{\tau} = \{\boldsymbol{x} \in \mathbf{M} \mid \boldsymbol{\tau} \cdot \boldsymbol{x} = \mathbf{0}\}$. ■

The elements of M and of $\mathbf{M}$ are called world points and world vectors, respectively.

For occurrences x and y (elements of M) $\boldsymbol{\tau} \cdot (y - x)$ is *the absolute time elapsed between x and y*. x is *earlier* than y and *later* than y if $\boldsymbol{\tau} \cdot (y - x) > 0$ and $\boldsymbol{\tau} \cdot (y - x) < 0$, respectively. (Recall that $\mathbb{T}$ is oriented, thus it makes sense to speak about its positive and negative elements, see IV.5.3.)

x and y are *absolutely simultaneous* if $\boldsymbol{\tau} \cdot (y - x) = 0$.

Elements of $\mathbf{S}$ are called *absolutely spacelike.*

World vectors outside $\mathbf{S}$ are called *timelike.* The set of timelike elements consists of two disjoint open subsets:

$$T^{\to} := \{\boldsymbol{x} \in \mathbf{M} \mid \boldsymbol{\tau} \cdot \boldsymbol{x} > 0\}, \qquad T^{\leftarrow} := \{\boldsymbol{x} \in \mathbf{M} \mid \boldsymbol{\tau} \cdot \boldsymbol{x} < 0\}.$$

Vectors in $T^{\to}$ and in $T^{\leftarrow}$ are called *future directed* and *past directed,* respectively.

We often illustrate the world vectors in the plane of the page:

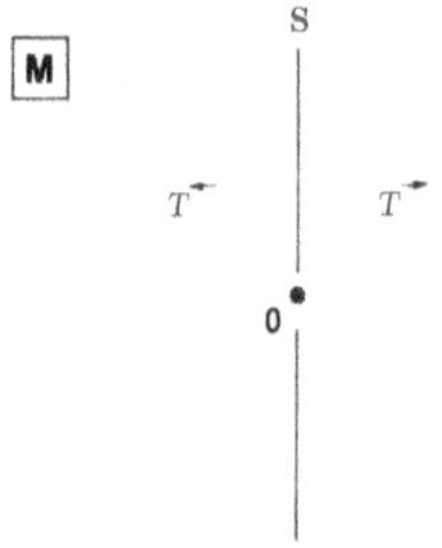

1.2.2. In this spacetime model there is a single synchronization (see later in 3.3.), given by absolute simultaneity. x and y are simultaneous if and only if $y - x \in \mathbf{S}$ which is equivalent to $y \in x + \mathbf{S}$ or $x \in y + \mathbf{S}$. Thus, a set of simultaneous world points is an affine hyperplane of M, directed by $\mathbf{S}$.

To be simultaneous is an equivalence relation on M. An equivalence class, a hyperplane directed by $\mathbf{S}$ is conceived to be an *absolute time point* or *instant;*

the set of equivalence classes is $\mathrm{T} := \mathrm{M}/\mathbf{S}$, the *absolute time.* We define the subtraction $\mathrm{T} \times \mathrm{T} \to \mathbb{T}$ by

$$t - s := \boldsymbol{\tau} \cdot (y - x) \qquad (y \in t,\ x \in s),$$

in another form,

$$(y + \mathbf{S}) - (x + \mathbf{S}) := \boldsymbol{\tau} \cdot (y - x).$$

It is not hard to see that this subtraction is well-defined and turns T into an affine space over $\mathbb{T}$. Then the *time evaluation*

$$\tau \colon \mathrm{M} \to \mathrm{T}, \qquad x \mapsto x + \mathbf{S}$$

becomes an affine map over $\boldsymbol{\tau}$.

We say that $t \in \mathrm{T}$ is *later* than $s \in \mathrm{T}$ (or s is *earlier* than t) and we write $s < t$ if $t - s$ is a positive element of $\mathbb{T}$.

Spacetime, too, will be illustrated in the plane of the page. Then vertical lines stand for the instants (hyperplanes of simultaneous world points). A line standing to the right of another is taken to be later.

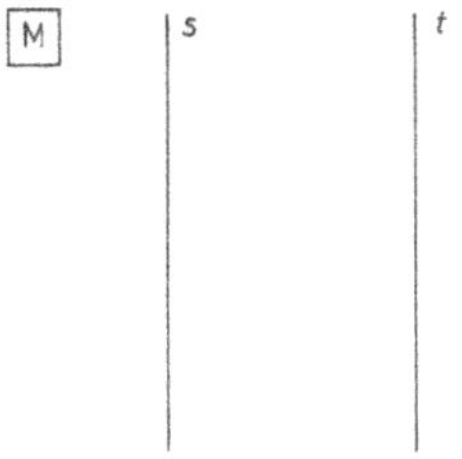

If x is a world point, $x + T^{\rightarrow}$ and $x + T^{\leftarrow}$ are called the *futurelike* and *pastlike part* of M, with respect to x.

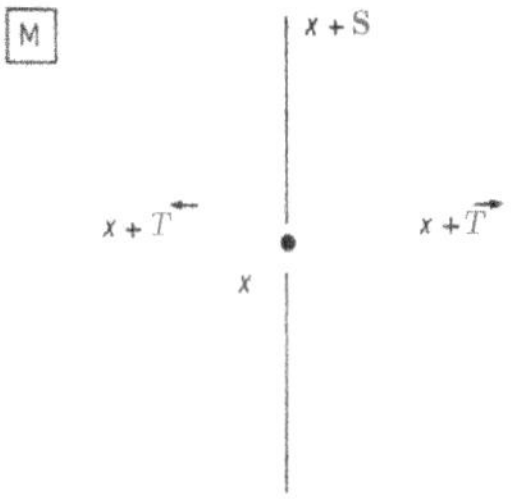

1.3. Structure of world vectors and covectors

1.3.1. There are spacetime and time in our nonrelativistic spacetime model and there is no space. However, there is something spacelike: the linear subspace $\mathbf{S}$ of $\mathbf{M}$. Later we see what the spacelike feature of $\mathbf{S}$ consists in. We find an important 'complementary' connection between $\mathbf{S}$ and $\mathbb{T}$. Let

$$\boldsymbol{i}: \mathbf{S} \to \mathbf{M}$$

denote the canonical injection (embedding; if $\boldsymbol{q} \in \mathbf{S}$, then $\boldsymbol{i}\cdot\boldsymbol{q}$ equals $\boldsymbol{q}$ regarded as an element of $\mathbf{M}$; evidently, $\boldsymbol{i}$ is linear). Then we can draw the diagram

$$\mathbf{S} \xrightarrow{\boldsymbol{i}} \mathbf{M} \xrightarrow{\boldsymbol{\tau}} \mathbb{T} \ ;$$

$\boldsymbol{i}$ is injective, $\boldsymbol{\tau}$ is surjective, and $\operatorname{Ran}\boldsymbol{i} = \operatorname{Ker}\boldsymbol{\tau}$, thus $\boldsymbol{\tau}\cdot\boldsymbol{i} = \mathbf{0}$.

$\mathbf{M}^*$, the dual of $\mathbf{M}$ will play an important role. Though it is also a four-dimensional oriented vector space, there is no canonical isomorphism between $\mathbf{M}$ and $\mathbf{M}^*$; these vector spaces are different.

A diagram similar to the previous one is drawn for the transposed maps:

$$\mathbb{T}^* \xrightarrow{\boldsymbol{\tau}^*} \mathbf{M}^* \xrightarrow{\boldsymbol{i}^*} \mathbf{S}^* \ .$$

$\boldsymbol{\tau}^*$ is injective, $\boldsymbol{i}^*$ is surjective (see IV.1.4) and $\operatorname{Ran}\boldsymbol{\tau}^* = \operatorname{Ker}\boldsymbol{i}^*$, thus $\boldsymbol{i}^*\cdot\boldsymbol{\tau}^* = \mathbf{0}$.

It is worth mentioning that for $\boldsymbol{k} \in \mathbf{M}^*$, $\boldsymbol{i}^*\cdot\boldsymbol{k} = \boldsymbol{k}\cdot\boldsymbol{i}$ is the restriction of $\boldsymbol{k}$ onto $\mathbf{S}$: $\boldsymbol{k}\cdot\boldsymbol{i} = \boldsymbol{k}|_{\mathbf{S}}$.

1.3.2. Since $\boldsymbol{\tau}^*$ is injective, its range is a one-dimensional linear subspace of $\mathbf{M}^*$ which will play an important role:

$$\operatorname{Ran}\boldsymbol{\tau}^* = \{\boldsymbol{\tau}^*\cdot\boldsymbol{e} \mid \boldsymbol{e} \in \mathbb{T}^*\} = \{\boldsymbol{e}\cdot\boldsymbol{\tau} \mid \boldsymbol{e} \in \mathbb{T}^*\} = \mathbb{T}^*\cdot\boldsymbol{\tau}.$$

Observe that $\boldsymbol{k} \in \mathbf{M}^*$ is in $\mathbb{T}^*\cdot\boldsymbol{\tau}$ if and only if $\boldsymbol{i}^*\cdot\boldsymbol{k} = \boldsymbol{k}\cdot\boldsymbol{i} = \mathbf{0}$, thus

$$\mathbb{T}^*\cdot\boldsymbol{\tau} = \{\boldsymbol{k} \in \mathbf{M}^* \mid \boldsymbol{k}\cdot\boldsymbol{q} = 0 \quad \text{for all} \quad \boldsymbol{q} \in \mathbf{S}\}.$$

We say that $\mathbb{T}^*\cdot\boldsymbol{\tau}$ is *the annihilator* of $\mathbf{S}$.

Illustrating $\mathbf{M}^*$ on the plane of the page, we draw a horizontal line for the one-dimensional linear subspace $\mathbb{T}^*\cdot\boldsymbol{\tau}$.

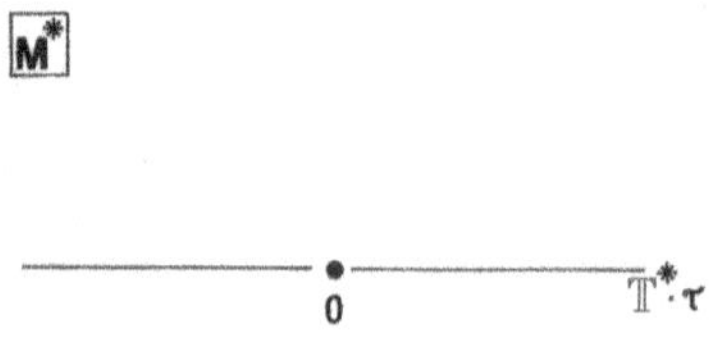

As usual, the elements of $\mathbf{M}^*$ are called *covectors.* The covectors in the linear subspace $\mathbb{T}^* \cdot \boldsymbol{\tau}$ are *timelike,* and the other ones are *spacelike.*

1.3.3. It will be often convenient to use tensorial forms of the above linear maps. According to IV.3.4 and IV.1.2 we have

$$\boldsymbol{\tau} \in \mathbb{T} \otimes \mathbf{M}^*, \qquad \boldsymbol{i} \in \mathbf{M} \otimes \mathbf{S}^*,$$
$$\boldsymbol{\tau}^* \in \mathbf{M}^* \otimes \mathbb{T}, \qquad \boldsymbol{i}^* \in \mathbf{S}^* \otimes \mathbf{M}.$$

1.3.4. With the aid of $\boldsymbol{\tau}$, the orientations of $\mathbf{M}$ and $\mathbb{T}$ determine a unique orientation of $\mathbf{S}$.

Proposition. If $(\boldsymbol{e}_1, \boldsymbol{e}_2, \boldsymbol{e}_3)$ is an ordered basis of $\mathbf{S}$, then $(\boldsymbol{x}, \boldsymbol{e}_1, \boldsymbol{e}_2, \boldsymbol{e}_3)$ and $(\boldsymbol{y}, \boldsymbol{e}_1, \boldsymbol{e}_2, \boldsymbol{e}_3)$ are equally oriented for all $\boldsymbol{x}, \boldsymbol{y} \in T^{\rightarrow}$.

Proof. Evidently, $(\boldsymbol{x}, \boldsymbol{e}_1, \boldsymbol{e}_2, \boldsymbol{e}_3)$ and $(\alpha\boldsymbol{x}, \boldsymbol{e}_1, \boldsymbol{e}_2, \boldsymbol{e}_3)$ are equally oriented if $\alpha \in \mathbb{R}^+$, hence we can suppose that $\boldsymbol{\tau} \cdot \boldsymbol{y} = \boldsymbol{\tau} \cdot \boldsymbol{x}$, i.e. $\boldsymbol{q} := \boldsymbol{y} - \boldsymbol{x} \in \mathbf{S}$. Then

$$\boldsymbol{y} \wedge \boldsymbol{e}_1 \wedge \boldsymbol{e}_2 \wedge \boldsymbol{e}_3 = (\boldsymbol{x} + \boldsymbol{q}) \wedge \boldsymbol{e}_1 \wedge \boldsymbol{e}_2 \wedge \boldsymbol{e}_3 = \boldsymbol{x} \wedge \boldsymbol{e}_1 \wedge \boldsymbol{e}_2 \wedge \boldsymbol{e}_3,$$

hence the statement is true by IV.5.1. ■

Definition. An ordered basis $(\boldsymbol{e}_1, \boldsymbol{e}_2, \boldsymbol{e}_3)$ of $\mathbf{S}$ is called *positively oriented* if $(\boldsymbol{x}, \boldsymbol{e}_1, \boldsymbol{e}_2, \boldsymbol{e}_3)$ is a positively oriented ordered basis of $\mathbf{M}$ for some (hence for all) $\boldsymbol{x} \in T^{\rightarrow}$. ■

1.3.5. $(\mathbf{S}, \mathbb{L}, \boldsymbol{h})$ is a three-dimensional Euclidean vector space, $\mathbf{S}$ and $\mathbb{L}$ are oriented. An important relation is the identification

$$\frac{\mathbf{S}}{\mathbb{L} \otimes \mathbb{L}} \equiv \mathbf{S}^*.$$

We shall use the notation

$$\mathbf{N} := \frac{\mathbf{S}}{\mathbb{L}}$$

and all the results of section V.3.

In particular, we use a dot product notation instead of $\boldsymbol{h}$:

$$\boldsymbol{q} \cdot \boldsymbol{q}' := \boldsymbol{h}(\boldsymbol{q}, \boldsymbol{q}') \in \mathbb{L} \otimes \mathbb{L} \qquad (\boldsymbol{q}, \boldsymbol{q}' \in \mathbf{S}).$$

The *length* of $\boldsymbol{q} \in \mathbf{S}$ is

$$|\boldsymbol{q}| := \sqrt{\boldsymbol{q} \cdot \boldsymbol{q}} \in \mathbb{L}_0^+,$$

and the *angle between* the nonzero elements q and q' of $\mathbf{S}$ is

$$\arg(q, q') := \arccos \frac{q \cdot q'}{|q||q'|}.$$

The dot product can be defined between spacelike vectors of different types (see later, Section 1.4) as well; e.g. if $\mathbb{A}$ and $\mathbb{B}$ are measure lines, for $w \in \frac{\mathbf{S}}{\mathbb{A}}$ and $z \in \frac{\mathbf{S}}{\mathbb{B}}$ we have

$$w \cdot z \in \frac{\mathbb{L} \otimes \mathbb{L}}{\mathbb{A} \otimes \mathbb{B}}, \quad \arg(w, z) := \arccos \frac{w \cdot z}{|w||z|},$$

$$|w| := \sqrt{w \cdot w} \in \frac{\mathbb{L}}{\mathbb{A}}.$$

1.3.6. Do not forget that timelike vectors (elements of $\mathbf{M}$ outside $\mathbf{S}$) have no length, no angles between them.

$\mathbb{T}^* \cdot \tau$ is an oriented one-dimensional vector space, hence the absolute value of its elements makes sense; thus a length (absolute value) can be assigned to a timelike covector. However, the length of spacelike covectors (elements of $\mathbf{M}^*$ outside $\mathbb{T}^* \cdot \tau$) and the angle between two covectors are not meaningful.

1.3.7. The Euclidean structure of our space is deeply fixed in our mind, therefore we must be careful when dealing with $\mathbf{M}$ which has no Euclidean structure; especially when illustrating it in the Euclidean plane of the page. Keep in mind that vectors out of $\mathbf{S}$ have no length, do not form angles. The following considerations help us to take in the situation.

Recall that the linear map $\tau : \mathbf{M} \to \mathbb{T}$ can be applied to element of $\frac{\mathbf{M}}{\mathbb{T}}$ and then it has values in $\frac{\mathbb{T}}{\mathbb{T}} \equiv \mathbb{R}$ (see V.2.1). Put

$$V(1) := \left\{ u \in \frac{\mathbf{M}}{\mathbb{T}} \;\middle|\; \tau \cdot u = 1 \right\}.$$

According to VI.2.2, $V(1)$ is an affine subspace of $\frac{\mathbf{M}}{\mathbb{T}}$ over $\frac{\mathbf{S}}{\mathbb{T}}$. It is illustrated as follows:

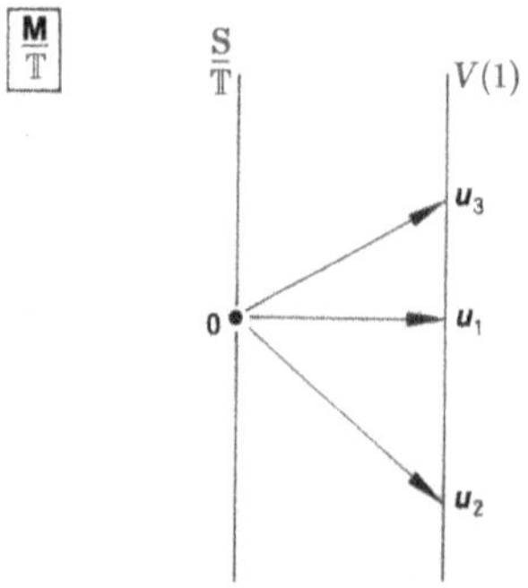

Three elements of $V(1)$ appear in the figure. Observe that it makes no sense to say that

— $\boldsymbol{u}_1$ is orthogonal to $\frac{\mathbf{S}}{\mathbb{T}}$ (there are no vectors orthogonal to $\frac{\mathbf{S}}{\mathbb{T}}$),

— the angle between $\boldsymbol{u}_1$ and $\boldsymbol{u}_3$ is less than the angle between $\boldsymbol{u}_1$ and $\boldsymbol{u}_2$ (there is no angle between the elements of $V(1)$),

— $\boldsymbol{u}_2$ is longer than $\boldsymbol{u}_1$ (the elements of $V(1)$ have no length).

We shall see in 2.1.2 that the elements of $V(1)$ can be interpreted as *velocity values*.

1.3.8. Since there is no vector orthogonal to $\mathbf{S}$, the orthogonal projection of vectors onto $\mathbf{S}$ makes no sense. Of course, we can project onto $\mathbf{S}$ in many equivalent ways; the following projections will play an important role.

Let $\boldsymbol{u}$ be an element of $V(1)$. Then $\boldsymbol{u}\otimes\mathbb{T} := \{\boldsymbol{u}\boldsymbol{t} \mid \boldsymbol{t}\in\mathbb{T}\}$ is a one-dimensional linear subspace of $\mathbf{M}$; $\boldsymbol{u}\otimes\mathbb{T}$ and $\mathbf{S}$ are complementary subspaces, thus every vector $\boldsymbol{x}$ can be uniquely decomposed into the sum of components in $\boldsymbol{u}\otimes\mathbb{T}$ and in $\mathbf{S}$, respectively:

$$\boldsymbol{x} = \boldsymbol{u}(\boldsymbol{\tau}\cdot\boldsymbol{x}) + (\boldsymbol{x} - \boldsymbol{u}(\boldsymbol{\tau}\cdot\boldsymbol{x}))\,.$$

The linear map

$$\boldsymbol{\pi}_{\boldsymbol{u}} : \mathbf{M}\to\mathbf{S}, \quad \boldsymbol{x}\to\boldsymbol{x} - \boldsymbol{u}(\boldsymbol{\tau}\cdot\boldsymbol{x})$$

is the *projection onto* $\mathbf{S}$ *along* $\boldsymbol{u}$. It is illustrated as follows:

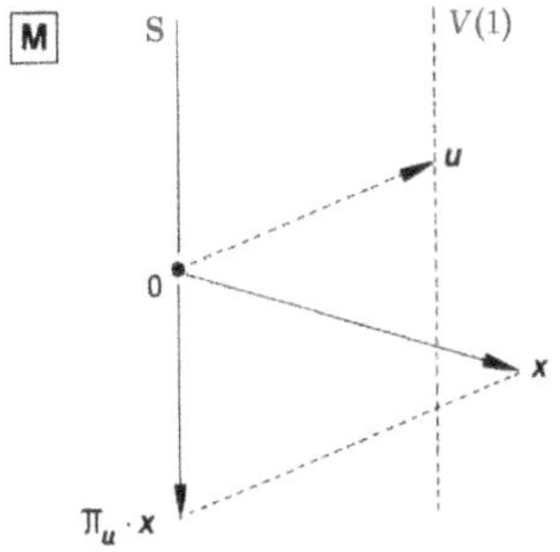

$V(1)$ is represented by a dashed line expressing that $V(1)$ is in fact a subset of $\frac{\mathbf{M}}{\mathbb{T}}$.

Observe that

$$\boldsymbol{\pi}_{\boldsymbol{u}}\cdot\boldsymbol{i} = \mathbf{1}_{\mathbf{S}}$$

and in a tensorial form $\boldsymbol{\pi}_{\boldsymbol{u}}\in\mathbf{S}\otimes\mathbf{M}^*$.

1.3.9. Proposition. Let $\boldsymbol{u}\in V(1)$. Then

$$\xi_{\boldsymbol{u}} := (\boldsymbol{\tau},\boldsymbol{\pi}_{\boldsymbol{u}}) : \mathbf{M}\to\mathbb{T}\times\mathbf{S} \qquad \boldsymbol{x}\mapsto(\boldsymbol{\tau}\cdot\boldsymbol{x},\ \boldsymbol{\pi}_{\boldsymbol{u}}\cdot\boldsymbol{x})$$

is an orientation preserving linear bijection, and

$$\boldsymbol{\xi}_u^{-1}(\boldsymbol{t},\boldsymbol{q}) = \boldsymbol{u}\boldsymbol{t} + \boldsymbol{q} \qquad (\boldsymbol{t} \in \mathbb{T}, \boldsymbol{q} \in \mathbf{S}).$$

1.4. The arithmetic spacetime model

1.4.1. Let us number the coordinates of elements of $\mathbb{R}^4$ from 0 to 3 : $(\xi^0, \xi^1, \xi^2, \xi^3) \in \mathbb{R}^4$. The canonical projection onto the zeroth coordinate,

$$pr^0 : \mathbb{R}^4 \to \mathbb{R}, \quad (\xi^0, \xi^1, \xi^2, \xi^3) \mapsto \xi^0$$

is a linear map whose kernel is $\{0\} \times \mathbb{R}^3$ which we *identify* with $\mathbb{R}^3$. Let $\boldsymbol{H}$ denote the usual inner product on $\mathbb{R}^3$: $\boldsymbol{H}(\boldsymbol{x}, \boldsymbol{y}) = \sum_{i=1}^{3} x^i y^i$. Endow $\mathbb{R}$ and $\mathbb{R}^4$ with the standard orientation.

It is quite evident that $\left(\mathbb{R}^4, \mathbb{R}, \mathbb{R}, pr^0, \boldsymbol{H}\right)$ is a nonrelativistic spacetime model which we call the *arithmetic nonrelativistic spacetime model.*

In the arithmetic spacetime model we have:

$$\mathrm{M} = \mathbf{M} = \mathbb{R}^4, \qquad \mathrm{T} = \mathbb{R}, \qquad \mathbb{L} = \mathbb{R},$$

$$\tau = pr^0,$$

$$\mathbf{S} = \{0\} \times \mathbb{R}^3 \equiv \mathbb{R}^3, \qquad \boldsymbol{h} = \boldsymbol{H}.$$

$$\mathrm{T} \equiv \mathbb{R} \times \{(0,0,0)\}, \quad \boldsymbol{\tau} = pr^0.$$

Then

$$\boldsymbol{i} \colon \mathbf{S} \to \mathbf{M} \quad \text{equals} \quad \mathbb{R}^3 \to \mathbb{R}^4, \quad (x^1, x^2, x^3) \mapsto (0, x^1, x^2, x^3).$$

The usual identification yields $\mathbf{M}^* = (\mathbb{R}^4)^* \equiv \mathbb{R}^4$; the covectors are indexed in subscripts: $(k_0, k_1, k_2, k_3) \in (\mathbb{R}^4)^*$ (see IV.1.4).

In the same way, $\mathbb{T}^* = \mathbb{R}^* \equiv \mathbb{R}$, but here we cannot make distinction with the aid of indices.

Then

$$\boldsymbol{i}^* : \mathbf{M}^* \mapsto \mathbf{S}^* \quad \text{equals} \quad (\mathbb{R}^4)^* \to (\mathbb{R}^3)^*, \quad (k_0, k_1, k_2, k_3) \mapsto (k_1, k_2, k_3)$$

and

$$\boldsymbol{\tau}^* : \mathbb{T}^* \to \mathbf{M}^* \quad \text{equals} \quad \mathbb{R}^* \to (\mathbb{R}^4)^*, \quad e \mapsto (e, 0, 0, 0).$$

1.4.2. It is an unpleasant feature of the arithmetic spacetime model that the same object, $\mathbb{R}^4$, represents the affine space of world points and the vector space of world vectors and even the vector space of covectors. For a clear distinction we shall write Greek letters indicating world points (affine space elements) and Latin letters indicating world vectors or covectors.

Another shortcoming of this spacetime model is that spacetime seems to be the Cartesian product of time and space. Then considering $\mathbb{R}^4 \equiv \mathbb{R} \times \mathbb{R}^3$, we write $(\alpha, \boldsymbol{\xi})$ or $(\xi^0, \boldsymbol{\xi})$ and $(t, \boldsymbol{q})$ for its elements; similarly, $(e, \boldsymbol{p})$ denotes an element of $(\mathbb{R} \times \mathbb{R}^3)^* \equiv \mathbb{R}^* \times (\mathbb{R}^3)^*$. Then

$$
\begin{aligned}
\tau\colon \mathbb{R} \times \mathbb{R}^3 \to \mathbb{R}, &\qquad (\alpha, \boldsymbol{\xi}) \mapsto \alpha,\\
\boldsymbol{\tau}\colon \mathbb{R} \times \mathbb{R}^3 \to \mathbb{R}, &\qquad (\boldsymbol{t}, \boldsymbol{q}) \mapsto \boldsymbol{t},\\
\boldsymbol{i}\colon \mathbb{R}^3 \to \mathbb{R} \times \mathbb{R}^3, &\qquad \boldsymbol{q} \mapsto (0, \boldsymbol{q}),\\
\boldsymbol{i}^*\colon (\mathbb{R} \times \mathbb{R}^3)^* \to (\mathbb{R}^3)^*, &\qquad (e, \boldsymbol{p}) \mapsto \boldsymbol{p},\\
\boldsymbol{\tau}^*\colon \mathbb{R}^* \to (\mathbb{R} \times \mathbb{R}^3)^*, &\qquad e \mapsto (e, \boldsymbol{0}).
\end{aligned}
$$

The last formula means that $\mathbb{T}^* \cdot \boldsymbol{\tau}$ now equals $\mathbb{R} \times \{\boldsymbol{0}\}$.

Of course, τ and $\boldsymbol{\tau}$ are equal though we have written the same formula with different symbols. This is a trick similar to that of subscripts and superscripts: we wish to distinguish between different objects that appear in the same form.

1.4.3. Now we have $\frac{\mathrm{M}}{\mathbb{T}} = \frac{\mathbb{R}^4}{\mathbb{R}} \equiv \mathbb{R}^4$, and

$$
V(1) = \left\{(v^0, \boldsymbol{v}) \in \mathbb{R} \times \mathbb{R}^3 \,\middle|\, v^0 = 1\right\} = \{1\} \times \mathbb{R}^3.
$$

Here, too, we find a misleading feature of this spacetime model: $V(1)$ seems to have a distinguished, simplest element, namely $(1, \boldsymbol{0})$.

For $(1, \boldsymbol{v}) \in V(1)$ we easily derive that

$$
\boldsymbol{\pi}_{(1,\boldsymbol{v})}\colon \mathbb{R} \times \mathbb{R}^3 \to \mathbb{R}^3, \qquad (t, \boldsymbol{q}) \mapsto \boldsymbol{q} - \boldsymbol{v}t.
$$

In particular, $\boldsymbol{\pi}_{(1,\boldsymbol{0})}$ is the canonical projection from $\mathbb{R} \times \mathbb{R}^3$ onto $\mathbb{R}^3$.

1.5. Classification of physical quantities

1.5.1. In physics one usually says e.g. that (relative) velocity and acceleration are three-dimensional vectors and are considered as triplets of real numbers. Although both are taken as elements of $\mathbb{R}^3$, they cannot be added because they have 'different physical dimensions'. The framework of our spacetime model assures a precise meaning of these notions.

A physical dimension is represented by a measure line. Let $\mathbb{A}$ be a measure line. Then the elements of

$$\begin{array}{rll} \mathbb{A} & \text{are called} & \textit{scalars of type } \mathbb{A}, \\ \mathbb{A}\otimes\mathbf{M} & \text{are called} & \textit{vectors of type } \mathbb{A}, \\ \dfrac{\mathbf{M}}{\mathbb{A}} & \text{are called} & \textit{vectors of cotype } \mathbb{A}, \\ \mathbb{A}\otimes(\mathbf{M}\otimes\mathbf{M}) & \text{are called} & \textit{tensors of type } \mathbb{A}, \\ \dfrac{\mathbf{M}\otimes\mathbf{M}}{\mathbb{A}} & \text{are called} & \textit{tensors of cotype } \mathbb{A}. \end{array}$$

Covectors of type $\mathbb{A}$, etc. are defined similarly with $\mathbf{M}^*$ instead of $\mathbf{M}$.

In the case $\mathbb{A}=\mathbb{R}$ we omit the term 'of type $\mathbb{R}$'. In particular, the elements of $\mathbf{M}\otimes\mathbf{M}$ and $\mathbf{M}^*\otimes\mathbf{M}^*$ are called *tensors* and *cotensors*, respectively; the elements of $\mathbf{M}^*\otimes\mathbf{M}$ and $\mathbf{M}\otimes\mathbf{M}^*$ are *mixed tensors*.

Recall the identifications $\mathbb{A}\otimes\mathbf{M}\equiv\mathbf{M}\otimes\mathbb{A}$ etc. (see IV.3.6).

Because of the identification $\frac{\mathbf{M}}{\mathbb{A}}\equiv\mathbf{M}\otimes\mathbb{A}^*$ the vectors of cotype $\mathbb{A}$ coincide with the vectors of type $\mathbb{A}^*$.

1.5.2. The vectors and tensors of type $\mathbb{A}$ in the subspaces $\mathbb{A}\otimes\mathbf{S}$ and $\mathbb{A}\otimes(\mathbf{S}\otimes\mathbf{S})$, respectively, are called *spacelike.*

The covectors of type $\mathbb{A}$ in the subspace $\mathbb{A}\otimes(\mathbb{T}^*\cdot\boldsymbol{\tau})$ are called *timelike.*

According to our convention (V.2.1 and V.2.2), the dot product of covectors and vectors of different types makes sense; e.g.

$$\text{for}\quad \boldsymbol{k}\in\mathbb{B}\otimes\mathbf{M}^*\quad\text{and}\quad \boldsymbol{z}\in\mathbb{A}\otimes\mathbf{M}\equiv\mathbf{M}\otimes\mathbb{A}\quad\text{we have}\quad \boldsymbol{k}\cdot\boldsymbol{z}\in\mathbb{B}\otimes\mathbb{A}.$$

In particular,

$$\text{for}\quad \boldsymbol{\tau}\in\mathbb{T}\otimes\mathbf{M}^*\quad\text{and}\quad \boldsymbol{z}\in\mathbb{A}\otimes\mathbf{M}\quad\text{we have}\quad \boldsymbol{\tau}\cdot\boldsymbol{z}\in\mathbb{T}\otimes\mathbb{A};$$

similarly,

$$\begin{array}{rll} \text{for}\quad \boldsymbol{w}\in\dfrac{\mathbf{M}}{\mathbb{A}} & \text{we have} & \boldsymbol{\tau}\cdot\boldsymbol{w}\in\dfrac{\mathbb{T}}{\mathbb{A}}; \\ \text{for}\quad \boldsymbol{T}\in\mathbb{A}\otimes(\mathbf{M}\otimes\mathbf{M}) & \text{we have} & \boldsymbol{\tau}\cdot\boldsymbol{T}\in\mathbb{T}\otimes\mathbb{A}\otimes\mathbf{M}. \end{array}$$

Evidently, $\boldsymbol{z}\in\mathbb{A}\otimes\mathbf{M}$ is spacelike if and only if $\boldsymbol{\tau}\cdot\boldsymbol{z}=\mathbf{0}$.

In the same way, $\boldsymbol{i}^*:\mathbf{M}^*\to\mathbf{S}^*$ is lifted to covectors of type $\mathbb{A}$, etc. i.e.

$$\text{for}\quad \boldsymbol{i}^*\in\mathbf{S}^*\otimes\mathbf{M}\quad\text{and}\quad \boldsymbol{k}\in\mathbb{A}\otimes\mathbf{M}^*\quad\text{we have}\quad \boldsymbol{i}^*\cdot\boldsymbol{k}\in\mathbb{A}\otimes\mathbf{S}^*\quad\text{etc.}$$

Evidently, $\boldsymbol{k}\in\mathbb{A}\otimes\mathbf{M}^*$ is timelike if and only if $\boldsymbol{i}^*\cdot\boldsymbol{k}=\mathbf{0}$.

1.5.3. In nonrelativistic physics one usually introduces the notion of scalars, three-dimensional vectors, three-dimensional pseudovectors and pseudoscalars as quantities having some prescribed transformation properties. One is forced to adapt such a definition because only coordinates are considered, only numbers and triplets of numbers are used, and one must know whether a triplet of numbers is the set of coordinates of a vector, or not. Of course, vectors can have different 'physical dimensions'.

Now we formulate the corresponding notion in the framework of our nonrelativistic spacetime model. The elements of

$$\begin{aligned} \mathbb{R} \quad & \text{are the scalars,} \\ \mathbf{S} \quad & \text{are the spacelike vectors,} \\ \mathbf{S}\wedge\mathbf{S} \quad & \text{are the spacelike pseudovectors of type } \mathbb{L}, \\ \mathbf{S}\wedge\mathbf{S}\wedge\mathbf{S} \quad & \text{are the pseudoscalars of type } \overset{3}{\otimes}\mathbb{L}. \end{aligned}$$

The first and the second names do not require an explanation. The third and fourth names are based on the fact that we have canonical linear bijections $\mathbf{S}\wedge\mathbf{S} \to \mathbf{S}\otimes\mathbb{L}$ and $\mathbf{S}\wedge\mathbf{S}\wedge\mathbf{S} \to \mathbb{L}\otimes\mathbb{L}\otimes\mathbb{L}$ (see V.3.17); pseudovectors are 'similar' to spacelike vectors of type $\mathbb{L}$, and pseudoscalars are 'similar' to scalars of type $\overset{3}{\otimes}\mathbb{L}$.

Having the notion of vectors of type $\mathbb{A}$, it is evident how we shall define spacelike pseudovectors and pseudoscalars of diverse types. For the sake of simplicity, we consider now 'physically dimensionless' quantities: $\mathbb{R}$, $\mathbf{N}$, $\mathbf{N}\wedge\mathbf{N}$, $\mathbf{N}\wedge\mathbf{N}\wedge\mathbf{N}$. Then we have the linear bijections $\boldsymbol{j}: \mathbf{N}\wedge\mathbf{N} \to \mathbf{N}$ and $\mathrm{j_o}: \mathbf{N}\wedge\mathbf{N}\wedge \mathbf{N} \to \mathbb{R}$.

Let $\boldsymbol{R}: \mathbf{S} \to \mathbf{S}$ be an orthogonal map which is considered to be an orthogonal map $\mathbf{N} \to \mathbf{N}$ as well. We say that $\boldsymbol{R}$ is a rotation if it has positive determinant. The determinant of the inversion $\boldsymbol{P} := -\mathbf{1}_{\mathbf{S}}$ is negative.

By definition, $\overset{0}{\otimes}\boldsymbol{R} := \overset{0}{\otimes}\boldsymbol{P} := \mathbf{1}_{\mathbb{R}}$; the scalars are invariant.

Vectors are transformed under $\boldsymbol{R}$ and $\boldsymbol{P}$ according to the definition of these operations.

Pseudovectors are transformed by $\boldsymbol{R}\wedge\boldsymbol{R}$ and $\boldsymbol{P}\wedge\boldsymbol{P}$ (IV.3.21.); formulae in V.3.16 say that

$$\boldsymbol{j}\circ(\boldsymbol{R}\wedge\boldsymbol{R}) = \boldsymbol{R}\circ\boldsymbol{j}, \qquad \boldsymbol{j}\circ(\boldsymbol{P}\wedge\boldsymbol{P}) = -\boldsymbol{P}\circ\boldsymbol{j} = \boldsymbol{j}$$

which means that the pseudovectors are transformed by rotations like vectors but they are invariant under the inversion.

Similarly we have that

$$\mathrm{j_o}\circ(\boldsymbol{R}\wedge\boldsymbol{R}\wedge\boldsymbol{R}) = \mathrm{j_o}, \qquad \mathrm{j_o}\circ(\boldsymbol{P}\wedge\boldsymbol{P}\wedge\boldsymbol{P}) = -\mathrm{j_o},$$

the pseudoscalars are invariant under rotations and they change sign by the inversion.

1.6. Comparison of spacetime models

1.6.1. A spacetime model is defined as a mathematical structure. It is an interesting question both from mathematical and from physical points of view: how many 'different' nonrelativistic spacetime models exist?

To answer, first we must define what the 'difference' and the 'similarity' between two spacetime models mean. We proceed as it is usual in mathematics; for instance, one defines the linear structure (vector space) and then the linear maps as the tool of comparison between linear structures; two vector spaces are of the same kind if there is a linear bijection between them, in other words, if they are isomorphic.

Definition. The nonrelativistic spacetime model $(\mathrm{M}, \mathbb{T}, \mathbb{L}, \boldsymbol{\tau}, \boldsymbol{h})$ is *isomorphic* to the nonrelativistic spacetime model $(\mathrm{M}', \mathbb{T}', \mathbb{L}', \boldsymbol{\tau}', \boldsymbol{h}')$ if there are

(*i*) an orientation preserving affine bijection $F\colon \mathrm{M} \to \mathrm{M}'$, over the linear bijecion $\boldsymbol{F}$.

(*ii*) an orientation preserving linear bijection $\boldsymbol{B}\colon \mathbb{T} \to \mathbb{T}'$,

(*iii*) an orientation preserving linear bijection $\boldsymbol{Z}\colon \mathbb{L} \to \mathbb{L}'$ such that

(I) $\boldsymbol{\tau}' \circ \boldsymbol{F} = \boldsymbol{B} \circ \boldsymbol{\tau}$,

(II) $\boldsymbol{h}' \circ (\boldsymbol{F} \times \boldsymbol{F}) = (\boldsymbol{Z} \otimes \boldsymbol{Z}) \circ \boldsymbol{h}$.

The triplet $(F, \boldsymbol{B}, \boldsymbol{Z})$ is an *isomorphism* between the two spacetime models.

If the two models coincide, isomorphism is called *automorphism.* An automorphism $(F, \boldsymbol{B}, \boldsymbol{Z})$ of $(\mathrm{M}, \mathbb{T}, \mathbb{L}, \boldsymbol{\tau}, \boldsymbol{h})$ is *strict* if $\boldsymbol{B} = \mathbf{1}_{\mathbb{T}}$ and $\boldsymbol{Z} = \mathbf{1}_{\mathbb{L}}$. ■

Two commutative diagrams illustrate the isomorphism:

$$
\begin{array}{ccc}
\mathrm{M} & \xrightarrow{\boldsymbol{\tau}} & \mathbb{T} \\
F\big\downarrow & & \big\downarrow \boldsymbol{B} \\
\mathrm{M}' & \xrightarrow[\boldsymbol{\tau}']{} & \mathbb{T}'
\end{array}
\qquad\qquad
\begin{array}{ccc}
\mathbf{S} \times \mathbf{S} & \xrightarrow{\boldsymbol{h}} & \mathbb{L} \otimes \mathbb{L} \\
\boldsymbol{F} \times \boldsymbol{F}\big\downarrow & & \big\downarrow \boldsymbol{Z} \otimes \boldsymbol{Z}\,. \\
\mathbf{S}' \times \mathbf{S}' & \xrightarrow[\boldsymbol{h}']{} & \mathbb{L}' \otimes \mathbb{L}'
\end{array}
$$

The definition is quite natural and simple. It is worth mentioning that (I) implies that for $\boldsymbol{q} \in \mathbf{S}$ we have $\boldsymbol{\tau}' \cdot \boldsymbol{F} \cdot \boldsymbol{q} = \boldsymbol{B} \cdot \boldsymbol{\tau} \cdot \boldsymbol{q} = \mathbf{0}$ which means that $\boldsymbol{F}$ maps $\mathbf{S}$ into (and even onto) $\mathbf{S}'$; hence the requirement in (II) is meaningful.

It is evident that $(F^{-1}, \boldsymbol{B}^{-1}, \boldsymbol{Z}^{-1})$, the inverse of $(F, \boldsymbol{B}, \boldsymbol{Z})$, is an isomorphism as well. Moreover, if $(\boldsymbol{F}', \boldsymbol{B}', \boldsymbol{Z}')$ is an isomorphism between nonrelativistic spacetime models $(\mathrm{M}', \mathbb{T}', \mathbb{L}', \boldsymbol{\tau}', \boldsymbol{h}')$ and $(\mathrm{M}'', \mathbb{T}'', \mathbb{L}'', \boldsymbol{\tau}'', \boldsymbol{h}'')$, then $(\boldsymbol{F}' \circ \boldsymbol{F}, \boldsymbol{B}' \circ \boldsymbol{B}, \boldsymbol{Z}' \circ \boldsymbol{Z})$ is an isomorphism, too.

1.6.2. Proposition. The nonrelativistic spacetime model $(\mathrm{M}, \mathbb{T}, \mathbb{L}, \boldsymbol{\tau}, \boldsymbol{h})$ is isomorphic to the arithmetic spacetime model.

Proof. Take
(*i*) a positive element $\boldsymbol{s}$ of $\mathbb{T}$,
(*ii*) a positive element $\boldsymbol{m}$ of $\mathbb{L}$,
(*iii*) an element $\boldsymbol{e}_0$ of $T^{\rightarrow}$ such that $\boldsymbol{\tau}\cdot\boldsymbol{e}_0 = \boldsymbol{s}$,
(*iv*) a positively oriented orthogonal basis $(\boldsymbol{e}_1, \boldsymbol{e}_2, \boldsymbol{e}_3)$, normed to $\boldsymbol{m}$, of $\mathbf{S}$,
(*v*) an element o of M.
Then $\boldsymbol{u} := \frac{\boldsymbol{e}_0}{\boldsymbol{s}}$ is in $V(1)$ and it is not hard to see that

$$F\colon \mathrm{M}\to\mathbb{R}^4, \quad x\mapsto \left(\frac{\boldsymbol{\tau}\cdot(x-o)}{\boldsymbol{s}}, \left(\frac{\boldsymbol{e}_\alpha\cdot\boldsymbol{\pi}_{\boldsymbol{u}}\cdot(x-o)}{\boldsymbol{m}^2}\right)_{\alpha=1,2,3}\right),$$

$$B\colon \mathbb{T}\to\mathbb{R}, \quad t\mapsto \frac{t}{\boldsymbol{s}},$$

$$Z\colon \mathbb{L}\to\mathbb{R}, \quad d\mapsto \frac{d}{\boldsymbol{m}}$$

is an isomorphism. ■

Observe that $(\boldsymbol{e}_0, \boldsymbol{e}_1, \boldsymbol{e}_2, \boldsymbol{e}_3)$ is a positively oriented basis in $\mathbf{M}$, and F is the affine coordinatization of M corresponding to o and that basis.

The isomorphism above has the inverse

$$\mathbb{R}^4\to\mathrm{M}, \quad (\xi^0,\xi^1,\xi^2,\xi^3)\mapsto o+\sum_{i=o}^{3}\xi^i\boldsymbol{e}_i,$$

$$\mathbb{R}\to\mathbb{T}, \quad \alpha\mapsto\alpha\boldsymbol{s},$$

$$\mathbb{R}\to\mathbb{L}, \quad \delta\mapsto\delta\boldsymbol{m}.$$

1.6.3. An important consequence of the previous result is that *two arbitrary nonrelativistic spacetime models are isomorphic,* i.e. are of the same kind. The nonrelativistic spacetime model as a mathematical structure is unique. This means that there is a unique 'nonrelativistic physics'.

Please, note: the nonrelativistic spacetime models are of the same kind, but, in general, are not identical. They are isomorphic, but, in general, there is no 'canonical' isomorphism between them, we cannot identify them by a distinguished isomorphism. It is a situation similar to that well known in the theory of vector spaces: all N-dimensional vector spaces are isomorphic to $\mathbb{K}^N$ but, in general, there is no canonical isomorphism between them.

Since all nonrelativistic spacetime models are isomorphic, we can use an arbitrary one for investigation and application. However, an actual model can have additional structures. For instance, in the arithmetic model, spacetime and time are vector spaces, time is canonically embedded into spacetime as $\mathbb{R}\times\{\mathbf{0}\}$, $V(1)$ has a distinguished element, $(1,\mathbf{0})$. This model tempts us to multiply world points by real numbers (although this has no physical meaning and that is why it is not meaningful in the abstract spacetime structure), to consider spacetime to be the Cartesian product of time and space (but space does not exist!), to say that the distinguished element of $V(1)$ is orthogonal to the space (such an orthogonality makes no sense in the abstract spacetime structure), etc.

To avoid such confusions, we should keep away from similar specially constructed models for theoretical investigation and application of the nonrelativistic spacetime model. However, for solving special problems, for executing some particular calculations, we can choose a convenient concrete model. In the same way as in the theory of vector spaces where a coordinatization—i.e. the use of $\mathbb{K}^N$ —may help us to perform our task.

1.6.4. In present day physics one uses tacitly the arithmetic spacetime model. One represents time points by real numbers, space points by triplets of real numbers. To arrive at such representations, one chooses tacitly a unit of measurement for time and an initial time point, a unit of measurement for distance and an initial space point (origin) and an orthogonal spatial basis whose elements have unit length.

However, all the previous notions have merely a heuristic sense. Take a glance at the isomorphism established in 1.6.2 to recognize that the nonrelativistic spacetime model will give these notions a mathematically precise meaning. Evidently, $\boldsymbol{s}$ and $\boldsymbol{m}$ are the units of time period and distance, respectively, $\{\boldsymbol{e}_1, \boldsymbol{e}_2, \boldsymbol{e}_3\}$ is the orthogonal spatial basis whose elements have unit length; $\tau(o)$ is the initial time point and o includes somehow the origin of space as well. At present only the sense of $\boldsymbol{e}_0$ is not clear; later we shall see that it determines the space in question, because we know that absolute space does not exist; $\boldsymbol{e}_0$ characterizes an observer which realizes a space.

1.7. The split spacetime model

1.7.1. As we have said, the arithmetic spacetime model is useful for solving particular problems, for executing practical calculations. Moreover, at present, one usually expounds theories, too, in the frame of the arithmetic spacetime model, so we ought to 'translate' every notion in the arithmetic language. However, the arithmetic spacetime model is a little ponderous; that is why we introduce an 'intermediate' spacetime model between the abstract and the arithmetic ones, a more terse model which has all the essential features of the arithmetic spacetime model.

1.7.2. Let $(\mathrm{M}, \mathbb{T}, \mathbb{L}, \boldsymbol{\tau}, \boldsymbol{h})$ be a nonrelativistic spacetime model, and use the notations introduced in this chapter. Let $pr_{\mathbb{T}} : \mathbb{T} \times \mathbf{S} \to \mathbb{T}$ be the canonical projection $(\boldsymbol{t}, \boldsymbol{q}) \mapsto \boldsymbol{t}$.

Then $(\mathbb{T} \times \mathbf{S}, \mathbb{T}, \mathbb{L}, pr_{\mathbb{T}}, \boldsymbol{h})$ is a nonrelativistic spacetime model, called the *split nonrelativistic spacetime model* corresponding to $(\mathrm{M}, \mathbb{T}, \mathbb{L}, \boldsymbol{\tau}, \boldsymbol{h})$.

It is quite obvious that for all $o \in \mathrm{M}$ and $\boldsymbol{u} \in V(1)$,

$$\begin{aligned} \mathrm{M} \to \mathbb{T} \times \mathbf{S}, &\qquad x \mapsto \boldsymbol{\xi}_{\boldsymbol{u}} \cdot (x - o) \\ \mathbb{T} \to \mathbb{T}, &\qquad \boldsymbol{t} \to \boldsymbol{t} \\ \mathbb{L} \to \mathbb{L}, &\qquad \boldsymbol{d} \to \boldsymbol{d} \end{aligned}$$

is an isomorphism of the two nonrelativistic spacetime models where $\boldsymbol{\xi}_{\boldsymbol{u}}$ is defined in 1.3.9.

1.7.3. In the split spacetime model

$$\begin{aligned} \boldsymbol{\tau} : \mathbb{T} \times \mathbf{S} \to \mathbb{T}, &\qquad (\boldsymbol{t}, \boldsymbol{q}) \mapsto \boldsymbol{t}, \\ \boldsymbol{i} : \mathbf{S} \to \mathbb{T} \times \mathbf{S}, &\qquad \boldsymbol{q} \mapsto (\mathbf{0}, \boldsymbol{q}). \end{aligned}$$

With the usual identification (see IV.1.3) we have that in the split spacetime model the covectors are elements of $\mathbb{T}^* \times \mathbf{S}^*$, correspondingly,

$$\begin{aligned} \boldsymbol{\tau}^* : \mathbb{T}^* \to \mathbb{T}^* \times \mathbf{S}^*, &\qquad \boldsymbol{e} \mapsto (\boldsymbol{e}, \mathbf{0}), \\ \boldsymbol{i}^* : \mathbb{T}^* \times \mathbf{S}^* \to \mathbf{S}^*, &\qquad (\boldsymbol{e}, \boldsymbol{p}) \mapsto \boldsymbol{p}. \end{aligned}$$

As a consequence, $\mathbb{T}^* \cdot \boldsymbol{\tau} = \mathbb{T}^* \times \{0\}$.

In this model

$$V(1) = \{1\} \times \frac{\mathbf{S}}{\mathbb{T}}$$

and we easily derive for $(1, \boldsymbol{v}) \in V(1)$:

$$\boldsymbol{\pi}_{(1,\boldsymbol{v})} : \mathbb{T} \times \mathbf{S} \to \mathbf{S}, \qquad (\boldsymbol{t}, \boldsymbol{q}) \mapsto \boldsymbol{q} - \boldsymbol{v}\boldsymbol{t}.$$

1.8. Exercises

1. Let $\{\mathbf{e}_0, \mathbf{e}_1, \mathbf{e}_2, \mathbf{e}_3\}$ be a basis in $\mathbf{M}$ such that $\{\mathbf{e}_1, \mathbf{e}_2, \mathbf{e}_3\}$ is an orthogonal basis in $\mathbf{S}$, normed to $\boldsymbol{m} \in \mathbb{L}^+$. Put $\boldsymbol{s} := \boldsymbol{\tau} \cdot \mathbf{e}_0$, $\boldsymbol{u} := \frac{\mathbf{e}_0}{\boldsymbol{s}}$. Then $\left\{\frac{\boldsymbol{\tau}}{\boldsymbol{s}}, \left(\frac{\boldsymbol{\pi}_{\boldsymbol{u}}^* \cdot \mathbf{e}_i}{\boldsymbol{m}^2}\right)_{i=1,2,3}\right\}$ is the dual of the basis in question.

2. (*i*) Let $(\boldsymbol{e}_0, \boldsymbol{e}_1, \boldsymbol{e}_2, \boldsymbol{e}_3)$ be a positively oriented basis in $\mathbf{M}$ such that $(\boldsymbol{e}_1, \boldsymbol{e}_2, \boldsymbol{e}_3)$ is a positively oriented orthogonal basis in $\mathbf{S}$, normed to $\boldsymbol{m} \in \mathbb{L}^+$. Put $\boldsymbol{s} := \boldsymbol{\tau} \cdot \boldsymbol{e}_0$. Take another 'primed' basis with the same properties. Then

$$\boldsymbol{\epsilon} := \frac{\overset{3}{\underset{i=0}{\wedge}} \boldsymbol{e}_i}{\boldsymbol{s}\boldsymbol{m}^3} = \frac{\overset{3}{\underset{i=0}{\wedge}} \boldsymbol{e}'_i}{\boldsymbol{s}'\boldsymbol{m}'^3} \in \frac{\overset{4}{\wedge}\mathbf{M}}{\mathbb{T}\otimes\mathbb{L}^{\otimes 3}},$$

which is called the *Levi-Civita tensor* of the nonrelativistic spacetime model.

In other words, if $\boldsymbol{u} \in V(1)$ and $(\boldsymbol{n}_1, \boldsymbol{n}_2, \boldsymbol{n}_3)$ is a positively oriented orthonormal basis in $\mathbf{N} = \frac{\mathbf{S}}{\mathbb{L}}$, then

$$\boldsymbol{\epsilon} = \boldsymbol{u} \wedge \overset{3}{\underset{\alpha=1}{\wedge}} \boldsymbol{n}_\alpha.$$

(*ii*) Let $(\boldsymbol{k}^0, \boldsymbol{k}^1, \boldsymbol{k}^2, \boldsymbol{k}^3))$ and $(\boldsymbol{k}'^0, \boldsymbol{k}'^1, \boldsymbol{k}'^2, \boldsymbol{k}'^3)$ be the dual of the bases in question (see the previous exercise). Then

$$\overline{\boldsymbol{\epsilon}} := \boldsymbol{s}\boldsymbol{m}^3 \overset{3}{\underset{i=0}{\wedge}} \boldsymbol{k}^i = \boldsymbol{s}'\boldsymbol{m}'^3 \overset{3}{\underset{i=0}{\wedge}} \boldsymbol{k}'^i \in \mathbb{T}\otimes\mathbb{L}^{\otimes 3} \otimes \overset{4}{\wedge}\mathbf{M}^*,$$

which is called the *Levi-Civita cotensor* of the nonrelativistic spacetime model.

In other words, if the elements $\boldsymbol{\eta}^\alpha \in \mathbb{L}\otimes\mathbf{M}^*$ are such that $\boldsymbol{i}^* \cdot \boldsymbol{\eta}^\alpha$ $(\alpha = 1, 2, 3)$ form a positively oriented orthonormal basis in $\mathbf{N} = \frac{\mathbf{S}}{\mathbb{L}}$, then

$$\overline{\boldsymbol{\epsilon}} = \boldsymbol{\tau} \wedge \overset{3}{\underset{\alpha=1}{\wedge}} \boldsymbol{\eta}^\alpha.$$

3. $\boldsymbol{\epsilon}$ and $\overline{\boldsymbol{\epsilon}}$ can be regarded as linear maps from $\mathbb{T}\otimes\mathbb{L}^{\otimes 3}$ into $\overset{4}{\wedge}\mathbf{M}$ and from $\overset{4}{\wedge}\mathbf{M}$ into $\mathbb{T}\otimes\mathbb{L}^{\otimes 3}$ (recall that $\overset{4}{\wedge}\mathbf{M}^* \equiv \left[\overset{4}{\wedge}\mathbf{M}\right]^*$). Prove that $\overline{\boldsymbol{\epsilon}}$ is the inverse of $\boldsymbol{\epsilon}$.

4. Take the arithmetic spacetime model and the usual matrix form of linear maps $\mathbb{R}^M \to \mathbb{R}^N$. Then

$$\boldsymbol{\tau} = (1\,0\,0\,0),$$

$$\boldsymbol{i} = \begin{pmatrix} 0 & 0 & 0 \\ 1 & 0 & 0 \\ 0 & 1 & 0 \\ 0 & 0 & 1 \end{pmatrix}, \qquad \boldsymbol{\pi}_{(1,\boldsymbol{v})} = \begin{pmatrix} -v^1 & 1 & 0 & 0 \\ -v^2 & 0 & 1 & 0 \\ -v^3 & 0 & 0 & 1 \end{pmatrix}.$$

2. World lines

2.1. History of a masspoint: world line

2.1.1. Let us consider a pointlike material body. As mentioned in the Introduction, and is well-known, its motion makes sense only relative to other material objects, i.e., motion is a relative notion. The motion is described usually by a function assigning space points (of an observer) to time points (of a synchronization), i.e. in a reference frame.

Our spacetime model allows us an absolute description (independent of reference frames, to de defined later). We have to recognize only that the *existence* or the *history*) of the body is an absolute notion and this history seems to be a motion to another material object.

The history of a material point can be described in the spacetime model by a function that assigns world points to instants; the world point assigned to an instant gives the instantaneous spacetime position of the existence of the material point. Of course, the instant of the assigned world point must coincide with the instant itself.

Definition. A function $r\colon \mathrm{T} \rightarrowtail \mathrm{M}$ is called a *world line function* if

(*i*) $\mathrm{Dom}\, r$ is an interval,

(*ii*) r is piecewise twice continuously differentiable,

(*iii*) $\tau\big(r(t)\big) = t$ for all $t \in \mathrm{Dom}\, r$.

A subset C of M is a *world line* if it is the range of a world line function.

The world line function r and the world line $\mathrm{Ran}\, r$ are *global* if $\mathrm{Dom}\, r = \mathrm{T}$.

■

It can be shown easily that a world line C uniquely determines the world line function r such that $\mathrm{C} = \mathrm{Ran}\, r$.

2.1.2. Let the world line function r be twice differentiable at t. Then $\dot r(t) \in \frac{\mathrm{M}}{\mathrm{T}}$ and $\ddot r(t) \in \frac{\mathrm{M}}{\mathrm{T}\otimes\mathrm{T}}$ (see VI.3.9); moreover,

$$\boldsymbol{\tau} \cdot \dot r(t) = \lim_{s\to t} \frac{\boldsymbol{\tau} \cdot \big(r(s) - r(t)\big)}{s - t} = \lim_{s\to t} \frac{\tau\big(r(s)\big) - \tau\big(r(t)\big)}{s - t} = \lim_{s\to t} \frac{s-t}{s-t} = 1$$

and similarly we deduce $\boldsymbol{\tau} \cdot \ddot r(t) = \mathbf{0}$; in other words,

$$\dot r(t) \in V(1), \qquad \ddot r(t) \in \frac{\mathbf{S}}{\mathbb{T}\otimes\mathbb{T}}.$$

The same is true for the right and left derivatives at instants t where r is not twice differentiable.

The functions $\dot r\colon \mathrm{T} \rightarrowtail V(1)$ and $\ddot r\colon \mathrm{T} \rightarrowtail \frac{\mathbf{S}}{\mathrm{T}\otimes\mathrm{T}}$ can be interpreted as the *absolute velocity* and the *absolute acceleration* of the material point whose history is described by r.

That is why we call the elements of $V(1)$ *absolute velocity values* and the elements of $\frac{\mathbf{S}}{\mathbb{T}\otimes\mathbb{T}}$ *absolute acceleration values.*

2.1.3. Recall that $V(1)$ is a three-dimensional affine space over $\frac{\mathbf{S}}{\mathbb{T}}$. The elements of $\frac{\mathbf{S}}{\mathbb{T}}$ will be called *relative velocity values;* later we shall see the motivation of this name.

We know that the Euclidean structure of $\mathbf{S}$ induces Euclidean structures on $\frac{\mathbf{S}}{\mathbb{T}}$ and on $\frac{\mathbf{S}}{\mathbb{T}\otimes\mathbb{T}}$ (see 1.2.5). The magnitude of a relative velocity value is a positive element of $\frac{\mathbb{L}}{\mathbb{T}}$; the magnitude of an acceleration value is a positive element of $\frac{\mathbb{L}}{\mathbb{T}\otimes\mathbb{T}}$.

$\mathbb{L}$ and $\mathbb{T}$ are the measure lines of distances and time periods, respectively. Choosing a positive element in $\mathbb{L}$ and in $\mathbb{T}$ we fix the unit of distances and the unit of time periods; for instance, (meter=) $\boldsymbol{m} \in \mathbb{L}$ and (second=) $\boldsymbol{s} \in \mathbb{T}$. Then the units of measurements of the relative velocity and the acceleration are $\frac{\boldsymbol{m}}{\boldsymbol{s}} \in \frac{\mathbb{L}}{\mathbb{T}}$ and $\frac{\boldsymbol{m}}{\boldsymbol{s}^2} := \frac{\boldsymbol{m}}{\boldsymbol{s}\otimes\boldsymbol{s}} \in \frac{\mathbb{L}}{\mathbb{T}\otimes\mathbb{T}}$, respectively.

We emphasize the following important facts.

(*i*) The absolute velocity values are timelike vectors of cotype $\mathbb{T}$, in particular they are future directed. They form a three-dimensional affine space which is not a vector space; in particular, there is no zero absolute velocity value. An absolute velocity value has no magnitude, absolute velocity values have no angles between themselves.

(*ii*) The relative velocity values are spacelike vectors of cotype $\mathbb{T}$, they form a three-dimensional Euclidean vector space; there is a zero relative velocity value. Magnitudes and angles make sense for relative velocity values.

(*iii*) The absolute acceleration values are spacelike vectors of cotype $\mathbb{T}\otimes\mathbb{T}$, they form a three-dimensional Euclidean vector space; the acceleration values have magnitudes and angles between themselves.

The absence of magnitudes of absolute velocity values means that 'quickness' makes no absolute sense; it is not meaningful that a material object exists more quickly than another. An absolute velocity value characterizes somehow the *tendency* of the history of a material point. Mass points can move slowly or quickly *relative to each other.*

2.1.4. A world line function in the arithmetic spacetime model is $r = (r^0, \boldsymbol{r}) : \mathbb{R} \rightarrowtail \mathbb{R} \times \mathbb{R}^3$ such that $r^0(t) = t$ for all $t \in \operatorname{Dom} r$. In other words, a world line function is given by a function $\boldsymbol{r} : \mathbb{R} \rightarrowtail \mathbb{R}^3$ in the form $t \mapsto \big(t, \boldsymbol{r}(t)\big)$.

The first and the second derivative of the world line function (i.e. velocity and acceleration) are $t \mapsto \big(1, \dot{\boldsymbol{r}}(t)\big)$ and $t \mapsto \big(0, \ddot{\boldsymbol{r}}(t)\big)$, respectively.

2.2. A characterization of world lines

World lines are special *curves* in M (for the notion of curves see VI.4.3).

It is evident that if C is a world line then $\mathrm{C} \cap t$ has at most one element for all $t \in \mathrm{T}$ (where T is identified with the affine subspaces in M, directed by $\mathbf{S}$, see 1.1.4). We shall use the symbol

$$\mathrm{C} \star t$$

for the unique element of $\mathrm{C} \cap t$ if this latter is not void. Then we have that the world line function r corresponding to C is given by

$$\begin{aligned} \mathrm{Dom}\, r &= \{t \in \mathrm{T} \mid \mathrm{C} \cap t \neq \emptyset\}, \\ r(t) &= \mathrm{C} \star t \qquad\qquad (t \in \mathrm{Dom}\, r). \end{aligned}$$

It is evident as well that a twice differentiable curve C for which $\mathrm{C} \cap t$ has at most one element for all $t \in \mathrm{T}$ need not be a world line: it can have a spacelike tangent vector.

Every nonzero tangent vector of a world line is timelike. The converse is true as well.

Proposition. Let C be a connected twice differentiable curve in M whose nonzero tangent vectors are timelike; then C is a world line.

Proof. Let $p\colon \mathbb{R} \rightarrowtail \mathrm{M}$ be a parametrization of C. Then $\boldsymbol{\tau} \cdot \big(\dot{p}(\alpha)\big) \neq \mathbf{0}$ for all $\alpha \in \mathrm{Dom}\, p$. The function $\tau \circ p\colon \mathbb{R} \rightarrowtail \mathrm{T}$ is defined in an interval, is twice continuously differentiable, its derivative $\boldsymbol{\tau} \cdot \dot{p}$ is nowhere zero; hence it is strictly monotonous, its inverse $z := (\tau \circ p)^{-1}$ is twice continuously differentiable as well and $\dot{z}(t) = 1/\big(\boldsymbol{\tau} \cdot \dot{p}\big(z(t)\big)\big)$, as it is well known. It is obvious then that $r := p \circ z$ is a world line function and $\mathrm{Ran}\, r = \mathrm{C}$.

2.3. Classification of world lines

Definition. The twice continuously differentiable world line function r and the corresponding world line are called

(*i*) *inertial* if $\ddot{r} = \mathbf{0}$,

(*ii*) *uniformly accelerated* if $\ddot{r}$ is constant,

(*iii*) *twist-free* if $\ddot{r}(s)$ is parallel to $\ddot{r}(t)$ for all $t, s \in \mathrm{Dom}\, r$.

Proposition. The twice continuously differentiable world line function r is

(*i*) inertial if and only if there are $x_o \in \mathrm{M}$ and $\boldsymbol{u}_o \in V(1)$ such that

$$r(t) = x_o + \boldsymbol{u}_o\big(t - \tau(x_o)\big) \qquad (t \in \mathrm{Dom}\, r);$$

(*ii*) uniformly accelerated if and only if there are $x_o \in M$, $\boldsymbol{u}_o \in V(1)$ and $\boldsymbol{a}_o \in \frac{\mathbf{S}}{\mathbb{T}\otimes\mathbb{T}}$ such that

$$r(t) = x_o + \boldsymbol{u}_o\big(t - \tau(x_o)\big) + \frac{1}{2}\boldsymbol{a}_o\big(t - \tau(x_o)\big)^2 \qquad (t \in \mathrm{Dom}\, r);$$

(*iii*) twist-free if and only if there exist $x_o \in M$, $\boldsymbol{u}_o \in V(1)$, $\boldsymbol{0} \neq \boldsymbol{a}_o \in \frac{\mathbf{S}}{\mathbb{T}\otimes\mathbb{T}}$ and a twice continuously differentiable function $\boldsymbol{h}\colon \mathbb{T} \rightarrowtail \mathbb{T}\otimes\mathbb{T}$ for which $\boldsymbol{h}(\boldsymbol{0}) = \boldsymbol{0}$, $\dot{\boldsymbol{h}}(\boldsymbol{0}) = \boldsymbol{0}$ and

$$r(t) = x_o + \boldsymbol{u}_o\big(t - \tau(x_o)\big) + \boldsymbol{a}_o\boldsymbol{h}\big(t - \tau(x_o)\big) \qquad (t \in \mathrm{Dom}\, r).$$

Proof. The validity of the assertions comes from the theory of differential equations; (*i*) and (*ii*) are quite trivial. For (*iii*) observe that r is twist-free if and only if there is a nonzero acceleration value $\boldsymbol{a}_o$ and a continuous function $\alpha\colon \mathrm{T} \rightarrowtail \mathbb{R}$ (which can be zero) such that $\ddot{r}(t) = \boldsymbol{a}_o\alpha(t)$. If x_o is a point in the range of r, we define $\chi\colon \mathbb{T} \rightarrowtail \mathbb{R}$ by $\chi(\boldsymbol{t}) := \alpha\big(\tau(x_o) + \boldsymbol{t}\big)$ which means that $\chi\big(t - \tau(x_o)\big) = \alpha(t)$. Then $\boldsymbol{h}$ will be the function whose second derivative is χ and that satisfies the initial condition given above. ■

Observe that a twice continuously differentiable world line function r is twist-free if and only if $\ddot{r}/|\ddot{r}|$ is constant on each interval where the second derivative is not zero.

An inertial world line is uniformly accelerated (with zero acceleration) and a uniformly accelerated world line is twist-free (with constant acceleration).

A world line is inertial if and only if it is a straight line segment.

2.4. Newtonian equation

2.4.1. We shall say some words about the Newtonian equation though it does not belong to the subject of this volume; the Newtonian equation motivates the notion of force fields and potentials which will make us understand the importance of splitting of vectors and covectors (see Section 6).

First of all we have to say something about mass. One usually introduces the unit of mass, $\boldsymbol{kg}$, as a unit independent of the unit of distances, $\boldsymbol{m}$, and of the unit of time periods, $\boldsymbol{s}$. This means in our language that we introduce the measure line $\mathbb{M}$ of mass as a measure line 'independent' of $\mathbb{L}$ and $\mathbb{T}$. We shall do so in another book where we wish to treat physical theories in a form suitable for applications, so in a form which applies the SI physical dimensions. However, for the present purposes we choose another possibility.

The results of quantum mechanics showed that Nature establishes a relation among the measure lines $\mathbb{L}$, $\mathbb{T}$ and $\mathbb{M}$. Namely, it is discovered, that the values of angular momentum are integer multiples of a given quantum denoted by $h/4\pi$ where h is known as the Planck constant. Hence we can choose $\mathbb{R}$ for the measure line of angular momentum; a real number (more precisely an integer) n represents the angular momentum value $nh/4\pi$. As it is known, angular momentum is the product of mass, position and velocity; thus its measure line is $\mathbb{M}\otimes\mathbb{L}\otimes\frac{\mathbb{L}}{\mathbb{T}}$ which is identified with $\mathbb{R}$; consequently, $\mathbb{M}\equiv\frac{\mathbb{T}}{\mathbb{L}\otimes\mathbb{L}}$.

In this book, for easier theoretical considerations, we take $\frac{\mathbb{T}}{\mathbb{L}\otimes\mathbb{L}}$ as the measure line of masses. If $\boldsymbol{m}$ is the unit distance and $\boldsymbol{s}$ is the unit time period then $\frac{\boldsymbol{s}}{\boldsymbol{m}^2}$ is the unit mass. One finds the experimental data

$$h/4\pi = 1.05\ldots\cdot 10^{-34}\frac{\boldsymbol{m}^2\,\boldsymbol{kg}}{\boldsymbol{s}},$$

hence if we take it equal to the real number one we arrive at the definition

$$\boldsymbol{kg} := 9.4813\ldots\cdot 10^{33}\frac{\boldsymbol{s}}{\boldsymbol{m}^2}.$$

2.4.2. Since acceleration values are elements of $\frac{\mathbf{S}}{\mathbb{T}\otimes\mathbb{T}}$ and 'the product of mass and acceleration equals the force', the force values are elements of $\frac{\mathbb{T}}{\mathbb{L}\otimes\mathbb{L}}\otimes\frac{\mathbf{S}}{\mathbb{T}\otimes\mathbb{T}}\equiv\frac{\mathbf{S}}{\mathbb{T}\otimes\mathbb{L}\otimes\mathbb{L}}\equiv\frac{\mathbf{S}^*}{\mathbb{T}}$; moreover, 'a force can depend on time, on space and on velocity'. Thus we accept that a *force field* is a differentiable mapping

$$\boldsymbol{f}\colon \mathrm{M}\times V(1)\rightarrowtail\frac{\mathbf{S}^*}{\mathbb{T}}$$

and *the history of the material point with mass m under the action of the force field $\boldsymbol{f}$* is given by the Newtonian equation

$$m\ddot{x} = \boldsymbol{f}(x,\dot{x}),$$

i.e. the world line modelling the history is a solution of this differential equation.

2.4.3. The most important force fields can be derived from potentials; e.g. the gravitational field and the electromagnetic field. Usually the gravitational field is the gradient of a scalar potential and the electromagnetic field is given by the gradient of a scalar potential and the curl of a vector potential. The gravitational force acting on a material point depends only on the spacetime position of the masspoint, the electromagnetic force depends on the velocity of the masspoint as well. To introduce the notion of potential in the spacetime model, we have to rely on these facts. Now we give the convenient definition and we shall show in Section 6 that it is suitable indeed.

A *potential* is a twice differentiable mapping

$$\boldsymbol{K}: \mathrm{M} \rightarrowtail \mathbf{M}^*$$

(in other words, a potential is a twice differentiable covector field).

The *field strength* corresponding to $\boldsymbol{K}$ is $\mathrm{D} \wedge \boldsymbol{K}: \mathrm{M} \rightarrowtail \mathbf{M}^* \wedge \mathbf{M}^*$ (the antisymmetric or exterior derivative of $\boldsymbol{K}$, see VI.3.6).

The force field $\boldsymbol{f}$ has a potential (is derived from a potential) if
— there is an open subset $\mathrm{O} \subset \mathrm{M}$ such that $\mathrm{Dom}\,\boldsymbol{f} = \mathrm{O} \times V(1)$,
— there is a potential $\boldsymbol{K}$ defined on O such that

$$\boldsymbol{f}(x, \dot{x}) = \boldsymbol{i}^* \cdot \boldsymbol{F}(x)\dot{x} \qquad \big(x \in \mathrm{O},\ \dot{x} \in V(1)\big),$$

where $\boldsymbol{F} := \mathrm{D} \wedge \boldsymbol{K}$ and $\boldsymbol{i}: \mathbf{S} \to \mathbf{M}$ is the embedding. Checking this formula, the reader can seize the opportunity to practice using the dot product.

2.5. Exercises

1. Let r_1 and r_2 be world line functions. Characterize the function $r_1 - r_2$.
2. Another formulation of the preceding exercise: give necessary and sufficient conditions for a function $\boldsymbol{z}$ that $r + \boldsymbol{z}$ be a world line for all world lines r.
3. Describe the world line function s in the split spacetime model (cf. 2.1.4).

3. Observers

3.1. The notions of an observer and its space

3.1.1. Observers played an important role in our heuristics treated in the Introduction; an observer in the real word is a material object or a set of material objects, e.g the earth, the houses on it form an observer, the car is another observer.

Now we are in position to give an exact definition in the spacetime model.

We can imagine that an observer is a collection of material points existing 'in close proximity' to each other. The existence of a masspoint in spacetime is described by a world line. Thus an observer would be modelled by a collection of world lines that fill 'continuously' a domain of spacetime. How can a convenient notion of such continuity be defined? To all points of every world line of the observer we assign the corresponding absolute velocity value; in this way we define an absolute velocity field: a function defined for some world points and having values in $V(1)$. Conversely, given an absolute velocity field (with convenient mathematical properties), we can recover the world lines of the observers:

world lines having everywhere the velocity value prescribed by the velocity field. We shall see that the velocity field is extremely suitable for our purposes, hence we prefer it to the collection of world lines.

Definition. An *observer* is a smooth map $\boldsymbol{U}\colon \mathrm{M} \rightarrowtail V(1)$ whose domain is connected.

If $\operatorname{Dom} \boldsymbol{U} = \mathrm{M}$, the observer is called *global.* ■

We emphasize that we are dealing with mathematical models; an observer as it is defined is a mathematical model for a physical object. To underline this fact we might use the term 'observer model' instead of 'observer' but we wish to avoid ponderousness. If necessary, we shall say physical observer for the material objects in question.

3.1.2. Let $\boldsymbol{U}$ be an observer. The integral curves of the differential equation

$$(x\colon \mathrm{T} \rightarrowtail \mathrm{M})? \qquad \dot{x} = \boldsymbol{U}(x)$$

have exclusively timelike tangent vectors, thus they are world lines (see 2.2).

The maximal integral curves of this differential equation will be called *$\boldsymbol{U}$-lines.*

If the world line function r is a solution of the above differential equation—i.e. $\operatorname{Ran} r$ is an integral curve of $\boldsymbol{U}$ —then $\dot{r}(t) = \boldsymbol{U}(r(t))$ and so $\ddot{r}(t) = \mathrm{D}\boldsymbol{U}(r(t)) \cdot \dot{r}(t) = \mathrm{D}\boldsymbol{U}(r(t)) \cdot \boldsymbol{U}(r(t))$ for all $t \in \operatorname{Dom} r$. This motivates that

$$\boldsymbol{A_U}\colon \mathrm{M} \rightarrowtail \frac{\mathbf{S}}{\mathrm{T}\otimes\mathrm{T}}, \qquad x \mapsto \mathrm{D}\boldsymbol{U}(x)\cdot\boldsymbol{U}(x)$$

is called the *acceleration field* corresponding to the observer $\boldsymbol{U}$.

3.1.3. Definition. An observer $\boldsymbol{U}$ is called *fit* if all the world line functions giving the $\boldsymbol{U}$-lines have the same domain; this uniquely determined interval of T is the *lifetime* of the observer. ■

It may happen that the maximal integral curves of a global observer are not global world lines (see Exercise 3.4). A global observer $\boldsymbol{U}$ is fit if and only if all $\boldsymbol{U}$-lines are global.

3.1.4. In the arithmetic spacetime model an observer is given by a function $\boldsymbol{V}\colon \mathbb{R}\times\mathbb{R}^3 \rightarrowtail \mathbb{R}^3$ in the form $(1,\boldsymbol{V}) = (1,\ V^1,\ V^2,\ V^3)$. If we denote the partial derivatives corresponding to $\mathbb{R}$ and $\mathbb{R}^3$ by ∂_0 and $\nabla = (\partial_1,\ \partial_2,\ \partial_3)$, respectively, then the acceleration field of the observer is $(0,\ \partial_0\boldsymbol{V} + \boldsymbol{V}\cdot\nabla\boldsymbol{V}) = \left(0, (\partial_0 V^i + \sum_{k=1}^{3} V^k\partial_k V^i)_{i=1,2,3}\right)$.

3.1.5. As it is stated in the Introduction, a physical observer—a material object—establishes space for itself. The points of its space are just the material points that it consists of. In our model these points correspond to the maximal integral curves of the observer. Thus the space of an observer $\boldsymbol{U}$ is just the collection of $\boldsymbol{U}$-lines:

Definition. Let $\boldsymbol{U}$ be an observer and let S_U denote the set of maximal integral curves of $\boldsymbol{U}$. S_U is called *the space of the observer* $\boldsymbol{U}$, or the $\boldsymbol{U}$-space. ■

The elements of the $\boldsymbol{U}$-space are world lines. We have to get accustomed to this situation, which is strange at first sight but common in mathematics: the elements of a set are sets themselves.

A maximal integral curve of $\boldsymbol{U}$ will be called a $\boldsymbol{U}$*-line* if considered to be a subset of M and will be called a $\boldsymbol{U}$*-space point* if considered to be an element of S_U.

We measure distances in our physical space, we know what is near, what is far. We define limit procedures regarding our space. These notions must appear in the model, too.

It can be shown (see [1]) that, in general, the $\boldsymbol{U}$-space can be endowed with a smooth structure in a natural way, thus limits, differentiability etc. will make sense. However, in this book we avoid the general theory of smooth manifolds, that is why, in general, we do not deal with the structure of observer spaces. Later the spaces of some special observers, important from the point of view of applications, will be treated in detail.

3.1.6. Recall from the theory of differential equations that different integral curves of $\boldsymbol{U}$ do not intersect (VI.6.2). Let us introduce the map $C_U : \mathrm{Dom}\,\boldsymbol{U} \to \mathrm{S}_U$ in such a way that $C_U(x)$ is the (unique) $\boldsymbol{U}$-line passing through x.

We shall say as well that $C_U(x)$ is the $\boldsymbol{U}$-space point that the world point x is *incident* with.

3.2. Classification of observers

3.2.1. We have considered the room and the car as examples of physical observers. However, much 'worse' material objects can be observers as well. For instance, the stormy sea: the distance of its space points (which are the molecules of the water) and even the direction of their mutual positions are not constant. A ship on the stormy sea is a little better because it does not change its shape, it is rigid. However, it rotates, i.e. the directions of relative positions of its space points vary with time. The slightly waving water is better than the stormy one because it does not whirl. These examples show from what point of view we should classify observers in our spacetime model.

We mention that physical observers, in reality, are never rigid and rotation-free; at least molecular motion contradicts these properties. Besides, a physical observer is never global, it cannot fill all the spacetime. All these notions, as all models, are idealizations, extrapolations for a convenient mathematical description.

Recall the notation introduced in 2.2.

Definition. A fit observer $\boldsymbol{U}$ is called

(i) *rigid* if for all $q_1, q_2 \in \mathrm{S}_U$ the distance between $q_1 \star t$ and $q_2 \star t$, in other words $|q_1 \star t - q_2 \star t|$, is the same for all t in the lifetime of $\boldsymbol{U}$;

(ii) *rotation-free* if for all $q_1, q_2 \in \mathrm{S}_U$ the direction of the vector $q_1 \star t - q_2 \star t$ is the same for all t in the lifetime of $\boldsymbol{U}$;

(iii) *twist-free* if all $\boldsymbol{U}$-space points are twist-free;

(iv) *inertial* if $\boldsymbol{U}$ is a constant function; in other words, if the $\boldsymbol{U}$-lines are parallel straight line segments in spacetime. ■

Except the inertial observers, it is difficult to give a good illustration of these types of observers. The following figure tries to show a rigid or rotation-free observer.

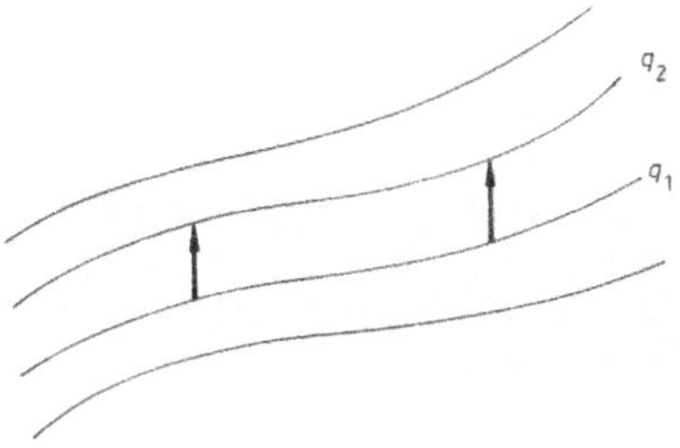

Suppose q_1 runs in the plane of the sheet. Letting q_2 bend below the plane of the sheet in such a way that its points have the same distances from the corresponding points of q_1, we can draw a picture of a rigid observer which is not rotation-free.

Letting q_2 bend in the plane of the sheet we can draw a picture of a rotation-free observer which is not rigid.

3.2.2. We call attention to the fact that a fit observer whose space points are all inertial (i.e. straight line segments) is not necessarily inertial: it may occur that its integral curves are not parallel (see Exercise 5.4.1).

Evidently, an inertial observer is rigid, rotation-free and twist-free. The converse is not true: see 5.2.

A fit observer $\boldsymbol{U}$ is rigid and rotation-free if and only if for all q_1, $q_2 \in \mathrm{S}_U$, $q_1 \star t - q_2 \star t$ is the same for all t in the lifetime of $\boldsymbol{U}$.

3.3. Reference frames, splitting of spacetime

3.3.1. Synchronization means defining simultaneity 'continuously'. Evidently, if we want to give somehow a synchronization, then we must adhere to the following rules:

1. every world point x must be simultaneous with itself,
2. if x is simultaneous with y then y is simultaneous with x,
3. if x is simultaneous with y and y is simultaneous with z then x is simultaneous with z.

Thus, a synchronization is an equivalence relation. Moreover, a further evident requirement is that

4. different occurrences of any world line cannot be simultaneous.

Thus, x and y cannot be simultaneous if $y - x$ is timelike.

Further, we expect that the equivalence classes be 'good' sets; all these lead us to the following definition,

Definition. A *synchronization* is an equivalence relation defined in a connected open subset of spacetime, such that every equivalence class is a three-dimensional submanifold whose every tangent space is transversal to all timelike vectors. ■

In other words, the tangent spaces must be spacelike. It is trivial that *there is only one global (i.e. everywhere defined) synchronization in the nonrelativistic spacetime model:* the equivalence class of x is $x + \mathbf{S}$.

3.3.2. The following definition seems superfluous but it is necessary for better comprehending the relativistic case.

Definition. An observer and a synchronization together form a *reference frame.* ■

Evidently, an observer now uniquely determines a reference frame. Thus, in the nonrelativistic spacetime model, but *only here, we can say observer instead of reference frame.*

3.3.3. Given an observer $\boldsymbol{U}$, then the map

$$\xi_U : \mathrm{Dom}\,\boldsymbol{U} \to \mathrm{T} \times \mathrm{S}_U, \qquad x \mapsto \big(\tau(x),\ C_U(x)\big)$$

is clearly injective, its inverse is

$$(t, q) \mapsto q \star t \qquad \big((t, q) \in \mathrm{Ran}\,\xi_U \subset \mathrm{T} \times \mathrm{S}_U\big)$$

where the notation introduced in 2.2 is used.

In this way spacetime points in the domain of $\boldsymbol{U}$ are represented by pairs of time points and $\boldsymbol{U}$-space points. We say that the observer $\boldsymbol{U}$ *splits spacetime* into time and $\boldsymbol{U}$-space with the aid of ξ_U.

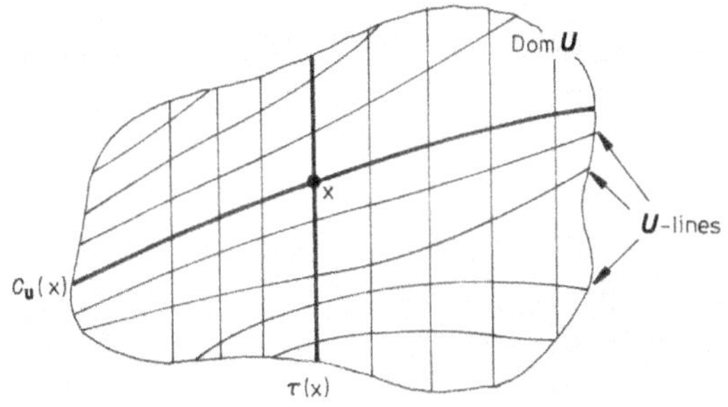

Definition. ξ_U is the *splitting* of spacetime *according* to $\boldsymbol{U}$. ■

If S_U is endowed with the smooth structure mentioned previously then ξ_U will be smooth. Its properties will be clarified in special cases.

3.4. Exercise

The observer

$$(\xi^0,\ \xi^1,\ \xi^2,\ \xi^3) \mapsto (1,\ -(\xi^1)^2,\ 0,\ 0)$$

in the arithmetic spacetime model is global, its maximal integral curve passing through $(\xi^0,\ \xi^1,\ \xi^2,\ \xi^3)$ is

$$\{(t,\ 0,\ \xi^2,\ \xi^3) \mid t \in \mathbb{R}\} \qquad \text{if} \qquad \xi^1 = 0,$$

$$\left\{(t,\ \frac{1}{t-\xi^0+1/\xi^1},\ \xi^2,\ \xi^3) \,\middle|\, t > \xi^0 - 1/\xi^1\right\} \qquad \text{if} \qquad \xi^1 > 0,$$

$$\{(t,\ \frac{1}{t-\xi^0+1/\xi^1},\ \xi^2,\ \xi^3) | t < \xi^0 - 1/\xi^1\} \qquad \text{if} \qquad \xi^1 < 0.$$

Consequently, most of the maximal integral curves of the observer are not global.

4. Rigid observers

4.1. Inertial observers

4.1.1. Let us consider a global inertial observer with constant absolute velocity value $\boldsymbol{u}$; we shall refer to this observer by $\boldsymbol{u}$ and we omit the epithet global.

Recall the linear map $\boldsymbol{\pi_u}$—the projection onto $\mathbf{S}$ along $\boldsymbol{u}\otimes\mathbb{T}$—defined in 1.3.8.

The observer space $\mathrm{S}_{\boldsymbol{u}}$ is the set of straight lines directed by $\boldsymbol{u}$; more closely,

$$C_{\boldsymbol{u}}(x) = x + \boldsymbol{u}\otimes\mathbb{T} := \{x + \boldsymbol{u}t \mid t \in \mathbb{T}\}.$$

Note that

$$(x + \boldsymbol{u}\otimes\mathbb{T}) \star t = x + \boldsymbol{u}(t - \tau(x)).$$

As a consequence, the inertial observer is rigid and rotation-free:

$$\begin{aligned}(x_2 + \boldsymbol{u}\otimes\mathbb{T}) \star t - (x_1 + \boldsymbol{u}\otimes\mathbb{T}) \star t = \\ &= \big(x_2 + \boldsymbol{u}(t - \tau(x_2))\big) - \big(x_1 + \boldsymbol{u}(t - \tau(x_1))\big) = \\ &= x_2 - x_1 - \boldsymbol{u}\big(\boldsymbol{\tau}\cdot(x_2 - x_1)\big) = \\ &= \boldsymbol{\pi_u}\cdot(x_2 - x_1).\end{aligned}$$

4.1.2. According to the previous formula, if q_2 and q_1 are $\boldsymbol{u}$-space points then $q_2 \star t - q_1 \star t$ is the same vector in $\mathbf{S}$ for all $t \in \mathrm{T}$: more closely, it equals $\boldsymbol{\pi_u}\cdot(x_2 - x_1)$ where x_1 and x_2 are arbitrary elements of q_1 and q_2, respectively. Regarding this vector as the difference of the $\boldsymbol{u}$-space points, we define an affine structure on $\mathrm{S}_{\boldsymbol{u}}$ in a natural way.

Proposition. $\mathrm{S}_{\boldsymbol{u}}$, endowed with the subtraction

$$q_2 - q_1 := \boldsymbol{\pi_u}\cdot(x_2 - x_1) \qquad (q_1,\ q_2 \in \mathrm{S}_{\boldsymbol{u}},\ x_1 \in q_1,\ x_2 \in q_2)$$

is an affine space over $\mathbf{S}$. ■

Observe that if $x_1 \in q_1$, $x_2 \in q_2$ and $\tau(x_1) = \tau(x_2)$, then $q_2 - q_1 = x_2 - x_1$.

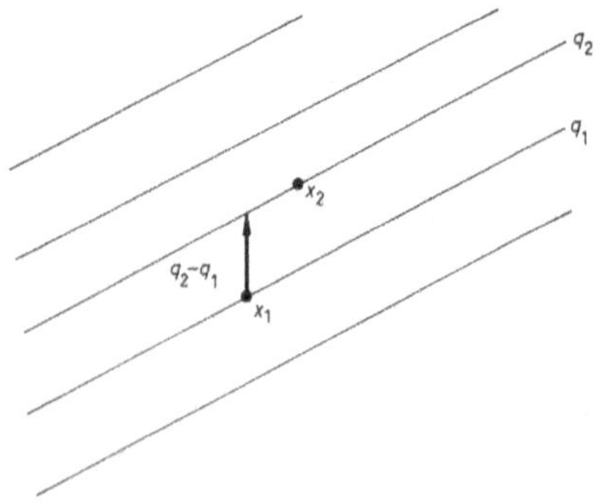

It is worth remarking that

$$(x + \boldsymbol{q}) + \boldsymbol{u} \otimes \mathbb{T} = (x + \boldsymbol{u} \otimes \mathbb{T}) + \boldsymbol{q} \qquad (x \in \mathrm{M},\ \boldsymbol{q} \in \mathbf{S}),$$

which is not trivial because here the same sign $+$ denotes different operations: the first one refers to the addition between elements of M and **M**; the second and the third ones denote a set addition between elements of M and **M**; the fourth one indicates the addition between elements of $\mathrm{S}_{\boldsymbol{u}}$ and **S**. This formula has the generalization

$$(x + \boldsymbol{x}) + \boldsymbol{u} \otimes \mathbb{T} = (x + \boldsymbol{u} \otimes \mathbb{T}) + \boldsymbol{\pi}_{\boldsymbol{u}} \cdot \boldsymbol{x} \qquad (x \in \mathrm{M},\ \boldsymbol{x} \in \mathbf{M}).$$

4.1.3. The space of any inertial observer is a three-dimensional oriented Euclidean affine space (over **S**). In this way we regain our experience regarding our physical space from the spacetime model (see the Introduction).

Now we see why the vectors in **S** are called spacelike.

Note that *the spaces of different inertial observers are* **different affine spaces** *over the* **same vector space.** The following assertion is proved without any difficulty.

Proposition. The splitting of spacetime according to an inertial observer with velocity value $\boldsymbol{u}$,

$$\mathrm{M} \to \mathrm{T} \times \mathrm{S}_{\boldsymbol{u}}, \qquad x \mapsto \xi_{\boldsymbol{u}}(x) := \big(\tau(x),\, C_{\boldsymbol{u}}(x)\big) = (x + \mathbf{S},\ x + \boldsymbol{u} \otimes \mathbb{T})$$

is an orientation preserving affine bijection having $\boldsymbol{\xi}_{\boldsymbol{u}} = (\boldsymbol{\tau}, \boldsymbol{\pi}_{\boldsymbol{u}})$ as its underlying linear map. ∎

4.1.4. We have to get accustomed to the fact that a physical notion which seems 'structureless', 'as simple as possible' (e.g. a space point of an observer) is modelled by a less simple, structured mathematical object (by a line). In mathematics it is customary that the *elements* of a set are themselves *sets or functions.*

However, we have a tool to reduce some of our mathematical objects to simpler ones. This tool is the vectorization of affine spaces: choosing an arbitrary element ('reference origin') in an affine space, we can represent every element of the affine space by a vector.

An inertial observer with velocity value $\boldsymbol{u}$, taking a $t_o \in \mathrm{T}$ and a $q_o \in \mathrm{S}_{\boldsymbol{u}}$, can establish the vectorization of time and $\boldsymbol{u}$-space:

$$V_o \colon \mathrm{T} \times \mathrm{S}_{\boldsymbol{u}} \to \mathbb{T} \times \mathbf{S}, \qquad (t, q) \to (t - t_o,\ q - q_o)$$

by which, in particular, we represent $\boldsymbol{u}$-space points by vectors in **S** that are simpler objects than straight lines in M.

Notice that choosing t_o and q_o is equivalent to choosing a *spacetime reference origin* $o \in \mathrm{M}$: $o := q_o \star t_o$, $t_o = \tau(o)$, $q_o = C_{\boldsymbol{u}}(o)$.

Definition. An *inertial observer with origin* is a pair $(\boldsymbol{u}, o)$ where $\boldsymbol{u}$ is a the constant velocity value of an inertial observer and o is a world point.

The *vectorized splitting* of spacetime corresponding to $(\boldsymbol{u}, o)$ is the map

$$\xi_{\boldsymbol{u},o} := V_o \circ \xi_{\boldsymbol{u}} : \mathrm{M} \to \mathbb{T} \times \mathbf{S}, \quad x \to \big(\tau(x) - \tau(o),\ C_{\boldsymbol{u}}(x) - C_{\boldsymbol{u}}(o)\big) = \\ = \big(\boldsymbol{\tau} \cdot (x - o),\ \boldsymbol{\pi}_{\boldsymbol{u}} \cdot (x - o)\big). \quad \blacksquare$$

Note that

$$\xi_{\boldsymbol{u},o} = \boldsymbol{\xi}_{\boldsymbol{u}} \circ O_o,$$

where $\boldsymbol{\xi}_{\boldsymbol{u}} = (\boldsymbol{\tau}, \boldsymbol{\pi}_{\boldsymbol{u}})$ and O_o is the vectorization of M with origin o:, i.e. $O_o : \mathrm{M} \to \mathbf{M}$, $x \mapsto x - o$.

4.1.5. Let us consider the arithmetic spacetime model and the inertial observer with constant value $(1, \boldsymbol{v})$. The space point of the observer that $(\alpha, \boldsymbol{\xi})$ is incident with is the straight line $(\alpha, \boldsymbol{\xi}) + (1, \boldsymbol{v})\mathbb{R} = \{(\alpha + t, \boldsymbol{\xi} + \boldsymbol{v}t) \mid t \in \mathbb{R}\}$.

As concerns the affine structure of the set of such lines we have

$$[(\alpha, \boldsymbol{\xi}) + (1 + \boldsymbol{v})\mathbb{R}] - [(\beta, \boldsymbol{\zeta}) + (1, \boldsymbol{v})\mathbb{R}] = \boldsymbol{\xi} - \boldsymbol{\zeta} - \boldsymbol{v}(\alpha - \beta) \in \mathbb{R}^3.$$

Let the observer in question choose $(0, \mathbf{0})$ as reference origin. Then the observer space will be represented by $\mathbb{R}^3$; the space point $(\alpha, \boldsymbol{\xi}) + (1, \boldsymbol{v})\mathbb{R}$ will correspond to the difference of this straight line and that passing through $(0, \mathbf{0})$ which is $(1, \boldsymbol{v})\mathbb{R}$; this difference is exactly $\boldsymbol{\xi} - \boldsymbol{v}\alpha$.

Consequently, the vectorized splitting of spacetime due to this observer is

$$\mathbb{R} \times \mathbb{R}^3 \to \mathbb{R} \times \mathbb{R}^3, \qquad (\alpha, \boldsymbol{\xi}) \mapsto (\alpha, \boldsymbol{\xi} - \boldsymbol{v}\alpha).$$

In particular, the splitting of spacetime according to the 'basic observer'—the one whose value is the basic velocity value $(1, \mathbf{0})$ —with reference origin $(0, \mathbf{0})$ is the identity of $\mathbb{R} \times \mathbb{R}^3$: the arithmetic spacetime model is the Cartesian product of vectorized time and vectorized space relative to the basic observer.

In other words, the observer with reference origin makes the correspondence that previously has been accepted as a natural identification. The vectorized splitting of spacetime is described by the formula above.

4.2. Characterization of rigid observers *

4.2.1. Now we derive some mathematical results to characterize some properties of observers. Simple but important relations for deducing our results are the following.

Recall that $C_{\boldsymbol{U}}(x)$ denotes the $\boldsymbol{U}$-line passing through x. Then $t \mapsto C_{\boldsymbol{U}}(x) \star t$ is the corresponding world line function. So we have

$$C_{\boldsymbol{U}}(x) \star \tau(x) = x$$

and

$$\frac{\mathrm{d}}{\mathrm{d}t}\big(C_{\boldsymbol{U}}(x) \star t\big) = \boldsymbol{U}\big(C_{\boldsymbol{U}}(x) \star t\big).$$

Proposition. Let $\boldsymbol{U}$ be a fit global observer.
(i) $\boldsymbol{U}$ is rigid if and only if

$$\big(\boldsymbol{U}(x + \boldsymbol{q}) - \boldsymbol{U}(x)\big) \cdot \boldsymbol{q} = \boldsymbol{0} \qquad (x \in \mathrm{M},\ \boldsymbol{q} \in \mathbf{S}).$$

(ii) $\boldsymbol{U}$ is rigid and rotation-free if and only if

$$\boldsymbol{U}(x + \boldsymbol{q}) - \boldsymbol{U}(x) = \boldsymbol{0} \qquad (x \in \mathrm{M},\ \boldsymbol{q} \in \mathbf{S}),$$

which is equivalent to the existence of a smooth map $\boldsymbol{V} : \mathrm{T} \rightarrowtail V(1)$ such that

$$\boldsymbol{U} = \boldsymbol{V} \circ \tau.$$

Proof. Let $q_1, q_2 \in \mathrm{S}_{\boldsymbol{U}}$.
(i) The function

$$t \mapsto |q_1 \star t - q_2 \star t|^2$$

is constant if and only if its derivative

$$t \mapsto 2\big(\boldsymbol{U}(q_1 \star t) - \boldsymbol{U}(q_2 \star t)\big) \cdot (q_1 \star t - q_2 \star t)$$

is zero.

Putting $x := q_2 \star t$, $\boldsymbol{q} := q_1 \star t - q_2 \star t$ in the derivative we infer that the derivative is zero if and only if the equality in the assertion holds (every $x \in \mathrm{M}$ is of the form $q_2 \star t$ for some q_2 and t and every $\boldsymbol{q} \in \mathbf{S}$ is of the form $q_1 \star t - q_2 \star t$ for some q_1).
(ii) The function

$$t \mapsto q_1 \star t - q_2 \star t$$

is constant if and only if its derivative

$$t \mapsto \boldsymbol{U}(q_1 \star t) - \boldsymbol{U}(q_2 \star t)$$

is zero.

Reasoning as previously we get the desired result.

4.2.2. Let U be a global rigid observer. For $t_o, t \in T$ let us define

$$R_U(t,t_o) : \mathbf{S} \to \mathbf{S}, \qquad q \mapsto C_U(x_o + q) \star t \; - \; C_U(x_o) \star t,$$

where x_o is an arbitrary element of t_o (i.e. $x_o \in M$ and $\tau(x_o) = t_o$).

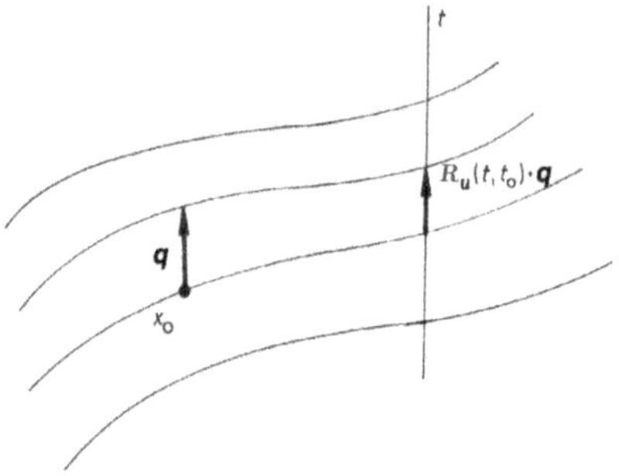

Proposition. If U is a global rigid observer then $R_U(t,t_o)$ is a rotation in $\mathbf{S}$ (a linear orthogonal map with determinant 1) for all $t_o, t \in T$. Moreover, $R_U(t,t_o)$ is independent of x_o appearing in its definition.

The global rigid observer U is rotation-free if and only if $R_U(t,t_o) = \mathbf{1_S}$ for all $t_o, t \in \mathbb{T}$.

Proof. Evidently,

$$R_U(t,t_o)(\mathbf{0}) = \mathbf{0}.$$

Moreover, since U is rigid, for all $q_1, q_2 \in \mathbf{S}$ we have

$$\begin{aligned}
&| R_U(t,t_o)(q_1) - R_U(t,t_o)(q_2) | = \\
&= \big| \big(C_U(x_o + q_1) \star t - C_U(x_o) \star t\big) - \big(C_U(x_o + q_2) \star t - C_U(x_o) \star t\big) \big| = \\
&= \big| C_U(x_o + q_1) \star t - C_U(x_o + q_2) \star t \big| = \big| (x_o + q_1) - (x_o + q_2) \big| = \\
&= | q_1 - q_2 |.
\end{aligned}$$

As a consequence, $R_U(t,t_o)$ is a linear orthogonal map (see V.3.7).

For fixed t_o and fixed $q \in \mathbf{S}$, the function $T \to \mathbf{S}$, $t \mapsto R_U(t,t_o) \cdot q$ is smooth since it is the difference of two solutions of the smooth differential equation $\dot{x} = U(x)$. Consequently, $t \mapsto \det R_U(t,t_o)$ is a smooth function. Since the determinants in question can be 1 or -1 only and

$$R_U(t_o,t_o) = \mathbf{1_S},$$

all the determinants in question equal 1.

If y_o is another element of t_o, then with the notation $\boldsymbol{q}_o := y_o - x_o \in \mathbf{S}$ we deduce

$$C_U(y_o + \boldsymbol{q}) \star t - C_U(y_o) \star t = C_U(x_o + \boldsymbol{q}_o + \boldsymbol{q}) \star t - C_U(x_o + \boldsymbol{q}_o) \star t =$$
$$= \big(C_U(x_o + \boldsymbol{q}_o + \boldsymbol{q}) \star t - C_U(x_o) \star t\big) - \big(C_U(x_o + \boldsymbol{q}_o) \star t - C_U(x_o \star t)\big) =$$
$$= \boldsymbol{R}_U(t, t_o) \cdot (\boldsymbol{q}_o + \boldsymbol{q}) - \boldsymbol{R}_U(t, t_o) \cdot \boldsymbol{q}_o = \boldsymbol{R}_U(t, t_o) \cdot \boldsymbol{q},$$

which means that the definition of $\boldsymbol{R}_U(t, t_o)$ is independent of x_o.

4.2.3. Proposition. For all $t_o, t, s \in \mathrm{T}$ we have

(*i*) $\boldsymbol{R}_U(t_o, t_o) = \mathbf{1}_{\mathbf{S}}$,

(*ii*) $\boldsymbol{R}_U(t, t_o)^{-1} = \boldsymbol{R}_U(t_o, t)$,

(*iii*) $\boldsymbol{R}_U(t, t_o) = \boldsymbol{R}_U(t, s) \cdot \boldsymbol{R}_U(s, t_o)$.

Proof. (*i*) is trivial.

The defining formula of $\boldsymbol{R}_U(t, t_o)$ can be rewritten in the following form: if q, q_o are $\boldsymbol{U}$-space points then

$$q \star t - q_o \star t = \boldsymbol{R}_U(t, t_o) \cdot (q \star t_o - q_o \star t_o) \qquad (t_o, t \in \mathrm{T}). \quad (*)$$

Interchanging t and t_o we get

$$q \star t_o - q_o \star t_o = \boldsymbol{R}_U(t_o, t) \cdot (q \star t - q_o \star t)$$

from which we infer (*ii*).

In a similar way we obtain (*iii*.) ■

Observe that $(*)$ implies that if $\boldsymbol{R}_U(t, t_o)$ is known for a t_o and for all t then every $\boldsymbol{U}$-space point q can be deduced from an arbitrarily chosen q_o.

4.2.4. Let $\boldsymbol{U}$ be a global rigid observer. For fixed $t_o \in \mathrm{T}$, the function $\mathrm{T} \to \mathbf{S} \otimes \mathbf{S}^*$, $t \mapsto \boldsymbol{R}_U(t, t_o)$ is smooth (because for all $\boldsymbol{q} \in \mathbf{S}$, $t \mapsto \boldsymbol{R}_U(t, t_o) \cdot \boldsymbol{q}$ is smooth); we introduce

$$\dot{\boldsymbol{R}}_U(t, t_o) := \frac{\mathrm{d}\boldsymbol{R}_U(t, t_o)}{\mathrm{d}t} \in \frac{\mathbf{S} \otimes \mathbf{S}^*}{\mathbb{T}} \qquad (t, t_o \in \mathrm{T}),$$

which can be regarded as a linear map

$$\dot{\boldsymbol{R}}_U(t, t_o) \colon \mathbf{S} \to \frac{\mathbf{S}}{\mathbb{T}}, \qquad \boldsymbol{q} \mapsto \frac{\mathrm{d}}{\mathrm{d}t} \boldsymbol{R}_U(t, t_o) \cdot \boldsymbol{q}$$

(VI.3.11). We deduce from the defining formula of $\boldsymbol{R}_U(t, t_o)$ that

$$\dot{\boldsymbol{R}}_U(t, t_o) \cdot \boldsymbol{q} = \boldsymbol{U}\big(C_U(x_o + \boldsymbol{q}) \star t\big) - \boldsymbol{U}\big(C_U(x_o) \star t\big) =$$
$$= \boldsymbol{U}\big(C_U(x_o) \star t + \boldsymbol{R}_U(t, t_o) \cdot \boldsymbol{q}\big) - \boldsymbol{U}\big(C_U(x_o) \star t\big) =$$
$$= \boldsymbol{U}\big(q \star t + \boldsymbol{R}_U(t, t_o) \cdot \boldsymbol{q}\big) - \boldsymbol{U}(q \star t),$$

where x_o is an arbitrary element of t_o and q is an arbitrary element of S_U.

Substituting $\boldsymbol{R}_U(t,t_o)^{-1}\cdot\boldsymbol{q}$ for $\boldsymbol{q}$ and introducing the linear map

$$\boldsymbol{\Omega}_U(t) := \dot{\boldsymbol{R}}_U(t,t_o)\cdot\boldsymbol{R}_U(t,t_o)^{-1} : \mathbf{S}\to\frac{\mathbf{S}}{\mathbb{T}}$$

for $t\in\mathrm{T}$, we obtain

$$\boldsymbol{\Omega}_U(t)\cdot\boldsymbol{q} = \boldsymbol{U}(q\star t+\boldsymbol{q}) - \boldsymbol{U}(q\star t) \qquad (t\in\mathrm{T}, \boldsymbol{q}\in\mathbf{S}).$$

We know that $\boldsymbol{\Omega}_U(t)$ is antisymmetric (see 11.1.10). Since $q\star t$ can be an arbitrary world point, we have proved:

Proposition. If $\boldsymbol{U}$ is a global rigid observer then $\boldsymbol{\Omega}_U(t)$ is an antisymmetric linear map for all $t\in\mathrm{T}$; it is independent of t_o appearing in its definition. Moreover,

$$\boldsymbol{U}(x+\boldsymbol{q}) - \boldsymbol{U}(x) = \boldsymbol{\Omega}_U\big(\tau(x)\big)\cdot\boldsymbol{q} \qquad (x\in\mathrm{M},\ \boldsymbol{q}\in\mathbf{S}). \quad (**)$$

The global rigid observer $\boldsymbol{U}$ is rotation-free if and only if $\boldsymbol{\Omega}_U(t)=\mathbf{0}$ for all $t\in\mathrm{T}$. ■

Notice that the restriction of $\boldsymbol{U}$ to an arbitrary simultaneous hyperplane t is an affine map whose underlying linear map is $\boldsymbol{\Omega}_U(t)$.

$\boldsymbol{\Omega}_U(t)$ can be interpreted as the *angular velocity* of the observer at the instant t (see 11.1.10).

4.2.5. For arbitrarily fixed $t_o\in\mathrm{T}$, the function $t\mapsto\boldsymbol{R}_U(t,t_o)$ defines the function $t\mapsto\boldsymbol{\Omega}_U(t)$ according to the preceding paragraph. Conversely, if the function $t\mapsto\boldsymbol{\Omega}_U(t)$ is known, then $t\mapsto\boldsymbol{R}_U(t,t_o)$ is determined as the unique solution of the differential equation

$$(\boldsymbol{X}:\mathrm{T}\to\mathbf{S}\otimes\mathbf{S}^*)? \qquad \dot{\boldsymbol{X}} = \boldsymbol{\Omega}_U\cdot\boldsymbol{X}$$

with the initial condition

$$\boldsymbol{X}(t_o) = \mathbf{1}_{\mathbf{S}}.$$

4.2.6. We see from the formula (**) of 4.2.4 that the rigid observer $\boldsymbol{U}$ is completely determined by an arbitrarily chosen $\boldsymbol{U}$-space point q_o and by the angular velocity of the observer, i.e. by the function $t\mapsto\boldsymbol{\Omega}_U(t)$. Indeed, putting $\boldsymbol{q} := q_o\star\tau(x) - x$ in that formula we obtain

$$\boldsymbol{U}(x) = \boldsymbol{U}\big(q_o\star\tau(x)\big) + \boldsymbol{\Omega}_U\big(\tau(x)\big)\cdot\big(x - q_o\star\tau(x)\big) \qquad (x\in\mathrm{M})$$

and we know that the values of $\boldsymbol{U}$ on q_o coincide with the derivative of the world line function $t\mapsto q_o\star t$.

4.3. About the spaces of rigid observers*

4.3.1. Proposition. Let $\boldsymbol{U}$ be a fit global observer. $\boldsymbol{U}$ is rigid and rotation-free if and only if S_U, equipped with the subtraction

$$q_1 - q_2 := q_1 \star t - q_2 \star t \qquad (q_1, q_2 \in \mathrm{S}_U,\ t \in \mathrm{T})$$

is an affine space over $\mathbf{S}$.

Proof. If $\boldsymbol{U}$ is rigid and rotation-free then, for all $q_1, q_2 \in \mathrm{S}_U$, $q_1 \star t - q_2 \star t$ is the same element of $\mathbf{S}$ for all $t \in \mathrm{T}$. It is not hard to see that the subtraction in the assertion satisfies the requirements listed in the definition of affine spaces.

Conversely, if S_U is an affine space over $\mathbf{S}$ with the given subtraction then, in particular, $q_1 \star t - q_2 \star t$ is independent of t for all $q_1, q_2 \in \mathrm{S}_U$, hence $\boldsymbol{U}$ is rigid and rotation-free.

4.3.2. If $\boldsymbol{U}$ is a global rigid and rotation-free observer, then S_U is an affine space, thus the differentiability of the splitting of spacetime according to $\boldsymbol{U}$ makes sense.

Proposition. Let U be a global rigid and rotation-free observer. Then the splitting

$$\xi_U : \mathrm{M} \to \mathrm{T} \times \mathrm{S}_U, \qquad x \mapsto \big(\tau(x),\ C_U(x)\big)$$

is a smooth bijection,

$$\mathrm{D}\xi_U(x) = \big(\boldsymbol{\tau}, \boldsymbol{\pi}_{\boldsymbol{U}(x)}\big) \qquad (x \in \mathrm{M}),$$

and the inverse of ξ_U is smooth as well.

Proof. For $x \in \mathrm{M}$ and $t \in \mathrm{T}$ we have $C_U(x) \star t = x + \boldsymbol{U}(x)\big(t - \tau(x)\big) + \mathrm{ordo}\big(t - \tau(x)\big)$ (VI.3.3). Thus for all $y, x \in \mathrm{M}$ (see Exercise 4.5.1),

$$\begin{aligned} C_U(y) - C_U(x) &= y - C_U(x) \star \tau(y) = \\ &= y - x + \boldsymbol{U}(x)\big(\tau(y) - \tau(x)\big) + \mathrm{ordo}\big(\tau(y) - \tau(x)\big) \end{aligned}$$

and so

$$\begin{aligned} \xi_U(y) - \xi_U(x) &= \big(\tau(y) - \tau(x),\quad C_U(y) - C_U(x)\big) = \\ &= \big(\boldsymbol{\tau} \cdot (y - x),\quad \boldsymbol{\pi}_{\boldsymbol{U}(x)} \cdot (y - x)\big) + \mathrm{ordo}\big(\boldsymbol{\tau}(y - x)\big). \end{aligned}$$

Hence ξ_U is differentiable, its derivative is the one given in the proposition. As a consequence, we see that ξ_U is smooth; its inverse is smooth by the inverse mapping theorem.

4.3.3. The space of a rigid and rotation-free global observer, endowed with a natural subtraction, is an affine space over $\mathbf{S}$. The space of another observer is *not* affine space with *that subtraction* (in fact that subtraction makes no sense for other observers). This does not mean that the space of other observers cannot be endowed with an affine structure in some other way.

Let us consider a fit global observer $\boldsymbol{U}$. For every instant t we can define the *instantaneous* affine structure on S_U by the subtraction $q_1 - q_2 := q_1 \star t - q_2 \star t$. In general, different instants determine different instantaneous affine structures and all instants have the same 'right' for establishing an affine structure on the $\boldsymbol{U}$-space. There is no natural way to select an instant and to use the corresponding instantaneous affine structure as the affine structure of S_U.

Nevertheless, we can define a natural affine structure on the spaces of *rigid global* observers.

4.3.4. Though the Earth rotates, we experience on it an affine structure independent of time. A stick on the earth represents a vector. Evidently, the stick rotates together with the earth. The stick will be represented in the following reasoning by two points (the extremities of the stick) in the observer space. Now we wish to define that two points in the space of a rigid observer determine a vector (rotating together with the observer).

Let $\boldsymbol{U}$ be a rigid global observer. If q_1 and q_2 are points in the observer space S_U then for all $t, t' \in \mathrm{T}$

$$q_1 \star t - q_2 \star t = \boldsymbol{R}_U(t, t') \cdot (q_1 \star t' - q_2 \star t').$$

Let us introduce

$$\mathbf{S}_U := \{\boldsymbol{s} \colon \mathrm{T} \to \mathbf{S} \mid \boldsymbol{s} \text{ is smooth, } \boldsymbol{s}(t) = \boldsymbol{R}_U(t, t') \cdot \boldsymbol{s}(t') \text{ for all } t, t' \in \mathrm{T}\}.$$

It is a routine to check that $\mathbf{S}_U$, endowed with the usual pointwise addition and pointwise multiplication by real numbers, is a vector space; it is three-dimensional, because $\mathbf{S}_U \to \mathbf{S}$, $\boldsymbol{s} \mapsto \boldsymbol{s}(t)$ is a linear bijection for arbitrary $t \in \mathrm{T}$ (which means in particular, that the function $\boldsymbol{s}$ is completely determined by a single one of its values). Moreover, if $\boldsymbol{s}_1$ and $\boldsymbol{s}_2$ are elements of $\mathbf{S}_U$, then $\boldsymbol{s}_1(t) \cdot \boldsymbol{s}_2(t)$ is the same for all instants t, thus

$$\mathbf{S}_U \times \mathbf{S}_U \to \mathbb{L} \otimes \mathbb{L}, \qquad (\boldsymbol{s}_1, \boldsymbol{s}_2) \mapsto \boldsymbol{s}_1 \cdot \boldsymbol{s}_2 : = \boldsymbol{s}_1(t) \cdot \boldsymbol{s}_2(t)$$

is a positive definite symmetric bilinear map which turns $\mathbf{S}_U$ into a Euclidean vector space.

Now it is quite evident that S_U, endowed with the subtraction

$$q_1 - q_2 := \big(\mathrm{T} \to \mathbf{S},\ t \mapsto (q_1 \star t - q_2 \star t)\big)$$

will be an affine space over $\mathbf{S}_U$. In other words, the difference of two $\boldsymbol{U}$-space points is exactly the difference of the corresponding world line functions, as the difference of functions is defined.

If $\boldsymbol{U}$ is rotation-free, then $\mathbf{S}_U$ consists of the constant functions from T into $\mathbf{S}$ which can be identified with $\mathbf{S}$. So we get back our previous result that the space of a global rigid and rotation-free observer is an affine space over $\mathbf{S}$ in a natural way.

If $\boldsymbol{U}$ is not rotation-free then S_U is a three-dimensional Euclidean affine space in a natural way, but the underlying vector space is not $\mathbf{S}$; in fact the underlying vector space $\mathbf{S}_U$ depends on the observer itself.

4.3.5. The space of a global rigid observer is an affine space, thus the differentiability of the splitting of spacetime according to the observer makes sense. This question, reduced to a simpler affine structure, will be studied in the next section.

4.4. Observers with origin*

4.4.1. The vectorization of observer spaces simplifies some formulae for inertial observers and it will be a powerful tool for noninertial rigid observers.

Let $\boldsymbol{U}$ be a global rigid and rotation-free observer. Choosing an instant t_o and a $\boldsymbol{U}$-space point q_o, we give the corresponding *vectorization* of time and $\boldsymbol{U}$-space:

$$V_o\colon \mathrm{T}\times\mathrm{S}_U \to \mathbb{T}\times\mathbf{S}, \qquad (t,q)\mapsto (t-t_o,\ \ q-q_o).$$

We see that in this way $\boldsymbol{U}$-space points (curves in M) are represented by spacelike vectors (points in $\mathbf{S}$).

Notice that choosing t_o and q_o is equivalent to choosing a 'spacetime reference origin' $o\in\mathrm{M}$: $o := q_o\star t_o,\ t_o=\tau(o),\ q_o = C_U(o)$. That is why we have used the symbol V_o for the vectorization which can be written in the following form, too:

$$V_o\colon \mathrm{T}\times\mathrm{S}_U \to \mathbb{T}\times\mathbf{S}, \qquad (t,q,)\mapsto (t-\tau(o),\ \ q\star\tau(o)-o),$$

since $q-q_o = q\star t - q_o\star t$ for all $t\in\mathrm{T}$, in particular for $t:=\tau(o)$.

If $\boldsymbol{U}$ is not rotation-free, the result of a similar vectorization

$$V_o\colon \mathrm{T}\times\mathrm{S}_U \to \mathbb{T}\times\mathbf{S}_U, \qquad (t,q)\mapsto (t-t_o,\ \ q-q_o)$$

is not simple enough because the elements of $\mathbf{S}_U$ are functions. That is why we make a further step by the linear bijection

$$\boldsymbol{L}_o\colon \mathbf{S}_U\to\mathbf{S}, \qquad \boldsymbol{s}\mapsto\boldsymbol{s}(t_o).$$

Since $\boldsymbol{L}_o\cdot(q-q_o) = (q-q_o)(t_o) = q\star t_o - q_o\star t_o$, we get the *double vectorization* of time and $\boldsymbol{U}$-space:

$$W_o := (\mathbf{1}_\mathbb{T}\times\boldsymbol{L}_o)\circ V_o\colon \mathrm{T}\times\mathrm{S}_U\to\mathbb{T}\times\mathbf{S}, \qquad (t,q)\mapsto(t-\tau(o),\ \ q\star\tau(o)-o),$$

which coincides formally with the vectorization of time and space of a rigid and rotation-free observer.

4.4.2. Definition. A *rigid observer with reference origin* is a pair $(\boldsymbol{U}, o)$ where $\boldsymbol{U}$ is a global rigid observer and o is a world point.

If $\boldsymbol{U}$ is rotation-free, the *vectorized splitting* of spacetime corresponding to $(\boldsymbol{U}, o)$ is the map

$$\xi_{\boldsymbol{U},o} := V_o \circ \xi_{\boldsymbol{U}} : \mathrm{M} \to \mathbb{T} \times \mathbf{S}, \qquad x \mapsto \big(\tau(x) - \tau(o),\ \ C_{\boldsymbol{U}}(x) - C_{\boldsymbol{U}}(o)\big) = \big(\tau(x) - \tau(o),\ \ C_{\boldsymbol{U}}(x) \star \tau(o) - o\big),$$

and if $\boldsymbol{U}$ is not rotation-free then the *double vectorized splitting* of spacetime is the map

$$\xi_{\boldsymbol{U},o} := W_o \circ \xi_{\boldsymbol{U}} : \mathrm{M} \to \mathbb{T} \times \mathbf{S}, \qquad x \mapsto \big(\tau(x) - \tau(o),\ \ C_{\boldsymbol{U}}(x) \star \tau(o) - o\big).$$

4.4.3. Proposition. Let $(\boldsymbol{U}, o)$ be an observer with reference origin.

If $\boldsymbol{U}$ is rotation-free then the vectorized splitting is a smooth bijection whose inverse is smooth as well and

$$\mathrm{D}\xi_{\boldsymbol{U},o}(x) = (\boldsymbol{\tau}, \boldsymbol{\pi}_{\boldsymbol{U}(x)}) \qquad (x \in \mathrm{M}).$$

If $\boldsymbol{U}$ is not rotation-free, the double vectorized splitting is a smooth bijection whose inverse is smooth as well and

$$\mathrm{D}\xi_{\boldsymbol{U},o}(x) = \Big(\boldsymbol{\tau}, \boldsymbol{R}_{\boldsymbol{U}}(\tau(x), t_o)^{-1} \cdot \boldsymbol{\pi}_{\boldsymbol{U}(x)}\Big) \qquad (x \in \mathrm{M})$$

where $t_o := \tau(o)$.

Proof. For rotation-free observers the assertion is trivial because of 4.3.2 and because the derivative of V_o is the identity of $\mathbb{T} \times \mathbf{S}$.

For the double vectorization we argue as follows: the map $\mathrm{M} \to \mathbf{S}$, $x \mapsto C_{\boldsymbol{U}}(x) \star t_o - o = \boldsymbol{R}_{\boldsymbol{U}}(\tau(x), t_o)^{-1} \cdot \big(x - C_{\boldsymbol{U}}(o) \star \tau(x)\big)$ (see formula $(*)$ in 4.2.3.) is clearly differentiable, its derivative is the linear map (see Exercise 11.2.2)

$$\mathbf{M} \to \mathbf{S}, \qquad \boldsymbol{x} \mapsto - \boldsymbol{R}_{\boldsymbol{U}}(\tau(x), t_o)^{-1} \cdot \boldsymbol{\Omega}_{\boldsymbol{U}}(\tau(x)) \cdot \big(x - C_{\boldsymbol{U}}(o) \star \tau(x)\big)\boldsymbol{\tau} \cdot \boldsymbol{x} + {} + \boldsymbol{R}_{\boldsymbol{U}}(\tau(x), t_o)^{-1} \cdot \big(\boldsymbol{x} - \boldsymbol{U}\big(C_{\boldsymbol{U}}(o) \star \tau(x)\big)\boldsymbol{\tau} \cdot \boldsymbol{x}\big) = {} = \boldsymbol{R}_{\boldsymbol{U}}(\tau(x), t_o)^{-1} \cdot \boldsymbol{\pi}_{\boldsymbol{U}(x)} \cdot \boldsymbol{x}. \qquad \blacksquare$$

Since W_o is an affine bijection, it follows that the splitting $\xi_{\boldsymbol{U}} : \mathrm{M} \to \mathbb{T} \times \mathbf{S}_{\boldsymbol{U}}$ is smooth and has a smooth inverse as well (cf. 4.3.5.).

4.4.4. Dealing with observers in the arithmetic spacetime model it is extremely convenient to consider observers with reference origin where the reference origin coincides with the origin $(0, \mathbf{0})$ of $\mathbb{R} \times \mathbb{R}^3$. Namely, in this case the (double) vectorized observer spaces are $\mathbb{R}^3$ and the (double) vectorized splitting is a linear map $\mathbb{R} \times \mathbb{R}^3 \to \mathbb{R} \times \mathbb{R}^3$ whose zeroth component is the zeroth projection.

4.5. Exercises

1. If S_U is the affine space over $\mathbf{S}$ with the subtraction given in 4.3.1, then

$$\begin{aligned} C_U(x + \boldsymbol{q}) &= C_U(x) + \boldsymbol{q}, \\ C_U(y) - C_U(x) &= C_U(y) \star \tau(x) - x = \\ &= y - C_U(x) \star \tau(y). \end{aligned}$$

for all $x, y \in \mathrm{M}, \boldsymbol{q} \in \mathbf{S}$.

2. Prove that

$$\boldsymbol{\Omega}_U(t) = \dot{\boldsymbol{R}}_U(t,t) \qquad (t \in \mathrm{T})$$

(see 4.2.4).

3. We know that the derivative at a point of a double vectorization is of the form $(\boldsymbol{\tau}, \boldsymbol{R}^{-1} \cdot \boldsymbol{\pi_u}) \colon \mathbf{M} \to \mathbb{T} \times \mathbf{S}$ where $\boldsymbol{u} \in V(1)$ and $\boldsymbol{R}$ is an orthogonal map $\mathbf{S} \to \mathbf{S}$, i.e. $\boldsymbol{R}^{\circledast} = \boldsymbol{R}^{-1}$ (see 4.4.3). Recall that the adjoint $\boldsymbol{R}^{\circledast}$ is identified with the transpose $\boldsymbol{R}^*$ due to the identification $\frac{\mathbf{S}}{\mathbb{L} \otimes \mathbb{L}} \equiv \mathbf{S}^*$. Thus we have $\boldsymbol{R}^* \equiv \boldsymbol{R}^{\circledast} = \boldsymbol{R}^{-1}$ and so $(\boldsymbol{i} \cdot \boldsymbol{R})^* = \boldsymbol{R}^{-1} \cdot \boldsymbol{i}^*$. Prove that

$$(\boldsymbol{\tau}, \boldsymbol{R}^{-1} \cdot \boldsymbol{\pi_u})^{*^{-1}} = (\boldsymbol{u}, \boldsymbol{R}^{-1} \cdot \boldsymbol{i}^*).$$

4. Let $\boldsymbol{U}$ be a global rigid observer. Using Proposition 4.2.1. prove that $\mathrm{D}\boldsymbol{U}(x)|_{\mathbf{S}}$ is antisymmetric for all $x \in \mathrm{M}$ (which is proved in 4.2.4 in another way).

5. Let $\boldsymbol{U}$ be a fit global observer. Demonstrate that $\boldsymbol{U}$ is rotation-free if and only if there is a smooth map $\alpha \colon \mathrm{T} \times \mathrm{M} \times \mathbf{S} \to \mathbb{R}$ such that

(*i*) $C_U(x + \boldsymbol{q}) \star t - C_U(x) \star t = \alpha(t, x, \boldsymbol{q})\boldsymbol{q}$ $\qquad (t \in \mathrm{T}, x \in \mathrm{M}, \boldsymbol{q} \in \mathbf{S})$;

(*ii*) $\alpha(\tau(x), x, \boldsymbol{q}) = 1$ $\qquad (x \in \mathrm{M}, \boldsymbol{q} \in \mathbf{S})$;

(*iii*) $\alpha(t, x, 0) = 1$ $\qquad (t \in \mathrm{T}, x \in \mathrm{M})$.

6. Using the previous result prove that if $\boldsymbol{U}$ is a global rigid and rotation-free observer then there is a smooth map $\beta \colon \mathrm{M} \to \frac{\mathbb{R}}{\mathbb{T}}$ such that $\mathrm{D}\boldsymbol{U}(x)|_{\mathbf{S}} = \beta(x)\mathbf{1}_{\mathbf{S}}$ for all $x \in \mathrm{M}$.

5. Some special observers

5.1. Why inertial observers are better than others

5.1.1. We know that the space of a rigid and rotation-free global observer, even if it is not inertial, is an affine space over $\mathbf{S}$. However, the splitting of spacetime according to noninertial observers is not affine.

Proposition. Let $\boldsymbol{U}$ be a rigid and rotation-free global observer. The splitting of spacetime according to $\boldsymbol{U}$,

$$\xi_U : \mathrm{M} \to \mathrm{T} \times \mathrm{S}_U, \qquad x \mapsto \big(\tau(x), C_U(x)\big)$$

is an affine map if and only if $\boldsymbol{U}$ is inertial.

Proof. We have seen that if $\boldsymbol{U}$ is inertial then ξ_U is affine.

We know that ξ_U is differentiable, $\mathrm{D}\xi_U(x) = (\boldsymbol{\tau}, \boldsymbol{\pi}_{U(x)})$ (see 4.3.2). If ξ_U is affine, then $\mathrm{D}\xi_U(x)$ is the same for all $x \in \mathrm{M}$. This means that $\boldsymbol{\pi}_{U(x)}$ does not depend on x which implies that $\boldsymbol{U}$ is a constant map as well.

5.1.2. We can say that if S_U is affine but $\boldsymbol{U}$ is not inertial then the affine structures of M and $\mathrm{T}\times\mathrm{S}_U$ —though they are mathematically isomorphic—are not related from a physical point of view.

If S_U is affine, then $(\mathrm{T}\times\mathrm{S}_U, \mathbb{T}, \mathbb{L}, \boldsymbol{\tau}, \boldsymbol{h})$ is a nonrelativistic spacetime model and so it is isomorphic to the spacetime model $(\mathrm{M}, \mathbb{T}, \mathbb{L}, \boldsymbol{\tau}, \boldsymbol{h})$; however, the physically meaningful triplet $(\xi_U, \mathbf{1}_{\mathbb{T}}, \mathbf{1}_{\mathbb{L}})$ is an isomorphism between them if and only if $\boldsymbol{U}$ is inertial.

This shows that global inertial observers play an important role in applications. Let $\boldsymbol{U}$ be a global inertial observer and suppose an assertion is formulated for some objects related to $\mathrm{T}\times\mathrm{S}_U$; then the assertion concerns an absolute fact if it uses only the affine structure of $\mathrm{T}\times\mathrm{S}_U$. The assertion has not necessarily an absolute content if it uses other properties of $\mathrm{T}\times\mathrm{S}_U$; for instance, the Cartesian product structure or the affine structure of S_U alone.

5.2. Uniformly accelerated observer

5.2.1. The rigid global observer $\boldsymbol{U}$ is called *uniformly accelerated* if its acceleration field is a nonzero constant, i.e. there is a $\mathbf{0} \neq \boldsymbol{a} \in \frac{\mathbf{S}}{\mathbb{T}\otimes\mathbb{T}}$ such that

$$\boldsymbol{A}_U(x) := \mathrm{D}\boldsymbol{U}(x)\cdot\boldsymbol{U}(x) = \boldsymbol{a}. \qquad (x \in \mathrm{M}).$$

Equivalently, for all $\boldsymbol{U}$-space points q, $\frac{\mathrm{d}^2}{\mathrm{d}t^2}(q \star t) = \boldsymbol{a}$ $(t \in \mathrm{T})$.

We have for all $x \in \mathrm{M}$ and $t \in \mathrm{T}$ that

$$C_{\boldsymbol{U}}(x) \star t = x + \boldsymbol{U}(x)\big(t - \tau(x)\big) + \frac{1}{2}\boldsymbol{a}\big(t - \tau(x)\big)^2$$

and

$$\boldsymbol{U}\big(C_{\boldsymbol{U}}(x) \star t\big) = \frac{\mathrm{d}}{\mathrm{d}t}\big(C_{\boldsymbol{U}}(x) \star t\big) = \boldsymbol{U}(x) + \boldsymbol{a}\big(t - \tau(x)\big). \qquad (*)$$

Now it follows that for all $x \in \mathrm{M}$, $\boldsymbol{q} \in \mathbf{S}$ and $t \in \mathrm{T}$

$$C_{\boldsymbol{U}}(x + \boldsymbol{q}) \star t - C_{\boldsymbol{U}}(x) \star t = \boldsymbol{q} + \big(\boldsymbol{U}(x + \boldsymbol{q}) - \boldsymbol{U}(x)\big)\big(t - \tau(x)\big).$$

Since $\boldsymbol{U}$ is rigid, the length of this vector is independent of t, so it equals the length of $\boldsymbol{q}$. Then assertion (i) in Proposition 4.2.1 implies that

$$\boldsymbol{U}(x + \boldsymbol{q}) - \boldsymbol{U}(x) = \mathbf{0} \qquad (x \in \mathrm{M},\ \boldsymbol{q} \in \mathbf{S})$$

which means, according to the quoted proposition, that $\boldsymbol{U}$ is rotation-free.

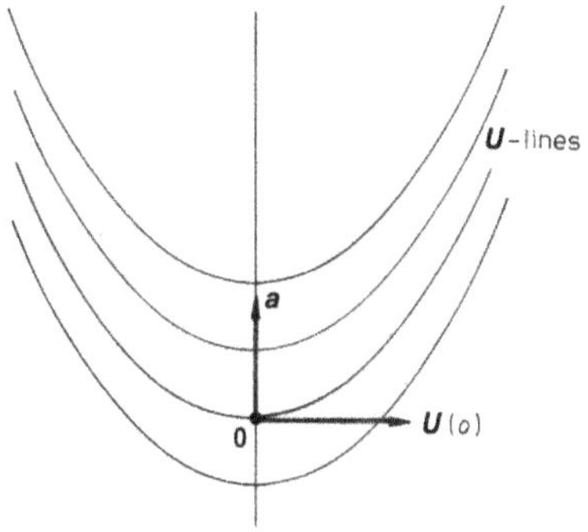

5.2.2. $\boldsymbol{U}$ is constant on the simultaneous hyperplanes. Thus $\boldsymbol{U}\big(C_{\boldsymbol{U}}(x) \star \tau(y)\big) = \boldsymbol{U}(y)$ for all $x, y \in \mathrm{M}$ and we infer from $(*)$ that

$$\boldsymbol{U}(y) = \boldsymbol{U}(x) + \boldsymbol{a}\big(\boldsymbol{\tau}(y - x)\big) \qquad (x, y \in \mathrm{M}).$$

As a corollary, the uniformly accelerated observer $\boldsymbol{U}$ is uniquely determined by a single value of $\boldsymbol{U}$ at an arbitrary world point and by the constant value of the acceleration field of $\boldsymbol{U}$.

We see as well that the uniformly accelerated observer is an affine map from M into $V(1)$ whose underlying linear map is $\boldsymbol{a} \cdot \boldsymbol{\tau}$.

5.2.3. Let the previous observer choose a reference origin o. Then

$$C_U(x) - C_U(o) = x - C_U(o) \star \tau(x) =$$
$$= x - o - \boldsymbol{U}(o)\,\big(\tau(x) - \tau(o)\big) - \frac{1}{2}\boldsymbol{a}\big(\tau(x) - \tau(o)\big)^2.$$

As a consequence, the vectorized splitting of spacetime is

$$\mathrm{M} \to \mathbb{T} \times \mathbf{S}, \quad x \mapsto \Big(\boldsymbol{\tau} \cdot (x - o), \quad \boldsymbol{\pi}_{U(o)} \cdot (x - o) - \frac{1}{2}\boldsymbol{a}(\boldsymbol{\tau} \cdot (x - o))^2\Big).$$

5.2.4. For $\alpha > 0$, the observer

$$(\xi^0, \xi^1, \xi^2, \xi^3) \mapsto (1, \alpha\xi^0, 0, 0)$$

in the arithmetic spacetime model is uniformly accelerated. Its maximal integral curve passing through $(\xi^0, \xi^1, \xi^2, \xi^3)$ is

$$\Big\{\Big(t, \xi^1 + \alpha\xi^0(t - \xi^0) + \frac{1}{2}\alpha(t - \xi^0)^2, \xi^2, \xi^3\Big) \,\Big|\, t \in \mathbb{R}\Big\} =$$
$$= \Big\{\Big(t, \xi^1 + \frac{1}{2}\alpha t^2 - \frac{1}{2}\alpha(\xi^0)^2, \xi^2, \xi^3\Big) \,\Big|\, t \in \mathbb{R}\Big\}.$$

Accordingly, if the observer chooses $(0, \mathbf{0})$ as a reference origin then the vectorized splitting becomes

$$\mathbb{R} \times \mathbb{R}^3 \to \mathbb{R} \times \mathbb{R}^3, \qquad (\xi^0, \xi^1, \xi^2, \xi^3) \mapsto \Big(\xi^0, \xi^1 - \frac{1}{2}\alpha(\xi^0)^2, \xi^2, \xi^3\Big).$$

5.3. Uniformly rotating observer

5.3.1. The global observer $\boldsymbol{U}$ is called *uniformly rotating* if there is a nonzero antisymmetric linear map $\boldsymbol{\Omega}\colon \mathbf{S} \to \frac{\mathbf{S}}{\mathbb{T}}$ (in other words, $\boldsymbol{\Omega} \in \frac{\mathbf{N} \wedge \mathbf{N}}{\mathbb{T}}$, $\mathbf{N} := \frac{\mathbf{S}}{\mathbb{L}}$), called the *angular velocity,* such that

$$\boldsymbol{U}(x + \boldsymbol{q}) - \boldsymbol{U}(x) = \boldsymbol{\Omega} \cdot \boldsymbol{q} \qquad (x \in \mathrm{M}, \boldsymbol{q} \in \mathbf{S}).$$

Proposition 4.2.1 (i) implies that $\boldsymbol{U}$ is rigid. Moreover, we easily obtain that

$$\boldsymbol{R}_U(t, t_o) = e^{(t - t_o)\boldsymbol{\Omega}} \qquad (t_o, t \in \mathrm{T}),$$

because this is the (necessarily unique) solution of the initial value problem given in 4.2.5.

Consequently, 4.2.2. yields that if $o, x \in \mathrm{M}$, $\tau(o) = t_o$ and $t \in \mathrm{T}$ then

$$C_U(x) \star t = C_U(o) \star t + e^{(t-t_o)\boldsymbol{\Omega}} \cdot (C_U(x) \star t_o - o)\,. \qquad (*)$$

Every $\boldsymbol{U}$-line is obtained from a given one and from $\boldsymbol{\Omega}$. This formula becomes simpler if we consider $x \in t_o$:

$$C_U(x) \star t = C_U(o) \star t + e^{(t-t_o)\boldsymbol{\Omega}} \cdot (x - o).$$

$\boldsymbol{U}$ itself is determined by its values on a given $\boldsymbol{U}$-line q_o and by $\boldsymbol{\Omega}$ (4.2.6):

$$\boldsymbol{U}(x) = \boldsymbol{U}\left(q_o \star \tau(x)\right) + \boldsymbol{\Omega} \cdot \left(x - q_o \star \tau(x)\right) \qquad (x \in \mathrm{M}).$$

5.3.2. Reformulating the previous result we can say that a uniformly rotating observer can be given by the history of a point of the observer (by a space point of the observer) and by its angular velocity. If we deal with a uniformly rotating observer then we are to look for its 'best' space points to have a simple description of the observer. Even if the observer is given by one of its space points and by its angular velocity, it may happen that we find a 'better' space point than the given one.

Now we shall examine a uniformly rotating observer $\boldsymbol{U}$ that has an inertial spacepoint. Then there is an $o \in \mathrm{M}$ and a $\boldsymbol{c} \in V(1)$ such that $q_o := o + \boldsymbol{c} \otimes \mathrm{T}$ is a $\boldsymbol{U}$-line. $\boldsymbol{U}$ equals $\boldsymbol{c}$ on q_o, thus

$$\boldsymbol{U}(x) = \boldsymbol{c} + \boldsymbol{\Omega} \cdot \boldsymbol{\pi}_{\boldsymbol{c}} \cdot (x - o) \qquad (x \in \mathrm{M}).$$

We see that $\boldsymbol{U}$ is an affine map, the underlying linear map is $\boldsymbol{\Omega} \cdot \boldsymbol{\pi}_{\boldsymbol{c}}$ whose range coincides with the range of $\boldsymbol{\Omega}$ which is a two- dimensional linear subspace in $\frac{\mathbf{S}}{\mathrm{T}}$.

We know that the kernel of $\boldsymbol{\Omega}$ is one-dimensional and orthogonal to Ran $\boldsymbol{\Omega}$ (see V.3.9). If $\boldsymbol{e} \in \operatorname{Ker}\boldsymbol{\Omega}$, then $\boldsymbol{U}(o + \boldsymbol{e} + \boldsymbol{c}\boldsymbol{t}) = \boldsymbol{c}$ for all $\boldsymbol{t} \in \mathbb{T}$, i.e. $\boldsymbol{U}$ is constant on the inertial world line $o + \boldsymbol{e} + \boldsymbol{c} \otimes \mathbb{T}$ as well. Thus it is a maximal integral curve of $\boldsymbol{U}$, parallel to q_o. It is an easy task to show that

$$\{x \in \mathrm{M} \mid \boldsymbol{U}(x) = \boldsymbol{c}\} = o + \operatorname{Ker}\boldsymbol{\Omega} + \boldsymbol{c} \otimes \mathbb{T}.$$

The observer has the acceleration field

$$\begin{aligned} \boldsymbol{A}_U(x) &= \mathrm{D}\boldsymbol{U}(x) \cdot \boldsymbol{U}(x) = \boldsymbol{\Omega} \cdot \boldsymbol{\pi}_{\boldsymbol{c}} \cdot \boldsymbol{U}(x) = \boldsymbol{\Omega} \cdot \left(\boldsymbol{U}(x) - \boldsymbol{c}\right) = \\ &= \boldsymbol{\Omega} \cdot \boldsymbol{\Omega} \cdot \boldsymbol{\pi}_{\boldsymbol{c}} \cdot (x - o) \qquad (x \in \mathrm{M}). \end{aligned}$$

Since $\mathrm{Ker}(\boldsymbol{\Omega}^2) = \mathrm{Ker}\,\boldsymbol{\Omega}$ (Exercise V.3.21.2), the set of acceleration-free world points is $\{x \in \mathrm{M} \mid \boldsymbol{\pi}_{\boldsymbol{c}} \cdot (x - o) \in \mathrm{Ker}\,\boldsymbol{\Omega}\}$ which equals $o + \mathrm{Ker}\,\boldsymbol{\Omega} + \boldsymbol{c} \otimes \mathbb{T}$.

Thus for all $\boldsymbol{e} \in \mathrm{Ker}\,\boldsymbol{\Omega}$, $o + \boldsymbol{e} + \boldsymbol{c} \otimes \mathbb{T}$ is an inertial $\boldsymbol{U}$-space point and there are no other inertial $\boldsymbol{U}$-space points. The inertial $\boldsymbol{U}$-space points corresponding to different elements of $\mathrm{Ker}\,\boldsymbol{\Omega}$ are different. The set

$$\{o + \boldsymbol{e} + \boldsymbol{c} \otimes \mathbb{T} \mid \boldsymbol{e} \in \mathrm{Ker}\,\boldsymbol{\Omega}\}$$

in $\mathrm{S}_{\boldsymbol{U}}$ is called the *axis of rotation.*

5.3.3. The axis of rotation makes sense for arbitrary uniformly rotating observers (see Exercise 5.4.4).

The Earth can be modelled by a uniformly rotating observer. Note that the angle between the axis of rotation and the direction of progression makes no absolute sense. The direction of progression is the direction of the relative velocity with respect to the Sun. The axis of rotation ($\mathrm{Ker}\,\boldsymbol{\Omega}$, an oriented one-dimensional linear subspace in $\mathbf{S}$) and a relative velocity value (an element of $\frac{\mathbf{S}}{\mathbb{T}}$ as we shall see in Section 6.2) make an angle; however, $\mathrm{Ker}\,\boldsymbol{\Omega}$ and an absolute velocity value ($\boldsymbol{c}$ in the former treatment) form no angle.

5.3.4. Let the previous observer choose o as a reference origin. Then formula $(*)$ in 5.3.1 yields that

$$C_{\boldsymbol{U}}(x) \star t_o - o = e^{-(\tau(x) - \tau(o))\boldsymbol{\Omega}} \cdot \big(x - (o + \boldsymbol{c}(\tau(x) - \tau(o)))\big),$$

thus the double vectorized splitting of spacetime becomes

$$\mathrm{M} \to \mathbb{T} \times \mathbf{S}, \qquad x \mapsto \Big(\boldsymbol{\tau} \cdot (x - o), e^{-\boldsymbol{\tau} \cdot (x-o)\boldsymbol{\Omega}} \cdot \boldsymbol{\pi}_{\boldsymbol{c}} \cdot (x - o)\Big).$$

5.3.5. For $\omega > 0$, the observer

$$(\xi^0,\ \xi^1,\ \xi^2,\ \xi^3) \mapsto (1,\ -\omega\xi^2,\ \omega\xi^1,\ 0)$$

in the arithmetic spacetime model is uniformly rotating. Its maximal integral curve passing through $(\xi^0,\ \xi^1,\ \xi^2,\ \xi^3)$ is

$$\begin{aligned}\big\{\big(t,\ &\xi^1 \cos\omega(t - \xi^0) - \xi^2 \sin\omega(t - \xi^0),\\ &\xi^1 \sin\omega(t - \xi^0) + \xi^2 \cos\omega(t - \xi^0), \xi^3\big) \bigm| t \in \mathbb{R}\big\}.\end{aligned}$$

If the observer chooses $(0, \mathbf{0})$ as a reference origin, the double vectorized splitting will be

$$\begin{aligned}&\mathbb{R} \times \mathbb{R}^3 \to \mathbb{R} \times \mathbb{R}^3,\\ &(\xi^0,\ \xi^1,\ \xi^2,\ \xi^3) \mapsto (\xi^0,\ \xi^1 \cos\omega\xi^0 + \xi^2 \sin\omega\xi^0,\ -\xi^1 \sin\omega\xi^0 + \xi^2 \cos\omega\xi^0,\ \xi^3).\end{aligned}$$

5.4. Exercises

1. Let $\boldsymbol{U}$ be a global observer. Demonstrate that the following assertions are equivalent:

(i) the acceleration field of $\boldsymbol{U}$ is zero,

(ii) all the integral curves of $\boldsymbol{U}$ are straight lines.

Such an observer need not be inertial. Consider the observer

$$(\xi^0,\ \xi^1,\ \xi^2,\ \xi^3) \mapsto (1,\ 0,\ \xi^1,\ 0)$$

in the arithmetic spacetime model. Give its maximal integral curves. Show that the observer is not rigid.

2. Let $\boldsymbol{U}$ be a global observer. Demonstrate that the following assertions are equivalent:

(i) the acceleration field of $\boldsymbol{U}$ is a nonzero constant,

(ii) all the integral curves of $\boldsymbol{U}$ are uniformly accelerated with the same nonzero acceleration.

Such an observer need not be uniformly accelerated. Consider the observer

$$(\xi^0,\ \xi^1,\ \xi^2,\ \xi^3) \mapsto (1, 0, \xi^0 + \xi^1, 0)$$

in the arithmetic spacetime model.

3. Prove that a global rigid observer whose integral curves are straight lines is inertial.

4. Define the axis of rotation for an arbitrary uniformly rotating observer.

5. Find the axis of rotation of the observer given in 5.3.5.

6. Since M and $V(1)$ are affine spaces, it makes sense that a global observer $\boldsymbol{U}\colon \mathrm{M} \to V(1)$ is affine; let $\mathrm{D}\boldsymbol{U}\colon \mathrm{M} \to \frac{\mathbf{S}}{\mathbb{T}}$ be the underlying linear map (the derivative of $\boldsymbol{U}$ at every point equals the linear map under $\boldsymbol{U}$). The restriction of $\mathrm{D}\boldsymbol{U}$ onto $\mathbf{S}$ will be denoted by $\boldsymbol{\Omega}_U$; it is a linear map from $\mathbf{S}$ into $\frac{\mathbf{S}}{\mathbb{T}}$. Prove that for all $x \in \mathrm{M}$ the world line function

$$\mathrm{T} \to \mathrm{M}, \quad t \mapsto x + \boldsymbol{U}(x)\big(t - \tau(x)\big) + \frac{1}{2}\mathrm{D}\boldsymbol{U} \cdot \boldsymbol{U}(x)\big(t - \tau(x)\big)^2 + \\ + \sum_{n=3}^{\infty} \frac{1}{n!}\Big(\big(t - \tau(x)\big)\boldsymbol{\Omega}_U\Big)^{n-2} \cdot \mathrm{D}\boldsymbol{U} \cdot \boldsymbol{U}(x)\big(t - \tau(x)\big)^2$$

gives the maximal integral curve passing through x.

7. Let $\boldsymbol{U}$ be an affine observer. Then $\boldsymbol{\Omega}_U := \mathrm{D}\,\boldsymbol{U}|_{\mathbf{S}} : \mathbf{S} \to \frac{\mathbf{S}}{\mathbb{T}}$ is a linear map. Prove that

(i) $$C_U(x + \boldsymbol{q}) = C_U(x) + \boldsymbol{q} \qquad (x \in \mathrm{M},\ \boldsymbol{q} \in \mathbf{S})$$

if and only if $\boldsymbol{q} \in \operatorname{Ker} \boldsymbol{\Omega}_U$;

(ii) $\boldsymbol{U}$ is rigid if and only if $\boldsymbol{\Omega}_U$ is antisymmetric.

8. Let $\boldsymbol{U}$ be a rigid affine observer. Then, according to the previous exercise, $\boldsymbol{\Omega}_U$ is antisymmetric.

We distinguish four cases:

$$(i)\quad \boldsymbol{\Omega}_U \neq \mathbf{0},\quad \mathrm{D}\boldsymbol{U}\cdot\boldsymbol{u} \neq \mathbf{0} \quad \text{for all} \quad \boldsymbol{u} \in V(1);$$

$$(ii)\quad \boldsymbol{\Omega}_U \neq \mathbf{0},\quad \mathrm{D}\boldsymbol{U}\cdot\boldsymbol{c} = \mathbf{0} \quad \text{for some} \quad \boldsymbol{c} \in V(1);$$

$$(iii)\quad \boldsymbol{\Omega}_U = \mathbf{0},\quad \mathrm{D}\boldsymbol{U}\cdot\boldsymbol{u} \neq \mathbf{0} \quad \text{for all} \quad \boldsymbol{u} \in V(1);$$

$$(iv)\quad \boldsymbol{\Omega}_U = \mathbf{0},\quad \mathrm{D}\boldsymbol{U}\cdot\boldsymbol{c} = \mathbf{0} \quad \text{for some} \quad \boldsymbol{c} \in V(1) \ \text{(i.e. } \mathrm{D}\boldsymbol{U} = \mathbf{0}).$$

Demonstrate that

(iv) is an inertial observer,

(iii) is a uniformly accelerated observer,

(ii) is a uniformly rotating observer having an inertial space point,

(i) is a uniformly rotating observer having a uniformly accelerated space point.

(Hint: the kernel of $\boldsymbol{\Omega}_U \neq \mathbf{0}$ is one-dimensional, $\boldsymbol{U}$ and $\mathrm{D}\boldsymbol{U}$ are surjections. Hence there is a $\boldsymbol{c} \in V(1)$ such that $\boldsymbol{a} := \mathrm{D}\boldsymbol{U}\cdot\boldsymbol{c}$ is in the kernel of $\boldsymbol{\Omega}_U$. Consequently, there is a world point o such that for all world points x

$$\boldsymbol{U}(x) = \boldsymbol{U}(o) + \mathrm{D}\boldsymbol{U}\cdot(x-o) = \boldsymbol{c} + \boldsymbol{\Omega}_U\cdot\boldsymbol{\pi}_{\boldsymbol{c}}\cdot(x-o) + \boldsymbol{a}\boldsymbol{\tau}\cdot(x-o)$$

and so the observer has the acceleration field

$$\boldsymbol{A}_U(x) = \boldsymbol{a} + \boldsymbol{\Omega}_U\cdot\boldsymbol{\Omega}_U\cdot\boldsymbol{\pi}_{\boldsymbol{c}}\cdot(x-o).)$$

9. Take an $o \in \mathrm{M}$ and define the observer

$$\boldsymbol{U}(x) := \frac{x-o}{\boldsymbol{\tau}\cdot(x-o)} \qquad (x \in o + T^{\rightarrow}).$$

Prove that

(i) every $\boldsymbol{U}$-space point is inertial, more closely,

$$C_U(x) \star t = o + \frac{x-o}{\boldsymbol{\tau}\cdot(x-o)}\big(t-\tau(o)\big) \qquad (x \in \mathrm{Dom}\,\boldsymbol{U},\ t > \tau(o));$$

(ii) the acceleration field corresponding to $\boldsymbol{U}$ is zero which follows from

$$\mathrm{D}\boldsymbol{U}(x) = \frac{\boldsymbol{\pi}_{\boldsymbol{U}(x)}}{\boldsymbol{\tau}\cdot(x-o)};$$

(iii) $\boldsymbol{U}$ is not rigid; the distance between two $\boldsymbol{U}$-space points increases as time passes.

10. Take an $o \in \mathrm{M}$, a $\boldsymbol{u} \in V(1)$, an $\boldsymbol{s} \in \mathbb{T}^+$ and define the observer

$$\boldsymbol{U}(x) := \boldsymbol{u} + \frac{\boldsymbol{\pi_u} \cdot (x - o)}{\boldsymbol{s}} \qquad (x \in \mathrm{M}).$$

Demonstrate that

(i) $\boldsymbol{U}$ is an affine observer, more closely

$$\mathrm{D}\boldsymbol{U}(x) = \frac{\boldsymbol{\pi_u}}{\boldsymbol{s}} \qquad \text{for all} \qquad x \in \mathrm{M};$$

(ii) the acceleration field corresponding to $\boldsymbol{U}$ is

$$x \mapsto \frac{\boldsymbol{\pi_u} \cdot (x - o)}{\boldsymbol{s}^2} = \frac{\boldsymbol{U}(x) - \boldsymbol{u}}{\boldsymbol{s}};$$

(iii) $C_{\boldsymbol{U}}(o + \boldsymbol{q}) \star t = o + \boldsymbol{u}\big(t - \tau(o)\big) + e^{(t-\tau(o))/\boldsymbol{s}}\boldsymbol{q} \qquad (\boldsymbol{q} \in \mathbf{S})$.

(iv) $\boldsymbol{U}$ is rotation-free and is not rigid: the distance between two $\boldsymbol{U}$-space points increases with time.

6. Kinematics

6.1. The history of a masspoint is observed as a motion

6.1.1. The motion of a material point relative to an observer is described by a function assigning to an instant the space point where the material point is at that instant.

Now we are able to give how an observer determines the motion from the history of a material point.

Definition. Let $\boldsymbol{U}$ be a fit observer and let r be a world line function, $\mathrm{Ran}\, r \subset \mathrm{Dom}\, \boldsymbol{U}$. Then

$$r_{\boldsymbol{U}} : \mathrm{T} \rightarrowtail \mathrm{S}_{\boldsymbol{U}}, \qquad t \mapsto C_{\boldsymbol{U}}\big(r(t)\big)$$

is called the *motion relative to* $\boldsymbol{U}$, or the $\boldsymbol{U}$*-motion*, corresponding to the world line function r. ∎

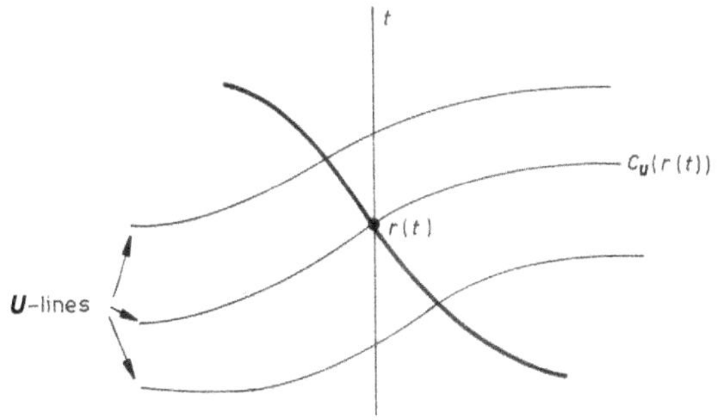

6.1.2. If $\boldsymbol{U}$ is a global rigid observer, then $\mathrm{S}_{\boldsymbol{U}}$ is an affine space thus the differentiability of $r_{\boldsymbol{U}}$ makes sense and $r_{\boldsymbol{U}}$ is piecewise twice differentiable.

Given a rigid and rotation-free global observer $\boldsymbol{U}$ and a motion relative to $\boldsymbol{U}$, i.e. a piecewise twice differentiable function $m\colon \mathrm{T} \rightarrowtail \mathrm{S}_{\boldsymbol{U}}$, we can regain the history, i.e. the world line function r for which $r_{\boldsymbol{U}} = m$ holds. Indeed, for every t, $m(t)$ is a $\boldsymbol{U}$-space point, i.e. a maximal integral curve of $\boldsymbol{U}$; then $r(t)$ will be the unique element in $t \cap m(t)$. In other words, using the splitting $\xi_{\boldsymbol{U}}$ we have

$$r(t) = {\xi_{\boldsymbol{U}}}^{-1}\big(t, m(t)\big) = m(t) \star t.$$

Similar considerations can be made for a general global rigid observer.

6.1.3. Let us consider the arithmetic spacetime model. As we know (see 2.1.4), a world line function r in it is given by a function $\boldsymbol{r}\colon \mathbb{R} \rightarrowtail \mathbb{R}^3$ in the form $r(t) = \big(t, \boldsymbol{r}(t)\big)$. Paragraph 4.1.5 shows that $\boldsymbol{r}$ is the corresponding motion relative to the basic observer. We see that the history is regained very simply from the motion (in view of the previous considerations it is a consequence of the fact that, for the basic observer $(1, \boldsymbol{0})$, $\xi_{(1,\boldsymbol{0})}$ is the identity the fact that for the basic observer $(1, \boldsymbol{0})$, the splitting $\xi_{(1,\boldsymbol{0})}$ is the identity of $\mathbb{R} \times \mathbb{R}^3$).

Thus if $\boldsymbol{r}\colon \mathbb{R} \rightarrowtail \mathbb{R}^3$ describes the motion relative to the basic observer then

$$r(t) = \big(t, \boldsymbol{r}(t)\big) \qquad (t \in \mathrm{Dom}\,\boldsymbol{r})$$

is the corresponding world line function.

6.2. Relative velocities

6.2.1. Proposition. Let $\boldsymbol{U}$ be a global rigid and rotation-free observer; if the world line function r is twice differentiable then $r_{\boldsymbol{U}}$ is twice differentiable as well and

$$\dot{r}_{\boldsymbol{U}}(t) = \dot{r}(t) - \boldsymbol{U}\big(r(t)\big), \qquad \ddot{r}_{\boldsymbol{U}}(t) = \ddot{r}(t) - \boldsymbol{A}_{\boldsymbol{U}}\big(r(t)\big).$$

Proof. Taking into account the relations

$$C_{\boldsymbol{U}}\big(r(s)\big) - C_{\boldsymbol{U}}\big(r(t)\big) = C_{\boldsymbol{U}}\big(r(s)\big) \star s - C_{\boldsymbol{U}}\big(r(t)\big) \star s = \\ = r(s) - r(t) - \Big[C_{\boldsymbol{U}}\big(r(t)\big) \star s - C_{\boldsymbol{U}}\big(r(t)\big) \star t\Big]$$

we deduce

$$\dot{r}_{\boldsymbol{U}}(t) = \lim_{s \to t} \frac{C_{\boldsymbol{U}}\big(r(s)\big) - C_{\boldsymbol{U}}\big(r(t)\big)}{s - t} = \dot{r}(t) - \boldsymbol{U}\big(r(t)\big),$$

from which $\ddot{r}_U(t) = \ddot{r}(t) - \mathrm{D}\boldsymbol{U}(r(t)) \cdot \dot{r}(t)$ follows immediately. Since now $\boldsymbol{U} = \boldsymbol{V} \circ \tau$ (see 4.2.1 (*ii*)), we have $\mathrm{D}\boldsymbol{U} = (\mathrm{D}\boldsymbol{V} \circ \tau) \cdot \boldsymbol{\tau}$ and $\boldsymbol{A}_U = \mathrm{D}\boldsymbol{V} \circ \tau$; thus the equality regarding the relative acceleration is verified. ■

The first and the second derivative of r_U is accepted as the *relative velocity* and the *relative acceleration* of r with respect to the global rigid and rotation-free observer $\boldsymbol{U}$, respectively.

6.2.2. The preceding result motivates the following definition.

Definition. Let $\boldsymbol{u}$ and $\boldsymbol{u}'$ be elements of $V(1)$. Then

$$\boldsymbol{v}_{\boldsymbol{u}'\boldsymbol{u}} := \boldsymbol{u}' - \boldsymbol{u}$$

is called the *relative velocity of* $\boldsymbol{u}'$ *with respect to* $\boldsymbol{u}$.

Proposition. Suppose $\boldsymbol{u}$, $\boldsymbol{u}'$ and $\boldsymbol{u}''$ are elements of $V(1)$. Then
(*i*) $\boldsymbol{v}_{\boldsymbol{u}'\boldsymbol{u}}$ is in $\frac{\mathbf{S}}{\mathbb{T}}$,
(*ii*) $\boldsymbol{v}_{\boldsymbol{u}'\boldsymbol{u}} = -\boldsymbol{v}_{\boldsymbol{u}\boldsymbol{u}'}$,
(*iii*) $\boldsymbol{v}_{\boldsymbol{u}''\boldsymbol{u}} = \boldsymbol{v}_{\boldsymbol{u}''\boldsymbol{u}'} + \boldsymbol{v}_{\boldsymbol{u}'\boldsymbol{u}}$. ■

These relations are very simple and they are in accordance with our everyday experience:
(*i*) the relative velocity values form a three-dimensional Euclidean vector space, the length of a relative velocity is in $\frac{\mathbb{L}}{\mathbb{T}}$;
(*ii*) if a body moves with a given relative velocity with respect to another body then the second body moves relative to the first one with the opposite velocity.
(*iii*) the sum of relative velocity values in a given order yields the resulting relative velocity value.

6.2.3. Let us imagine that a car is going on a straight road and it is raining. The raindrops hit the road and the car at different angles. What is the relation between the two angles?

Let $\boldsymbol{u}$ and $\boldsymbol{u}'$ be two different elements of $V(1)$ (the absolute velocity values of the road and of the car, respectively). If $\boldsymbol{w}$ is an element of $V(1)$, too, $\boldsymbol{w} \neq \boldsymbol{u}$, $\boldsymbol{w} \neq \boldsymbol{u}'$ (the absolute velocity value of the raindrops),

$$\theta(\boldsymbol{w}) := \arccos \frac{\boldsymbol{v}_{\boldsymbol{w}\boldsymbol{u}} \cdot \boldsymbol{v}_{\boldsymbol{u}'\boldsymbol{u}}}{|\boldsymbol{v}_{\boldsymbol{w}\boldsymbol{u}}||\boldsymbol{v}_{\boldsymbol{u}'\boldsymbol{u}}|}, \qquad \theta'(\boldsymbol{w}) := \arccos \frac{\boldsymbol{v}_{\boldsymbol{w}\boldsymbol{u}'} \cdot (-\boldsymbol{v}_{\boldsymbol{u}\boldsymbol{u}'})}{|\boldsymbol{v}_{\boldsymbol{w}\boldsymbol{u}'}||\boldsymbol{v}_{\boldsymbol{u}\boldsymbol{u}'}|}$$

are the angle formed by the relative velocity values $\boldsymbol{v}_{\boldsymbol{w}\boldsymbol{u}}$ and $\boldsymbol{v}_{\boldsymbol{u}'\boldsymbol{u}}$ and the angle formed by the relative velocity values $\boldsymbol{v}_{\boldsymbol{w}\boldsymbol{u}'}$ and $-\boldsymbol{v}_{\boldsymbol{u}'\boldsymbol{u}} = \boldsymbol{v}_{\boldsymbol{u}\boldsymbol{u}'}$, respectively (the angles at which the raindrops hit the road and the car, respectively).

A simple calculation yields that

$$\cos\theta(\boldsymbol{w}) = \frac{|\boldsymbol{v}_{\boldsymbol{w}\boldsymbol{u}'}|}{|\boldsymbol{v}_{\boldsymbol{w}\boldsymbol{u}}|}\cos\theta'(\boldsymbol{w}) + \frac{|\boldsymbol{v}_{\boldsymbol{u}'\boldsymbol{u}}|}{|\boldsymbol{v}_{\boldsymbol{w}\boldsymbol{u}}|}.$$

We call attention to an interesting limit case. Suppose $\boldsymbol{u}$ and $\boldsymbol{u}'$ are fixed and $\boldsymbol{w}$ tends to infinity, i.e. it varies in such a way that $|\boldsymbol{v_{wu}}|$ tends to infinity; then $|\boldsymbol{v_{wu'}}|$ tends to infinity as well and the quotient of these quantities tends to the number 1:

$$\lim_{w\to\infty} \cos\theta(\boldsymbol{w}) = \lim_{w\to\infty} \cos\theta'(\boldsymbol{w}),$$

which implies $\lim_{w\to\infty} \theta(\boldsymbol{w}) = \lim_{w\to\infty} \theta'(\boldsymbol{w})$.

Roughly speaking, the raindrops arriving with an 'infinitely big' relative velocity hit the road and the car at the same angle. Replacing 'raindrops with infinitely big relative velocity' by a 'light beam' we get that nonrelativistically there is *no aberration of light:* a light beam forms the same angle with the road and the car moving on the road.

We have spoken intuitively; of course the question arises at once: what is the model of a light beam in the nonrelativistic spacetime model? What mathematical object in the nonrelativistic spacetime model will correspond to a light beam? We shall see that none. A light beam cannot be modelled in the present spacetime model.

6.2.4. We can obtain the results of 6.2.1 by choosing a reference origin o in M, too, for the global rigid and rotation-free observer $\boldsymbol{U}$. Let us put $t_o := \tau(o)$, $q_o := C_U(o)$. Evidently, the derivative of the *vectorized motion*

$$\boldsymbol{r_U} : \mathrm{T} \rightarrowtail \mathbf{S}, \qquad t \mapsto r_U(t) - q_o$$

equals the derivative of r_U. Since

$$r_U(t) - q_o = C_U\big(r(t)\big) - q_o = C_U\big(r(t)\big) \star t - q_o \star t =$$
$$= r(t) - q_o \star t,$$

we get immediately

$$\dot{r}_U(t) = \dot{r}(t) - \boldsymbol{U}(q_o \star t) = \dot{r}(t) - \boldsymbol{U}\big(r(t)\big), \qquad (t \in \mathrm{T}),$$

because $\boldsymbol{U}$ is constant on the simultaneous hyperplanes.

We mention, that in practice it is more convenient to use the vectorized motion in such a form that time is vectorized, too:

$$\mathbb{T} \rightarrowtail \mathbf{S}, \qquad \boldsymbol{t} \mapsto r_U(t_o + \boldsymbol{t}) - q_o = r(t_o + \boldsymbol{t}) - q_o \star (t_o + \boldsymbol{t}).$$

6.3. Motions relative to a rigid observer*

6.3.1. Recall that the space S_U of a rigid global observer $\boldsymbol{U}$ is an affine space over $\mathbf{S}_U$ consisting of functions $\mathrm{T} \to \mathbf{S}$ whose values 'rotate together with the observer'. Thus, in general, it is somewhat complicated to control the affine structure based on these vectors; we can simplify the calculations by performing a double vectorization of the observer space, corresponding to a chosen reference origin o in M. Let $t_o := \tau(o)$, $q_o := C_U(o)$ and let $\boldsymbol{L}_o\colon \mathbf{S}_U \to \mathbf{S}$ be the linear bijection introduced in 4.4.1.

Let us take the motion r_U corresponding to the world line function r and let us consider the *double vectorized motion*

$$\boldsymbol{r}_U\colon \mathrm{T} \rightarrowtail \mathbf{S}\,, \qquad t \mapsto \boldsymbol{L}_o \cdot \big(r_U(t) - q_o\big) = C_U\big(r(t)\big) \star t_o - o = \\ = \boldsymbol{R}_U(t,t_o)^{-1} \cdot \big(r(t) - q_o \star t\big).$$

For the sake of simplicity, we shall use the notations $\boldsymbol{r}(t) := \boldsymbol{r}_U(t)$, $R(t) := \boldsymbol{R}_U(t,t_o)$, $\boldsymbol{\Omega}(t) := \boldsymbol{\Omega}_U(t)$, $\boldsymbol{u}_o(t) := \boldsymbol{U}(q_o \star t)$, $\boldsymbol{a}_o(t) := \boldsymbol{A}_U(q_o \star t)$.

Then the previous formula can be written in the form

$$R(t) \cdot \boldsymbol{r}(t) = r(t) - q_o \star t;$$

differentiating with respect to t and then omitting t from the notation we obtain

$$\dot{R} \cdot \boldsymbol{r} + R \cdot \dot{\boldsymbol{r}} = \dot{r} - \boldsymbol{u}_o$$

yielding

$$R \cdot \dot{\boldsymbol{r}} = -\boldsymbol{\Omega} \cdot R \cdot \boldsymbol{r} + \dot{r} - \boldsymbol{u}_o. \tag{$*$}$$

A second differentiation gives

$$\dot{R} \cdot \dot{\boldsymbol{r}} + R \cdot \ddot{\boldsymbol{r}} = -\dot{\boldsymbol{\Omega}} \cdot R \cdot \boldsymbol{r} - \boldsymbol{\Omega} \cdot \dot{R} \cdot \boldsymbol{r} - \boldsymbol{\Omega} \cdot R \cdot \dot{\boldsymbol{r}} + \ddot{r} - \boldsymbol{a}_o$$

from which we infer

$$R \cdot \ddot{\boldsymbol{r}} = -2\boldsymbol{\Omega} \cdot R \cdot \dot{\boldsymbol{r}} - \boldsymbol{\Omega} \cdot \boldsymbol{\Omega} \cdot R \cdot \boldsymbol{r} - \dot{\boldsymbol{\Omega}} \cdot R \cdot \boldsymbol{r} + \ddot{r} - \boldsymbol{a}_o.$$

6.3.2. Let us introduce the notation

$$\omega(t) := R(t)^{-1} \cdot \boldsymbol{\Omega}(t) \cdot R(t) = R(t)^{-1} \cdot \dot{R}(t) \qquad (t \in \mathrm{T}).$$

From $R \cdot \omega = \boldsymbol{\Omega} \cdot R$ we derive that $\dot{R} \cdot \omega + R \cdot \dot{\omega} = \dot{\boldsymbol{\Omega}} \cdot R + \boldsymbol{\Omega} \cdot \dot{R}$, which implies $\boldsymbol{\Omega} \cdot R \cdot \omega + R \cdot \dot{\omega} = \dot{\boldsymbol{\Omega}} \cdot R + \boldsymbol{\Omega} \cdot \boldsymbol{\Omega} \cdot R$; then we can state that

$$\dot{\omega} = R^{-1} \cdot \dot{\boldsymbol{\Omega}} \cdot R.$$

Consequently, the last formula in the preceding paragraph can be written in the form

$$\ddot{r} = -2\omega \cdot \dot{r} - \omega \cdot \omega \cdot \boldsymbol{r} - \dot{\omega} \cdot \boldsymbol{r} + R^{-1}(\ddot{r} - \boldsymbol{a}_o).$$

$-2\omega \cdot \dot{r}$ and $-\omega \cdot \omega \cdot \boldsymbol{r}$ are called the *Coriolis acceleration* and the *centrifugal acceleration* with respect to the observer.

6.3.3. Recall that $\boldsymbol{r} : \mathrm{T} \rightarrowtail \mathbf{S}$ denotes the double vectorized motion: $\ddot{r}(t) = L_o \cdot (r_U(t) - q_o)$; consequently, the relative velocity value at the instant t, $\dot{r}_U(t) = \boldsymbol{L}_o^{-1} \cdot \dot{\boldsymbol{r}}(t)$ is in $\mathbf{S}_U$, i.e. it is a function from T into $\mathbf{S}$ which is uniquely determined by an arbitrary one of its values:

$$\dot{r}_U(t)(s) = \boldsymbol{R}_U(s, t_o) \cdot \dot{\boldsymbol{r}}(t) \qquad (s \in \mathrm{T}).$$

Since $\boldsymbol{\Omega}(t) \cdot (r(t) - q_o \star t) = \boldsymbol{U}(r(t)) - \boldsymbol{U}(q_o \star t)$, formula $(*)$ in 6.3.1 gives

$$\dot{r}_U(t)(t) = \dot{r}(t) - \boldsymbol{U}\big(r(t)\big).$$

The expression on the right-hand side coincides with that for the relative velocity with respect to a rotation-free observer. However, keep in mind that now this expression is only a convenient representative (a value) of the relative velocity and not the relative velocity itself.

6.4. Some motions relative to an inertial observer

6.4.1. In this paragraph $\boldsymbol{u}$ denotes a global inertial observer.

Suppose r is an inertial world line function, use the notations of 2.3.1(*iii*) and put $t_o := \tau(x_o)$:

$$r(t) = x_o + \boldsymbol{u}_o(t - t_o). \qquad (*)$$

Applying one of the formulae in 4.1.1, we get the corresponding motion relative to $\boldsymbol{u}$:

$$r_{\boldsymbol{u}}(t) = \big(x_o + \boldsymbol{u}_o(t - t_o)\big) + \boldsymbol{u} \otimes \mathbb{T} = (x_o + \boldsymbol{u} \otimes \mathbb{T}) + (\boldsymbol{u} - \boldsymbol{u}_o)(t - t_o) =$$
$$= q_{x_o} + \boldsymbol{v}_{\boldsymbol{u}_o \boldsymbol{u}}(t - t_o)$$

where $q_{x_o} := x_o + \boldsymbol{u} \otimes \mathbb{T}$ is the $\boldsymbol{u}$-space point that x_o is incident with.

This is a uniform motion along a straight line.

Conversely, suppose that we are given a uniform motion relative to the inertial observer $\boldsymbol{u}$, i.e. there is a $q_o \in \mathrm{S}_{\boldsymbol{u}}$, a $t_o \in \mathrm{T}$ and a $\boldsymbol{v}_o \in \frac{\mathbf{S}}{\mathbb{T}}$ such that

$$r_{\boldsymbol{u}}(t) = q_o + \boldsymbol{v}_o(t - t_o) \qquad (t \in \mathrm{T}).$$

Then the corresponding history is inertial; putting $x_o := q_o \star t_o$, $\boldsymbol{u}_o := \boldsymbol{u} + \boldsymbol{v}_o$, we get the world line function $(*)$ which gives rise to the given motion.

6.4.2. Let r be a twist-free world line function (see 2.3.1(*iii*)):

$$r(t) = x_o + \boldsymbol{u}_o(t - t_o) + \boldsymbol{a}_o\boldsymbol{h}(t - t_o).$$

Then

$$r_{\boldsymbol{u}}(t) = q_{x_o} + \boldsymbol{v}_{\boldsymbol{u}_o\boldsymbol{u}}(t - t_o) + \boldsymbol{a}_o\boldsymbol{h}(t - t_o),$$

where $q_{x_o} := x_o + \boldsymbol{u} \otimes \mathbb{T}$.

If the world line function is not inertial, i.e. $\ddot{\boldsymbol{h}} \neq \boldsymbol{0}$, then the motion is not uniform. The motion is rectilinear relative to the observer if and only if $\boldsymbol{v}_{\boldsymbol{u}_o\boldsymbol{u}}$ is parallel to $\boldsymbol{a}_o$.

6.4.3. Now we see that the property 'rectilinear' of a motion is not absolute, in general. The same history can appear as a rectilinear motion to an observer and as a nonrectilinear one to another observer; exceptions are the uniformly rectilinear motions, i.e. the inertial histories.

Recall that assertion involving the inertial observer $\boldsymbol{u}$ is absolute if and only if it can be formulated exclusively with the aid of the affine structure of $\mathrm{T} \times \mathrm{S}_{\boldsymbol{u}}$.

Let $r_{\boldsymbol{u}} : \mathrm{T} \to \mathrm{S}_{\boldsymbol{u}}$ be a motion. Saying that the motion is rectilinear we state that the range of $r_{\boldsymbol{u}}$ is a straight line in the observer space, i.e. we involve the affine structure of $\mathrm{S}_{\boldsymbol{u}}$ only. This is not an absolute property.

Saying the motion is rectilinear and uniform we state that $\{(t, r_{\boldsymbol{u}}(t)) \mid t \in \mathrm{T}\}$ is a straight line in $\mathrm{T} \times \mathrm{S}_{\boldsymbol{u}}$; this is an absolute property.

6.4.4. Suppose that the inertial observer $\boldsymbol{u}$ chooses a reference origin o. Then, $q_o := o + \boldsymbol{u} \otimes \mathbb{T}$ is the $\boldsymbol{u}$-space point that o is incident with; hence the vectorized motion corresponding to the world line function r becomes

$$\mathrm{T} \rightarrowtail \mathbf{S}, \qquad t \mapsto r(t) - \big(o + \boldsymbol{u}(t - t_o)\big),$$

or

$$\mathbb{T} \rightarrowtail \mathbf{S}, \qquad \boldsymbol{t} \mapsto r(t_o + \boldsymbol{t}) - (o + \boldsymbol{u}\boldsymbol{t}),$$

where $t_o := \tau(o)$.

In particular, if r is the twist-free world line function treated in 6.4.2. and $\tau(x_o) = t_o$ (which can be assumed without loss of generality) then the vectorized motion is

$$\mathbb{T} \to \mathbf{S}, \qquad \boldsymbol{t} \mapsto \boldsymbol{q}_o + \boldsymbol{v}_{\boldsymbol{u}_o\boldsymbol{u}}\boldsymbol{t} + \boldsymbol{a}_o\boldsymbol{h}(\boldsymbol{t}),$$

where $\boldsymbol{q}_o := x_o - o$.

Since $\boldsymbol{q}_o = q_{x_o} - q_o$ holds as well, comparing our present result with that of 6.4.2, evidently we have—as it must be by definition—that the vectorized motion equals to $\boldsymbol{t} \mapsto r_{\boldsymbol{u}}(t_o + \boldsymbol{t}) - q_o$. The advantage of the vectorized motion is that it is easier to calculate.

6.5. Some motions relative to a uniformly accelerated observer

6.5.1. Let r be the previous twist-free world line function and let us examine the corresponding motion relative to a uniformly accelerated observer $\boldsymbol{U}$ with constant acceleration $\boldsymbol{a}$. We easily obtain by 5.2.1 that

$$C_{\boldsymbol{U}}\big(r(t)\big) \star s = x_o + \boldsymbol{u}_o(t - t_o) + \boldsymbol{a}_o\boldsymbol{h}(t - t_o) + \boldsymbol{U}\big(r(t)\big)(s - t) + \frac{1}{2}\boldsymbol{a}(s - t)^2.$$

Then 5.2.2 helps us to transform this expression:

$$\boldsymbol{U}\big(r(t)\big) = \boldsymbol{U}(x_o) + \boldsymbol{a}(t - t_o)$$

and so

$$C_{\boldsymbol{U}}\big(r(t)\big) \star s = x_o + \boldsymbol{U}(x_o)(s - t_o) + \frac{1}{2}\boldsymbol{a}(s - t_o)^2 + \\ + \big(\boldsymbol{u}_o - \boldsymbol{U}(x_o)\big)(t - t_o) + \left(\boldsymbol{a}_o\boldsymbol{h}(t - t_o) - \frac{1}{2}\boldsymbol{a}(t - t_o)^2\right).$$

Denoting by q_{x_o} the $\boldsymbol{U}$-space point that x_o is incident with and putting $\boldsymbol{v}_o := \boldsymbol{u}_o - \boldsymbol{U}(x_o)$, we can write:

$$r_{\boldsymbol{U}}(t) = q_{x_o} + \boldsymbol{v}_o(t - t_o) + \left(\boldsymbol{a}_o\boldsymbol{h}(t - t_o) - \frac{1}{2}\boldsymbol{a}(t - t_o)^2\right).$$

In particular, it is a uniformly accelerated motion, if $\ddot{\boldsymbol{h}} =$ const., i.e. if r is inertial or uniformly accelerated.

6.5.2. Let the previous uniformly accelerated observer $\boldsymbol{U}$ choose a reference origin o. Then the $\boldsymbol{U}$-space point that o is incident with is given by the world line function $t \mapsto q_o \star t := o + \boldsymbol{U}(o)(t - t_o) + \frac{1}{2}\boldsymbol{a}(t - t_o)^2$; hence the vectorized motion corresponding to the world line function r becomes

$$\mathrm{T} \rightarrowtail \mathbf{S}, \qquad t \mapsto r(t) - q_o \star t$$

or

$$\mathbb{T} \rightarrowtail \mathbf{S}, \qquad \boldsymbol{t} \mapsto r(t_o + \boldsymbol{t}) - q_o \star (t_o + \boldsymbol{t}).$$

In particular, the vectorized motion corresponding to the twist-free world line function r treated above is

$$\mathbb{T} \to \mathbf{S}, \qquad \boldsymbol{t} \mapsto \boldsymbol{q}_o + \boldsymbol{v}_o\boldsymbol{t} + \left(\boldsymbol{a}_o\boldsymbol{h}(\boldsymbol{t}) - \frac{1}{2}\boldsymbol{a}\boldsymbol{t}^2\right),$$

where $\boldsymbol{q}_o := x_o - o$ and $\boldsymbol{v}_o := \boldsymbol{u}_o - \boldsymbol{U}(o) = \boldsymbol{u}_o - \boldsymbol{U}(x_o)$ (recall that $\boldsymbol{U}$ is constant on simultaneous hyperplanes).

We see in this case, too, that the vectorized motion is $\boldsymbol{t} \mapsto r_U(t_o + \boldsymbol{t}) - q_o$, as it must be, but it is more complicated to determine the motion r_U and then the vectorized motion than to calculate the vectorized motion directly.

6.6. Some motions relative to a uniformly rotating observer*

6.6.1. Let the uniformly rotating observer $\boldsymbol{U}$ choose a reference origin o. If q_o is the $\boldsymbol{U}$-space point that o is incident with and $\boldsymbol{\Omega}$ is the constant angular velocity of the observer, then the double vectorized motion is

$$\mathrm{T} \to \mathbf{S}, \qquad t \mapsto e^{-(t-t_o)\boldsymbol{\Omega}} \cdot \big(r(t) - q_o \star t\big).$$

In particular, if q_o is an inertial world line, $q_o = o + \boldsymbol{c}\otimes\mathbb{T}$, and r is an inertial world line function, $r(t) = x_o + \boldsymbol{u}_o(t - t_o)$, where we supposed without loss of generality that $\tau(x_o) = \tau(o) = t_o$, then the double vectorized motion becomes

$$\mathrm{T} \to \mathbf{S}, \qquad t \mapsto e^{-(t-t_o)\boldsymbol{\Omega}} \cdot \big(\boldsymbol{q}_o + \boldsymbol{v}_{\boldsymbol{u}_o\boldsymbol{c}}(t - t_o)\big)$$

where $\boldsymbol{q}_o := x_o - o$; again it is more convenient to use vectorized time:

$$\mathbb{T} \to \mathbf{S}, \qquad \boldsymbol{t} \mapsto e^{-\boldsymbol{t}\boldsymbol{\Omega}} \cdot (\boldsymbol{q}_o + \boldsymbol{v}_{\boldsymbol{u}_o\boldsymbol{c}}\boldsymbol{t}).$$

If $\boldsymbol{v}_{\boldsymbol{u}_o\boldsymbol{c}} = \boldsymbol{0}$, i.e. the relative velocity of the material point with respect to the axis of rotation is zero, then the motion relative to the observer is a simple rotation around the axis. If $\boldsymbol{v}_{\boldsymbol{u}_o\boldsymbol{c}} \neq \boldsymbol{0}$, then the motion is the 'rotation of a uniform motion'. Anyway, the observed rotation of the inertial masspoint is opposite to the rotation of the observer (take into account the negative sign in the exponent).

6.6.2. In the case of inertial observers and uniformly accelerated observers, the vectorized motion can be deduced a little easier than motion. On the other hand, for uniformly rotated observers, it is significantly simpler to get the double vectorized motion than motion itself, as it will be seen from the following calculation.

Let $\boldsymbol{U}$ and r be as in the preceding paragraph. Then $q_o := C_U(o) = o+\boldsymbol{c}\otimes\mathbb{T}$ and so

$$\begin{aligned} C_U\big(r(t)\big)\star s &= q_o\star s + e^{(s-t_o)\boldsymbol{\Omega}}\cdot\big(C_U\big(r(t)\big)\star t_o - o\big) = \\ &= q_o\star s + e^{(s-t_o)\boldsymbol{\Omega}}\cdot e^{-(t-t_o)\boldsymbol{\Omega}}\cdot\big(C_U\big(r(t)\big)\star t - q_o\star t\big) = \\ &= q_o\star s + e^{-(t-t_o)\boldsymbol{\Omega}}\cdot e^{(s-t_o)\boldsymbol{\Omega}}\cdot\big(x_o - o + \boldsymbol{v}_{\boldsymbol{u}_o\boldsymbol{c}}(t-t_o)\big). \end{aligned}$$

The functions

$$\mathrm{T}\to\mathbf{S},\qquad s\mapsto \boldsymbol{q}_o(s) := e^{(s-t_o)\boldsymbol{\Omega}}\cdot(x_o - o)$$

$$\mathrm{T}\to\frac{\mathbf{S}}{\mathbb{T}},\qquad s\mapsto \boldsymbol{v}_o(s) := e^{(s-t_o)\boldsymbol{\Omega}}\cdot\boldsymbol{v}_{\boldsymbol{u}_o\boldsymbol{c}}$$

are in $\mathbf{S}_U$ and in $\frac{\mathbf{S}_U}{\mathbb{T}}$ (they are a vector and a vector of cotype $\mathbb{T}$ in the observer space), respectively. Thus we have got for the motion that

$$r_U(t) = q_o + e^{-(t-t_o)\boldsymbol{\Omega}}\cdot(\boldsymbol{q}_o + \boldsymbol{v}_o(t-t_o)) \qquad (t\in I).$$

Originally, the exponent of $\boldsymbol{\Omega}$ is a linear map from $\mathbf{S}$ into $\mathbf{S}$. Here it is regarded as a linear map from $\mathbf{S}_U$ into $\mathbf{S}_U$ defined by

$$\left(e^{-(t-t_o)\boldsymbol{\Omega}}\cdot\boldsymbol{s}\right)(s) := e^{-(t-t_o)\boldsymbol{\Omega}}\cdot\boldsymbol{s}(s) \qquad (\boldsymbol{s}\in\mathbf{S}_U,\ s\in\mathrm{T}).$$

6.7. Exercise

Let $\boldsymbol{U}$ be a uniformly rotating observer that has an inertial space point. Use the notations of Section 5.3. For $\boldsymbol{q}_o\in\mathbf{S}$ and $\boldsymbol{v}_o\in\frac{\mathbf{S}}{\mathbb{T}}$ define the world line function

$$t\mapsto o + \boldsymbol{c}(t-t_o) + e^{(t-t_o)\boldsymbol{\Omega}}\cdot(\boldsymbol{q}_o + \boldsymbol{v}_o(t-t_o)).$$

Prove that the corresponding motion relative to the observer $\boldsymbol{U}$ is a uniform straight line motion.

7. Some kinds of observation

7.1. Vectors observed by inertial observers

7.1.1. Let C_1 and C_2 be two world lines defined over the same time interval J. The *vector* between C_1 and C_2 at the instant $t\in\mathrm{J}$ is $\mathrm{C}_2\star t - \mathrm{C}_1\star t$. The *distance* at t between the two world lines is $|\mathrm{C}_2\star t - \mathrm{C}_1\star t|$.

The two world lines represent the history of two material points. An inertial observer $\boldsymbol{u}$ observes the two material points describing their history by the corresponding motions $r_{1,\boldsymbol{u}}$ and $r_{2,\boldsymbol{u}}$. Hence, the *vector observed* by the inertial observer between the material points at the instant t is evidently

$$r_{2,\boldsymbol{u}}(t) - r_{1,\boldsymbol{u}}(t) = (\mathrm{C}_2 \star t + \boldsymbol{u} \otimes \mathbb{T}) - (\mathrm{C}_1 \star t + \boldsymbol{u} \otimes \mathbb{T}) = \mathrm{C}_2 \star t - \mathrm{C}_1 \star t.$$

The observed vector coincides with the (absolute) vector; consequently, the observed distance, too, coincides with the (absolute) distance.

7.1.2. The question arises how a straight line segment in the space of an inertial observer is observed by another observer. The question and the answer are formulated correctly as follows.

Let us consider two inertial observers $\boldsymbol{u}_o$ and $\boldsymbol{u}$. Let H_o be a subset (a geometrical figure) in the $\boldsymbol{u}_o$-space. The corresponding figure observed by $\boldsymbol{u}$ at the instant t —called the trace of H_o at t in $\mathrm{S}_{\boldsymbol{u}}$ —is the set of $\boldsymbol{u}$-space points that coincide at t with the points of H_o:

$$\{q \star t + \boldsymbol{u} \otimes \mathbb{T} \mid q \in \mathrm{H}_o\}.$$

Introducing the mapping

$$P_t : \mathrm{S}_{\boldsymbol{u}_o} \to \mathrm{S}_{\boldsymbol{u}}, \qquad q \mapsto q \star t + \boldsymbol{u} \otimes \mathbb{T},$$

we see that the trace of H_o at t equals $P_t[\mathrm{H}_o]$. It is quite easy to see (recall the definition of subtraction in observer spaces) that

$$P_t(q_2) - P_t(q_1) = q_2 \star t - q_1 \star t = q_2 - q_1$$

for all $q_1, q_2 \in \mathrm{S}_{\boldsymbol{u}_o}$. Thus P_t is an affine map whose underlying linear map is the identity of $\mathbf{S}$.

We can say that the observed figure and the original figure are *congruent.* Evidently, every figure in the $\boldsymbol{u}_o$-space is of the form $q_o + \mathbf{H}_o$, where $q_o \in \mathrm{S}_{\boldsymbol{u}_o}$ and $\mathbf{H}_o \subset \mathbf{S}$; then $P_t\,[q_o + \mathbf{H}_o] = P_t(q_o) + \mathbf{H}_o$.

In particular, a straight line segment in the $\boldsymbol{u}_o$-space observed at an arbitrary instant by the observer $\boldsymbol{u}$ is a straight line segment parallel to the original one. Moreover, the original and the observed segments have the same length; the original and the observed angle between two segments are equal as well.

7.1.3. It is an important fact that the spaces of different global inertial observers are *different* affine spaces over the *same* vector space $\mathbf{S}$. Thus, though the observer spaces are different, it makes sense that a vector in the space of an inertial observer coincides with a vector in the space of another inertial observer.

Evidently, the coincidence of vectors in different observer spaces is a symmetric and transitive relation (if 'your' vector coincides with 'my' vector then 'mine' coincides with 'yours'; if, moreover, 'his' vector coincides with 'yours' then it coincides with 'mine' as well.)

This is a trivial fact here that does not hold in the relativistic spacetime model.

7.2. Measuring rods

7.2.1. A physical observer makes measurements in his space: measures the distance between two points, the length of a line, etc. In practice such measurements are based on measuring rods: one takes a rod, carries it to the figure to be measured, puts it consecutively at convenient places ... One supposes that during all this procedure the rod is *absolutely rigid:* it remains a straight line segment and its length does not change.

We are interested in whether the nonrelativistic spacetime model allows such measuring rods, i.e. whether we can permit in it the existence of such an absolutely rigid rod.

As we shall see, the answer is positive (in contradistinction to the relativistic case).

7.2.2. The existence of an absolutely rigid rod—if it is meaningful—can be determined uniquely by the history of its extremities. Two world lines C_0 and C_1 correspond to the two extremities of a measuring rod if and only if they are defined on the same interval J and their distance at every instant is the same: $|\mathrm{C}_1 \star t - \mathrm{C}_0 \star t| = \boldsymbol{d}$ for all $t \in \mathrm{J}$.

Then for all $\alpha \in [0,1]$ we can define the world line C_α as follows:

$$\mathrm{C}_\alpha \star t := \mathrm{C}_0 \star t + \alpha\,(\mathrm{C}_1 \star t - \mathrm{C}_0 \star t) \qquad (t \in \mathrm{J}).$$

It is quite evident that the set of world lines, $\{\mathrm{C}_\alpha \mid \alpha \in [0,1]\}$ gives an existence of a rigid rod: at every instant $t \in \mathrm{J}$, $\{\mathrm{C}_\alpha \star t \mid \alpha \in [0,1]\}$ is a straight line segment in $\mathbf{S}$, having the length $\boldsymbol{d}$.

8. Vector splittings

8.1. What is a splitting?

Recall what has been said in 3.1.1: in the experience of a physical observer relative to a phenomenon, and in the notions deduced from experience, properties of the phenomenon are mixed with properties of the observer. Our aim is to find the *absolute notions* that model some properties or aspects of phenomena independently of observers and then to give how the observers derive *relative notions* from the absolute ones (how the absolute objects are observed).

We know already how spacetime is observed as space and time and how the history of a mass point is observed as a motion. In the following, the splitting of force fields, potentials etc. will be treated: such splittings describe somehow the observed form of force fields, potentials, etc. We begin with the splitting of vectors and covectors according to velocity values and then we define the splitting of vector fields and covector fields according to observers.

8.2. Splitting of vectors

8.2.1. For $\boldsymbol{u} \in V(1)$ we have already defined

$$\boldsymbol{\pi}_{\boldsymbol{u}} : \mathbf{M} \to \mathbf{S}, \qquad \boldsymbol{x} \mapsto \boldsymbol{x} - (\boldsymbol{\tau} \cdot \boldsymbol{x})\boldsymbol{u}$$

and the linear bijection

$$\boldsymbol{\xi}_{\boldsymbol{u}} := (\boldsymbol{\tau}, \boldsymbol{\pi}_{\boldsymbol{u}}) : \mathbf{M} \to \mathbb{T} \times \mathbf{S}, \qquad \boldsymbol{x} \mapsto (\boldsymbol{\tau} \cdot \boldsymbol{x}, \boldsymbol{\pi}_{\boldsymbol{u}} \cdot \boldsymbol{x})$$

having the inverse

$$(t, \boldsymbol{q}) \mapsto \boldsymbol{u}t + \boldsymbol{q}$$

(1.2.8). Thus

$$\boldsymbol{\pi}_{\boldsymbol{u}} = \mathbf{1}_{\mathbf{M}} - \boldsymbol{u} \otimes \boldsymbol{\tau}, \qquad \boldsymbol{\pi}_{\boldsymbol{u}}^* = \mathbf{1}_{\mathbf{M}^*} - \boldsymbol{\tau} \otimes \boldsymbol{u}.$$

Moreover,

$$\boldsymbol{\tau} \cdot \boldsymbol{\pi}_{\boldsymbol{u}} = \mathbf{0}, \qquad \boldsymbol{\pi}_{\boldsymbol{u}} \cdot \boldsymbol{i} = \mathbf{1}_{\mathbf{S}}, \qquad \boldsymbol{\pi}_{\boldsymbol{u}} \cdot \boldsymbol{u} = \mathbf{0}.$$

8.2.2. Definition. $\boldsymbol{\tau} \cdot \boldsymbol{x}$ and $\boldsymbol{\pi}_{\boldsymbol{u}} \cdot \boldsymbol{x}$ are called the *timelike component* and the $\boldsymbol{u}$*-spacelike component* of the vector $\boldsymbol{x}$. $(\boldsymbol{\tau} \cdot \boldsymbol{x},\ \boldsymbol{\pi}_{\boldsymbol{u}} \cdot \boldsymbol{x})$ is the $\boldsymbol{u}$*-split form* of $\boldsymbol{x}$. $\boldsymbol{\xi}_{\boldsymbol{u}} := (\boldsymbol{\tau}, \boldsymbol{\pi}_{\boldsymbol{u}})$ is the *splitting* of $\mathbf{M}$ corresponding to $\boldsymbol{u}$, or the $\boldsymbol{u}$*-splitting* of $\mathbf{M}$. ■

Note that $\boldsymbol{\xi}_{\boldsymbol{u}} \cdot \boldsymbol{q} = (0, \boldsymbol{q})$ for all $\boldsymbol{q} \in \mathbf{S}$. In other words, $\mathbf{S}$ is split into $\{0\} \times \mathbf{S}$ trivially. In applications it is convenient to identify $\{0\} \times \mathbf{S}$ with $\mathbf{S}$ and to assume that the split form of a spacelike vector $\boldsymbol{q}$ is itself.

8.2.3. If $\mathbb{A}$ is a measure line then $\mathbb{A} \otimes \mathbf{M}$ is split into $(\mathbb{A} \otimes \mathbb{T}) \times (\mathbb{A} \otimes \mathbf{S})$ by $\boldsymbol{\xi}_{\boldsymbol{u}}$; similarly, $\frac{\mathbf{M}}{\mathbb{A}}$ is split into $\frac{\mathbb{T}}{\mathbb{A}} \times \frac{\mathbf{S}}{\mathbb{A}}$. Correspondingly, the timelike component and the $\boldsymbol{u}$-spacelike component of a vector of type $\mathbb{A}$ are in $\mathbb{A} \otimes \mathbb{T}$ and $\mathbb{A} \otimes \mathbf{S}$, respectively, and the timelike component and the $\boldsymbol{u}$-spacelike component of a vector of cotype $\mathbb{A}$ are in $\frac{\mathbb{T}}{\mathbb{A}}$ and $\frac{\mathbf{S}}{\mathbb{A}}$, respectively.

In particular, ξ_u splits $\frac{\mathbf{M}}{\mathbb{T}}$ into $\mathbb{R} \times \frac{\mathbf{S}}{\mathbb{T}}$ and for all $u' \in V(1)$

$$\xi_u \cdot u' = (1, u' - u) = (1, v_{u'u});$$

the u-spacelike component of the velocity value u' is the relative velocity of u' with respect to u.

Thus $V(1)$ is split into $\{1\} \times \frac{\mathbf{S}}{\mathbb{T}}$; in applications it is often convenient to omit the trivial component $\{1\}$, and to regard only π_u instead of ξ_u as the splitting of $V(1)$:

$$V(1) \to \frac{\mathbf{S}}{\mathbb{T}}, \qquad u' \mapsto u' - u = v_{u'u}.$$

8.2.4. The timelike component of a vector is independent of the velocity value u producing the splitting but the u-spacelike components vary with u, except when the vector is spacelike (an element of $\mathbf{S}$); then the timelike component is zero and the u-spacelike component is the vector itself for all $u \in V(1)$.

The transformation rule that shows how the u-spacelike components of a vector vary with u can be well seen from the following formula giving the u'-spacelike component of the vector having the timelike component t and the u-spacelike component q.

Definition. Let $u, u' \in V(1)$. Then

$$\xi_{u'u} := \xi_{u'} \cdot \xi_u^{-1} : \mathbb{T} \times \mathbf{S} \to \mathbb{T} \times \mathbf{S}$$

is called the *vector transformation law* from u-splitting into u'-splitting.

Proposition.

$$\xi_{u'u} \cdot (t, q) = (t, -v_{u'u} t + q) \qquad (t \in \mathbb{T}, q \in \mathbf{S}). \quad \blacksquare$$

Using the matrix form of the linear maps $\mathbb{T} \times \mathbf{S} \to \mathbb{T} \times \mathbf{S}$ (see IV.3.7), we can write

$$\xi_{u'u} = \begin{pmatrix} \mathbf{1}_{\mathbb{T}} & 0 \\ -v_{u'u} & \mathbf{1}_{\mathbf{S}} \end{pmatrix}.$$

According to the identification $\mathrm{Lin}(\mathbb{T}) \equiv \mathbb{R}$ we have $\mathbf{1}_{\mathbb{T}} \equiv 1$. Moreover, applying the usual convention that the identity of a vector space is denoted by $\mathbf{1}$ (the identity is the operation of multiplication by 1), we obtain

$$\xi_{u'u} = \begin{pmatrix} 1 & 0 \\ -v_{u'u} & \mathbf{1} \end{pmatrix}.$$

In the lower left position of the matrix a linear map $\mathbb{T} \to \mathbf{S}$ must appear; recall that $v_{u'u} \in \frac{\mathbf{S}}{\mathbb{T}} \equiv \mathrm{Lin}(\mathbb{T}, \mathbf{S})$.

8.2.5. Let us give the transformation rule in a form which is more usual in the literature.

Let $(\boldsymbol{t}, \boldsymbol{q})$ and $(\boldsymbol{t}', \boldsymbol{q}')$ be the $\boldsymbol{u}$-split form and the $\boldsymbol{u}'$-split form of the same vector, respectively. Let $\boldsymbol{v}$ denote the relative velocity of $\boldsymbol{u}'$ with respect to $\boldsymbol{u}$. Then

$$\boldsymbol{t}' = \boldsymbol{t}, \qquad \boldsymbol{q}' = \boldsymbol{q} - \boldsymbol{v}\boldsymbol{t}.$$

Usually one calls this formula—or, rather, a similar formula in the arithmetic spacetime model—the Galilean transformation rule and one even defines Galilean transformations by it.

The transformation rule is a mapping from $\mathbb{T} \times \mathbf{S}$ into $\mathbb{T} \times \mathbf{S}$. A (special) Galilean transformation is to be defined on spacetime vectors, i.e. as a mapping from $\mathbf{M}$ into $\mathbf{M}$. Thus the transformation rule and a Galilean transformation cannot be equal. In the split spacetime model $\mathbb{T} \times \mathbf{S}$ stands for both spacetime vectors and spacetime. Thus, using the split model (or, similarly, the arithmetic spacetime model) one can confuse the transformation rule with a mapping defined on spacetime vectors or on spacetime. This indicates very well that we must not use the split model or the arithmetic model for the composition of general ideas.

Of course, there is some connection between transformation rules and Galilean transformations. We shall see (11.3.7) that there is a special Galilean transformation $\boldsymbol{L}(\boldsymbol{u}, \boldsymbol{u}') : \mathbf{M} \to \mathbf{M}$ such that

$$\xi_{\boldsymbol{u}'\boldsymbol{u}} = \xi_{\boldsymbol{u}} \cdot \boldsymbol{L}(\boldsymbol{u}, \boldsymbol{u}') \cdot \xi_{\boldsymbol{u}}^{-1}.$$

8.3. Splitting of covectors

8.3.1. For $\boldsymbol{u} \in V(1)$, $\mathbf{M}^*$ is split by the transpose of the inverse of $\xi_{\boldsymbol{u}}$:

$$\eta_{\boldsymbol{u}} := \left(\xi_{\boldsymbol{u}}^{-1}\right)^* : \mathbf{M}^* \to (\mathbb{T} \times \mathbf{S})^* \equiv \mathbb{T}^* \times \mathbf{S}^*,$$

where we used the identification described in IV.1.3. Then for all $\boldsymbol{k} \in \mathbf{M}^*$, $(\boldsymbol{t}, \boldsymbol{q}) \in \mathbb{T} \times \mathbf{S}$ we have

$$(\eta_{\boldsymbol{u}} \cdot \boldsymbol{k}) \cdot (\boldsymbol{t}, \boldsymbol{q}) = \boldsymbol{k} \cdot \xi_{\boldsymbol{u}}^{-1} \cdot (\boldsymbol{t}, \boldsymbol{q}) = \boldsymbol{k} \cdot (\boldsymbol{u}\boldsymbol{t} + \boldsymbol{q}) = (\boldsymbol{k} \cdot \boldsymbol{u})\boldsymbol{t} + \boldsymbol{k} \cdot \boldsymbol{q}.$$

Of course, in the last term $\boldsymbol{k}$ can be replaced by $\boldsymbol{k}|_{\mathbf{S}} = \boldsymbol{i}^* \cdot \boldsymbol{k}$, where $\boldsymbol{i} : \mathbf{S} \to \mathbf{M}$ is the canonical embedding. Furthermore, recall that $\boldsymbol{k} \cdot \boldsymbol{u} \in \frac{\mathbb{R}}{\mathbb{T}} \equiv \mathbb{T}^*$ and $(\boldsymbol{k} \cdot \boldsymbol{u})\boldsymbol{t}$ stand for the tensor product of $\boldsymbol{k} \cdot \boldsymbol{u}$ and $\boldsymbol{t}$. Then, in view of our convention regarding the duals of one-dimensional vector spaces (IV.3.8), we can state that $\eta_{\boldsymbol{u}} \cdot \boldsymbol{k} = (\boldsymbol{k} \cdot \boldsymbol{u},\ \boldsymbol{i}^* \cdot \boldsymbol{k})$. Recall that $\boldsymbol{i}^* \cdot \boldsymbol{k} = \boldsymbol{k} \cdot \boldsymbol{i}$; moreover, our dot notation convention allows us to interchange the order of $\boldsymbol{k}$ and $\boldsymbol{u}$ to have the more suitable forms

$$\eta_{\boldsymbol{u}} \cdot \boldsymbol{k} = (\boldsymbol{k} \cdot \boldsymbol{u}, \boldsymbol{k} \cdot \boldsymbol{i}) = (\boldsymbol{u} \cdot \boldsymbol{k}, \boldsymbol{i}^* \cdot \boldsymbol{k}) \qquad (\boldsymbol{k} \in \mathbf{M}^*).$$

Definition. $\boldsymbol{u} \cdot \boldsymbol{k}$ and $\boldsymbol{i}^* \cdot \boldsymbol{k}$ are called the $\boldsymbol{u}$*-timelike component* and the *spacelike component* of the covector $\boldsymbol{k}$. $(\boldsymbol{u} \cdot \boldsymbol{k},\ \boldsymbol{i}^* \cdot \boldsymbol{k})$ is the $\boldsymbol{u}$*-split form* of the covector $\boldsymbol{k}$. $\eta_{\boldsymbol{u}}$ is the splitting of $\mathbf{M}^*$ corresponding to $\boldsymbol{u}$, or the $\boldsymbol{u}$*-splitting* of $\mathbf{M}^*$. ■

Note that $\eta_{\boldsymbol{u}} \cdot (\boldsymbol{e}\boldsymbol{\tau}) = (\boldsymbol{e}, \mathbf{0})$ for all $\boldsymbol{e} \in \mathbb{T}^*$. In other words, $\mathbb{T}^* \cdot \boldsymbol{\tau}$ is split into $\mathbb{T}^* \times \{\mathbf{0}\}$ trivially. In applications it is convenient to identify $\mathbb{T}^* \times \{\mathbf{0}\}$ with $\mathbb{T}^*$ and consider that the split form of $\boldsymbol{e}\boldsymbol{\tau}$ is simply $\boldsymbol{e}$.

8.3.2. The spacelike component of a covector is independent of the velocity value $\boldsymbol{u}$ establishing the splitting, but the $\boldsymbol{u}$-timelike components vary with $\boldsymbol{u}$, except when the covector is timelike (an element of $\mathbb{T}^* \cdot \boldsymbol{\tau}$; then the spacelike component is zero and the $\boldsymbol{u}$-timelike component coincides with the corresponding element of $\mathbb{T}^*$). The transformation rule that shows how the $\boldsymbol{u}$-timelike components of a covector vary with $\boldsymbol{u}$ can be well seen from the following formula giving the $\boldsymbol{u}'$-timelike component of the covector having the $\boldsymbol{u}$-timelike component $\boldsymbol{e}$ and the spacelike component $\boldsymbol{p}$.

Definition. Let $\boldsymbol{u}, \boldsymbol{u}' \in V(1)$. Then

$$\eta_{\boldsymbol{u}'\boldsymbol{u}} := \eta_{\boldsymbol{u}'} \cdot \eta_{\boldsymbol{u}}^{-1} : \mathbb{T}^* \times \mathbf{S}^* \to \mathbb{T}^* \times \mathbf{S}^*$$

is called the *covector transformation law* from $\boldsymbol{u}$-splitting into $\boldsymbol{u}'$-splitting.

Proposition.

$$\eta_{\boldsymbol{u}'\boldsymbol{u}} \cdot (\boldsymbol{e}, \boldsymbol{p}) = (\boldsymbol{e} + \boldsymbol{p} \cdot \boldsymbol{v}_{\boldsymbol{u}'\boldsymbol{u}}, \boldsymbol{p}) \qquad (\boldsymbol{e} \in \mathbb{T}^*,\ \boldsymbol{p} \in \mathbf{S}^*).$$

Proof. It is not hard to see that

$$\eta_{\boldsymbol{u}}^{-1}(\boldsymbol{e}, \boldsymbol{p}) = \boldsymbol{e}\boldsymbol{\tau} + \boldsymbol{\pi}_{\boldsymbol{u}}^* \cdot \boldsymbol{p}$$

from which we easily obtain the desired formula. ■

Using the matrix form of the linear maps $\mathbb{T}^* \times \mathbf{S}^* \to \mathbb{T}^* \times \mathbf{S}^*$, we can write

$$\eta_{\boldsymbol{u}'\boldsymbol{u}} = \begin{pmatrix} \mathbf{1}_{\mathbb{T}^*} & \boldsymbol{v}_{\boldsymbol{u}'\boldsymbol{u}} \\ \mathbf{0} & \mathbf{1}_{\mathbf{S}^*} \end{pmatrix} \equiv \begin{pmatrix} 1 & \boldsymbol{v}_{\boldsymbol{u}'\boldsymbol{u}} \\ 0 & 1 \end{pmatrix}.$$

In the upper right position a linear map $\mathbf{S}^* \to \mathbb{T}^*$ must appear. The identifications $\mathrm{Lin}(\mathbf{S}^*, \mathbb{T}^*) \equiv \mathbb{T}^* \otimes \mathbf{S} \equiv \frac{\mathbf{S}}{\mathbb{T}}$ justify that $\boldsymbol{v}_{\boldsymbol{u}'\boldsymbol{u}}$ stands in that position.

The definitions imply that $\eta_{u'u} = \left(\xi_{uu'}^{-1}\right)^*$, which is reflected in the matrix form as well.

8.3.3. In 1.2.8 we have drawn a good picture how vectors are split. Now we give an illustration for splitting of covectors.

Recall that for all $u \in V(1)$, the surjection $\pi_u : \mathbf{M} \to \mathbf{S}$ is the left inverse of the canonical embedding $i : \mathbf{S} \to \mathbf{M}$, i.e. $\pi_u \cdot i = \mathbf{1}_{\mathbf{S}}$. As a consequence, the injection $\pi_u^* : \mathbf{S}^* \to \mathbf{M}^*$ is the right inverse of the surjection $i^* : \mathbf{M}^* \to \mathbf{S}^*$:

$$i^* \cdot \pi_u^* = \mathbf{1}_{\mathbf{S}^*}.$$

Since $\mathbb{T}^* \cdot \tau = \operatorname{Ker} i^*$ (see 1.2.1),

$$\mathbf{S}^* \cdot \pi_u = \operatorname{Ran} \pi_u^*$$

is a three-dimensional linear subspace in $\mathbf{M}^*$, complementary to $\mathbb{T}^* \cdot \tau$. Evidently, the restriction of i^* is a linear bijection from $\mathbf{S}^* \cdot \pi_u$ onto $\mathbf{S}^*$.

Moreover, we easily find that

$$\mathbf{S}^* \cdot \pi_u = \{k \in \mathbf{M}^* \mid k \cdot u = 0\};$$

in other words, $\mathbf{S}^* \cdot \pi_u$ is the annihilator of $u \otimes \mathbb{T}$.

Then the splitting of covectors according to u is illustrated as follows:

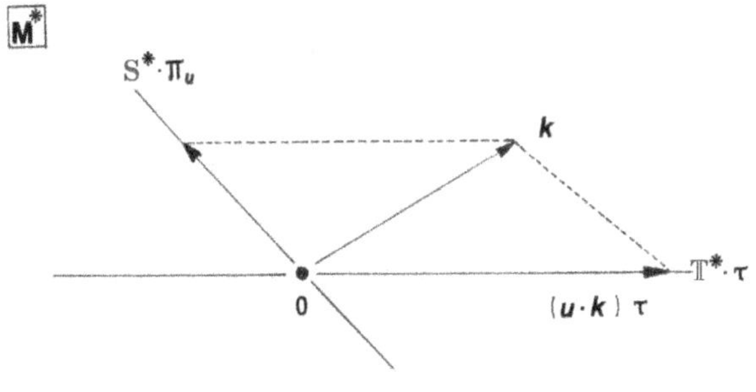

$k - (u \cdot k) \cdot \tau$ is in $\mathbf{S}^* \cdot \pi_u$, its image by i^* is the spacelike component of k.

8.4. Vectors and covectors are split in a different way

The splitting of vectors and the splitting of covectors according to $u \in V(1)$ are essentially different. The timelike component of vectors is independent of u, whereas the spacelike component of covectors is independent of u. The transformation laws for vectors and covectors are essentially different as well.

The reason of these differences lies in the fact that there is no one-dimensional vector space $\mathbb{A}$ in such a way that $\mathbf{M}^*$ could be canonically identified with $\frac{\mathbf{M}}{\mathbb{A}}$, in contradistinction to the relativistic case.

8.5. Splitting of vector fields and covector fields according to inertial observers

8.5.1. In applications, vectors and covectors appear in two ways: first, as values of functions defined in time; secondly, as values of functions defined in spacetime. The first case can be reduced to the second one: a function defined in time can be considered a function defined in spacetime that is constant on the simultaneous hyperplanes. Thus we shall study *vector fields* and *covector fields,* i.e. functions $\boldsymbol{X}: \mathrm{M} \rightarrowtail \mathbf{M}$ and $\boldsymbol{K}: \mathrm{M} \rightarrowtail \mathbf{M}^*$, respectively.

An inertial observer $\boldsymbol{u}$ splits vector fields and covector fields in two steps by splitting the range and then the domain of these functions.

First, at every world point x the values of the fields, $\boldsymbol{X}(x)$ and $\boldsymbol{K}(x)$, are split according to the velocity value $\boldsymbol{u}$ of the observer; thus the *half $\boldsymbol{u}$-split form* of the fields will be

$$\boldsymbol{\xi}_{\boldsymbol{u}} \cdot \boldsymbol{X}: \mathrm{M} \rightarrowtail \mathbb{T} \times \mathbf{S}, \qquad x \mapsto \big(\boldsymbol{\tau} \cdot \boldsymbol{X}(x),\ \boldsymbol{\pi}_{\boldsymbol{u}} \cdot \boldsymbol{X}(x)\big),$$
$$\boldsymbol{\eta}_{\boldsymbol{u}} \cdot \boldsymbol{K}: \mathrm{M} \rightarrowtail \mathbb{T}^* \times \mathbf{S}^*, \qquad x \mapsto \big(\boldsymbol{u} \cdot \boldsymbol{K}(x),\ \boldsymbol{i}^* \cdot \boldsymbol{K}(x)\big).$$

Secondly, the observer splits spacetime as well (the observer regards spacetime as time and space); accordingly, instead of world points, instants and $\boldsymbol{u}$-space points will be introduced to get the *completely $\boldsymbol{u}$-split form* of the fields:

$$\boldsymbol{\xi}_{\boldsymbol{u}} \cdot \boldsymbol{X} \circ \xi_{\boldsymbol{u}}^{-1}: \mathrm{T} \times \mathrm{S}_{\boldsymbol{u}} \rightarrowtail \mathbb{T} \times \mathbf{S}, \qquad (t,q) \mapsto \big(\boldsymbol{\tau} \cdot \boldsymbol{X}(q \star t),\ \boldsymbol{\pi}_{\boldsymbol{u}} \cdot \boldsymbol{X}(q \star t)\big),$$
$$\boldsymbol{\eta}_{\boldsymbol{u}} \cdot \boldsymbol{K} \circ \xi_{\boldsymbol{u}}^{-1}: \mathrm{T} \times \mathrm{S}_{\boldsymbol{u}} \rightarrowtail \mathbb{T}^* \times \mathbf{S}^*, \qquad (t,q) \mapsto \big(\boldsymbol{u} \cdot \boldsymbol{K}(q \star t),\ \boldsymbol{i}^* \cdot \boldsymbol{K}(q \star t)\big)$$

where $q \star t := \xi_{\boldsymbol{u}}^{-1}(t,q)$ (see 3.2.2).

8.5.2. Let us examine more closely a covector field $\boldsymbol{K}$: it has the half split form

$$(-V_{\boldsymbol{u}}, \boldsymbol{A}_{\boldsymbol{u}}) := \boldsymbol{\eta}_{\boldsymbol{u}} \cdot \boldsymbol{K}: \mathrm{M} \rightarrowtail \mathbb{T}^* \times \mathbf{S}^*, \qquad x \mapsto \big(\boldsymbol{u} \cdot \boldsymbol{K}(x),\ \boldsymbol{i}^* \cdot \boldsymbol{K}(x)\big),$$

and the completely split form

$$\boldsymbol{\eta}_{\boldsymbol{u}} \cdot \boldsymbol{K} \circ \xi_{\boldsymbol{u}}^{-1}: \mathrm{T} \times \mathrm{S}_{\boldsymbol{u}} \rightarrowtail \mathbb{T}^* \times \mathbf{S}^*, \qquad (t,q) \mapsto \big(-V_{\boldsymbol{u}}(q \star t), \boldsymbol{A}_{\boldsymbol{u}}(q \star t)\big).$$

A covector field $\boldsymbol{K}$ is a potential (see 2.4.3). $(t,q) \mapsto V_{\boldsymbol{u}}(q \star t)$ and $(t,q) \mapsto \boldsymbol{A}_{\boldsymbol{u}}(q \star t)$ are called the corresponding *scalar potential* and *vector potential* according to $\boldsymbol{u}$.

If $\boldsymbol{u}'$ is another inertial observer then, in view of 8.3.2,

$$V_{\boldsymbol{u}'} = V_{\boldsymbol{u}} - \boldsymbol{v}_{\boldsymbol{u}'\boldsymbol{u}} \cdot \boldsymbol{A}_{\boldsymbol{u}}, \qquad \boldsymbol{A}_{\boldsymbol{u}'} = \boldsymbol{A}_{\boldsymbol{u}}.$$

Introducing $V := V_{\boldsymbol{u}}$, $V' := V_{\boldsymbol{u}'}$, $\boldsymbol{A} := \boldsymbol{A}_{\boldsymbol{u}}$, $\boldsymbol{A}' := \boldsymbol{A}_{\boldsymbol{u}'}$, $\boldsymbol{v} := \boldsymbol{v}_{\boldsymbol{u}'\boldsymbol{u}}$, we get the formulae

$$V' = V - \boldsymbol{v} \cdot \boldsymbol{A}, \qquad \boldsymbol{A}' = \boldsymbol{A},$$

which are the well-known nonrelativistic transformation law for scalar and vector potentials in electromagnetism.

This supports our choice that (absolute) potentials are cotensor fields.

The reader is asked to bear the following remark in mind. One usually says that if an observer perceives scalar potential V and vector potential $\boldsymbol{A}$, then another observer moving with relative velocity $\boldsymbol{v}$ perceives scalar potential $V - \boldsymbol{v} \cdot \boldsymbol{A}$ and vector potential $\boldsymbol{A}$, i.e. 'the vector potential is not transformed'. However, an observer $\boldsymbol{u}$ perceives spacetime as time and $\boldsymbol{u}$-space, perceives the potentials to be functions depending on time and $\boldsymbol{u}$-space; thus, in fact, an observer works with the completely split form of the potentials. In particular, if $\boldsymbol{u}' \neq \boldsymbol{u}$, then $\boldsymbol{A} \circ \xi_{\boldsymbol{u}}^{-1} \neq \boldsymbol{A} \circ \xi_{\boldsymbol{u}'}^{-1}$: *the observed vector potentials are different!* Remember, usually one does not distinguish between the half split forms and the completely split forms.

8.5.3. Similarly, one usually says that force is not transformed, a force field is the same for all observers. Of course, this is true for the half split form of force fields; the completely split forms of force fields—which are actually observed—depend on the observers.

A force field

$$\boldsymbol{f} : \mathrm{M} \times \mathrm{V}(1) \rightarrowtail \frac{\mathbf{S}}{\mathbb{T} \otimes \mathbb{L} \otimes \mathbb{L}}$$

has exclusively spacelike values, thus its half split form is $\boldsymbol{f}$ itself for all global inertial observers. On the other hand, $\boldsymbol{f}$ has the completely split form

$$\mathrm{T} \times \mathrm{S}_{\boldsymbol{u}} \times \frac{\mathbf{S}}{\mathbb{T}} \rightarrowtail \frac{\mathbf{S}}{\mathbb{T} \otimes \mathbb{L} \otimes \mathbb{L}}, \qquad (t, q, \boldsymbol{v}) \mapsto \boldsymbol{f}(q \star t, \boldsymbol{u} + \boldsymbol{v})$$

strongly depending on $\boldsymbol{u}$.

8.5.4. Let the inertial observer $\boldsymbol{u}$ choose a reference origin o; then $(\boldsymbol{u}, o)$ performs another splitting using $\xi_{\boldsymbol{u},o}$ instead of $\xi_{\boldsymbol{u}}$. The half split form of vector fields and covector fields according to $(\boldsymbol{u}, o)$ is the same as the half split form according to $\boldsymbol{u}$; on the other hand, the observer with reference origin obtains functions $\mathbb{T} \times \mathbf{S} \rightarrowtail \mathbb{T} \times \mathbf{S}$ and $\mathbb{T} \times \mathbf{S} \rightarrowtail \mathbb{T}^* \times \mathbf{S}^*$ for the completely split forms of the fields.

8.6. Splitting of vector fields and covector fields according to rigid observers

8.6.1. Recall that the space S_U of a global rigid observer $\boldsymbol{U}$ is an affine space over $\mathbf{S}$ or $\mathbf{S}_U$ (Section 4.3), depending on whether $\boldsymbol{U}$ is rotation-free or not. The corresponding splitting of spacetime, $\xi_U : M \to T \times S_U$ is a smooth bijection whose inverse is smooth as well.

The splitting of a vector field $\boldsymbol{X}$ according to $\boldsymbol{U}$ is defined by the corresponding formula of coordinatization: at every world point x, the value of the field, $\boldsymbol{X}(x)$, is split—i.e. is mapped from $\mathbf{M}$ into $\mathbb{T} \times \mathbf{S}$ or $\mathbb{T} \times \mathbf{S}_U$ —by $D\xi_U(x)$. Similarly, the covector field $\boldsymbol{K}$, is split in such a way that at every world point x the value of the field, $\boldsymbol{K}(x)$ is split—i.e. is mapped from $\mathbf{M}^*$ into $\mathbb{T}^* \times \mathbf{S}^*$ or $\mathbb{T}^* \times \mathbf{S}_U^*$ —by $\left((D\xi_U(x))^*\right)^{-1}$. Thus the *half $\boldsymbol{U}$-split forms* of such fields are

$$M \rightarrowtail \mathbb{T} \times \mathbf{S} \quad (\text{or } \mathbb{T} \times \mathbf{S}_U), \qquad x \mapsto D\xi_U(x) \cdot \boldsymbol{X}(x),$$
$$M \rightarrowtail \mathbb{T}^* \times \mathbf{S}^* \quad (\text{or } \mathbb{T}^* \times \mathbf{S}_U^*), \qquad x \mapsto \left((D\xi_U(x))^*\right)^{-1} \cdot \boldsymbol{K}(x).$$

We get the *completely U-split* forms by substituting $\xi_U^{-1}(t,q) = q \star t$ for x in these formulae.

8.6.2. If $\boldsymbol{U}$ is rotation-free, then, in view of 4.3.2, the half split forms of the fields are

$$M \rightarrowtail \mathbb{T} \times \mathbf{S}, \qquad x \mapsto \left(\boldsymbol{\tau} \cdot \boldsymbol{X}(x),\ \boldsymbol{\pi}_{U(x)} \cdot \boldsymbol{X}(x)\right),$$
$$M \rightarrowtail \mathbb{T}^* \times \mathbf{S}^*, \qquad x \mapsto \left(\boldsymbol{U}(x) \cdot \boldsymbol{K}(x),\ \boldsymbol{i}^* \cdot \boldsymbol{K}(x)\right).$$

The values of the fields at x are split by the corresponding value $\boldsymbol{U}(x)$ of the observer.

8.6.3. If $\boldsymbol{U}$ is not rotation-free, S_U is an inconvenient object; that is why we let the global rigid observer choose a reference origin o and use the double vectorization $\xi_{U,o}$ of spacetime.

Then the half split forms become

$$M \rightarrowtail \mathbb{T} \times \mathbf{S}, \quad x \mapsto D\xi_{U,o}(x) \cdot \boldsymbol{X}(x) = \left(\boldsymbol{\tau} \cdot \boldsymbol{X}(x), R(x)^{-1} \cdot \boldsymbol{\pi}_{U(x)} \cdot \boldsymbol{X}(x)\right),$$

$$M \rightarrowtail \mathbb{T}^* \times \mathbf{S}^*, \quad x \mapsto \left((D\xi_{U,o}(x))^*\right)^{-1} \cdot \boldsymbol{K}(x) = \\ = \left(\boldsymbol{U}(x) \cdot \boldsymbol{K}(x),\ R(x)^{-1} \cdot \boldsymbol{i}^* \cdot \boldsymbol{K}(x)\right),$$

where

$$R(x) := R_U\left(\tau(x), t_o\right),$$

(see 4.4.3 and Exercise 4.5.3).

8.7. Exercises

1. Give the split form of vector fields and covector fields that depend only on time; more closely, if $\boldsymbol{\chi}: \mathrm{T} \rightarrowtail \mathbf{M}$ and $\boldsymbol{\kappa}: \mathrm{T} \rightarrowtail \mathbf{M}^*$, consider the splitting of the fields $\boldsymbol{X} := \boldsymbol{\chi} \circ \tau$ and $\boldsymbol{K} := \boldsymbol{\kappa} \circ \tau$.

2. We know, it has an absolute meaning that a function $\boldsymbol{s}$ defined in spacetime depends only on time: if $\boldsymbol{s}$ is constant on the simultaneous hyperplanes.

On the other hand, it does not have an absolute meaning that $\boldsymbol{s}$ depends only on space (absolute space does not exist). If $\boldsymbol{U}$ is an observer, it makes sense that $\boldsymbol{s}$ depends only on $\boldsymbol{U}$-space, in other words, $\boldsymbol{s}$ is $\boldsymbol{U}$-static: if $\boldsymbol{s}$ is constant in the $\boldsymbol{U}$-space points (on the $\boldsymbol{U}$-lines), i.e. if the completely split form of $\boldsymbol{s}$ depends only on the elements of S_U.

Let $o \in \mathrm{M}$, $\boldsymbol{c} \in V(1)$, $\boldsymbol{C}: \mathbf{S} \to \mathbf{M}$, and let $\boldsymbol{U}$ be the global inertial observer with velocity value $\boldsymbol{u}$. Prove that the vector field $x \mapsto \boldsymbol{C}\big(\boldsymbol{\pi}_{\boldsymbol{c}} \cdot (x-o)\big)$ is $\boldsymbol{U}$-static if and only if $\boldsymbol{u} = \boldsymbol{c}$.

3. Take the arithmetic spacetime model. Give the completely split form of the vector fields

$$(\xi^0, \boldsymbol{\xi}) \mapsto (|\boldsymbol{\xi}|, \mathbf{0}),$$
$$(\xi^0, \boldsymbol{\xi}) \mapsto (\xi^0 + |\boldsymbol{\xi}|,\ \xi^2 + \xi^3,\ 1 + \xi^1, 0)$$

according to the inertial observer with velocity value $(1, \boldsymbol{v})$.

Consider the previous mappings to be covector fields and give their completely split form.

4. Take the arithmetic spacetime model. Give the completely split form of the vector field

$$(\xi^0, \boldsymbol{\xi}) \mapsto (\xi^1 + \xi^2, \cos(\xi^0 - \xi^3), 0, 0)$$

according to the uniformly accelerated observer with reference origin treated in 5.2.4 and to the uniformly rotating observer with reference origin treated in 5.3.5. [It is easy to obtain the composition of this vector field and the inverse of the splitting if we use different symbols for the variables; e.g. the splitting due to the uniformly accelerated observer has the inverse $(\zeta^0, \boldsymbol{\zeta}) \mapsto \left(\zeta^0, \zeta^1 + \frac{1}{2}\alpha(\zeta^0)^2, \zeta^2, \zeta^3\right)$.]

5. In the split spacetime model the splitting of vectors according to the basic velocity value $(1, \mathbf{0})$ is the identity of $\mathbb{T} \times \mathbf{S}$. The splitting according to $(1, \boldsymbol{v})$ is

$$\mathbb{T} \times \mathbf{S} \to \mathbb{T} \times \mathbf{S}, \qquad (\boldsymbol{t}, \boldsymbol{q}) \mapsto (\boldsymbol{t}, \boldsymbol{q} - \boldsymbol{v}\boldsymbol{t}),$$

which coincides with the transformation rule from $(1, \mathbf{0})$ into $(1, \boldsymbol{v})$. Because of the special structure of the split spacetime model a splitting and a transformation rule—which are in fact different objects in principle—can be equal. To deal with fundamental ideas do not use the split spacetime model or the arithmetic one.

6. In the split spacetime model the splitting of covectors according to the basic velocity value $(1, \mathbf{0})$ is the identity of $\mathbb{T}^* \times \mathbf{S}^*$ and the splitting according to $(1, \boldsymbol{v})$ equals

$$\mathbb{T}^* \times \mathbf{S}^* \to \mathbb{T}^* \times \mathbf{S}^*, \qquad (e, \boldsymbol{p}) \mapsto (e + \boldsymbol{p} \cdot \boldsymbol{v},\ \boldsymbol{p}).$$

Again we see that the splitting coincides with the transformation rule from $(1, \mathbf{0})$ into $(1, \boldsymbol{v})$.

9. Tensor splittings

9.1. Splitting of tensors, cotensors, etc.

9.1.1. The various tensors are split according to $\boldsymbol{u} \in V(1)$ by the maps

$$\begin{aligned}
\boldsymbol{\xi_u} \otimes \boldsymbol{\xi_u} &: \mathbf{M} \otimes \mathbf{M} \to (\mathbb{T} \times \mathbf{S}) \otimes (\mathbb{T} \times \mathbf{S}) = \\
&= (\mathbb{T} \otimes \mathbb{T}) \times (\mathbb{T} \otimes \mathbf{S}) \times (\mathbf{S} \otimes \mathbb{T}) \times (\mathbf{S} \otimes \mathbf{S}), \\
\boldsymbol{\xi_u} \otimes \boldsymbol{\eta_u} &: \mathbf{M} \otimes \mathbf{M}^* \to (\mathbb{T} \times \mathbf{S}) \otimes (\mathbb{T}^* \times \mathbf{S}^*) = \\
&= (\mathbb{T} \otimes \mathbb{T}^*) \times (\mathbb{T} \otimes \mathbf{S}^*) \times (\mathbf{S} \otimes \mathbb{T}^*) \times (\mathbf{S} \otimes \mathbf{S}^*), \\
\boldsymbol{\eta_u} \otimes \boldsymbol{\xi_u} &: \mathbf{M}^* \otimes \mathbf{M} \to (\mathbb{T}^* \times \mathbf{S}^*) \otimes (\mathbb{T} \times \mathbf{S}) = \\
&= (\mathbb{T}^* \otimes \mathbb{T}) \times (\mathbb{T}^* \otimes \mathbf{S}) \times (\mathbf{S}^* \otimes \mathbb{T}) \times (\mathbf{S}^* \otimes \mathbf{S}), \\
\boldsymbol{\eta_u} \otimes \boldsymbol{\eta_u} &: \mathbf{M}^* \otimes \mathbf{M}^* \to (\mathbb{T}^* \times \mathbf{S}^*) \otimes (\mathbb{T}^* \times \mathbf{S}^*) = \\
&= (\mathbb{T}^* \otimes \mathbb{T}^*) \times (\mathbb{T}^* \otimes \mathbf{S}^*) \times (\mathbf{S}^* \otimes \mathbb{T}^*) \times (\mathbf{S}^* \otimes \mathbf{S}^*).
\end{aligned}$$

Since we know $\boldsymbol{\xi_u}$ and $\boldsymbol{\eta_u}$, our task is only to determine the above splittings in a perspicuous way. First recall that the elements of the Cartesian products on the right-hand sides can be well given in a matrix form (see IV.3.7). Second, with the aid of the usual identifications, consider $\boldsymbol{\xi_u} = (\boldsymbol{\tau}, \boldsymbol{\pi_u}) \in (\mathbb{T} \times \mathbf{S}) \otimes \mathbf{M}^*$, $\boldsymbol{\eta_u} = (\boldsymbol{u}, \boldsymbol{i}^*) \in (\mathbb{T}^* \times \mathbf{S}^*) \otimes \mathbf{M}$, take into account the identifications $\mathbb{T} \otimes \mathbf{M}^* \equiv \mathbf{M}^* \otimes \mathbb{T}$, $\boldsymbol{\tau} \equiv \boldsymbol{\tau}^*$ and $\mathbb{T}^* \otimes \mathbf{M} \equiv \mathbf{M} \otimes \mathbb{T}^*$, $\boldsymbol{u} \equiv \boldsymbol{u}^*$ (see IV.3.6), and apply the dot products to have

for $\boldsymbol{T} \in \mathbf{M} \otimes \mathbf{M}$:

$$(\boldsymbol{\xi_u} \otimes \boldsymbol{\xi_u})(\boldsymbol{T}) = \boldsymbol{\xi_u} \cdot \boldsymbol{T} \cdot \boldsymbol{\xi_u^*} = \boldsymbol{\xi_u} \cdot \boldsymbol{T} \cdot \boldsymbol{\eta_u^{-1}} = \begin{pmatrix} \boldsymbol{\tau} \cdot \boldsymbol{T} \cdot \boldsymbol{\tau} & \boldsymbol{\tau} \cdot \boldsymbol{T} \cdot \boldsymbol{\pi_u^*} \\ \boldsymbol{\pi_u} \cdot \boldsymbol{T} \cdot \boldsymbol{\tau} & \boldsymbol{\pi_u} \cdot \boldsymbol{T} \cdot \boldsymbol{\pi_u^*} \end{pmatrix} =$$

$$= \begin{pmatrix} \boldsymbol{\tau} \cdot \boldsymbol{T} \cdot \boldsymbol{\tau} & \boldsymbol{\tau} \cdot \boldsymbol{T} - \boldsymbol{u}(\boldsymbol{\tau} \cdot \boldsymbol{T} \cdot \boldsymbol{\tau}) \\ \boldsymbol{T} \cdot \boldsymbol{\tau} - \boldsymbol{u}(\boldsymbol{\tau} \cdot \boldsymbol{T} \cdot \boldsymbol{\tau}) & \boldsymbol{T} - \boldsymbol{u} \otimes (\boldsymbol{\tau} \cdot \boldsymbol{T}) - (\boldsymbol{T} \cdot \boldsymbol{\tau}) \otimes \boldsymbol{u} + \boldsymbol{u} \otimes \boldsymbol{u}(\boldsymbol{\tau} \cdot \boldsymbol{T} \cdot \boldsymbol{\tau}) \end{pmatrix},$$

for $\boldsymbol{L} \in \mathbf{M} \otimes \mathbf{M}^*$:

$$(\boldsymbol{\xi_u} \otimes \boldsymbol{\eta_u})(\boldsymbol{L}) = \boldsymbol{\xi_u} \cdot \boldsymbol{L} \cdot \boldsymbol{\eta_u^*} = \boldsymbol{\xi_u} \cdot \boldsymbol{L} \cdot \boldsymbol{\xi_u^{-1}} = \begin{pmatrix} \boldsymbol{\tau} \cdot \boldsymbol{L} \cdot \boldsymbol{u} & \boldsymbol{\tau} \cdot \boldsymbol{L} \cdot \boldsymbol{i} \\ \boldsymbol{\pi_u} \cdot \boldsymbol{L} \cdot \boldsymbol{u} & \boldsymbol{\pi_u} \cdot \boldsymbol{L} \cdot \boldsymbol{i} \end{pmatrix} =$$

$$= \begin{pmatrix} \boldsymbol{\tau} \cdot \boldsymbol{L} \cdot \boldsymbol{u} & \boldsymbol{\tau} \cdot \boldsymbol{L} \cdot \boldsymbol{i} \\ \boldsymbol{L} \cdot \boldsymbol{u} - \boldsymbol{u}(\boldsymbol{\tau} \cdot \boldsymbol{L} \cdot \boldsymbol{u}) & \boldsymbol{L} \cdot \boldsymbol{i} - \boldsymbol{u} \otimes (\boldsymbol{\tau} \cdot \boldsymbol{L} \cdot \boldsymbol{i}) \end{pmatrix},$$

for $\boldsymbol{P} \in \mathbf{M}^* \otimes \mathbf{M}$:

$$(\eta_u \otimes \xi_u)(\boldsymbol{P}) = \eta_u \cdot \boldsymbol{P} \cdot \xi_u^* = \eta_u \cdot \boldsymbol{P} \cdot \eta_u^{-1} = \begin{pmatrix} \boldsymbol{u} \cdot \boldsymbol{P} \cdot \boldsymbol{\tau} & \boldsymbol{u} \cdot \boldsymbol{P} \cdot \pi_u^* \\ \boldsymbol{i}^* \cdot \boldsymbol{P} \cdot \boldsymbol{\tau} & \boldsymbol{i}^* \cdot \boldsymbol{P} \cdot \pi_u^* \end{pmatrix} =$$

$$= \begin{pmatrix} \boldsymbol{u} \cdot \boldsymbol{P} \cdot \boldsymbol{\tau} & \boldsymbol{u} \cdot \boldsymbol{P} - (\boldsymbol{u} \cdot \boldsymbol{P} \cdot \boldsymbol{\tau})\boldsymbol{u} \\ \boldsymbol{i}^* \cdot \boldsymbol{P} \cdot \boldsymbol{\tau} & \boldsymbol{i}^* \cdot \boldsymbol{P} - (\boldsymbol{i}^* \cdot \boldsymbol{P} \cdot \boldsymbol{\tau}) \otimes \boldsymbol{u} \end{pmatrix},$$

for $\boldsymbol{F} \in \mathbf{M}^* \otimes \mathbf{M}^*$:

$$(\eta_u \otimes \eta_u)(\boldsymbol{F}) = \eta_u \cdot \boldsymbol{F} \cdot \eta_u^* = \eta_u \cdot \boldsymbol{F} \cdot \xi_u^{-1} = \begin{pmatrix} \boldsymbol{u} \cdot \boldsymbol{F} \cdot \boldsymbol{u} & \boldsymbol{u} \cdot \boldsymbol{F} \cdot \boldsymbol{i} \\ \boldsymbol{i}^* \cdot \boldsymbol{F} \cdot \boldsymbol{u} & \boldsymbol{i}^* \cdot \boldsymbol{F} \cdot \boldsymbol{i} \end{pmatrix}.$$

(To see, e.g. that $\pi_u \cdot \boldsymbol{T} \cdot \boldsymbol{\tau} = \boldsymbol{T} \cdot \boldsymbol{\tau} - \boldsymbol{u}(\boldsymbol{\tau} \cdot \boldsymbol{T} \cdot \boldsymbol{\tau})$, take $\boldsymbol{T} = \boldsymbol{x} \otimes \boldsymbol{y}$.)

9.1.2. The splittings corresponding to different velocity values $\boldsymbol{u}$ and $\boldsymbol{u}'$ are different. To compare the different splittings we can deduce transformation rules by giving

$$\xi_{u'u} \cdot \begin{pmatrix} \alpha & b \\ a & A \end{pmatrix} \cdot {\xi_{u'u}}^*,$$

where $\alpha \in \mathbb{T} \otimes \mathbb{T}$, $\boldsymbol{a} \in \mathbf{S} \otimes \mathbb{T}$, $\boldsymbol{b} \in \mathbb{T} \otimes \mathbf{S}$, $\boldsymbol{A} \in \mathbf{S} \otimes \mathbf{S}$, and using similar formulae for the other three cases as well. In general, the transformation rules are rather complicated. We shall study them for antisymmetric tensors and cotensors.

9.2. Splitting of antisymmetric tensors

9.2.1. If the tensor $\boldsymbol{T}$ is antisymmetric—i.e. $\boldsymbol{T} \in \mathbf{M} \wedge \mathbf{M}$ —then $\boldsymbol{\tau} \cdot \boldsymbol{T} \cdot \boldsymbol{\tau} = 0$, $\boldsymbol{\tau} \cdot \boldsymbol{T} \cdot \pi_u^* = -(\pi_u \cdot \boldsymbol{T} \cdot \boldsymbol{\tau})^*$ and $\pi_u \cdot \boldsymbol{T} \cdot \pi_u^* \in \mathbf{S} \wedge \mathbf{S}$, which (of course) means that the split forms of $\boldsymbol{T}$ are antisymmetric as well. Thus splittings map the elements of $\mathbf{M} \wedge \mathbf{M}$ into elements of the form

$$\begin{pmatrix} 0 & -\boldsymbol{a}^* \\ \boldsymbol{a} & \boldsymbol{A} \end{pmatrix} \equiv \begin{pmatrix} 0 & -\boldsymbol{a} \\ \boldsymbol{a} & \boldsymbol{A} \end{pmatrix},$$

where $\boldsymbol{a} \in \mathbf{S} \otimes \mathbb{T}$, $\boldsymbol{A} \in \mathbf{S} \wedge \mathbf{S}$; $\boldsymbol{a}^* \in \mathbb{T} \otimes \mathbf{S}$ is the transpose of $\boldsymbol{a}$, which is identified with $\boldsymbol{a}$ in the usual identification $\mathbb{T} \otimes \mathbf{S} \equiv \mathbf{S} \otimes \mathbb{T}$. We shall find convenient to write

$$(\mathbf{S} \otimes \mathbb{T}) \times (\mathbf{S} \wedge \mathbf{S}) \equiv (\mathbb{T} \times \mathbf{S}) \wedge (\mathbb{T} \times \mathbf{S}),$$

$$(\boldsymbol{a}, \boldsymbol{A}) \equiv \begin{pmatrix} 0 & -\boldsymbol{a} \\ \boldsymbol{a} & \boldsymbol{A} \end{pmatrix}.$$

The corresponding formula in 9.1.1 gives us for $\boldsymbol{T} \in \mathbf{M} \wedge \mathbf{M}$

$$\xi_u \cdot \boldsymbol{T} \cdot \xi_u^* = (\boldsymbol{T} \cdot \boldsymbol{\tau},\ \boldsymbol{T} - (\boldsymbol{T} \cdot \boldsymbol{\tau}) \wedge \boldsymbol{u}).$$

Definition. $\boldsymbol{T}\cdot\boldsymbol{\tau}$ and $\boldsymbol{T}-(\boldsymbol{T}\cdot\boldsymbol{\tau})\wedge\boldsymbol{u}$ are called the *timelike component* and the $\boldsymbol{u}$*-spacelike component* of the antisymmetric tensor $\boldsymbol{T}$.

9.2.2. Notice the similarity between splittings of vectors and splittings of antisymmetric tensors. The timelike component of $\boldsymbol{T}$ is independent of $\boldsymbol{u}$, the $\boldsymbol{u}$-spacelike component varies with $\boldsymbol{u}$ except when $\boldsymbol{T}$ is spacelike, i.e. is in $\mathbf{S}\wedge\mathbf{S}$; then the timelike component is zero and the $\boldsymbol{u}$-spacelike component is $\boldsymbol{T}$ itself for all $\boldsymbol{u}$.

The following *transformation rule* shows well how the splittings depend on the velocity values.

Proposition. Let $\boldsymbol{u},\boldsymbol{u}'\in V(1)$. Then

$$\xi_{\boldsymbol{u}'\boldsymbol{u}}\cdot(\boldsymbol{a},\boldsymbol{A})\cdot{\xi_{\boldsymbol{u}'\boldsymbol{u}}}^{*}=(\boldsymbol{a},-\boldsymbol{a}\wedge\boldsymbol{v}_{\boldsymbol{u}'\boldsymbol{u}}+\boldsymbol{A})\qquad(\boldsymbol{a}\in\mathbf{S}\otimes\mathbb{T},\ \boldsymbol{A}\in\mathbf{S}\wedge\mathbf{S}).$$

Proof. Use the matrix forms:

$$\begin{pmatrix}1 & 0\\ -\boldsymbol{v}_{\boldsymbol{u}'\boldsymbol{u}} & 1\end{pmatrix}\begin{pmatrix}0 & -\boldsymbol{a}\\ \boldsymbol{a} & \boldsymbol{A}\end{pmatrix}\begin{pmatrix}1 & -\boldsymbol{v}_{\boldsymbol{u}'\boldsymbol{u}}\\ 0 & 1\end{pmatrix}=\begin{pmatrix}0 & -\boldsymbol{a}\\ \boldsymbol{a} & -\boldsymbol{a}\wedge\boldsymbol{v}_{\boldsymbol{u}'\boldsymbol{u}}+\boldsymbol{A}\end{pmatrix}.$$

9.3. Splitting of antisymmetric cotensors

9.3.1. If $\boldsymbol{F}\in\mathbf{M}^*\wedge\mathbf{M}^*$ then $\boldsymbol{u}\cdot\boldsymbol{F}\cdot\boldsymbol{u}=0$, $\boldsymbol{u}\cdot\boldsymbol{F}\cdot\boldsymbol{i}=-(\boldsymbol{i}^*\cdot\boldsymbol{F}\cdot\boldsymbol{u})^*$ and $\boldsymbol{i}^*\cdot\boldsymbol{F}\cdot\boldsymbol{i}\in\mathbf{S}^*\wedge\mathbf{S}^*$; the split forms of $\boldsymbol{F}$ are antisymmetric as well. Thus splitting maps the elements of $\mathbf{M}^*\wedge\mathbf{M}^*$ into elements of the form

$$\begin{pmatrix}0 & -\boldsymbol{z}^*\\ \boldsymbol{z} & \boldsymbol{Z}\end{pmatrix}\equiv\begin{pmatrix}0 & -\boldsymbol{z}\\ \boldsymbol{z} & \boldsymbol{Z}\end{pmatrix}\equiv(\boldsymbol{z},\boldsymbol{Z})\in(\mathbf{S}^*\otimes\mathbb{T}^*)\times(\mathbf{S}^*\wedge\mathbf{S}^*),$$

where we used notations similar to those in 9.2.1.

The corresponding formula in 9.1.1 gives for $\boldsymbol{F}\in\mathbf{M}^*\wedge\mathbf{M}^*$:

$$\eta_{\boldsymbol{u}}\cdot\boldsymbol{F}\cdot\eta_{\boldsymbol{u}}^*=(\boldsymbol{i}^*\cdot\boldsymbol{F}\cdot\boldsymbol{u},\ \boldsymbol{i}^*\cdot\boldsymbol{F}\cdot\boldsymbol{i}).$$

Definition. $\boldsymbol{i}^*\cdot\boldsymbol{F}\cdot\boldsymbol{u}$ and $\boldsymbol{i}^*\cdot\boldsymbol{F}\cdot\boldsymbol{i}$ are called the $\boldsymbol{u}$*-timelike component* and the *spacelike component* of the antisymmetric cotensor $\boldsymbol{F}$.

9.3.2. Notice the similarity between splittings of covectors and splittings of antisymmetric cotensors. The spacelike component of $\boldsymbol{F}$ is independent of $\boldsymbol{u}$, the $\boldsymbol{u}$-timelike component varies with $\boldsymbol{u}$ except when $\boldsymbol{F}$ is in $\mathbf{M}^*\wedge(\mathbb{T}^*\cdot\boldsymbol{\tau}):=\{\boldsymbol{k}\wedge(\boldsymbol{e}\cdot\boldsymbol{\tau})\mid\boldsymbol{k}\in\mathbf{M}^*,\ \boldsymbol{e}\in\mathbb{T}^*\}$; then the spacelike component is zero and the $\boldsymbol{u}$-timelike component is the same for all $\boldsymbol{u}$.

The following *transformation rule* shows well, how the splittings depend on the velocity values.

Proposition. Let $\boldsymbol{u}, \boldsymbol{u}' \in V(1)$. Then

$$\eta_{\boldsymbol{u}'\boldsymbol{u}} \cdot (\boldsymbol{z}, \boldsymbol{Z}) \cdot {\eta_{\boldsymbol{u}'\boldsymbol{u}}}^* = (\boldsymbol{z} + \boldsymbol{Z} \cdot \boldsymbol{v}_{\boldsymbol{u}'\boldsymbol{u}}, \boldsymbol{Z}) \qquad (\boldsymbol{z} \in \mathbb{T}^* \otimes \mathbf{S}^*,\ \boldsymbol{Z} \in \mathbf{S}^* \wedge \mathbf{S}^*).$$

Proof. Use the matrix forms:

$$\begin{pmatrix} 1 & \boldsymbol{v}_{\boldsymbol{u}'\boldsymbol{u}} \\ 0 & 1 \end{pmatrix} \begin{pmatrix} 0 & -\boldsymbol{z} \\ \boldsymbol{z} & \boldsymbol{Z} \end{pmatrix} \begin{pmatrix} 1 & 0 \\ \boldsymbol{v}_{\boldsymbol{u}'\boldsymbol{u}} & 1 \end{pmatrix} = \begin{pmatrix} 0 & -(\boldsymbol{z} + \boldsymbol{Z} \cdot \boldsymbol{v}_{\boldsymbol{u}'\boldsymbol{u}}) \\ \boldsymbol{z} + \boldsymbol{Z} \cdot \boldsymbol{v}_{\boldsymbol{u}'\boldsymbol{u}} & \boldsymbol{Z} \end{pmatrix}.$$

9.4. Splitting of cotensor fields

9.4.1. A rotation-free rigid observer $\boldsymbol{U}$ splits various tensor fields in such a way that the value of the tensor field at the world point x is split according to $\boldsymbol{U}(x)$; for the sake of definiteness we shall consider cotensor fields. The *half split form* of the cotensor field $\boldsymbol{F} \colon \mathrm{M} \rightarrowtail \mathbf{M}^* \otimes \mathbf{M}^*$ according to $\boldsymbol{U}$ is

$$\mathrm{M} \rightarrowtail (\mathbb{T}^* \times \mathbf{S}^*) \otimes (\mathbb{T}^* \times \mathbf{S}^*), \qquad x \mapsto \eta_{\boldsymbol{U}(x)} \cdot \boldsymbol{F}(x) \cdot {\eta_{\boldsymbol{U}(x)}}^*.$$

The *completely split form* of $\boldsymbol{F}$ according to $\boldsymbol{U}$ is

$$\mathrm{T} \times \mathrm{S}_{\boldsymbol{U}} \rightarrowtail (\mathrm{T} \times \mathbf{S}^*) \otimes (\mathbb{T}^* \times \mathbf{S}^*), \qquad (t, q) \mapsto \eta_{\boldsymbol{U}(q \star t)} \cdot \boldsymbol{F}(q \star t) \cdot \eta_{\boldsymbol{U}(q \star t)},$$

where $q \star t = \xi_{\boldsymbol{U}}^{-1}(t, q)$.

In particular, if $\boldsymbol{F}$ is antisymmetric, then it has the half split form

$$\mathrm{M} \rightarrowtail (\mathbf{S}^* \otimes \mathbb{T}^*) \times (\mathbf{S}^* \wedge \mathbf{S}^*), \qquad x \mapsto \big(\boldsymbol{i}^* \cdot \boldsymbol{F}(x) \cdot \boldsymbol{U}(x), \boldsymbol{i}^* \cdot \boldsymbol{F}(x) \cdot \boldsymbol{i}\big).$$

9.4.2. Now let us suppose that $\boldsymbol{U}$ is a global inertial observer with constant velocity value $\boldsymbol{u}$. Then the antisymmetric cotensor field $\boldsymbol{F}$ has the half split form

$$(\boldsymbol{E}_{\boldsymbol{u}}, \boldsymbol{B}_{\boldsymbol{u}}) := \eta_{\boldsymbol{u}} \cdot \boldsymbol{F} \cdot \eta_{\boldsymbol{u}}^* \colon \mathrm{M} \rightarrowtail (\mathbf{S}^* \otimes \mathbb{T}^*) \times (\mathbf{S}^* \wedge \mathbf{S}^*),$$
$$x \mapsto \big(\boldsymbol{i}^* \cdot \boldsymbol{F}(x) \cdot \boldsymbol{u},\ \boldsymbol{i}^* \cdot \boldsymbol{F}(x) \cdot \boldsymbol{i}\big)$$

and the completely split form

$$\eta_{\boldsymbol{u}} \cdot \boldsymbol{F} \cdot \eta_{\boldsymbol{u}}^* \circ \xi_{\boldsymbol{u}}^{-1} \colon \mathrm{T} \times \mathrm{S}_{\boldsymbol{U}} \rightarrowtail (\mathbf{S}^* \otimes \mathbb{T}^*) \times (\mathbf{S}^* \wedge \mathbf{S}^*),$$
$$(t, q) \mapsto \big(\boldsymbol{E}_{\boldsymbol{u}}(q \star t), \boldsymbol{B}_{\boldsymbol{u}}(q \star t)\big).$$

If u' is another global inertial observer, then 9.3.2 gives

$$\boldsymbol{E}_{u'} = \boldsymbol{E}_u + \boldsymbol{B}_u \cdot \boldsymbol{v}_{u'u}, \qquad \boldsymbol{B}_{u'} = \boldsymbol{B}_u.$$

9.4.3. Introducing $\boldsymbol{E} := \boldsymbol{E}_u$, $\boldsymbol{E}' := \boldsymbol{E}_{u'}$, $\boldsymbol{B} := \boldsymbol{B}_u$, $\boldsymbol{B}' := \boldsymbol{B}_{u'}$, $\boldsymbol{v} := \boldsymbol{v}_{u'u}$, we get the formula

$$\boldsymbol{E}' = \boldsymbol{E} + \boldsymbol{B} \cdot \boldsymbol{v}, \qquad \boldsymbol{B}' = \boldsymbol{B}$$

which is the well-known nonrelativistic transformation law for the electric field $\boldsymbol{E}$ and magnetic field $\boldsymbol{B}$. (Here $\boldsymbol{B}$ is an antisymmetric spacelike tensor of cotype $\mathbb{L}^{\otimes 4}$, an element of $\mathbf{S}^* \wedge \mathbf{S}^* = \frac{\mathbf{S}\wedge\mathbf{S}}{\mathbb{L}\otimes\mathbb{L}\otimes\mathbb{L}\otimes\mathbb{L}}$, which can be identified with a vector of cotype $\mathbb{L}^{\otimes 3}$, an element of $\frac{\mathbf{S}}{\mathbb{L}\otimes\mathbb{L}\otimes\mathbb{L}} \equiv \frac{\mathbf{S}^*}{\mathbb{L}}$ (V.3.17.); with the aid of this identification magnetic field is regarded as a vector field and then instead of $\boldsymbol{B} \cdot \boldsymbol{v}$ one has a vectorial product.)

This supports the idea that (absolute) electromagnetic fields exist whose timelike and spacelike components according to an observer are the observed electric and magnetic fields, respectively.

One usually says that if an observer perceives electric field $\boldsymbol{E}$ and magnetic field $\boldsymbol{B}$, then another observer moving with the velocity $\boldsymbol{v}$ perceives electric field $\boldsymbol{E}+\boldsymbol{B}\cdot\boldsymbol{v}$ and magnetic field $\boldsymbol{B}$. However, an observer perceives spacetime as time and $\boldsymbol{u}$-space, perceives the fields as functions depending on time and $\boldsymbol{u}$-space; thus, in fact, an observer observes the completely split form of the fields, and we can repeat the remark at the end of 8.5.3.

9.4.4. Consider the completely split form of a potential $\boldsymbol{K}$ according to the inertial observer $\boldsymbol{u}$:

$$(-V_u^c, \boldsymbol{A}_u^c) := \boldsymbol{\eta}_u \cdot \left(\boldsymbol{K} \circ \xi_u^{-1}\right) : \mathrm{T} \times \mathrm{S}_u \rightarrowtail \mathbb{T}^* \times \mathbf{S}^*.$$

Its derivative is

$$\mathrm{D}(-V_u^c, \boldsymbol{A}_u^c) = \boldsymbol{\eta}_u \cdot (\mathrm{D}\boldsymbol{K} \circ \xi_u^{-1}) \cdot \boldsymbol{\eta}_u{}^*$$

having the transpose

$$\left(\mathrm{D}(-V_u^c, \boldsymbol{A}_u^c)\right)^* = \boldsymbol{\eta}_u \cdot \left((\mathrm{D}\boldsymbol{K} \circ \xi_u^{-1})\right)^* \cdot \boldsymbol{\eta}_u{}^*.$$

Consequently, for the exterior derivatives (see VI.3.6(i)) we have

$$\mathrm{D} \wedge (-V_u^c, \boldsymbol{A}_u^c) = \boldsymbol{\eta}_u \cdot \left((\mathrm{D} \wedge \boldsymbol{K}) \circ \xi_u^{-1}\right) \cdot \boldsymbol{\eta}_u{}^*.$$

Let $\boldsymbol{F} := \mathrm{D} \wedge \boldsymbol{K}$, use the notations of the previous paragraph and let ∂_0 and ∇ denote the partial derivations with respect to T and S_u, respectively. Then (see VI.3.7(ii)) the above equality yields

$$-\partial_0 \boldsymbol{A}_u^c - \nabla V_u^c = \boldsymbol{E}_u^c, \qquad \nabla \wedge \boldsymbol{A}_u^c = \boldsymbol{B}_u^c.$$

9.4.5. Let us consider the force field defined by the potential $\boldsymbol{K}$:

$$\boldsymbol{f}(x,\dot{x}) = \boldsymbol{i}^* \cdot \boldsymbol{F}(x) \cdot \dot{x} \qquad (x \in \mathrm{Dom}\boldsymbol{K}, \ \dot{x} \in V(1)).$$
$$(\boldsymbol{F} := \mathrm{D} \wedge \boldsymbol{K}).$$

According to 9.3.1, the value of the force field at $(x,\dot{x})$ is the $\dot{x}$-timelike component of the antisymmetric cotensor $\boldsymbol{F}(x)$.

A masspoint at the world point x having the instantaneous velocity value $\dot{x}$ 'feels' only the $\dot{x}$-timelike component of the field; a masspoint always 'feels' the time component of the field according to its instantaneous velocity value.

Consider now the inertial observer with velocity value $\boldsymbol{u}$ and use the notations of the previous paragraphs. Then

$$\begin{aligned}\boldsymbol{f}(x,\dot{x}) &= \boldsymbol{i}^* \cdot \boldsymbol{F}(x) \cdot \boldsymbol{u} + \boldsymbol{i}^* \cdot \boldsymbol{F}(x)(\dot{x} - \boldsymbol{u}) = \\ &= \boldsymbol{E}_{\boldsymbol{u}}(x) + \boldsymbol{B}_{\boldsymbol{u}}(x) \cdot \boldsymbol{v}_{\dot{x}\boldsymbol{u}},\end{aligned}$$

a well-known formula for the Lorentz force in electromagnetism.

9.4.6. If a potential $\boldsymbol{K}$ is timelike, i.e. has values in $\mathbb{T}^* \cdot \boldsymbol{\tau}$, (in fact $\boldsymbol{K}$ is a scalar field: there is a function $V : \mathrm{M} \rightarrowtail \mathbb{T}^*$ such that $\boldsymbol{K} = V \cdot \boldsymbol{\tau}$) then $\mathrm{D} \wedge \boldsymbol{K}$ takes values in $\mathbf{M}^* \wedge (\mathbb{T}^* \cdot \boldsymbol{\tau})$; consequently the corresponding force field does not depend on velocity values; the spacelike component of $\boldsymbol{K}$ is zero and the half split form of $\boldsymbol{K}$ is the same for all observers.

The possibility of (absolute) scalar potentials is a peculiar feature of the non-relativistic spacetime model in contradistinction to relativistic spacetime models. (Newtonian gravitational fields, elastic fields are modelled by such timelike potentials in nonrelativistic physics.)

9.4.7. Let us mention the case of a general (rotating) global rigid observer $\boldsymbol{U}$. Then it is convenient to choose a reference origin o for the observer and consider the corresponding double vectorization of spacetime.

We easily infer from the splitting of vector fields and covector fields that the half split forms of various tensor fields according to $(\boldsymbol{U},o)$ are obtained from the half split forms according to a rotation-free observer in such a way that $R_{\boldsymbol{U}}(\tau(x),t_o)^{-1} \cdot \boldsymbol{\pi}_{\boldsymbol{U}(x)}$ and $R_{\boldsymbol{U}}(\tau(x),t_o)^{-1} \cdot \boldsymbol{i}^*$ are substituted for $\boldsymbol{\pi}_{\boldsymbol{U}(x)}$ and $\boldsymbol{i}^*$, respectively (then $\boldsymbol{i} \cdot R_{\boldsymbol{U}}(\tau(x),t_o)$ is substituted for $\boldsymbol{i}$).

For instance, the half split form of an antisymmetric cotensor field $\boldsymbol{F}$ becomes

$$\begin{aligned}&\mathrm{M} \rightarrowtail (\mathbf{S}^* \otimes \mathbb{T}^*) \times (\mathbf{S}^* \wedge \mathbf{S}^*), \\ &\qquad x \mapsto \left(R(x)^{-1} \cdot \boldsymbol{i}^* \cdot \boldsymbol{F}(x) \cdot \boldsymbol{U}(x), \quad R(x)^{-1} \cdot \boldsymbol{i}^* \cdot \boldsymbol{F}(x) \cdot \boldsymbol{i} \cdot R(x)\right),\end{aligned}$$

where $R(x) := R_{\boldsymbol{U}}\big(\tau(x),t_o\big)$.

9.5. Exercises

1. Give the $\boldsymbol{u}$-split form of tensors in $\mathbf{S}\otimes\mathbf{S}$, $\mathbf{S}\otimes\mathbf{M}$, $\mathbf{M}\otimes\mathbf{S}$, $\mathbf{S}\otimes\mathbf{M}^*$, $\mathbf{M}^*\otimes\mathbf{S}$, $(\mathbb{T}^*\cdot\boldsymbol{\tau})\otimes\mathbf{M}$, $\mathbf{M}\otimes(\mathbb{T}^*\cdot\boldsymbol{\tau})$, $(\mathbb{T}^*\cdot\boldsymbol{\tau})\otimes\mathbf{M}^*$, $\mathbf{M}^*\otimes(\mathbb{T}^*\cdot\boldsymbol{\tau})$ and derive the transformation rules between their $\boldsymbol{u}'$-splitting and $\boldsymbol{u}$-splitting.

2. Derive the transformation rules for the splitting of arbitrary tensors.

3. A potential in the arithmetic spacetime model is a function $(-V,\boldsymbol{A})\colon\mathbb{R}\times\mathbb{R}^3\rightarrowtail(\mathbb{R}\times\mathbb{R}^3)^*$ which is the completely split form of the potential according to the basic observer.

The half split form of this potential according to the inertial observer with velocity value $(1,\boldsymbol{v})$ is $(-V+\boldsymbol{v}\cdot\boldsymbol{A},\boldsymbol{A})$.

Choose $(0,\mathbf{0})$ as a reference origin for the observer and give the completely split form of the potential.

4. An antisymmetric cotensor field in the arithmetic spacetime model is a function $(\boldsymbol{E},\boldsymbol{B}):\mathbb{R}\times\mathbb{R}^3\rightarrowtail(\mathbb{R}^3)^*\times\big((\mathbb{R}^3)^*\wedge(\mathbb{R}^3)^*\big)$, being the completely split form of the field according to the basic observer:

$$(\boldsymbol{E},\boldsymbol{B})=\begin{pmatrix}0&-E_1&-E_2&-E_3\\E_1&0&B_3&-B_2\\E_2&-B_3&0&B_1\\E_3&B_2&-B_1&0\end{pmatrix}.$$

The half split form of this field according to the inertial observer with velocity value $(1,\boldsymbol{v})$ is $(\boldsymbol{E}+\boldsymbol{B}\cdot\boldsymbol{v},\ \boldsymbol{B})$.

Choose $(0,\mathbf{0})$ as a reference origin for the observer and give the completely split form of the field.

5. Take the uniformly accelerated observer treated in 5.2.4.

The half split forms of the previous potential and field according to this observer are

$$(-V+\alpha tA_1,\ \boldsymbol{A})\qquad\text{and}\qquad((E_1,\ E_2-\alpha tB_3,\ E_3+\alpha tB_2),\ \boldsymbol{B}),$$

where t is the time evaluation: $\mathbb{R}\times\mathbb{R}^3\to\mathbb{R}$, $(\xi^0,\boldsymbol{\xi})\mapsto\xi^0$.

Choose $(0,\mathbf{0})$ as a reference origin for the observer and give the completely split forms.

6. Take the uniformly rotating observer treated in 5.3.5.

Let $(-V',\boldsymbol{A}')$ and $(\boldsymbol{E}',\boldsymbol{B}')$ denote the half split forms of the previous potential and field, respectively, according to this observer. Then

$$\begin{aligned}V'&=V+\omega(x^2A_1-x^1A_2),\\A_1'&=A_1\cos\omega t-A_2\sin\omega t,\\A_2'&=A_1\sin\omega t+A_2\cos\omega t,\\A_3'&=A_3\end{aligned}$$

and

$$
\begin{aligned}
E_1' &= (E_1 + \omega x^1 B_3)\cos\omega t - (E_2 + \omega x^2 B_3)\sin\omega t,\\
E_2' &= (E_1 + \omega x^1 B_3)\sin\omega t + (E_2 + \omega x^2 B_3)\cos\omega t,\\
E_3' &= E_3 - \omega(x^2 B_2 + x^1 B_1),\\
B_1' &= B_1\cos\omega t - B_2\sin\omega t,\\
B_2' &= B_1\sin\omega t + B_2\cos\omega t,\\
B_3' &= B_3,
\end{aligned}
$$

where t is the time evaluation $\mathbb{R}\times\mathbb{R}^3 \to \mathbb{R},\ (\xi^0, \boldsymbol{\xi}) \mapsto \xi^0$ and x^i is the evaluation of the i-th space coordinate: $\mathbb{R}\times\mathbb{R}^3 \to \mathbb{R},\ (\xi^0, \boldsymbol{\xi}) \mapsto \xi^i$.

10. Reference systems

10.1. The notion of a reference system

10.1.1. Observers, reference frames, reference systems, coordinatizations are fundamental notions of usual textbooks in describing physical phenomena. However, these notions are often applied in different (sometimes heuristic i.e. not precisely defined) senses.

Our intention is to give an *absolute description* of phenomena, i.e. a description free of reference frames and coordinates. Reference frames and coordinates have only a practical (not theoretical) importance: it is convenient and suitable to use observers or coordinates for solving concrete problems, for achieving numerical characterization of quantities.

We defined reference frames in 3.3.2. In nonrelativistic spacetime model an observer determines a reference frame uniquely. A reference frame can introduce coordinates which model how timepoints are indicated by a 'synchronometer' and how spacepoints are labelled by triplets of numbers.

The reader is assumed to be familiar with coordinatizations of affine spaces (see Section VI.5.).

10.1.2. Recall that an observer $\boldsymbol{U}$ makes the splitting $\xi_U = (\tau, C_U)\colon \mathrm{M} \rightarrowtail \mathrm{T}\times\mathrm{S}_U$.

Definition. A *reference system* is a triplet $(\boldsymbol{U}, T, S_U)$ where

(*i*) $\boldsymbol{U}$ is an observer,

(*ii*) $T\colon \mathrm{T} \rightarrowtail \mathbb{R}$ is a strictly monotone increasing mapping,

(*iii*) $S_U\colon \mathrm{S}_U \rightarrowtail \mathbb{R}^3$ is a mapping

such that $(T\times S_U)\circ\xi_U = (T\circ\tau,\ S_U\circ C_U)\colon \mathrm{M} \rightarrowtail \mathbb{R}\times\mathbb{R}^3$ is an orientation preserving (local) coordinatization of spacetime. ■

Note the difference between the symbols: T and S_U are sets whereas T and S_U are the corresponding mappings.

According to the definition, $T \circ \tau$ is smooth which implies by VI.3.5 that T is smooth as well. Because of (*ii*) the derivative of T —denoted by T' —is everywhere positive,

$$0 < T'(t) \in \mathbb{T}^* \equiv \frac{\mathbb{R}}{\mathbb{T}} \qquad (t \in \mathrm{Dom}\,T),$$

i.e. T is an (orientation preserving) *coordinatization of time.*

If $\boldsymbol{U}$ is a global rigid observer then S_U is an affine space and C_U is a smooth map (see 4.4.3); consequently, we can state that S_U is a coordinatization of S_U. On the contrary, since S_U, in general, is not an affine space, and we introduced the notion of coordinatization only for affine spaces, we cannot state that S_U is a coordinatization of $\boldsymbol{U}$-space; nevertheless it will be called the *coordinatization of $\boldsymbol{U}$-space.* (We mentioned that in any case S_U can be endowed with a smooth structure; in the framework of smooth structures S_U does become a coordinatization.)

10.1.3. Let us consider a coordinatization $K : \mathrm{M} \rightarrowtail \mathbb{R} \times \mathbb{R}^3$.

As usual, the coordinates in $\mathbb{R} \times \mathbb{R}^3$ are numbered from zero to three. Accordingly, we find it convenient to write a coordinatization of spacetime in the form $K = (\kappa^0, \boldsymbol{\kappa}) : \mathrm{M} \rightarrowtail \mathbb{R} \times \mathbb{R}^3$. Using the notations $pr^0 : \mathbb{R} \times \mathbb{R}^3 \to \mathbb{R}$ and $\boldsymbol{pr} : \mathbb{R} \times \mathbb{R}^3 \to \mathbb{R}^3$ for the canonical projections, we have $\kappa^0 = pr^0 \circ K$, $\boldsymbol{\kappa} = \boldsymbol{pr} \circ K$.

The following important relation holds for an arbitrary coordinatization K:

$$\mathrm{D}\boldsymbol{\kappa}(x) \cdot \partial_0 K^{-1}(K(x)) = \mathbf{0} \qquad (x \in \mathrm{Dom}\,K).$$

Indeed, according to the definition of partial derivatives (VI.3.8) and the rules of differentiation (VI.3.4), we have

$$\begin{aligned} \partial_0 K^{-1}(K(x)) &= \left(\mathrm{D}K^{-1}\right)\left(K(x)\right) \cdot (1, \mathbf{0}) = \mathrm{D}K(x)^{-1} \cdot (1, \mathbf{0}), \qquad (*) \\ \mathrm{D}\boldsymbol{\kappa}(x) &= \boldsymbol{pr} \cdot \mathrm{D}K(x), \end{aligned}$$

from which we infer the desired equality.

We say that a coordinatization K *is referencelike* or *corresponds to a reference system* if there is a reference system $(\boldsymbol{U}, T, S_U)$ such that $K = (T \times S_U) \circ \xi_U$. In that case

$$\kappa^0 = T \circ \tau, \qquad \boldsymbol{\kappa} = S_U \circ C_U$$

and

$$\left(\mathrm{D}\kappa^0\right)(x) = T'\left(\tau(x)\right)\boldsymbol{\tau}$$

from which we deduce

$$T'\big(\tau(x)\big) = \frac{1}{\boldsymbol{\tau}\cdot\partial_0 K^{-1}\big(K(x)\big)}$$

in the following way:

$$\begin{aligned} pr^0\cdot \mathrm{D}K(x) &= T'\big(\tau(x)\big)\boldsymbol{\tau}, \\ pr^0 &= T'\big(\tau(x)\big)\boldsymbol{\tau}\cdot \mathrm{D}K(x)^{-1}, \\ 1 &= T'\big(\tau(x)\big)\boldsymbol{\tau}\cdot \mathrm{D}K(x)^{-1}\cdot(1,\mathbf{0}), \\ 1 &= T'\big(\tau(x)\big)\boldsymbol{\tau}\cdot \partial_0 K^{-1}\big(K(x)\big). \end{aligned}$$

10.1.4. Proposition. A coordinatization $K = (\kappa^0, \boldsymbol{\kappa})\colon \mathrm{M} \rightarrowtail \mathbb{R}\times\mathbb{R}^3$ is referencelike if and only if

(i) K is orientation preserving,
(ii) $\partial_0 K^{-1}\big(K(x)\big)$ is a future directed timelike vector,
(iii) $\kappa^0(x) < \kappa^0(y)$ is equivalent to $\tau(x) < \tau(y)$ for all $x, y \in \mathrm{Dom}\,K$.

In this case K corresponds to the reference system given by

(1)
$$\boldsymbol{U}(x) = \frac{\partial_0 K^{-1}\big(K(x)\big)}{\boldsymbol{\tau}\cdot\partial_0 K^{-1}\big(K(x)\big)} = \partial_0 K^{-1}\big(K(x)\big)\cdot T'\big(\tau(x)\big) \qquad (x\in \mathrm{Dom}\,K),$$

(2)
$$T(t) = \kappa^0(x) \qquad (t\cap \mathrm{Dom}\,K \neq \emptyset,\ \ x\in t),$$

(3)
$$S_{\boldsymbol{U}}(q) = \boldsymbol{\kappa}(x) \qquad (q\in \mathrm{S}_{\boldsymbol{U}},\ \ x\in q).$$

Proof. If $K = (T\times S_{\boldsymbol{U}})\circ \xi_{\boldsymbol{U}}$, then ($i$) is trivial and ($iii$) follows from $\kappa^0 = T\circ\tau$ and the strictly monotonous character of T. As concerns (ii), note that a world line function r satisfies $\dot r(t) = \boldsymbol{U}\big(r(t)\big)$ and takes values in the domain of K if and only if $K\big(r(t)\big) = \big(T(t), \boldsymbol{\xi}\big)$, i.e. $r(t) = K^{-1}\big(T(t),\boldsymbol{\xi}\big)$ for a $\boldsymbol{\xi}\in\mathbb{R}^3$ and for all $t\in\mathrm{Dom}\,r$. As a consequence, we have

$$\begin{aligned} \boldsymbol{U}\big(r(t)\big) &= \frac{\mathrm{d}}{\mathrm{d}t}K^{-1}\big(T(t),\boldsymbol{\xi}\big) = \partial_0 K^{-1}\big(T(t),\boldsymbol{\xi}\big)\cdot T'(t) = \\ &= \partial_0 K^{-1}\Big(K\big(r(t)\big)\Big)\cdot T'(t) \end{aligned}$$

implying

$$\boldsymbol{U}(x) = \partial_0 K^{-1}\big(K(x)\big)\cdot T'\big(\tau(x)\big), \qquad (x\in\mathrm{Dom}\,K),$$

which proves (ii) since $T'\big(\tau(x)\big) > 0$. It proves equality (1) as well; equalities (2) and (3) are trivial.

Suppose now that $K = (\kappa^0, \boldsymbol{\kappa})$ is a coordinatization that fulfills conditions (*i*)–(*iii*).

Then condition (*ii*) implies that $\boldsymbol{U}$ *defined* by the first equality in (1) is an observer.

According to (*iii*), K is constant on simultaneous hyperplanes, thus T is well *defined* by the formula (2). Moreover, T is strictly monotone increasing.

If r is a world line function such that $\dot{r}(t) = \boldsymbol{U}\big(r(t)\big)$ then according to $(*)$ in the preceding paragraph

$$\frac{\mathrm{d}}{\mathrm{d}t}\big(\boldsymbol{\kappa}\big(r(t)\big)\big) = \mathrm{D}\boldsymbol{\kappa}\big(r(t)\big)\cdot \boldsymbol{U}\big(r(t)\big) = \mathrm{D}\boldsymbol{\kappa}\big(r(t)\big)\cdot \frac{\partial_0 K^{-1}\big(K\big(r(t)\big)\big)}{\boldsymbol{\tau}\cdot\partial_0 K^{-1}\big(K\big(r(t)\big)\big)} = \mathbf{0},$$

which means that $\boldsymbol{\kappa}\circ r$ is a constant mapping, in other words, $\boldsymbol{\kappa}$ is constant on the $\boldsymbol{U}$-lines; hence $S_{\boldsymbol{U}}$ is well *defined* by the formula (3).

Finally, it is evident that $K = (T \times S_{\boldsymbol{U}})\circ \xi_{\boldsymbol{U}}$. ■

It is suitable to use $P := K^{-1}$, the parametrization corresponding to K. Then—putting $\phi(P)$ instead of $\phi\circ P$ for any function ϕ —we can rewrite formula (1) in the proposition:

$$\boldsymbol{U}(P) = \frac{\partial_0 P}{\boldsymbol{\tau}\cdot\partial_0 P}.$$

10.1.5. Condition (*iii*) in the previous proposition can be replaced by

(*iii*)' for all $x\in \mathrm{Dom}\,K$ there is an $\mathsf{e}(x) \in (\mathbb{T}^*)^+$ such that

$$\mathrm{D}\kappa^0(x) = \mathsf{e}(x)\boldsymbol{\tau},$$

i.e. the derivative of κ^0 in every point is a positive multiple of $\boldsymbol{\tau}$.

Indeed, if K corresponds to a reference system then $\mathsf{e}(x) = T'\big(\tau(x)\big)$.

Conversely, if (*iii*)' holds, then the restriction of κ^0 onto every simultaneous hyperplane t has zero derivative: $\mathrm{D}\left(\kappa^0|_t\right)(x) = \mathrm{D}\kappa^0(x)|_{\mathbf{S}} = \mathbf{0}$ $(x\in t)$ thus κ^0 is constant on every simultaneous hyperplane which allows us to define T by the formula (1) in the previous proposition.

Moreover, Lagrange's mean value theorem implies that every x in the domain of K has a neighbourhood such that for all y in that neighbourhood there is a z on the straight line segment connecting x and y in such a way that

$$\kappa^0(y) - \kappa^0(x) = \mathrm{D}\kappa^0(z)\cdot(y-x) = \mathsf{e}(z)\boldsymbol{\tau}\cdot(y-x),$$

hence $\kappa^0(y)-\kappa^0(x) > 0$ is equivalent to $\boldsymbol{\tau}\cdot(y-x) > 0$ in the neighbourhood in question. Since the domain of K is connected, this relation holds globally as well.

10.2. Galilean reference systems

10.2.1. Now we are interested in what kinds of affine coordinatizations of spacetime can correspond to reference systems.

Let us take an affine coordinatization K of M. Then there are

— an $o \in \mathrm{M}$,

— an ordered basis $(\boldsymbol{x}_0, \boldsymbol{x}_1, \boldsymbol{x}_2, \boldsymbol{x}_3)$ of $\mathbf{M}$ such that

$$K(x) = \big(\boldsymbol{k}^i \cdot (x - o) \mid i = 0, 1, 2, 3\big) \qquad (x \in \mathrm{M}),$$

where $(\boldsymbol{k}^0,\ \boldsymbol{k}^1,\ \boldsymbol{k}^2,\ \boldsymbol{k}^3)$ is the dual of the basis in question.

Proposition. The affine coordinatization K corresponds to a reference system if and only if

(i) $(\boldsymbol{x}_0, \boldsymbol{x}_1, \boldsymbol{x}_2, \boldsymbol{x}_3)$ is a positively oriented basis,

(ii) $\boldsymbol{x}_0$ is a future directed timelike vector,

(iii) $\boldsymbol{x}_1, \boldsymbol{x}_2, \boldsymbol{x}_3$ are spacelike vectors.

Then the corresponding observer is global and inertial having the constant value

$$\boldsymbol{u} := \frac{\boldsymbol{x}_0}{\boldsymbol{s}},$$

and

$$K(x) = \left(\frac{\boldsymbol{\tau} \cdot (x - o)}{\boldsymbol{s}},\ \big(\boldsymbol{p}^\alpha \cdot \boldsymbol{\pi}_{\boldsymbol{u}} \cdot (x - o)\big)_{\alpha=1,2,3}\right) \qquad (x \in \mathrm{M}),$$

$$K^{-1}(\xi^0, \boldsymbol{\xi}) = o + \xi^0 \boldsymbol{s}\boldsymbol{u} + \sum_{\alpha=1}^{3} \xi^\alpha \boldsymbol{x}_\alpha \qquad \big((\xi^0, \boldsymbol{\xi}) \in \mathbb{R} \times \mathbb{R}^3\big)$$

where

$$\boldsymbol{s} := \boldsymbol{\tau} \cdot \boldsymbol{x}_0$$

and $\{\boldsymbol{p}^\alpha := \boldsymbol{k}^\alpha|_{\boldsymbol{E}} = \boldsymbol{k}^\alpha \cdot \boldsymbol{i} \mid (\alpha = 1, 2, 3)\}$ is the dual of the basis $\{\boldsymbol{x}_1,\ \boldsymbol{x}_2,\ \boldsymbol{x}_3\}$ of $\mathbf{S}$.

Proof. We show that the present conditions (i)–(iii) correspond to conditions (i)–(ii) listed in Proposition 10.1.4 and condition (iii)' in 10.1.5.

(i) The coordinatization is orientation preserving if and only if the corresponding basis is positively oriented;

(ii) $\partial_0 K^{-1}(K(x)) = \boldsymbol{x}_0$;

(iii)' $\mathrm{D}\kappa^0(x) = \boldsymbol{k}^0$ for all $x \in \mathrm{M}$. Since $\boldsymbol{k}^0 \cdot \boldsymbol{x}_\alpha = 0$, $\boldsymbol{k}^0 = \boldsymbol{e}\boldsymbol{\tau}$ for some $\boldsymbol{e} \in \mathbb{T}^*$ if and only if $\boldsymbol{x}_\alpha$-s $(\alpha = 1, 2, 3)$ are spacelike; then, because of $\boldsymbol{k}^0 \cdot \boldsymbol{x}_0 = 1 > 0$, $\boldsymbol{e} = \frac{1}{\boldsymbol{\tau} \cdot \boldsymbol{x}_0} > \mathbf{0}$. ∎

According to our result, an *affine reference system* will be given in the form $(\boldsymbol{u}, o, \boldsymbol{s}, \boldsymbol{x}_1, \boldsymbol{x}_2, \boldsymbol{x}_3)$.

Note that an affine coordinatization K corresponds to a reference system if and only if the restriction of $\boldsymbol{K}$ (the linear map under K) onto $\mathbf{S}$ is a linear bijection between $\mathbf{S}$ and $\{0\}\times\mathbb{R}^3$.

10.2.2. Definition. A coordinatization K is called *Galilean* if
— K is affine,
— $\boldsymbol{K}\cdot\boldsymbol{i}:\mathbf{S}\to\{0\}\times\mathbb{R}^3$ is $\boldsymbol{h}$–$\boldsymbol{H}$-orthogonal
where $\boldsymbol{H}$ is the usual inner product on $\mathbb{R}^3\equiv\{0\}\times\mathbb{R}^3$.

Proposition. A coordinatization K is Galilean if and only if there are
(*i*) an $o\in\mathrm{M}$,
(*ii*) an ordered basis $(\boldsymbol{e}_0,\ \boldsymbol{e}_1,\ \boldsymbol{e}_2,\ \boldsymbol{e}_3)$ of $\mathbf{M}$,
— $(\boldsymbol{e}_0,\ \boldsymbol{e}_1,\ \boldsymbol{e}_2,\ \boldsymbol{e}_3)$ is positively oriented,
— $\boldsymbol{s}:=\boldsymbol{\tau}\cdot\boldsymbol{e}_0>0$,
— $(\boldsymbol{e}_1,\ \boldsymbol{e}_2,\ \boldsymbol{e}_3)$ is a (necessarily positively oriented) orthogonal basis in $\mathbf{S}$, normed to an $\boldsymbol{m}\in\mathbb{L}^+$, such that

$$K(x)=\left(\frac{\boldsymbol{\tau}\cdot(x-o)}{\boldsymbol{s}},\ \left(\frac{\boldsymbol{e}_\alpha\cdot\boldsymbol{\pi_u}\cdot(x-o)}{\boldsymbol{m}^2}\right)_{\alpha=1,2,3}\right)\qquad (x\in\mathrm{M}),$$

where

$$\boldsymbol{u}:=\frac{\boldsymbol{e}_0}{\boldsymbol{s}}$$

is the constant value of the corresponding inertial observer.

Proof. It is quite evident that an affine coordinatization is Galilean if and only if the spacelike elements of the corresponding basis in $\mathbf{M}$ are orthogonal to each other and have the same length. We know that the dual of the basis $(\boldsymbol{e}_1,\ \boldsymbol{e}_2,\ \boldsymbol{e}_3)$ becomes $\left(\frac{\boldsymbol{e}_1}{\boldsymbol{m}^2},\ \frac{\boldsymbol{e}_2}{\boldsymbol{m}^2},\ \frac{\boldsymbol{e}_3}{\boldsymbol{m}^2}\right)$ in the identification $\mathbf{S}^*\equiv\frac{\mathbf{S}}{\mathbb{L}\otimes\mathbb{L}}$ which proves the equality regarding K. ■

According to our result, a *Galilean reference system* will be given in the form $(\boldsymbol{u},o,\boldsymbol{s},\boldsymbol{m},\boldsymbol{e}_1,\boldsymbol{e}_2,\boldsymbol{e}_3)$ and we shall use the following names: $\boldsymbol{u}$ is *its velocity value,* o is its *origin,* $\boldsymbol{s}$ is its *time unit,* $\boldsymbol{m}$ is its *distance unit,* $(\boldsymbol{e}_1,\ \boldsymbol{e}_2,\ \boldsymbol{e}_3)$ is its *space basis.* Moreover, putting $\boldsymbol{e}_0:=\boldsymbol{s}\boldsymbol{u}$, we call $(\boldsymbol{e}_0,\ \boldsymbol{e}_1,\ \boldsymbol{e}_2,\ \boldsymbol{e}_3)$ its *spacetime basis.*

10.2.3. Let K be a Galilean coordinatization and use the previous notations.

Recalling 1.5.2, we see that the Galilean coordinatization establishes an isomorphism between the spacetime model $(\mathrm{M},\mathbb{T},\mathbb{L},\boldsymbol{\tau},\boldsymbol{h})$ and the arithmetic spacetime model. More precisely, the coordinatization K and the mappings $\boldsymbol{B}:\mathbb{T}\to\mathbb{R},\ \boldsymbol{t}\mapsto\frac{\boldsymbol{t}}{\boldsymbol{s}}$ and $\boldsymbol{Z}:\mathbb{L}\to\mathbb{R},\ \boldsymbol{d}\mapsto\frac{\boldsymbol{d}}{\boldsymbol{m}}$ constitute an isomorphism.

This isomorphism transforms vectors, covectors and tensors, cotensors, etc. into vectors, covectors, etc. of the arithmetic spacetime model.

In particular,

$$\boldsymbol{K}\colon \mathbf{M} \to \mathbb{R} \times \mathbb{R}^3, \qquad \boldsymbol{x} \mapsto \left(\frac{\boldsymbol{\tau} \cdot \boldsymbol{x}}{\boldsymbol{s}}, \left(\frac{\boldsymbol{e}_\alpha \cdot \boldsymbol{\pi}_{\boldsymbol{u}} \cdot \boldsymbol{x}}{\boldsymbol{m}^2} \right)_{\alpha=1,2,3} \right),$$

is the coordinatization of vectors; note that it maps $\mathbf{S}$ onto $\{0\} \times \mathbb{R}^3$;

$$\left(\boldsymbol{K}^{-1}\right)^* : \mathbf{M}^* \times \mathbb{R} \times \mathbb{R}^3, \qquad \boldsymbol{k} \mapsto (\boldsymbol{k} \cdot \boldsymbol{e}_i \mid i = 0,1,2,3),$$

is the coordinatization of covectors; note that it maps $\mathbb{T}^* \cdot \boldsymbol{\tau}$ onto $\mathbb{R} \times \{\mathbf{0}\}$.

We can generalize the coordinatization for vectors (covectors) of type or cotype $\mathbb{A}$, i.e. for elements in $\mathbf{M} \otimes \mathbb{A}$ or $\frac{\mathbf{M}}{\mathbb{A}}$ $\left(\mathbf{M}^* \otimes \mathbb{A}, \frac{\mathbf{M}^*}{\mathbb{A}}\right)$, too, where $\mathbb{A}$ is a measure line. For instance, elements of $\frac{\mathbf{M}}{\mathbb{T}}$ or $\frac{\mathbf{M}}{\mathbb{L}\otimes\mathbb{L}}$ are coordinatized by the basis $\left(\frac{\boldsymbol{e}_i}{\boldsymbol{s}} \mid i = 0,1,2,3\right)$ and by the basis $\left(\frac{\boldsymbol{e}_i}{\boldsymbol{m}^2} \mid i = 0,1,2,3\right)$, respectively:

$$\frac{\mathbf{M}}{\mathbb{T}} \to \mathbb{R} \times \mathbb{R}^3, \qquad \boldsymbol{w} \mapsto \boldsymbol{s} \left(\frac{\boldsymbol{\tau} \cdot \boldsymbol{w}}{\boldsymbol{s}}, \left(\frac{\boldsymbol{e}_\alpha \cdot \boldsymbol{\pi}_{\boldsymbol{u}} \cdot \boldsymbol{w}}{\boldsymbol{m}^2} \right)_{\alpha=1,2,3} \right),$$

$$\frac{\mathbf{M}}{\mathbb{L} \otimes \mathbb{L}} \to \mathbb{R} \times \mathbb{R}^3, \qquad \boldsymbol{p} \mapsto \boldsymbol{m}^2 \left(\frac{\boldsymbol{\tau} \cdot \boldsymbol{p}}{\boldsymbol{s}}, \left(\frac{\boldsymbol{e}_\alpha \cdot \boldsymbol{\pi}_{\boldsymbol{u}} \cdot \boldsymbol{p}}{\boldsymbol{m}^2} \right)_{\alpha=1,2,3} \right).$$

10.3. Subscripts and superscripts

10.3.1. In textbooks one generally uses, without a precise definition, Galilean coordinatizations and the arithmetic spacetime model. Vectors, covectors and tensors, cotensors, etc. are given by coordinates relative to a spacetime basis. Let us survey the usual formalism from our point of view.

Let us take a Galilean coordinatization and let us use the previous notations. If $(\boldsymbol{k}^0, \boldsymbol{k}^1, \boldsymbol{k}^2, \boldsymbol{k}^3)$ is the dual of the basis $(\boldsymbol{e}_0, \boldsymbol{e}_1, \boldsymbol{e}_2, \boldsymbol{e}_3)$, then

$$x^i := \boldsymbol{k}^i \cdot \boldsymbol{x} \qquad (i = 0,1,2,3)$$

are the coordinates of the vector $\boldsymbol{x}$; we know that

$$x^0 := \frac{\boldsymbol{\tau} \cdot \boldsymbol{x}}{\boldsymbol{s}}, \qquad x^\alpha := \frac{\boldsymbol{e}_\alpha \cdot \boldsymbol{x}}{\boldsymbol{m}^2} \qquad (\alpha = 1,2,3).$$

The covector $\boldsymbol{k}$ has the coordinates

$$k_i := \boldsymbol{k} \cdot \boldsymbol{e}_i \qquad (i = 0,1,2,3).$$

Let us accept the convention that the coordinates of vectors are denoted by superscripts and the coordinates of covectors are denoted by subscripts, and we shall not indicate that the coordinates run from 0 to 3. Then the symbol $\boldsymbol{x} \sim x^i$ and $\boldsymbol{k} \sim k_i$ will mean that the vector $\boldsymbol{x}$ (covector $\boldsymbol{k}$) has the coordinates x^i (k_i).

We have $\boldsymbol{k} \cdot \boldsymbol{x} = \sum_{i=0}^{3} k_i x^i$. According to the Einstein summation convention we shall omit the symbol of summation as well: $\boldsymbol{k} \cdot \boldsymbol{x} = k_i x^i$.

The various tensors are given by coordinates with respect to the tensor products of the corresponding bases (e.g. $(\boldsymbol{e}_i \otimes \boldsymbol{e}_j \mid i,j = 0,1,2,3)$ or $(\boldsymbol{e}_i \otimes \boldsymbol{k}^j \mid i,j = 0,1,2,3)$), as the following symbols show:

$$\begin{aligned} &\boldsymbol{T} \in \mathbf{M} \otimes \mathbf{M}, && \boldsymbol{T} \sim T^{ij}, \\ &\boldsymbol{L} \in \mathbf{M} \otimes \mathbf{M}^*, && \boldsymbol{L} \sim L^i{}_j, \\ &\boldsymbol{P} \in \mathbf{M}^* \otimes \mathbf{M}, && \boldsymbol{P} \sim P_i^j, \\ &\boldsymbol{F} \in \mathbf{M}^* \otimes \mathbf{M}^*, && \boldsymbol{F} \sim F_{ij}. \end{aligned}$$

Applying the Einstein summation convention we can write, e.g. $\boldsymbol{T} \cdot \boldsymbol{k} \sim T^{ij} k_j$, $\boldsymbol{L} \cdot \boldsymbol{x} \sim L^i{}_j x^j$, $\boldsymbol{L} \cdot \boldsymbol{T} \sim L^i{}_j T^{jk}$, $\mathrm{Tr}\, \boldsymbol{L} = L^i{}_i$, etc.

We know that $\boldsymbol{x} \cdot \boldsymbol{y}$ makes no sense for $\boldsymbol{x}, \boldsymbol{y} \in \mathbf{M}$; in coordinates this means that $x^i y^i$ makes no sense. Similarly, $\boldsymbol{L} \cdot \boldsymbol{k}$ makes no sense for $\boldsymbol{L} \in \mathbf{M} \otimes \mathbf{M}^*$ and $\boldsymbol{k} \in \mathbf{M}^*$; in coordinates this means that $L^i{}_j k_j$ makes no sense. More precisely, $x^i y^i$, etc. do not *make an absolute sense.* Of course, the value of this expression can be computed, but it depends on the coordinatization: taking the coordinates x'^i and y'^i relative to another coordinatization and computing $x'^i y'^i$ we get a different value.

We can see that, in general, a summation makes an absolute sense only for equal subscripts and superscripts.

10.3.2. Recall that we have the identification $\mathbf{S}^* \equiv \frac{\mathbf{S}}{\mathbb{L} \otimes \mathbb{L}}$ and under this identification the dual of the orthogonal basis $(\boldsymbol{e}_1, \boldsymbol{e}_2, \boldsymbol{e}_3)$ becomes $(\frac{\boldsymbol{e}_\alpha}{m^2} \mid \alpha = 1,2,3)$. The coordinates of $\boldsymbol{p} \in \mathbf{S}^*$ are $p_\alpha := \boldsymbol{p} \cdot \boldsymbol{e}_\alpha$ $(\alpha = 1,2,3)$. If we consider $\boldsymbol{p}$ as an element of $\frac{\mathbf{S}}{\mathbb{L} \otimes \mathbb{L}}$ then it has the coordinates $p^\alpha := m^2 \left(\frac{\boldsymbol{e}_\alpha}{m^2} \cdot \boldsymbol{p}\right) = p_\alpha$ $(\alpha = 1,2,3)$.

Similarly, $\boldsymbol{q} \in \mathbf{S}$ has the coordinates $q^\alpha := \frac{\boldsymbol{e}_\alpha}{m^2} \cdot \boldsymbol{q}$ $(\alpha = 1,2,3)$. If we consider $\boldsymbol{q}$ as an element of $\mathbf{S}^* \otimes \mathbb{L} \otimes \mathbb{L}$, then its coordinates are $q_\alpha := \frac{1}{m^2}(\boldsymbol{q} \cdot \boldsymbol{e}_\alpha) = q^\alpha$ $(\alpha = 1,2,3)$.

Thus dealing exclusively with *spacelike vectors,* we need not distinguish between superscripts and subscripts. We know that $\boldsymbol{q} \cdot \boldsymbol{q}$ makes sense for a spacelike vector $\boldsymbol{q}$, and $\boldsymbol{q} \cdot \boldsymbol{q} \sim q^\alpha q^\alpha = q^\alpha q_\alpha = q_\alpha q_\alpha$.

We emphasize that this is true only if we use an orthogonal and normed basis in $\mathbf{S}$ (see V.3.20).

10.4. Reference systems associated with global rigid observers*

10.4.1. We know that the space S_{U} of a global rigid observer $\boldsymbol{U}$ is a three-dimensional affine space. Moreover, given $t_o \in \mathrm{T}$ and $q_o \in \mathrm{S}_U$, or, equivalently, given $o \in \mathrm{M}$ —called the *origin*—such that $o = q_o \star t_o$, $t_o = \tau(o)$, $q_o = C_U(o)$, we establish the (double) vectorization

$$\mathrm{T} \times \mathrm{S}_U \to \mathbb{T} \times \mathbf{S}, \qquad (t,q) \mapsto (t - t_o,\ q \star t_o - q_o \star t_o) = (t - \tau(o), q \star \tau(o) - o),$$

which is an orientation preserving affine bijection.

Then choosing an $\boldsymbol{s} \in \mathbb{T}^+$ (a positively oriented basis in $\mathbb{T}$) —called the *time unit*—and a positively oriented basis $(\boldsymbol{x}_1, \boldsymbol{x}_2, \boldsymbol{x}_3)$ in $\mathbf{S}$ —called the space basis—, we can establish coordinatizations of time and $\boldsymbol{U}$-space:

$$T(t) := \frac{t - \tau(o)}{\boldsymbol{s}} \qquad (t \in \mathrm{T}),$$

$$S_U(q) := (\boldsymbol{p}^\alpha \cdot (q \star \tau(o) - o) \mid \alpha = 1,2,3), \qquad (q \in \mathrm{S}_U),$$

where $(\boldsymbol{p}^1, \boldsymbol{p}^2, \boldsymbol{p}^3)$ is the dual of the basis in question.

Evidently, T and S_U are orientation preserving affine bijections; we know that ξ_U is an orientation preserving smooth bijection whose inverse is smooth as well (see 4.3.2 and 4.4.3), thus $(\boldsymbol{U}, T, S_U)$ is a reference system. For the corresponding coordinatizatioin $K := (T \times S_U) \circ \xi_U$ we have

$$K(x) = \left(\frac{\boldsymbol{\tau} \cdot (x - o)}{\boldsymbol{s}}, (\boldsymbol{p}^\alpha \cdot (C_U(x) \star \tau(o) - o))_{\alpha=1,2,3}\right) \qquad (x \in \mathrm{M}),$$

$$K^{-1}(\xi^0, \boldsymbol{\xi}) = C_U\left(o + \sum_{\alpha=1}^{3} \xi^\alpha \boldsymbol{x}_\alpha\right) \star (\tau(o) + \xi^0 \boldsymbol{s}) \qquad ((\xi^0, \boldsymbol{\xi}) \in \mathbb{R} \times \mathbb{R}^3).$$

T and S_U are affine coordinatizations of time and $\boldsymbol{U}$-space. Evidently, $K = (T \times S_U) \circ \xi_U$ is an affine coordinatization of spacetime if and only if ξ_U is an affine map which holds if and only if $\boldsymbol{U}$ is a global inertial observer (see 5.1).

10.4.2. Let us take a uniformly accelerated observer $\boldsymbol{U}$ having the constant acceleration value $\boldsymbol{a}$ (see 5.2).

Then, according to 5.2.3, for the coordinatization treated in 10.4.1 we have

$$K(x) = \left(\frac{\boldsymbol{\tau} \cdot (x - o)}{\boldsymbol{s}},\ \boldsymbol{p}^\alpha \cdot \left(\boldsymbol{\pi}_{U(o)} \cdot (x - o) - \frac{1}{2}\boldsymbol{a}(\boldsymbol{\tau} \cdot (x - o))^2\right)_{\alpha=1,2,3}\right)$$

$$(x \in \mathrm{M})$$

and

$$K^{-1}(\xi^0, \boldsymbol{\xi}) = o + \xi^0 \boldsymbol{s}\boldsymbol{U}(o) + \sum_{\alpha=1}^{3} \xi^\alpha \boldsymbol{x}_\alpha + \frac{1}{2}(\xi^0)^2 \boldsymbol{s}^2 \boldsymbol{a} \qquad ((\xi^0, \boldsymbol{\xi}) \in \mathbb{R} \times \mathbb{R}^3).$$

10.4.3. Let us take a uniformly rotating observer $\boldsymbol{U}$, and let o, $\boldsymbol{c}$ and $\boldsymbol{\Omega}$ be the quantities introduced in 5.3.

Then, according to 5.3.4, for the coordinatization treated in 10.4.1 we have

$$K(x) = \left(\frac{\boldsymbol{\tau} \cdot (x - o)}{\boldsymbol{s}}, \left(\boldsymbol{p}^\alpha \cdot e^{-(\boldsymbol{\tau} \cdot (x-o))\boldsymbol{\Omega}} \cdot \boldsymbol{\pi}_{\boldsymbol{c}} \cdot (x - o) \right)_{\alpha=1,2,3} \right) \qquad (x \in \mathrm{M}),$$

$$K^{-1}(\xi^0, \boldsymbol{\xi}) = C_{\boldsymbol{U}}(o) \star \left(\tau(o) + \xi^0 \boldsymbol{s} \right) + e^{\xi^0 \boldsymbol{s}\boldsymbol{\Omega}} \cdot \sum_{\alpha=1}^{3} \xi^\alpha \boldsymbol{x}_\alpha \qquad ((\xi^0, \boldsymbol{\xi}) \in \mathbb{R} \times \mathbb{R}^3).$$

10.4.4. To summarize the results from this subsection, we can say:

Galilean reference system = global inertial observer + measuring time with respect to an initial instant and a time unit + introducing orthogonal (Cartesian) coordinates in the observer space.

Affine reference system = global inertial observer + measuring time with respect to an initial instant and a time unit + introducing (oblique-angled) rectilinear coordinates in the observer space.

Other reference systems treated previously = global rigid observer + measuring time with respect to an initial instant and a time unit + introducing rectilinear coordinates in the observer space.

For the solution of some practical problems we often use reference systems in which curvilinear coordinates (e.g. spherical coordinates or cylindrical coordinates) are introduced in the observer space.

10.5. Equivalent reference systems

10.5.1. In textbooks one usually formulates the principle—without a precise definition—that the Galilean reference systems are equivalent with respect to the description of phenomena. It is very important that then one takes tacitly into consideration Galilean reference systems with the same time unit and the same distance unit.

Reference systems as we defined them are mathematical objects. The physical object modelled by them will be called here a physical reference systems. When could we consider two physical reference systems to be equivalent? The answer is: if the experiments prepared in the same way in the reference systems give the same results. Let us see some illustrative examples.

Take two physical Galilean reference systems in which the time units and distance units are different and perform the following experiment in both systems: let an iron ball of unit diameter moving with unit relative velocity hit a sheet of glass of unit width perpendicularly. It may happen that the ball bounces in one of the reference systems, the glass breaks in the other. The two reference systems are not equivalent.

Take an affine reference system in which the first space basis element is perpendicular to the other two basis elements; take another affine reference system in which the first space basis element is not perpendicular to the other two basis elements. Perform the following experiment in both systems: let a ball moving parallel to the first space axis hit a plane parallel to the other two axes. The ball returns to its initial position in one of the reference systems and does not in the other. The two reference systems are not equivalent.

10.5.2. Recall the notion of automorphisms of the spacetime model (1.5.4). An automorphism is a transformation that leaves invariant (preserves) the structure of the spacetime model. Strict automorphisms do not change time periods and distances.

It is quite natural that two objects transformed into each other by a strict automorphism of the spacetime model are considered equivalent (i.e. identical from a physical point of view).

In the next paragraph we shall study the Noether transformations that involve the strict automorphisms of the spacetime model. Now we recall the basic facts.

Let $\mathcal{SO}(\boldsymbol{h})$ denote the set of linear maps $\boldsymbol{R}\colon \mathbf{S} \to \mathbf{S}$ that preserve the Euclidean structure and the orientation of $\mathbf{S}$: $\boldsymbol{h} \circ (\boldsymbol{R} \times \boldsymbol{R}) = \boldsymbol{h}$ and $\det \boldsymbol{R} = 1$ (see 11.1.2).

Let us introduce the notation

$$\mathcal{N}^{+\rightarrow} := \{L\colon \mathrm{M} \to \mathrm{M} \mid L \quad \text{is affine}, \quad \boldsymbol{\tau} \cdot \boldsymbol{L} = \boldsymbol{\tau}, \quad \boldsymbol{L}|_{\mathbf{S}} \in \mathcal{SO}(\boldsymbol{h})\}$$

and let us call the elements of $\mathcal{N}^{+\rightarrow}$ *proper Noether transformations.* It is quite evident that $(L, 1_{\mathrm{T}}, \mathbf{1}_{\mathbb{L}})$ is a strict automorphism of the spacetime model if and only if L is a proper Noether transformation (11.6.4).

An affine map $\mathrm{ti}L\colon \mathrm{T} \to \mathrm{T}$ can be assigned to every proper Noether transformation L in such a way that $\tau \circ L = (\mathrm{ti}L) \circ \tau$ (see 11.6.3).

10.5.3. Definition. The coordinatizations K and K' are called *equivalent* if there is a proper Noether transformation L such that

$$K' \circ L = K.$$

Two reference systems are *equivalent* if the corresponding coordinatizations are equivalent.

Proposition. The reference systems $(\boldsymbol{U}, T, S_U)$ and $(\boldsymbol{U}', T', S_{U'})$ are equivalent if and only if

(*i*) $\boldsymbol{L}^{-1} \cdot \boldsymbol{U}' \circ L = \boldsymbol{U}$, in other words, $\boldsymbol{L} \cdot \boldsymbol{U} = \boldsymbol{U}' \circ L$,

(*ii*) $T' \circ (\mathrm{ti}\, L) = T$, in other words, $T'^{-1} \circ T = \mathrm{ti}\, L$.

(*iii*) $\left(S_{U'}^{-1} \circ S_U\right) \circ C_U = C_{U'} \circ L$.

Proof. Let K and K' denote the corresponding coordinatizations. It is quite trivial that if the relations above hold, then K and K' are equivalent.

Let us suppose now that the two reference systems are equivalent. Then

(*i*) $K = K' \circ L$, $K^{-1} = L^{-1} \circ K'^{-1}$ hold for the coresponding coordinatizations. Then $\boldsymbol{\tau} \cdot \boldsymbol{L} = \boldsymbol{\tau}$ together with 10.1.4 imply

$$\boldsymbol{U}(x) = \frac{\partial_0 K^{-1}\big(K(x)\big)}{\boldsymbol{\tau} \cdot \partial_0 K^{-1}\big(K(x)\big)} = \frac{\boldsymbol{L}^{-1} \cdot \partial_0 K'^{-1}\big(K'(L(x))\big)}{\boldsymbol{\tau} \cdot \boldsymbol{L}^{-1} \cdot \partial_0 K'^{-1}\big(K'(L(x))\big)} = \boldsymbol{L}^{-1} \cdot \boldsymbol{U}'\big(L(x)\big).$$

(*ii*) The equalities

$$T \circ \tau = pr^0 \circ K = pr^0 \circ K' \circ L = T' \circ \tau \circ L = T' \circ (\mathrm{ti}\, L) \circ \tau$$

yield the desired relation immediately.

(*iii*) Consider the equalities

$$S_U \circ C_U = \boldsymbol{pr} \circ K = \boldsymbol{pr} \circ K' \circ L = S_{U'} \circ C_{U'} \circ L.$$

10.5.4. Now we shall see that our definition of equivalence of reference systems is in accordance with the intuitive notion expounded in 10.5.1.

Proposition. Two Galilean reference systems are equivalent if and only if they have the same time unit and distance unit, respectively.

Proof. Let the Galilean coordinatizations K and K' be defined by the origins o and o' and the spacetime bases $(\boldsymbol{e}_0, \boldsymbol{e}_1, \boldsymbol{e}_2, \boldsymbol{e}_3)$ and $(\boldsymbol{e}_0', \boldsymbol{e}_1', \boldsymbol{e}_2', \boldsymbol{e}_3')$, respectively.

Then $L := K'^{-1} \circ K : \mathrm{M} \to \mathrm{M}$ is the affine bijection determined by

$$L(o) = o', \qquad \boldsymbol{L} \cdot \boldsymbol{e}_i = \boldsymbol{e}_i' \qquad (i = 0, 1, 2, 3).$$

Evidently, L is orientation preserving. Moreover, $\boldsymbol{\tau} \cdot \boldsymbol{L} = \boldsymbol{\tau}$ if and only if $\boldsymbol{\tau} \cdot \boldsymbol{e}_0 = \boldsymbol{\tau} \cdot \boldsymbol{e}_0'$, and $\boldsymbol{L}|_{\mathbf{S}} \in \mathcal{SO}(\boldsymbol{h})$ if and only if $|\boldsymbol{e}_\alpha| = |\boldsymbol{e}_\alpha'|$ $(\alpha = 1, 2, 3)$.

10.6. Exercises

1. Reference systems give rise to coordinatizations, hence we can apply all the notions introduced in VI.5, e.g. the coordinatized form of vector fields.

Let K be a coordinatization corresponding to a reference system whose observer is $\boldsymbol{U}$. Demonstrate that the coordinatized form of $\boldsymbol{U}$ according to K is the constant mapping $(1, \mathbf{0})$. ($\boldsymbol{U}$ is a vector field of cotype $\mathbb{T}$, hence by definition, $(\mathrm{D}K \cdot \boldsymbol{U}) \circ K^{-1}$ is its coordinatized form according to K.)

2. Take a uniformly accelerated observer $\boldsymbol{U}$ having the acceleration value $\boldsymbol{a} \neq \mathbf{0}$. Define a Galilean reference system with arbitrary time unit $\boldsymbol{s}$ and distance unit $\boldsymbol{m}$, an origin o and with a spacetime basis such that $\boldsymbol{e}_0 := \boldsymbol{s}\boldsymbol{U}(o)$, $\boldsymbol{e}_1 := \boldsymbol{m}\frac{\boldsymbol{a}}{|\boldsymbol{a}|}$, $\boldsymbol{e}_2$ and $\boldsymbol{e}_3$ are arbitrary. Demonstrate that then $\boldsymbol{U}$ has the coordinatized form

$$(\xi^0, \xi^1, \xi^2, \xi^3) \mapsto (1, \alpha\xi^0, 0, 0),$$

where α is the number for which $|\boldsymbol{a}| = \alpha\frac{\boldsymbol{m}}{\boldsymbol{s}^2}$ holds.

The $\boldsymbol{U}$-line passing through $o + \sum_{i=0}^{3} \xi^i \boldsymbol{e}_i$ becomes

$$\left\{ \left(t, \xi^1 + \alpha\xi^0(t - \xi^0) + \frac{1}{2}\alpha(t - \xi^0)^2, \xi^2, \xi^3 \right) \middle| \ t \in \mathbb{R} \right\}.$$

3. Take a uniformly rotating observer $\boldsymbol{U}$ having the angular velocity $\boldsymbol{\Omega}$ and suppose there is an inertial $\boldsymbol{U}$-space point $q_o = o + \boldsymbol{c} \otimes \mathbb{T}$. Define a Galilean reference system with origin o, with arbitrary time unit $\boldsymbol{s}$ and distance unit $\boldsymbol{m}$ and $\boldsymbol{e}_0 := \boldsymbol{s}\boldsymbol{U}(o)$, $\boldsymbol{e}_3$ positively oriented in $\operatorname{Ker}\boldsymbol{\Omega}$, $|\boldsymbol{e}_3| = \boldsymbol{m}$, $\boldsymbol{e}_1$ and $\boldsymbol{e}_2$ being arbitrary. Demonstrate that then $\boldsymbol{U}$ has the coordinatized form

$$(\xi^0, \xi^1, \xi^2, \xi^3) \mapsto (1, -\omega\xi^2, \omega\xi^1, 0),$$

where ω is the number for which $|\boldsymbol{\Omega}| = \omega\frac{1}{\boldsymbol{s}}$ holds.

The $\boldsymbol{U}$-line passing through $o + \sum_{i=0}^{3} \xi^i \boldsymbol{e}_i$ becomes

$$\Big\{ \big(t, \ \xi^1 \cos\omega(t - \xi^0) - \xi^2 \sin\omega(t - \xi^0),$$
$$\xi^1 \sin\omega(t - \xi^0) + \xi^2 \cos\omega(t - \xi^0), \ \xi^3 \big) \mid t \in \mathbb{R} \Big\}.$$

4. Prove that two affine reference systems are equivalent if and only if they have the same time unit and the corresponding elements of the space bases have the same length and the same angles between themselves; in other words, the affine reference systems defined by the origins o and o' and the spacetime bases $(\boldsymbol{x}_0, \boldsymbol{x}_1, \boldsymbol{x}_2, x_3)$ and $(\boldsymbol{x}'_0, \boldsymbol{x}'_1, \boldsymbol{x}'_2, \boldsymbol{x}'_3)$, respectively, are equivalent if and only if

$$\boldsymbol{\tau} \cdot \boldsymbol{x}_0 = \boldsymbol{\tau} \cdot \boldsymbol{x}'_0$$

and

$$\boldsymbol{x}_\alpha \cdot \boldsymbol{x}_\beta = \boldsymbol{x}'_\alpha \cdot \boldsymbol{x}'_\beta \qquad (\alpha, \beta = 1, 2, 3).$$

5. Prove that two reference systems defined for uniformly accelerated observers in the form given in 10.4.2 are equivalent if and only if the two acceleration values have the same magnitude, the time units are equal, the corresponding elements of the space bases have the same length and the same angles between themselves, and the acceleration values incline in the same way to the basis elements; in other words, if $\boldsymbol{a}$ and $\boldsymbol{a}'$ are the acceleration values, $\boldsymbol{s}$ and $\boldsymbol{s}'$ are the time units, $(\boldsymbol{x}_1,\ \boldsymbol{x}_2,\ \boldsymbol{x}_3)$ and $(\boldsymbol{x}'_1,\ \boldsymbol{x}'_2,\ \boldsymbol{x}'_3)$ are the space bases, then the two reference systems are equivalent if and only if

$$|\boldsymbol{a}| = |\boldsymbol{a}'|, \qquad \boldsymbol{s} = \boldsymbol{s}',$$
$$\boldsymbol{x}_\alpha \cdot \boldsymbol{x}_\beta = \boldsymbol{x}'_\alpha \cdot \boldsymbol{x}'_\beta, \quad \frac{\boldsymbol{x}_\alpha \cdot \boldsymbol{a}}{|\boldsymbol{x}_\alpha||\boldsymbol{a}|} = \frac{\boldsymbol{x}'_\alpha \cdot \boldsymbol{a}'}{|\boldsymbol{x}'_\alpha||\boldsymbol{a}'|} \quad (\alpha, \beta = 1, 2, 3).$$

6. Prove that two reference systems defined for uniformly rotating observers in the form given in 10.4.3 are equivalent if and only if the angular velocities have the same magnitude, the time units are equal, the corresponding elements of the space bases have the same length and the same angles between themselves and the oriented kernels of the angular velocities incline in the same way to the basis elements.

7. Take a global inertial observer and construct a reference system by spherical (cylindrical) coordinatization of the observer space. Find necessary and sufficient conditions that two such reference systems be equivalent.

8. In all the treated reference systems time is coordinatized by an affine map. Construct a reference system based on a global inertial observer in which the time coordinatization is not affine.

11. Spacetime groups*

11.1. The three-dimensional orthogonal groups

11.1.1. $(\mathbf{S}, \mathbb{L}, \boldsymbol{h})$ is a three-dimensional oriented Euclidean vector space. Recall the notations (see V.2.7)

$$\mathrm{A}(\boldsymbol{h}) := \{\boldsymbol{A} \in \mathbf{S} \otimes \mathbf{S}^* \mid \boldsymbol{A}^* = -\boldsymbol{A}\} \equiv \frac{\mathbf{S}}{\mathbb{L}} \wedge \frac{\mathbf{S}}{\mathbb{L}},$$
$$\mathcal{O}(\boldsymbol{h}) := \{\boldsymbol{R} \in \mathbf{S} \otimes \mathbf{S}^* \mid \boldsymbol{R}^* = \boldsymbol{R}^{-1}\}.$$

$\mathrm{A}(\boldsymbol{h})$ is a three-dimensional subspace in $\mathbf{S} \otimes \mathbf{S}^*$ and $\mathcal{O}(\boldsymbol{h})$ is a three-dimensional Lie group having $\mathrm{A}(\boldsymbol{h})$ as its Lie algebra (VII.5).

11.1.2. We know that $|\det \boldsymbol{R}| = 1$ for $\boldsymbol{R} \in \mathcal{O}(\boldsymbol{h})$ (see V.2.8). We introduce the notations

$$\mathcal{SO}(\boldsymbol{h}) := \mathcal{O}(\boldsymbol{h})^+ := \{\boldsymbol{R} \in \mathcal{O}(\boldsymbol{h}) \mid \det \boldsymbol{R} = 1\},$$
$$\mathcal{O}(\boldsymbol{h})^- := \{\boldsymbol{R} \in \mathcal{O}(\boldsymbol{h}) \mid \det \boldsymbol{R} = -1\}.$$

The elements of $\mathcal{SO}(\boldsymbol{h})$ are called *rotations.*

Since the determinant is a continuous function, $\mathcal{O}(\boldsymbol{h})^+$ and $\mathcal{O}(\boldsymbol{h})^-$ are disjoint.

Evidently, $\mathbf{1_S} \in \mathcal{O}(\boldsymbol{h})^+$ and $-\mathbf{1_S} \in \mathcal{O}(\boldsymbol{h})^-$; moreover, $(-\mathbf{1_S}) \cdot \mathcal{O}(\boldsymbol{h})^+ = \mathcal{O}(\boldsymbol{h})^-$.

The determinant is a continuous function, hence both $\mathcal{O}(\boldsymbol{h})^+$ and $\mathcal{O}(\boldsymbol{h})^-$ are closed. Moreover, we know that $\boldsymbol{F} \mapsto \mathrm{Tr}(\boldsymbol{F}^* \cdot \boldsymbol{F})$ is an inner product (real-valued positive definite bilinear form) on $\mathbf{S} \otimes \mathbf{S}^*$ (see V.2.10). Since $\mathrm{Tr}(\boldsymbol{R}^* \cdot \boldsymbol{R}) = \mathrm{Tr}(\mathbf{1_S}) = 3$ for all $\boldsymbol{R} \in \mathcal{O}(\boldsymbol{h})$, $\mathcal{O}(\boldsymbol{h})$ is a bounded set.

Thus we can state, that $\mathcal{O}(\boldsymbol{h})$, $\mathcal{O}(\boldsymbol{h})^+$ and $\mathcal{O}(\boldsymbol{h})^-$ are compact (closed and bounded) sets.

11.1.3. Let $\boldsymbol{R} \in \mathcal{SO}(\boldsymbol{h})$. For all $\boldsymbol{x} \in \mathbf{S}$ we have $|\boldsymbol{R} \cdot \boldsymbol{x}| = |\boldsymbol{x}|$. As a consequence, $\boldsymbol{R} \cdot \boldsymbol{x} = \alpha \boldsymbol{x}$ implies $\alpha = \pm 1$.

Proposition. For every $\boldsymbol{R} \in \mathcal{SO}(\boldsymbol{h})$ there is a nonzero $\boldsymbol{x} \in \mathbf{S}$ such that $\boldsymbol{R} \cdot \boldsymbol{x} = \boldsymbol{x}$; moreover,

$$a_{\boldsymbol{R}} := \{\boldsymbol{x} \in \mathbf{S} \mid \boldsymbol{R} \cdot \boldsymbol{x} = \boldsymbol{x}\}$$

is a one-dimensional linear subspace if and only if $\boldsymbol{R} \neq \mathbf{1_S}$.

Proof. It is trivial that $a_{\boldsymbol{R}} = \mathbf{S}$ for $\boldsymbol{R} = \mathbf{1_S}$.

IV.3.18 and V.1.5 result in

$$\det(\boldsymbol{R} - \mathbf{1_S}) = \det(\boldsymbol{R} - \boldsymbol{R}^* \cdot \boldsymbol{R}) = \det(\mathbf{1_S} - \boldsymbol{R}^*)\det \boldsymbol{R} = -\det(\boldsymbol{R} - \mathbf{1_S}).$$

Consequently, $\det(\boldsymbol{R} - \mathbf{1_S}) = 0$, $\boldsymbol{R} - \mathbf{1_S}$ is not injective, there is a nonzero $\boldsymbol{x}$ such that $(\boldsymbol{R} - \mathbf{1_S}) \cdot \boldsymbol{x} = \mathbf{0}$.

Let us suppose $a_{\boldsymbol{R}}$ is not one-dimensional, i.e. $\boldsymbol{x}_1$ and $\boldsymbol{x}_2$ are not parallel vectors such that $\boldsymbol{R} \cdot \boldsymbol{x}_1 = \boldsymbol{x}_1$ and $\boldsymbol{R} \cdot \boldsymbol{x}_2 = \boldsymbol{x}_2$. Then for every element $\boldsymbol{x}$ in the plane spanned by $\boldsymbol{x}_1$ and $\boldsymbol{x}_2$ we have $\boldsymbol{R} \cdot \boldsymbol{x} = \boldsymbol{x}$. This means that the plane spanned by $\boldsymbol{x}_1$ and $\boldsymbol{x}_2$ is invariant under $\boldsymbol{R}$ and the restriction of $\boldsymbol{R}$ onto that plane is the identity. Let $\boldsymbol{y}$ be a nonzero vector orthogonal to the plane spanned by $\boldsymbol{x}_1$ and $\boldsymbol{x}_2$. Since $\boldsymbol{R}$ preserves orthogonality, $\boldsymbol{R} \cdot \boldsymbol{y}$ must be orthogonal to that plane, i.e. it is parallel to $\boldsymbol{y}$: $\boldsymbol{R} \cdot \boldsymbol{y} = \pm \boldsymbol{y}$. $\boldsymbol{R}$ is orientation preserving, thus $\boldsymbol{R} \cdot \boldsymbol{y} = \boldsymbol{y}$ must hold. This means that $\boldsymbol{R} = \mathbf{1_S}$. ■

For $\boldsymbol{R} \neq \mathbf{1_S}$, $a_{\boldsymbol{R}}$ is called the *axis of rotation* of $\boldsymbol{R}$.

11.1.4. For $\boldsymbol{R} \in \mathcal{SO}(\boldsymbol{h})$ the symbol $a_{\boldsymbol{R}}^{\perp}$ will stand for the orthogonal complement of $a_{\boldsymbol{R}}$:

$$a_{\boldsymbol{R}}^{\perp} := \{\boldsymbol{x} \in \mathbf{S} \mid \boldsymbol{x} \quad \text{is orthogonal to} \quad a_{\boldsymbol{R}}\}.$$

Evidently, $a_{\boldsymbol{R}}^{\perp} = \{0\}$ for $\boldsymbol{R} = \mathbf{1}_{\mathbf{S}}$ and $a_{\boldsymbol{R}}^{\perp}$ is a plane for $\boldsymbol{R} \neq \mathbf{1}_{\mathbf{S}}$. Moreover, $a_{\boldsymbol{R}}^{\perp}$ is invariant under $\boldsymbol{R}$.

The restriction of $\boldsymbol{R} \neq \mathbf{1}_{\mathbf{S}}$ onto $a_{\boldsymbol{R}}^{\perp}$ is a rotation in a plane which 'evidently' can be characterized by an angle of rotation. This is the content of the following proposition.

Proposition. If $\boldsymbol{x}$ and $\boldsymbol{y}$ are nonzero vectors in $a_{\boldsymbol{R}}^{\perp}$ then

$$\frac{\boldsymbol{x} \cdot \boldsymbol{R} \cdot \boldsymbol{x}}{|\boldsymbol{x}|^2} = \frac{\boldsymbol{y} \cdot \boldsymbol{R} \cdot \boldsymbol{y}}{|\boldsymbol{y}|^2}.$$

Proof. We can exclude the trivial cases $\boldsymbol{R} \cdot \boldsymbol{x} = \boldsymbol{x}$ and $\boldsymbol{R} \cdot \boldsymbol{x} = -\boldsymbol{x}$ for all $\boldsymbol{x} \in a_{\boldsymbol{R}}^{\perp}$ (note that the first case is $\boldsymbol{R} = \mathbf{1}_{\mathbf{S}}$).

It will be convenient to put $\boldsymbol{n} := \frac{\boldsymbol{x}}{|\boldsymbol{x}|}$, $\boldsymbol{k} := \frac{\boldsymbol{y}}{|\boldsymbol{y}|}$ and to consider $\boldsymbol{R}$ to be a linear map on $\frac{\mathbf{S}}{\mathbb{L}}$. Let us introduce the notation

$$S_{\boldsymbol{R}} := \left\{ \boldsymbol{n} \in \frac{\mathbf{S}}{\mathbb{L}} \mid \boldsymbol{n} \quad \text{is orthogonal to} \quad a_{\boldsymbol{R}}, \quad |\boldsymbol{n}| = 1 \right\}.$$

The proof consists of several simple steps whose details are left to the reader.

(*i*) Let $\boldsymbol{n}$ and $\boldsymbol{k}$ be elements of $S_{\boldsymbol{R}}$ orthogonal to each other. Then, excluding the trivial case,

$$\boldsymbol{n} \cdot \boldsymbol{R} \cdot \boldsymbol{k} \neq 0.$$

Indeed,

$$1 = \det \boldsymbol{R} = (\boldsymbol{n} \cdot \boldsymbol{R} \cdot \boldsymbol{n})(\boldsymbol{k} \cdot \boldsymbol{R} \cdot \boldsymbol{k}) - (\boldsymbol{n} \cdot \boldsymbol{R} \cdot \boldsymbol{k})(\boldsymbol{k} \cdot \boldsymbol{R} \cdot \boldsymbol{n})$$

and because of the Cauchy inequality (apart from the trivial case), $(\boldsymbol{n} \cdot \boldsymbol{R} \cdot \boldsymbol{n})(\boldsymbol{k} \cdot \boldsymbol{R} \cdot \boldsymbol{k}) < 1$.

(*ii*) $\boldsymbol{R} \cdot \boldsymbol{n} \neq \boldsymbol{R}^{-1} \cdot \boldsymbol{n}$. Indeed, suppose $\boldsymbol{R} \cdot \boldsymbol{n} = \boldsymbol{R}^{-1} \cdot \boldsymbol{n}$. Then we get from the previous formula that

$$\begin{aligned} 1 &= (\boldsymbol{n} \cdot \boldsymbol{R}^{-1} \cdot \boldsymbol{n})(\boldsymbol{k} \cdot \boldsymbol{R} \cdot \boldsymbol{k}) - (\boldsymbol{n} \cdot \boldsymbol{R} \cdot \boldsymbol{k})(\boldsymbol{k} \cdot \boldsymbol{R}^{-1} \cdot \boldsymbol{n}) = \\ &= (\boldsymbol{n} \cdot \boldsymbol{R} \cdot \boldsymbol{n})(\boldsymbol{k} \cdot \boldsymbol{R} \cdot \boldsymbol{k}) - (\boldsymbol{n} \cdot \boldsymbol{R} \cdot \boldsymbol{k})(\boldsymbol{n} \cdot \boldsymbol{R} \cdot \boldsymbol{k}), \end{aligned}$$

which implies $(\boldsymbol{n} \cdot \boldsymbol{R} \cdot \boldsymbol{n})(\boldsymbol{k} \cdot \boldsymbol{R} \cdot \boldsymbol{k}) > 1$ contradicting the Cauchy inequality.

(*iii*) $0 \neq \boldsymbol{R} \cdot \boldsymbol{n} - \boldsymbol{R}^{-1} \cdot \boldsymbol{n}$ is orthogonal to $\boldsymbol{n}$, hence it is parallel to $\boldsymbol{k}$.

(*iv*) $\boldsymbol{R}\cdot\boldsymbol{n}+\boldsymbol{R}^{-1}\cdot\boldsymbol{n}$ is orthogonal to $\boldsymbol{R}\cdot\boldsymbol{n}-\boldsymbol{R}^{-1}\cdot\boldsymbol{n}$, hence it is orthogonal to $\boldsymbol{k}$ as well. Consequently,

$$\boldsymbol{n}\cdot\boldsymbol{R}\cdot\boldsymbol{k}+\boldsymbol{k}\cdot\boldsymbol{R}\cdot\boldsymbol{n}=0.$$

(*v*) $\boldsymbol{R}\cdot\boldsymbol{n}=(\boldsymbol{n}\cdot\boldsymbol{R}\cdot\boldsymbol{n})\boldsymbol{n}+(\boldsymbol{k}\cdot\boldsymbol{R}\cdot\boldsymbol{n})\boldsymbol{k}$, and from a similar relation for $\boldsymbol{R}\cdot\boldsymbol{k}$ we have

$$0=(\boldsymbol{R}\cdot\boldsymbol{n})\cdot(\boldsymbol{R}\cdot\boldsymbol{k})=(\boldsymbol{n}\cdot\boldsymbol{R}\cdot\boldsymbol{n})(\boldsymbol{n}\cdot\boldsymbol{R}\cdot\boldsymbol{k})+(\boldsymbol{k}\cdot\boldsymbol{R}\cdot\boldsymbol{n})(\boldsymbol{k}\cdot\boldsymbol{R}\cdot\boldsymbol{k});$$

then we infer from (*i*) and (*iii*) that $\boldsymbol{n}\cdot\boldsymbol{R}\cdot\boldsymbol{n}=\boldsymbol{k}\cdot\boldsymbol{R}\cdot\boldsymbol{k}$.
(*vi*) If $\boldsymbol{m}\in S_{\boldsymbol{R}}$ then $\boldsymbol{m}=\alpha\boldsymbol{n}+\beta\boldsymbol{k}$, $\alpha^2+\beta^2=1$ and $\boldsymbol{m}\cdot\boldsymbol{R}\cdot\boldsymbol{m}=\boldsymbol{n}\cdot\boldsymbol{R}\cdot\boldsymbol{n}$.
■

Now let us return to $\boldsymbol{0}\neq\boldsymbol{x}\in\mathbf{S}$, orthogonal to $a_{\boldsymbol{R}}$. The Cauchy inequality gives $|\boldsymbol{x}\cdot\boldsymbol{R}\cdot\boldsymbol{x}|\leq|\boldsymbol{x}|^2$; thus

$$\alpha_{\boldsymbol{R}}:=\arccos\frac{\boldsymbol{x}\cdot\boldsymbol{R}\cdot\boldsymbol{x}}{|\boldsymbol{x}|^2}\in[0,\pi]$$

is meaningful, which is called the *angle of rotation of* $\boldsymbol{R}$.

Observe that

— $\alpha_{\boldsymbol{R}}=0$ if and only if $\boldsymbol{R}\cdot\boldsymbol{x}=\boldsymbol{x}$ for all $\boldsymbol{x}$ orthogonal to $a_{\boldsymbol{R}}$, i.e. $\boldsymbol{R}=\mathbf{1_S}$,
— $\alpha_{\boldsymbol{R}}=\pi$ if and only if $\boldsymbol{R}\cdot\boldsymbol{x}=-\boldsymbol{x}$ for all $\boldsymbol{x}$ orthogonal to $a_{\boldsymbol{R}}$.

11.1.5. Proposition. Let $\boldsymbol{R}\neq\mathbf{1_S}$ and $\alpha_{\boldsymbol{R}}\neq\pi$. Take an arbitrary nonzero $\boldsymbol{x}\in a_{\boldsymbol{R}}^{\perp}$. Let $\boldsymbol{y}\in a_{\boldsymbol{R}}^{\perp}$ be orthogonal to $\boldsymbol{x}$, $|\boldsymbol{y}|=|\boldsymbol{x}|$, and suppose $(\boldsymbol{x},\boldsymbol{y})$ and $(\boldsymbol{x},\boldsymbol{R}\cdot\boldsymbol{x})$ are equally oriented bases in $a_{\boldsymbol{R}}^{\perp}$. Then

$$\boldsymbol{R}\cdot\boldsymbol{x}=(\cos\alpha_{\boldsymbol{R}})\boldsymbol{x}+(\sin\alpha_{\boldsymbol{R}})\boldsymbol{y}.$$

Proof. Since $\boldsymbol{R}\cdot\boldsymbol{x}=\frac{\boldsymbol{x}\cdot\boldsymbol{R}\cdot\boldsymbol{x}}{|\boldsymbol{x}|^2}\boldsymbol{x}+\frac{\boldsymbol{y}\cdot\boldsymbol{R}\cdot\boldsymbol{x}}{|\boldsymbol{y}|^2}\boldsymbol{y}$, we easily find that $\cos^2\alpha_{\boldsymbol{R}}+\left(\frac{\boldsymbol{y}\cdot\boldsymbol{R}\cdot\boldsymbol{x}}{|\boldsymbol{y}|^2}\right)^2=1$. As a consequence of the equal orientation of $(\boldsymbol{x},\boldsymbol{y})$ and $(\boldsymbol{x},\boldsymbol{R}\cdot\boldsymbol{x})$, we have $\frac{\boldsymbol{y}\cdot\boldsymbol{R}\cdot\boldsymbol{x}}{|\boldsymbol{x}||\boldsymbol{y}|}>0$ which implies that this expression equals $\sin\alpha_{\boldsymbol{R}}$ (because $\alpha_{\boldsymbol{R}}$ is between 0 and π).

11.1.6. Let $\boldsymbol{R}\neq\mathbf{1_S}$ and $\alpha_{\boldsymbol{R}}\neq\pi$. Then $\boldsymbol{x}$ and $\boldsymbol{R}\cdot\boldsymbol{x}$ are linearly independent if $\boldsymbol{x}$ is a nonzero vector orthogonal to $a_{\boldsymbol{R}}$. It is not hard to see that if $\boldsymbol{y}$ is another nonzero vector orthogonal to $a_{\boldsymbol{R}}$, then the pairs $(\boldsymbol{x},\boldsymbol{R}\cdot\boldsymbol{x})$ and $(\boldsymbol{y},\boldsymbol{R}\cdot\boldsymbol{y})$ are equally oriented bases in $a_{\boldsymbol{R}}^{\perp}$. As a consequence, $(\boldsymbol{R}\cdot\boldsymbol{x})\wedge\boldsymbol{x}$ and $(\boldsymbol{R}\cdot\boldsymbol{y})\wedge\boldsymbol{y}$ are positive multiples of each other.

Since $|(\boldsymbol{R}\cdot\boldsymbol{x})\wedge\boldsymbol{x}|^2 = |\boldsymbol{x}|^4 - (\boldsymbol{x}\cdot\boldsymbol{R}\cdot\boldsymbol{x})^2 = |\boldsymbol{x}|^4\sin^2\alpha_{\boldsymbol{R}}$, we have that for $\boldsymbol{R}\neq\mathbf{1}_S$, $\alpha_{\boldsymbol{R}}\neq\pi$

$$\log\boldsymbol{R} := \frac{(\boldsymbol{R}\cdot\boldsymbol{x})\wedge\boldsymbol{x}}{|\boldsymbol{x}|^2\sin\alpha_{\boldsymbol{R}}}\alpha_{\boldsymbol{R}} \in \mathrm{A}(\boldsymbol{h}) \qquad (\mathbf{0}\neq\boldsymbol{x}\in a_{\boldsymbol{R}}^{\perp})$$

is independent of $\boldsymbol{x}$. Moreover, put

$$\log(\mathbf{1}_S) := \mathbf{0}\in\mathrm{A}(\boldsymbol{h}).$$

It is easy to see that

(*i*) $\mathrm{Ker}(\log\boldsymbol{R}) = a_{\boldsymbol{R}}$,

(*ii*) $|\log\boldsymbol{R}| = \alpha_{\boldsymbol{R}}$,

(*iii*) if $\boldsymbol{R}\neq\mathbf{1}_S$ then for an arbitrary nonzero $\boldsymbol{x}\in a_{\boldsymbol{R}}^{\perp}$, $(\boldsymbol{x},\ \boldsymbol{R}\cdot\boldsymbol{x})$ and $(\boldsymbol{x},\ (\log\boldsymbol{R})\cdot\boldsymbol{x})$ form equally oriented bases in $a_{\boldsymbol{R}}^{\perp}$.

In this way, assuming the notations

$$\mathrm{N} := \{\boldsymbol{R}\in\mathcal{SO}(\boldsymbol{h}) \mid \alpha_{\boldsymbol{R}}\neq\pi\}, \qquad \mathrm{P} := \{\boldsymbol{A}\in\mathrm{A}(\boldsymbol{h}) \mid |\boldsymbol{A}|<\pi\}$$

we defined a mapping $\log\colon \mathrm{N}\to\mathrm{P}$; we shall show that log is a bijection whose inverse is the restriction of the exponential mapping (see VII.3.7).

11.1.7. Proposition. For $\mathbf{0}\neq\boldsymbol{A}\in\mathrm{A}(\boldsymbol{h})$ putting $\alpha := |\boldsymbol{A}|$, $\boldsymbol{A}_o := \frac{\boldsymbol{A}}{|\boldsymbol{A}|}$, we have

$$\mathrm{e}^{\boldsymbol{A}} = -\boldsymbol{A}_o^{\,2}\cos\alpha + \boldsymbol{A}_o\sin\alpha + \left(\mathbf{1}_S + \boldsymbol{A}_o^{\,2}\right).$$

Proof. Recall that $\boldsymbol{A}^3 = -\alpha^2\boldsymbol{A}$ (see V.3.10); thus

$$\mathrm{e}^{\boldsymbol{A}} = \sum_{n=0}^{\infty}\frac{\boldsymbol{A}^n}{n!} = \mathbf{1}_S + \boldsymbol{A} + \frac{\boldsymbol{A}^2}{2!} + \frac{\boldsymbol{A}^3}{3!} + \frac{\boldsymbol{A}^4}{4!} + \frac{\boldsymbol{A}^5}{5!} + \frac{\boldsymbol{A}^6}{6!} + \frac{\boldsymbol{A}^7}{7!} + \ldots =$$

$$= \left(\mathbf{1}_S + \frac{\boldsymbol{A}^2}{\alpha^2}\right) - \frac{\boldsymbol{A}^2}{\alpha^2} + \frac{\boldsymbol{A}^2}{2!} - \frac{\alpha^2\boldsymbol{A}^2}{4!} + \frac{\alpha^4\boldsymbol{A}^2}{6!} - \ldots$$

$$+\boldsymbol{A} - \frac{\alpha^2\boldsymbol{A}}{3!} + \frac{\alpha^4\boldsymbol{A}}{5!} - \frac{\alpha^6\boldsymbol{A}}{7!} + \ldots$$

which yields the desired result by $\boldsymbol{A} = \alpha\boldsymbol{A}_o$. ∎

Note that for $\boldsymbol{A}\neq\mathbf{0}$, $\mathbf{1}_S + \boldsymbol{A}_o^2$ is the orthogonal projection onto the plane orthogonal to the kernel of $\boldsymbol{A}$.

As a consequence, if $\boldsymbol{A}\neq\mathbf{0}$ then

(*i*) $\mathrm{e}^{\boldsymbol{A}}\cdot\boldsymbol{x} = \boldsymbol{x}$ for $\boldsymbol{x}\in\mathrm{Ker}\,\boldsymbol{A}$ (the axis of rotation of $\mathrm{e}^{\boldsymbol{A}}$ is the kernel of $\boldsymbol{A}$);

(*ii*) $\mathrm{e}^{\boldsymbol{A}}\cdot\boldsymbol{x} = (\cos\alpha)\boldsymbol{x} + (\sin\alpha)\boldsymbol{A}_o\cdot\boldsymbol{x}$ for $\boldsymbol{x}$ orthogonal to $\mathrm{Ker}\,\boldsymbol{A}$ (the angle of rotation of $\mathrm{e}^{\boldsymbol{A}}$ is $\alpha := |\boldsymbol{A}|$);

11.1.8. Proposition. For $\boldsymbol{R} \in \mathrm{N}$

$$\mathrm{e}^{\log \boldsymbol{R}} = \boldsymbol{R}$$

and for $\boldsymbol{A} \in \mathrm{P}$

$$\log(\mathrm{e}^{\boldsymbol{A}}) = \boldsymbol{A}.$$

Proof. Evidently, for $\boldsymbol{R} = \mathbf{1}_{\mathrm{S}}$ and for $\boldsymbol{A} = \mathbf{0}$ the equalities hold.

If $\boldsymbol{R} \neq \mathbf{1}_{\mathrm{S}}$ and $\boldsymbol{x}$ is in $a_{\boldsymbol{R}}$ then, obviously, $\mathrm{e}^{\log \boldsymbol{R}} \cdot \boldsymbol{x} = \boldsymbol{x} = \boldsymbol{R} \cdot \boldsymbol{x}$. If $\boldsymbol{x}$ is orthogonal to the axis of rotation of $\boldsymbol{R}$, then

$$\mathrm{e}^{\log \boldsymbol{R}} \cdot \boldsymbol{x} = (\cos \alpha_{\boldsymbol{R}})\boldsymbol{x} + \sin \alpha_{\boldsymbol{R}} \frac{\log \boldsymbol{R}}{\alpha_{\boldsymbol{R}}} \cdot \boldsymbol{x} = \boldsymbol{R} \cdot \boldsymbol{x},$$

in view of 11.1.6 and 11.1.7.

According to the previous proposition, for $\boldsymbol{A} \neq \mathbf{0}$, the axis of rotation of $\mathrm{e}^{\boldsymbol{A}}$ is the kernel of $\boldsymbol{A}$; the angle of rotation of $\mathrm{e}^{\boldsymbol{A}}$ is $|\boldsymbol{A}|$. Thus if $\boldsymbol{x} \in \mathrm{Ker}\,\boldsymbol{A}$ then $\log(\mathrm{e}^{\boldsymbol{A}}) \cdot \boldsymbol{x} = \mathbf{0} = \boldsymbol{A} \cdot \boldsymbol{x}$. If $\boldsymbol{x}$ is orthogonal to the kernel of $\boldsymbol{A}$, then $\log(\mathrm{e}^{\boldsymbol{A}}) = \frac{(\mathrm{e}^{\boldsymbol{A}} \cdot \boldsymbol{x}) \wedge \boldsymbol{x}}{|\boldsymbol{x}|^2 \sin |\boldsymbol{A}|}$ and an easy calculation based on the formula in 11.1.6 yields that $\log(\mathrm{e}^{\boldsymbol{A}}) \cdot \boldsymbol{x} = \boldsymbol{A} \cdot \boldsymbol{x}$.

11.1.9. It is trivial that the closure of N is $\mathcal{SO}(\boldsymbol{h})$. It is not hard to see that exponential mapping $\mathrm{A}(\boldsymbol{h}) \to \mathcal{SO}(\boldsymbol{h})$ maps the closure of P onto $\mathcal{SO}(\boldsymbol{h})$. However, the exponential mapping on the closure of P is not injective: if $|\boldsymbol{A}| = \pi$ then $\mathrm{e}^{\boldsymbol{A}} = \mathrm{e}^{-\boldsymbol{A}}$.

Since the closure of P is connected and the exponential mapping is continuous, $\mathcal{SO}(\boldsymbol{h})$ is connected as well.

(However, $\mathcal{SO}(\boldsymbol{h})$ is not simply connected: it is homeomorphic to a set which is obtained from the closure of P by 'sticking' together antipodal points of the boundary of P.)

The one-parameter subgroup of $\mathcal{SO}(\boldsymbol{h})$ corresponding to $\boldsymbol{A} \in \mathrm{A}(\boldsymbol{h})$ is $\mathbb{R} \to \mathcal{SO}(\boldsymbol{h})$, $t \mapsto \mathrm{e}^{t\boldsymbol{A}}$. If $\boldsymbol{A} \neq \mathbf{0}$, then all the elements of the one-parameter subgroup are rotations around the same axis $\mathrm{Ker}\,\boldsymbol{A}$.

Since the exponential mapping is surjective, every element of $\mathcal{SO}(\boldsymbol{h})$ is in a one-parameter subgroup.

11.1.10. In physical applications we meet $\frac{\mathrm{A}(\boldsymbol{h})}{\mathbb{T}}$ instead of $\mathrm{A}(\boldsymbol{h})$. If $\boldsymbol{\Omega} \in \frac{\mathrm{A}(\boldsymbol{h})}{\mathbb{T}}$, then we can give a function $\boldsymbol{R}\colon \mathrm{T} \to \mathcal{SO}(\boldsymbol{h})$, $t \mapsto e^{(t-t_o)\boldsymbol{\Omega}}$, where t_o is a fixed element of T. Then every value of such a function is a rotation around the same axis; the angle of rotation of $\boldsymbol{R}(t)$ is $(t - t_o)|\boldsymbol{\Omega}|$. Thus $|\boldsymbol{\Omega}|$ is interpreted as the magnitude of the angular velocity and $\boldsymbol{\Omega}$ itself as the *angular velocity* of the rotation.

We know that $\boldsymbol{R}$ is differentiable, $\dot{\boldsymbol{R}} = \boldsymbol{\Omega} \cdot \boldsymbol{R}$, from which we infer that

$$\boldsymbol{\Omega} = \dot{\boldsymbol{R}} \cdot \boldsymbol{R}^{-1}.$$

In general, consider a differentiable function $\boldsymbol{R} \colon \mathrm{T} \rightarrowtail \mathcal{SO}(\boldsymbol{h}) \subset \mathbf{S} \otimes \mathbf{S}^*$. Its derivative at t, $\dot{\boldsymbol{R}}(t)$, is a linear map from T into $\mathbf{S} \otimes \mathbf{S}^*$ that takes values in the tangent space of $\mathcal{SO}(\boldsymbol{h})$ at $\boldsymbol{R}(t)$ which is $\boldsymbol{R}(t) \cdot \mathrm{A}(\boldsymbol{h}) = \{\boldsymbol{R}(t) \cdot \boldsymbol{A} \mid \boldsymbol{A} \in \mathrm{A}(\boldsymbol{h})\}$ (see VII.3.3). In other words, $\dot{\boldsymbol{R}}(t) \in \frac{\boldsymbol{R}(t) \cdot \mathrm{A}(\boldsymbol{h})}{\mathbb{T}}$, i.e. $\boldsymbol{R}(t)^{-1} \cdot \dot{\boldsymbol{R}}(t) \in \frac{\mathrm{A}(\boldsymbol{h})}{\mathbb{T}}$. Then V.2.11($ii$) implies that $\boldsymbol{R}(t) \cdot \left(\boldsymbol{R}(t)^{-1} \cdot \dot{\boldsymbol{R}}(t)\right) \cdot \boldsymbol{R}(t)^{-1}$ is in $\frac{\mathrm{A}(\boldsymbol{h})}{\mathbb{T}}$ as well;

$$\boldsymbol{\Omega}(t) := \dot{\boldsymbol{R}}(t) \cdot \boldsymbol{R}(t)^{-1} \in \frac{\mathrm{A}(\boldsymbol{h})}{\mathbb{T}}$$

is called the *angular velocity value* at t, and the function $\boldsymbol{\Omega} \colon \mathrm{T} \rightarrowtail \frac{\mathrm{A}(\boldsymbol{h})}{\mathbb{T}}$ is the *angular velocity*.

Evidently, $\boldsymbol{R}$ is the solution of the differential equation

$$(\boldsymbol{X} \colon \mathrm{T} \rightarrowtail \mathcal{SO}(\boldsymbol{h}))? \qquad \dot{\boldsymbol{X}} = \boldsymbol{\Omega} \cdot \boldsymbol{X}.$$

11.2. Exercises

1. Let us coordinatize $\mathcal{SO}(\boldsymbol{h})$ by the *Euler angles* as follows.

Let $(\boldsymbol{n}_1, \boldsymbol{n}_2, \boldsymbol{n}_3)$ be a positively oriented orthonormal basis in $\frac{\mathbf{S}}{\mathbb{L}}$. If $\boldsymbol{R} \cdot \boldsymbol{n}_3$ is not parallel to $\boldsymbol{n}_3$, put $\boldsymbol{n} := \frac{\boldsymbol{n}_3 \times (\boldsymbol{R} \cdot \boldsymbol{n}_3)}{|\boldsymbol{n}_3 \times (\boldsymbol{R} \cdot \boldsymbol{n}_3)|}$ and

$$\begin{aligned} \vartheta_{\boldsymbol{R}} &:= \arccos(\boldsymbol{n}_3 \cdot \boldsymbol{R} \cdot \boldsymbol{n}_3), \\ \psi_{\boldsymbol{R}} &:= \operatorname{sign}(\boldsymbol{n} \cdot \boldsymbol{n}_2) \arccos(\boldsymbol{n} \cdot \boldsymbol{n}_1), \\ \varphi_{\boldsymbol{R}} &:= \operatorname{sign}(\boldsymbol{n} \cdot \boldsymbol{R} \cdot \boldsymbol{n}_2) \arccos(\boldsymbol{n} \cdot \boldsymbol{R} \cdot \boldsymbol{n}_1) \end{aligned}$$

where $\operatorname{sign} x := \frac{x}{|x|}$ if $0 \neq x \in \mathbb{R}$ and $\operatorname{sign} 0 := 1$.

Prove that if R_i denotes the one-parameter subgroup of rotations around $\boldsymbol{n}_i$ $(i = 1, 2, 3)$ then

$$\boldsymbol{R} = R_3(\varphi_{\boldsymbol{R}}) \cdot R_1(\vartheta_{\boldsymbol{R}}) \cdot R_3(\psi_{\boldsymbol{R}}).$$

2. Let $\boldsymbol{R} \colon \mathrm{T} \rightarrowtail \mathcal{SO}(\boldsymbol{h})$ be a differentiable function and put $\boldsymbol{R}^{-1} \colon \mathrm{T} \rightarrowtail \mathcal{SO}(\boldsymbol{h})$, $t \mapsto \boldsymbol{R}(t)^{-1}$. Using $\boldsymbol{R} \cdot \boldsymbol{R}^{-1} = \mathbf{1}_{\mathbf{S}}$ prove that $\boldsymbol{R}^{-1}$ is also differentiable and

$$\left(\boldsymbol{R}^{-1}\right)^{\cdot} = -\boldsymbol{R}^{-1} \cdot \dot{\boldsymbol{R}} \cdot \boldsymbol{R}^{-1}.$$

3. Prove that for $\mathbf{0} \leq \boldsymbol{r} \in \mathbb{L}$, $\{\boldsymbol{x} \in \mathbf{S} \mid |\boldsymbol{x}| = \boldsymbol{r}\}$ is an orbit of $\mathcal{SO}(\boldsymbol{h})$ and all its orbits are of this kind.

11.3. The Galilean group

11.3.1. We shall deal with linear maps from $\mathbf{M}$ into $\mathbf{M}$, permanently using the identification $\mathrm{Lin}(\mathbf{M}) \equiv \mathbf{M} \otimes \mathbf{M}^*$. The restriction of a linear map $\boldsymbol{L}: \mathbf{M} \to \mathbf{M}$ ($\boldsymbol{L} \in \mathbf{M} \otimes \mathbf{M}^*$) onto $\mathbf{S}$ equals $\boldsymbol{L} \cdot \boldsymbol{i}$ where $\boldsymbol{i}: \mathbf{S} \to \mathbf{M}$ ($\boldsymbol{i} \in \mathbf{M} \otimes \mathbf{S}^*$) is the canonical embedding. The symbol $\boldsymbol{L} \cdot \boldsymbol{i} \in \boldsymbol{i} \cdot \mathcal{O}(\boldsymbol{h})$ means that the restriction of $\boldsymbol{L}$ onto $\mathbf{S}$ is in $\mathcal{O}(\boldsymbol{h})$, i.e. there is an $\boldsymbol{R} \in \mathcal{O}(\boldsymbol{h}) \subset \mathbf{S} \otimes \mathbf{S}^*$ such that $\boldsymbol{L} \cdot \boldsymbol{i} = \boldsymbol{i} \cdot \boldsymbol{R}$.

First we define the Galilean group and then studying it we find its physical meaning.

Definition.

$$\mathcal{G} := \{\boldsymbol{L} \in \mathbf{M} \otimes \mathbf{M}^* \mid \boldsymbol{\tau} \cdot \boldsymbol{L} = \pm\boldsymbol{\tau},\ \boldsymbol{L} \cdot \boldsymbol{i} \in \boldsymbol{i} \cdot \mathcal{O}(\boldsymbol{h})\}$$

is called the *Galilean group*; its elements are the *Galilean transformations.*

If $\boldsymbol{L}$ is a Galilean transformation then

$$\mathrm{ar}\,\boldsymbol{L} := \begin{cases} +1 & \text{if } \boldsymbol{\tau} \cdot \boldsymbol{L} = \boldsymbol{\tau} \\ -1 & \text{if } \boldsymbol{\tau} \cdot \boldsymbol{L} = -\boldsymbol{\tau} \end{cases}$$

is the *arrow* of $\boldsymbol{L}$ and

$$\mathrm{sign}\,\boldsymbol{L} := \begin{cases} +1 & \text{if } \boldsymbol{L} \cdot \boldsymbol{i} \in \boldsymbol{i} \cdot \mathcal{O}(\boldsymbol{h})^+ \\ -1 & \text{if } \boldsymbol{L} \cdot \boldsymbol{i} \in \boldsymbol{i} \cdot \mathcal{O}(\boldsymbol{h})^- \end{cases}$$

is the *sign* of $\boldsymbol{L}$.

Let us put

$$\begin{aligned}
\mathcal{G}^{+\to} &:= \{\boldsymbol{L} \in \mathcal{G} \mid \mathrm{sign}\,\boldsymbol{L} = \mathrm{ar}\,\boldsymbol{L} = 1\}, \\
\mathcal{G}^{+\leftarrow} &:= \{\boldsymbol{L} \in \mathcal{G} \mid \mathrm{sign}\,\boldsymbol{L} = -\mathrm{ar}\,\boldsymbol{L} = 1\}, \\
\mathcal{G}^{-\to} &:= \{\boldsymbol{L} \in \mathcal{G} \mid \mathrm{sign}\,\boldsymbol{L} = -\mathrm{ar}\,\boldsymbol{L} = -1\}, \\
\mathcal{G}^{-\leftarrow} &:= \{\boldsymbol{L} \in \mathcal{G} \mid \mathrm{sign}\,\boldsymbol{L} = \mathrm{ar}\,\boldsymbol{L} = -1\}.
\end{aligned}$$

$\mathcal{G}^{+\to}$ is called the *proper Galilean group.* ■

(*i*) The condition $\boldsymbol{\tau} \cdot \boldsymbol{L} = \pm\boldsymbol{\tau}$ implies that $\mathbf{S}$ is invariant under the linear map $\boldsymbol{L}: \mathbf{M} \to \mathbf{M}$.

(*ii*) The condition $\boldsymbol{L} \cdot \boldsymbol{i} \in \boldsymbol{i} \cdot \mathcal{O}(\boldsymbol{h})$ means that there is a (necessarily unique) $\boldsymbol{R_L}$ in $\mathcal{O}(\boldsymbol{h})$ such that

$$\boldsymbol{L} \cdot \boldsymbol{i} = \boldsymbol{i} \cdot \boldsymbol{R_L}.$$

(*iii*) The Galilean transformations are linear bijections: if $\boldsymbol{L} \cdot \boldsymbol{x} = \mathbf{0}$ then $\boldsymbol{\tau} \cdot \boldsymbol{x} = 0$, i.e. $\boldsymbol{x}$ is in $\mathbf{S}$; the restriction of $\boldsymbol{L}$ onto $\mathbf{S}$ is injective, thus $\boldsymbol{x} = \mathbf{0}$.

(*iv*) It is quite trivial that $\mathcal{G}$ is indeed a group: the product of its elements as well as the inverse of its elements are Galilean transformations.

11.3.2. Proposition. The Galilean group is a six-dimensional Lie group having the Lie algebra

$$\mathbf{La}(\mathcal{G}) = \{\boldsymbol{H} \in \mathbf{M} \otimes \mathbf{M}^* \mid \boldsymbol{\tau} \cdot \boldsymbol{H} = 0,\ \boldsymbol{H} \cdot \boldsymbol{i} \in \mathrm{A}(\boldsymbol{h})\}.$$

Proof. According to the previous remark, $\mathcal{G}$ is a subgroup of $\mathcal{GL}(\mathbf{M})$ which is sixteen-dimensional.

We have to show that the Galilean group is a six-dimensional smooth submanifold of $\mathcal{GL}(\mathbf{M})$.

Observe that if $\boldsymbol{L} \in \mathcal{G}$, then

$$\boldsymbol{\pi_u} \cdot \boldsymbol{L} \cdot \boldsymbol{i} = \boldsymbol{R_L}$$

for all $\boldsymbol{u} \in V(1)$, where $\boldsymbol{R_L}$ is given in the previous remark.

$S(\boldsymbol{h}) := \{\boldsymbol{S} \in \mathbf{S} \otimes \mathbf{S}^* \mid \boldsymbol{S}^* = \boldsymbol{S}\}$ is a six-dimensional linear subspace and

$$\phi_{\boldsymbol{u}} : \mathcal{GL}(\mathbf{M}) \to (\mathbb{T} \otimes \mathbf{M}^*) \times S(\boldsymbol{h}), \qquad \boldsymbol{L} \mapsto \big(\boldsymbol{\tau} \cdot \boldsymbol{L},\ (\boldsymbol{\pi_u} \cdot \boldsymbol{L} \cdot \boldsymbol{i})^* \cdot (\boldsymbol{\pi_u} \cdot \boldsymbol{L} \cdot \boldsymbol{i})\big)$$

is evidently a smooth map; $\mathcal{G}$ is the preimage of $\{(\pm\boldsymbol{\tau}, \mathbf{1_S})\}$ by $\phi_{\boldsymbol{u}}$.

The derivative of $\phi_{\boldsymbol{u}}$ at $\boldsymbol{L}$ is the linear map

$$\begin{aligned} \mathrm{D}\phi_{\boldsymbol{u}}(\boldsymbol{L}) :\ & \mathbf{M} \otimes \mathbf{M}^* \to (\mathbb{T} \otimes \mathbf{M}^*) \times S(\boldsymbol{h}), \\ & \boldsymbol{H} \mapsto \big(\boldsymbol{\tau} \cdot \boldsymbol{H},\ (\boldsymbol{\pi_u} \cdot \boldsymbol{H} \cdot \boldsymbol{i})^* \cdot (\boldsymbol{\pi_u} \cdot \boldsymbol{L} \cdot \boldsymbol{i}) + (\boldsymbol{\pi_u} \cdot \boldsymbol{L} \cdot \boldsymbol{i})^* \cdot (\boldsymbol{\pi_u} \cdot \boldsymbol{H} \cdot \boldsymbol{i})\big) \end{aligned}$$

which is surjective: $(\boldsymbol{k}, \boldsymbol{T}) \in (\mathbb{T} \otimes \mathbf{M}^*) \times S(\boldsymbol{h})$ is the image by $\mathrm{D}\phi_{\boldsymbol{u}}(\boldsymbol{L})$ of $\boldsymbol{u} \otimes \boldsymbol{k} + (1/2)(\boldsymbol{\pi_u} \cdot \boldsymbol{L} \cdot \boldsymbol{i})^{*^{-1}} \cdot \boldsymbol{T}$.

Thus, being a six-dimensional submanifold in $\mathcal{GL}(\mathbf{M})$, the Galilean group is a Lie group; its Lie algebra is $\mathrm{Ker}\,\mathrm{D}\phi_{\boldsymbol{u}}(\mathbf{1_M})$.

If $\mathrm{D}\phi_{\boldsymbol{u}}(\mathbf{1_M})(\boldsymbol{H}) = \mathbf{0}$, then $\boldsymbol{\tau} \cdot \boldsymbol{H} = 0$, and $\boldsymbol{\pi_u} \cdot \boldsymbol{H} \cdot \boldsymbol{i}$ is in $\mathrm{A}(\boldsymbol{h})$. Since the first condition means that $\boldsymbol{H} \in \mathbf{S} \otimes \mathbf{M}^*$, we have $\boldsymbol{\pi_u} \cdot \boldsymbol{H} \cdot \boldsymbol{i} = \boldsymbol{H} \cdot \boldsymbol{i}$. Hence the kernel of $\mathrm{D}\phi_{\boldsymbol{u}}(\mathbf{1_M})$ is the linear subspace given in our proposition.

11.3.3. The mappings $\mathcal{G} \to \{-1, 1\}$, $\boldsymbol{L} \mapsto \mathrm{ar}\,\boldsymbol{L}$ and $\mathcal{G} \to \{-1, 1\}$, $\boldsymbol{L} \mapsto \mathrm{sign}\,\boldsymbol{L}$ are continuous group homomorphisms. As a consequence, the Galilean group is disconnected. We shall see in 11.4.3 that the proper Galilean group $\mathcal{G}^{+\to}$ is connected. It is quite trivial that if $\boldsymbol{L} \in \mathcal{G}^{+\leftarrow}$ then $\boldsymbol{L} \cdot \mathcal{G}^{+\to} = \mathcal{G}^{+\leftarrow}$ and similar assertions hold for $\mathcal{G}^{-\to}$ and $\mathcal{G}^{-\leftarrow}$ as well. Consequently, the Galilean group has four connected components, the four subsets given in Definition 11.2.1.

From these four components only $\mathcal{G}^{+\to}$ —the proper Galilean group—is a subgroup; nevertheless, the union of an arbitrary component and of the proper Galilean group is a subgroup as well.

$\mathcal{G}^{\rightarrow} := \mathcal{G}^{+\rightarrow} \cup \mathcal{G}^{-\rightarrow}$ is called the *orthochronous Galilean group.*

If $\boldsymbol{L} \in \mathcal{G}$, then $\boldsymbol{L}$ preserves or reverses the 'orientation' of timelike vectors according to whether $\operatorname{ar} \boldsymbol{L} = 1$ or $\operatorname{ar} \boldsymbol{L} = -1$:

$$\begin{aligned} &\text{if} \quad \operatorname{ar} \boldsymbol{L} = 1 \qquad \text{then} \quad \boldsymbol{L}(T^{\rightarrow}) = T^{\rightarrow}, \quad \boldsymbol{L}(T^{\leftarrow}) = T^{\leftarrow}, \\ &\text{if} \quad \operatorname{ar} \boldsymbol{L} = -1 \qquad \text{then} \quad \boldsymbol{L}(T^{\rightarrow}) = T^{\leftarrow}, \quad \boldsymbol{L}(T^{\leftarrow}) = T^{\rightarrow}. \end{aligned}$$

Moreover, $\boldsymbol{L}$ preserves or reverses the orientation of $\mathbf{S}$ according to whether $\operatorname{sign} \boldsymbol{L} = 1$ or $\operatorname{sign} \boldsymbol{L} = -1$.

The orientation of $\mathbf{S}$ given in 1.2.4 shows that the elements of $\mathcal{G}^{+\rightarrow}$ and $\mathcal{G}^{-\leftarrow}$ preserve the orientation of $\mathbf{M}$, whereas the elements of $\mathcal{G}^{+\leftarrow}$ and $\mathcal{G}^{-\rightarrow}$ reverse the orientation.

11.3.4. $\mathbf{M}$ is of even dimension, thus $-\mathbf{1}_{\mathbf{M}}$ is orientation preserving. Evidently, $-\mathbf{1}_{\mathbf{M}}$ is in $\mathcal{G}^{-\leftarrow}$; it is called the *inversion of spacetime vectors.* We have that $\mathcal{G}^{-\leftarrow} = (-\mathbf{1}_{\mathbf{M}}) \cdot \mathcal{G}^{+\rightarrow}$.

We have seen previously that the elements of $\mathcal{G}^{+\leftarrow}$ invert in some sense the timelike vectors and do not invert the spacelike vectors; the elements of $\mathcal{G}^{-\rightarrow}$ invert in some sense the spacelike vectors and do not invert the timelike vectors. However, we cannot select an element of $\mathcal{G}^{+\leftarrow}$ and an element of $\mathcal{G}^{-\rightarrow}$ that we could consider the time inversion and the space inversion.

For each $\boldsymbol{u} \in V(1)$ we can give a $\boldsymbol{u}$-timelike inversion and a $\boldsymbol{u}$-spacelike inversion as follows.

The $\boldsymbol{u}$-timelike inversion $\boldsymbol{T}_{\boldsymbol{u}} \in \mathcal{G}^{+\leftarrow}$ inverts vectors parallel to $\boldsymbol{u}$ and leaves spacelike vectors invariant:

$$\boldsymbol{T}_{\boldsymbol{u}} \cdot \boldsymbol{u} := -\boldsymbol{u} \qquad \text{and} \qquad \boldsymbol{T}_{\boldsymbol{u}} \cdot \boldsymbol{q} := \boldsymbol{q} \quad \text{for} \quad \boldsymbol{q} \in \mathbf{S}.$$

In general,

$$\boldsymbol{T}_{\boldsymbol{u}} \cdot \boldsymbol{x} = -\boldsymbol{u}(\boldsymbol{\tau} \cdot \boldsymbol{x}) + \boldsymbol{\pi}_{\boldsymbol{u}} \cdot \boldsymbol{x} = -2\boldsymbol{u}(\boldsymbol{\tau} \cdot \boldsymbol{x}) + \boldsymbol{x} \qquad (\boldsymbol{x} \in \mathbf{M})$$

i.e.

$$\boldsymbol{T}_{\boldsymbol{u}} = \mathbf{1}_{\mathbf{M}} - 2\boldsymbol{u} \otimes \boldsymbol{\tau}.$$

The $\boldsymbol{u}$-spacelike inversion $\boldsymbol{P}_{\boldsymbol{u}} \in \mathcal{G}^{-\rightarrow}$ inverts spacelike vectors and leaves vectors parallel to $\boldsymbol{u}$ invariant:

$$\boldsymbol{P}_{\boldsymbol{u}} \cdot \boldsymbol{u} := \boldsymbol{u} \qquad \text{and} \qquad \boldsymbol{P}_{\boldsymbol{u}} \cdot \boldsymbol{q} := -\boldsymbol{q} \qquad \text{for} \qquad \boldsymbol{q} \in \mathbf{S}.$$

In general,

$$\boldsymbol{P}_{\boldsymbol{u}} \cdot \boldsymbol{x} = \boldsymbol{u}(\boldsymbol{\tau} \cdot \boldsymbol{x}) - \boldsymbol{\pi}_{\boldsymbol{u}} \cdot \boldsymbol{x} = 2\boldsymbol{u}(\boldsymbol{\tau} \cdot \boldsymbol{x}) - \boldsymbol{x} \qquad (\boldsymbol{x} \in \mathbf{M}),$$

i.e.

$$\boldsymbol{P}_{\boldsymbol{u}} = 2\boldsymbol{u} \otimes \boldsymbol{\tau} - \mathbf{1}_{\mathbf{M}}.$$

We easily deduce the following equalities:

$$\boldsymbol{T_u}^{-1} = \boldsymbol{T_u}, \qquad \boldsymbol{P_u}^{-1} = \boldsymbol{P_u}, \qquad -\boldsymbol{T_u} = \boldsymbol{P_u},$$

$$\boldsymbol{T_u} \cdot \boldsymbol{P_u} = \boldsymbol{P_u} \cdot \boldsymbol{T_u} = -\mathbf{1}_{\mathbf{M}}.$$

11.3.5. The three-dimensional orthogonal group *is not a subgroup* of the Galilean group: $\mathcal{O}(\boldsymbol{h})$ cannot be a subgroup of $\mathcal{G}$ because the elements of $\mathcal{G}$ are linear maps defined on $\mathbf{M}$ whereas the elements of $\mathcal{O}(\boldsymbol{h})$ are linear maps defined on $\mathbf{S}$ ($\mathbf{S} \otimes \mathbf{S}^*$ is not a subset of $\mathbf{M} \otimes \mathbf{M}^*$).

It is quite obvious that

$$\mathcal{G}^{\rightarrow} \to \mathcal{O}(\boldsymbol{h}), \qquad \boldsymbol{L} \mapsto \boldsymbol{R_L}$$

(where $\boldsymbol{L} \cdot \boldsymbol{i} = \boldsymbol{i} \cdot \boldsymbol{R_L}$) is a surjective Lie group homomorphism.

For every $\boldsymbol{u} \in V(1)$,

$$\mathcal{O}(\boldsymbol{h})_{\boldsymbol{u}} := \{\boldsymbol{L} \in \mathcal{G}^{\rightarrow} \mid \boldsymbol{L} \cdot \boldsymbol{u} = \boldsymbol{u}\},$$

called the group of $\boldsymbol{u}$*-spacelike orthogonal transformations,* is a subgroup of $\mathcal{G}^{\rightarrow}$; the restriction of the above Lie group homomorphism to $\mathcal{O}(\boldsymbol{h})_{\boldsymbol{u}}$ is a bijection between $\mathcal{O}(\boldsymbol{h})_{\boldsymbol{u}}$ and $\mathcal{O}(\boldsymbol{h})$.

Indeed, if $\boldsymbol{L} \cdot \boldsymbol{u} = \boldsymbol{u}$ and $\boldsymbol{R_L} = \mathbf{1}_{\mathbf{S}}$, then $\boldsymbol{L}$ is the identity on the complementary subspaces $\boldsymbol{u} \otimes \mathbb{T}$ and $\mathbf{S}$, thus $\boldsymbol{L} = \mathbf{1}_{\mathbf{M}}$: the group homomorphism from $\mathcal{O}(\boldsymbol{h})_{\boldsymbol{u}}$ into $\mathcal{O}(\boldsymbol{h})$ is injective.

If $\boldsymbol{R} \in \mathcal{O}(\boldsymbol{h})$ then

$$\boldsymbol{R_u} := \boldsymbol{u} \otimes \boldsymbol{\tau} + \boldsymbol{R} \cdot \boldsymbol{\pi_u}$$

is a Galilean transformation in $\mathcal{O}(\boldsymbol{h})_{\boldsymbol{u}}$ and $\boldsymbol{\pi_u} \cdot \boldsymbol{R_u} \cdot \boldsymbol{i} = \boldsymbol{R}$ (recall that $\boldsymbol{\pi_u} \cdot \boldsymbol{L} \cdot \boldsymbol{i} = \boldsymbol{R_L}$ for all Galilean transformations $\boldsymbol{L}$): the group homomorphism from $\mathcal{O}(\boldsymbol{h})_{\boldsymbol{u}}$ onto $\mathcal{O}(\boldsymbol{h})$ is surjective.

11.3.6. The kernel of the surjection $\mathcal{G}^{\rightarrow} \to \mathcal{O}(\boldsymbol{h})$, i.e.

$$\mathcal{V} := \{\boldsymbol{L} \in \mathcal{G}^{\rightarrow} \mid \boldsymbol{R_L} = \mathbf{1}_{\mathbf{S}}\} = \{\boldsymbol{L} \in \mathcal{G}^{\rightarrow} \mid \boldsymbol{L} \cdot \boldsymbol{i} = \boldsymbol{i}\}$$

is called the *special Galilean group.* Observe that $\mathcal{V}$ is a subgroup of $\mathcal{G}^{+\rightarrow}$.

The special Galilean group is a three-dimensional Lie group having the Lie algebra

$$\mathbf{La}(\mathcal{V}) = \{\boldsymbol{H} \in \mathbf{M} \otimes \mathbf{M}^* \mid \boldsymbol{\tau} \cdot \boldsymbol{H} = 0, \qquad \boldsymbol{H} \cdot \boldsymbol{i} = 0\}.$$

Proposition. If $\boldsymbol{L} \in \mathcal{V}$, then there is a unique $\boldsymbol{v_L} \in \frac{\mathbf{S}}{\mathbb{T}}$ such that

$$\boldsymbol{L} \cdot \boldsymbol{x} = \boldsymbol{v_L}(\boldsymbol{\tau} \cdot \boldsymbol{x}) + \boldsymbol{x} \qquad (\boldsymbol{x} \in \mathbf{M}),$$

i.e.

$$L = 1_M + v_L \otimes \tau.$$

The correspondence $\mathcal{V} \to \frac{S}{T}$, $L \mapsto v_L$ is a bijective group homomorphism regarding the additive structure of $\frac{S}{T}$ (i.e. $v_{L \cdot K} = v_L + v_K$ for all $L, K \in \mathcal{V}$).

Proof. Let L be an element of $\mathcal{V}$. Let us take an arbitrary $u \in V(1)$ and put $v_L := L \cdot u - u$. We claim that v_L does not depend on u. Indeed, if $u' \in V(1)$ then

$$(L \cdot u - u) - (L \cdot u' - u') = L \cdot (u - u') - (u - u') = 0,$$

because $L \cdot (u - u') = u - u'$. Moreover, $\tau \cdot (L \cdot u - u) = 0$, thus v_L is in $\frac{S}{T}$.

This means that $L \cdot u = u + v_L$ for all $u \in V(1)$.

Then we find that for $x \in M$

$$L \cdot x = L \cdot (u(\tau \cdot x) + \pi_u \cdot x) = (u + v_L)\tau \cdot x + \pi_u \cdot x = v_L(\tau \cdot x) + x.$$

This formula assures, too, that $L \mapsto v_L$ is a group homomorphism.

If $v_L = 0$ then $L \cdot u = u$ for all $u \in V(1)$ implying $L = 1_M$; thus the correspondence from $\mathcal{V}$ into $\frac{S}{T}$ is injective. Evidently, if v is in $\frac{S}{T}$ then $1_M + v \otimes \tau$ is a special Galilean transformation: the correspondence is surjective. ■

In view of our result, the special Galilean group is a three-dimensional commutative group.

11.3.7. (*i*) If $u, u' \in V(1)$, then the special Galilean transformation

$$L(u', u) := 1_M + (u' - u) \otimes \tau,$$

i.e. the one corresponding to $v_{u'u} = u' - u$ is the unique one with the property

$$L(u', u) \cdot u = u'.$$

Let us recall the splitting of M according to u and u'; then we easily find that

$$L(u', u) = \xi_{u'}{}^{-1} \cdot \xi_u.$$

(*ii*) The product of the u'-timelike inversion and the u-timelike inversion is a special Galilean transformation:

$$T_{u'} \cdot T_u = (1_M - 2u' \otimes \tau) \cdot (1_M - 2u \otimes \tau) = 1_M + 2v_{u'u} \otimes \tau.$$

We know that $T_u{}^{-1} = T_u = -P_u$; then we can assert that

$$T_{u'} \cdot T_u{}^{-1} = P_{u'} \cdot P_u{}^{-1} = L(u', u)^2 = \\ = L(u + 2v_{u'u}, u) = L(u - 2v_{uu'}, u).$$

11.3.8. Originally Galilean transformations are defined to be linear maps from $\mathbf{M}$ into $\mathbf{M}$. In the usual way, we can consider them to be linear maps from $\frac{\mathbf{M}}{\mathbb{T}}$ into $\frac{\mathbf{M}}{\mathbb{T}}$ as we already did in the preceding paragraphs as well.

$V(1)$ is invariant under orthochronous Galilean transformations. Moreover, the restriction of an orthochronous Galilean transformation $\boldsymbol{L}$ onto $V(1)$ is an affine bijection whose underlying linear map—which is the restriction of $\boldsymbol{L}$ onto $\frac{\mathbf{S}}{\mathbb{T}}$ —preserves the Euclidean structure.

Conversely, if $\boldsymbol{F}$ is a Euclidean transformation of $V(1)$ —an affine bijection whose underlying linear map preserves the Euclidean structure—then $\mathbf{M} \to \mathbf{M}$, $\boldsymbol{x} \mapsto \boldsymbol{F} \cdot (\boldsymbol{x}/\boldsymbol{\tau} \cdot \boldsymbol{x})\boldsymbol{\tau} \cdot \boldsymbol{x}$ is an orthochronous Galilean transformation whose restriction onto $V(1)$ coincides with $\boldsymbol{F}$.

Thus we can state that the orthochronous Galilean group is canonically isomorphic to the group of Euclidean transformations of $V(1)$.

11.4. The split Galilean group

11.4.1. The Galilean transformations, being elements of $\mathbf{M} \otimes \mathbf{M}^*$, are split by velocity values according to 8.1.1. Since $\boldsymbol{\tau} \cdot \boldsymbol{L} = (\operatorname{ar} \boldsymbol{L})\boldsymbol{\tau}$ and $\boldsymbol{\pi}_{\boldsymbol{u}} \cdot \boldsymbol{L} \cdot \boldsymbol{i} = \boldsymbol{R}_{\boldsymbol{L}}$ for a Galilean transformation $\boldsymbol{L}$ and for $\boldsymbol{u} \in V(1)$, we have

$$\boldsymbol{\xi}_{\boldsymbol{u}} \cdot \boldsymbol{L} \cdot \boldsymbol{\xi}_{\boldsymbol{u}}{}^{-1} = \begin{pmatrix} \operatorname{ar} \boldsymbol{L} & 0 \\ \boldsymbol{L} \cdot \boldsymbol{u} - (\operatorname{ar} \boldsymbol{L})\boldsymbol{u} & \boldsymbol{R}_{\boldsymbol{L}} \end{pmatrix}.$$

Writing $\boldsymbol{L} \cdot \boldsymbol{u} - (\operatorname{ar} \boldsymbol{L})\boldsymbol{u} = (\operatorname{ar} \boldsymbol{L}) \left((\operatorname{ar} \boldsymbol{L})\boldsymbol{L} \cdot \boldsymbol{u} - \boldsymbol{u} \right)$, we see that the following definition describes the split form of Galilean transformations.

Definition. The *split Galilean group* is

$$\left\{ \begin{pmatrix} \pm 1 & 0 \\ \boldsymbol{v} & \pm \boldsymbol{R} \end{pmatrix} \middle| \; \boldsymbol{v} \in \frac{\mathbf{S}}{\mathbb{T}}, \quad \boldsymbol{R} \in SO(\boldsymbol{h}) \right\}.$$

Its elements are called *split Galilean transformations.* ■

The split Galilean transformations can be regarded as linear maps $\mathbb{T} \times \mathbf{S} \to \mathbb{T} \times \mathbf{S}$; the one in the definition makes the correspondence

$$(\boldsymbol{t}, \boldsymbol{q}) \mapsto (\pm \boldsymbol{t}, \; \boldsymbol{v}\boldsymbol{t} \pm \boldsymbol{R} \cdot \boldsymbol{q}).$$

The split Galilean group is a six-dimensional Lie group having the Lie algebra

$$\left\{ \begin{pmatrix} 0 & 0 \\ \boldsymbol{v} & \boldsymbol{A} \end{pmatrix} \middle| \; \boldsymbol{v} \in \frac{\mathbf{S}}{\mathbb{T}}, \quad \boldsymbol{A} \in \mathrm{A}(\boldsymbol{h}) \right\}.$$

11.4.2. The splitting ξ_u according to u establishes a Lie-group isomorphism between the Galilean group and the split Galilean group. The isomorphisms corresponding to different u' and u are different.

The $\frac{\mathbb{S}}{\mathbb{T}}$ component in the split form of Galilean transformations, in general, varies according to the velocity value establishing the splitting.

The following *transformation rule* shows well how the splitting depends on the velocity values.

Let $u', u \in V(1)$. Recall the notation

$$\xi_{u'u} := \xi_{u'} \cdot \xi_u{}^{-1} = \begin{pmatrix} 1 & 0 \\ -v_{u'u} & 1_{\mathbb{S}} \end{pmatrix}.$$

Then

$$\xi_{u'u} \cdot \begin{pmatrix} \pm 1 & 0 \\ v & \pm R \end{pmatrix} \cdot \xi_{u'u}^{-1} = \begin{pmatrix} \pm 1 & 0 \\ (v - v_{u'u}) \pm R \cdot v_{u'u} & \pm R \end{pmatrix}. \quad \blacksquare$$

11.4.3. The splittings send the proper Galilean group into

$$\left\{ \begin{pmatrix} 1 & 0 \\ v & R \end{pmatrix} \middle| \, v \in \frac{\mathbb{S}}{\mathbb{T}}, \quad R \in \mathcal{SO}(h) \right\}$$

which is evidently a connected set. Since the splittings are Lie group isomorphisms, $\mathcal{G}^{+\to}$ is connected as well.

11.4.4. If L is a special Galilean transformation and v_L is the corresponding element of $\frac{\mathbb{S}}{\mathbb{T}}$, then L has the split form

$$\begin{pmatrix} 1 & 0 \\ v_L & 1_{\mathbb{S}} \end{pmatrix}$$

for all $u \in V(1)$: the splitting is independent of the velocity value. In other words, every $u \in V(1)$ makes the same bijection between the special Galilean group $\mathcal{V}$ and the group

$$\left\{ \begin{pmatrix} 1 & 0 \\ v & 1_{\mathbb{S}} \end{pmatrix} \middle| \, v \in \frac{\mathbb{S}}{\mathbb{T}} \right\}.$$

Observe that for all $u', u \in V(1)$, the vector transformation law is the split form of a special Galilean transformation:

$$\xi_{u'u} = \xi_{u'} \cdot \xi_u{}^{-1} = \xi_u \cdot L(u, u') \cdot \xi_u{}^{-1}.$$

11.4.5. The Lie algebra of the Galilean group, too, consists of elements of $\mathbf{M}\otimes\mathbf{M}^*$, thus they are split by velocity values in the same way as the Galilean transformations; evidently, their split forms will be different.

If $\boldsymbol{H}$ is in the Lie algebra of the Galilean group and $\boldsymbol{u}\in V(1)$, then

$$\xi_u\cdot\boldsymbol{H}\cdot\xi_u{}^{-1}=\begin{pmatrix}0 & 0\\ \boldsymbol{H}\cdot\boldsymbol{u} & \boldsymbol{H}\cdot\boldsymbol{i}\end{pmatrix}.$$

The splitting according to $\boldsymbol{u}$ establishes a Lie algebra isomorphism between the Lie algebra of the Galilean group and the Lie algebra of the split Galilean group. The isomorphisms corresponding to different $\boldsymbol{u}'$ and $\boldsymbol{u}$ are different:

$$\xi_{u'u}\cdot\begin{pmatrix}0 & 0\\ v & H\end{pmatrix}\cdot\xi_{u'u}{}^{-1}=\begin{pmatrix}0 & 0\\ v+H\cdot v_{u'u} & H\end{pmatrix}.$$

11.5. Exercises

1. Prove that for all $\mathbf{0}\neq\boldsymbol{t}\in\mathbb{T}$, $\{\boldsymbol{x}\in\mathbf{M}\mid\boldsymbol{\tau}\cdot\boldsymbol{x}=\boldsymbol{t}\}$ is an orbit of the special Galilean group as well as the proper Galilean group. What are the orbits for $\boldsymbol{t}=\mathbf{0}$?

2. Beside the trivial linear subspaces $\{\mathbf{0}\}$ and $\mathbf{M}$, $\mathbf{S}$ is invariant under all Galilean transformations and every linear subspace of $\mathbf{S}$ is invariant under all special Galilean transformations.

3. The transpose of a Galilean transformation is a linear bijection $\mathbf{M}^*\to\mathbf{M}^*$. Demonstrate that the *transposed Galilean group* $\{\boldsymbol{L}^*\mid\boldsymbol{L}\in\mathcal{G}\}$ leaves $\mathbb{T}^*\cdot\boldsymbol{\tau}$ invariant; more closely, if $\boldsymbol{L}\in\mathcal{G}$ and $\boldsymbol{e}\in\mathbb{T}^*\cdot\boldsymbol{\tau}$, then $\boldsymbol{L}^*\cdot\boldsymbol{e}=(\operatorname{ar}\boldsymbol{L})\boldsymbol{e}$.

Furthermore, if $\boldsymbol{k}\in\mathbf{M}^*$, and $\boldsymbol{L}^*\cdot\boldsymbol{k}$ is parallel to $\boldsymbol{k}$ for all Galilean transformations $\boldsymbol{L}$, then $\boldsymbol{k}$ is in $\mathbb{T}^*\cdot\boldsymbol{\tau}$.

4. The subgroup generated by $\{\boldsymbol{T_u}\mid\boldsymbol{u}\in V(1)\}$ is the special Galilean group.

5. Prove that

$$\xi_u\cdot\boldsymbol{T_u}\cdot\xi_u{}^{-1}=\begin{pmatrix}-1 & 0\\ 0 & 1_{\mathbf{S}}\end{pmatrix},\qquad \xi_u\cdot\boldsymbol{P_u}\cdot\xi_u{}^{-1}=\begin{pmatrix}1 & 0\\ 0 & -1_{\mathbf{S}}\end{pmatrix}.$$

Find $\xi_{u'}\cdot\boldsymbol{T_u}\cdot\xi_{u'}{}^{-1}$ and $\xi_{u'}\cdot\boldsymbol{P_u}\cdot\xi_{u'}{}^{-1}$.

6. The $\boldsymbol{u}$-splitting of the Galilean group sends the special Galilean group into the group

$$\left\{\begin{pmatrix}1 & 0\\ v & 1_{\mathbf{S}}\end{pmatrix}\,\middle|\, v\in\frac{\mathbf{S}}{\mathbb{T}}\right\}$$

whose Lie algebra is

$$\left\{\begin{pmatrix}0 & 0\\ v & 0\end{pmatrix}\,\middle|\, v\in\frac{\mathbf{S}}{\mathbb{T}}\right\}.$$

The $\boldsymbol{u}$-splitting of special Galilean transformations does not depend on $\boldsymbol{u}$.

7. The Lie algebra of $\mathcal{O}(\boldsymbol{h})_{\boldsymbol{u}}$ equals

$$\{\boldsymbol{H} \in \mathbf{La}(\mathcal{G}) \mid \boldsymbol{H} \cdot \boldsymbol{u} = \mathbf{0}\}.$$

8. The $\boldsymbol{u}$-splitting sends the subgroup $\mathcal{O}(\boldsymbol{h})_{\boldsymbol{u}}$ into the group

$$\left\{ \begin{pmatrix} 1 & \mathbf{0} \\ 0 & \boldsymbol{R} \end{pmatrix} \,\middle|\, \boldsymbol{R} \in \mathcal{O}(\boldsymbol{h}) \right\}$$

having the Lie algebra

$$\left\{ \begin{pmatrix} 0 & \mathbf{0} \\ 0 & \boldsymbol{A} \end{pmatrix} \,\middle|\, \boldsymbol{A} \in \mathrm{A}(\boldsymbol{h}) \right\}.$$

Find the $\boldsymbol{u}'$-splitting of $\mathcal{O}(\boldsymbol{h})_{\boldsymbol{u}}$ for $\boldsymbol{u}' \neq \boldsymbol{u}$.

9. Recall the notation introduced in 11.3.7 and prove that

— $\boldsymbol{L}(\boldsymbol{u}', \boldsymbol{u})^{-1} = \boldsymbol{L}(\boldsymbol{u}, \boldsymbol{u}')$,

— $\boldsymbol{L}(\boldsymbol{u}''\boldsymbol{u}') \cdot \boldsymbol{L}(\boldsymbol{u}', \boldsymbol{u}) = \boldsymbol{L}(\boldsymbol{u}'', \boldsymbol{u})$.

10. For all $\boldsymbol{u} \in V(1)$ and for all Galilean transformations $\boldsymbol{L}$ we have that

$$\boldsymbol{R}(\boldsymbol{L}, \boldsymbol{u}) := (\operatorname{ar} \boldsymbol{L})\boldsymbol{L}(\boldsymbol{u}, (\operatorname{ar} \boldsymbol{L}) \cdot \boldsymbol{u}) \cdot \boldsymbol{L} = (\operatorname{ar} \boldsymbol{L})\boldsymbol{L} + (\boldsymbol{u} - (\operatorname{ar} \boldsymbol{L})\boldsymbol{L} \cdot \boldsymbol{u}) \otimes \boldsymbol{\tau}$$

is in $\mathcal{O}(\boldsymbol{h})_{\boldsymbol{u}}$ and $\boldsymbol{R}\,(\boldsymbol{L}, \boldsymbol{u})|_{\mathbf{S}_{\boldsymbol{u}}} = \boldsymbol{R}_{\boldsymbol{L}}$. In other words, given an arbitrary $\boldsymbol{u} \in V(1)$, every Galilean transformation $\boldsymbol{L}$ is the product of a special Galilean transformation and a $\boldsymbol{u}$-spacelike orthogonal transformation, multiplied by the arrow of $\boldsymbol{L}$:

$$\boldsymbol{L} = (\operatorname{ar} \boldsymbol{L})\boldsymbol{L}((\operatorname{ar} \boldsymbol{L}(\boldsymbol{u}), \boldsymbol{u}) \cdot \boldsymbol{R}(\boldsymbol{L}, \boldsymbol{u}).$$

11.6. The Noether group

11.6.1. Now we shall deal with affine maps $L: \mathrm{M} \to \mathrm{M}$; as usual, the linear map under L is denoted by $\boldsymbol{L}$.

Definition.

$$\mathcal{N} := \{L: \mathrm{M} \to \mathrm{M} \mid L \quad \text{is affine,} \quad \boldsymbol{L} \in \mathcal{G}\}$$

is called the *Noether group*; its elements are the *Noether transformations.*

If L is a Noether transformation, then

$$\operatorname{ar} L := \operatorname{ar} \boldsymbol{L}, \qquad \operatorname{sign} L := \operatorname{sign} \boldsymbol{L}.$$

$\mathcal{N}^{+\to}$, $\mathcal{N}^{+\leftarrow}$, $\mathcal{N}^{-\to}$ and $\mathcal{N}^{-\leftarrow}$ are the subsets of $\mathcal{N}$ consisting of elements whose underlying linear maps belong to $\mathcal{G}^{+\to}$, $\mathcal{G}^{+\leftarrow}$, $\mathcal{G}^{-\to}$ and $\mathcal{G}^{-\leftarrow}$, respectively.

$\mathcal{N}^{+\to}$ is called the *proper Noether group.* ∎

The Noether group is the affine group over the Galilean group; according to VII.3.2(*ii*), we can state the following.

Proposition. The Noether group is a ten-dimensional Lie group; its Lie algebra consists of the affine maps $H: \mathrm{M} \to \mathbf{M}$ whose underlying linear map is in the Lie algebra of the Galilean group:

$$\mathbf{La}(\mathcal{N}) = \{H \in \mathrm{Aff}(\mathrm{M}, \mathbf{M}) \mid \boldsymbol{\tau} \cdot \boldsymbol{H} = \mathbf{0}, \quad \boldsymbol{H} \cdot \boldsymbol{i} \in \mathrm{A}(\boldsymbol{h})\}. \quad \blacksquare$$

The proper Noether group is a connected subgroup of the Noether group. As regards $\mathcal{N}^{+\leftarrow}$, etc. we can repeat what was said about the components of the Galilean group.

$\mathcal{N}^{\to} := \mathcal{N}^{+\to} \cup \mathcal{N}^{-\to}$ is called the *orthochronous Noether group.*

11.6.2. We can say that the elements of $\mathcal{N}^{-\leftarrow}$ invert spacetime in some sense but there is no element that we could call the spacetime inversion.

For every $o \in \mathrm{M}$ we can give the o*-centered* spacetime inversion in such a way that first M is vectorized by O_o, then the vectors are inverted ($-\mathbf{1}_{\mathbf{M}}$ is applied), finally the vectorization is removed:

$$I_o := O_o{}^{-1} \circ (-\mathbf{1}_{\mathbf{M}}) \circ O_o,$$

i.e.

$$I_o(x) := o - (x - o) \qquad (x \in \mathrm{M}).$$

Similarly, we can say that in some sense the elements of $\mathcal{N}^{-\to}$ contain spacelike inversion and do not contain timelike inversion; the elements of $\mathcal{N}^{+\leftarrow}$

contain timelike inversion and do not contain spacelike inversion. However, the space inversion and the time inversion do not exist.

For every $o \in \mathrm{M}$ and $\boldsymbol{u} \in V(1)$ we can give the *o-centered* $\boldsymbol{u}$*-timelike inversion* and the *o-centered* $\boldsymbol{u}$*-spacelike inversion* as follows:

$$T_{\boldsymbol{u},o}(x) := o + \boldsymbol{T}_{\boldsymbol{u}} \cdot (x - o), \qquad P_{\boldsymbol{u},o}(x) := o + \boldsymbol{P}_{\boldsymbol{u}} \cdot (x - o) \qquad (x \in \mathrm{M}).$$

11.6.3. Let L be a Noether transformation. If x and y are simultaneous then $L(x)$ and $L(y)$ are simultaneous as well:

$$\tau(L(x)) - \tau(L(y)) = \boldsymbol{\tau} \cdot \boldsymbol{L} \cdot (x - y) = (\mathrm{ar}\,\boldsymbol{L})\boldsymbol{\tau} \cdot (x - y) = \mathbf{0}.$$

Recall that T is identified with the set of hyperplanes of M directed by $\mathbf{S}$. Thus for a Noether transformation L we can define the mapping

$$\mathrm{ti}\,L \colon \mathrm{T} \to \mathrm{T}, \qquad t \mapsto L[t].$$

Observe that

$$(\mathrm{ti}\,L) \circ \tau = \tau \circ L$$

or, in other words,

$$(\mathrm{ti}\,L)(t) = \tau(L(x)) \qquad (x \in t),$$

from which we get immediately that

$$(\mathrm{ti}\,L)(t) - (\mathrm{ti}\,L)(s) = (\mathrm{ar}\,L)(t - s) \qquad (t, s \in \mathrm{T}).$$

Thus $\mathrm{ti}\,L$ is an affine map over $(\mathrm{ar}\,L)\mathbf{1}_{\mathbb{T}}$.

According to Exercises VI.2.5.6–7, if $\mathrm{ar}\,L = 1$, then $\mathrm{ti}\,L$ is a translation, i.e. there is a unique $\boldsymbol{t} \in \mathbb{T}$ such that $(\mathrm{ti}\,L)(t) = t + \boldsymbol{t}$; if $\mathrm{ar}\,L = -1$, then $\mathrm{ti}\,L$ is an inversion, i.e. there is a unique $t_o \in \mathrm{T}$ such that $(\mathrm{ti}\,L)(t) = t_o - (t - t_o)$.

11.6.4. Noether transformations are mappings of spacetime. They play a fundamental role because the proper Noether transformations can be considered to be the strict automorphisms of the spacetime model, according to Definition 1.6.1.

11.6.5. Let us denote the translation group of $\mathbb{T}$ by $\mathcal{T}n(\mathbb{T})$ and consider it as an affine transformation group of T: $\boldsymbol{t} \in \mathbb{T}$ acts as $\mathrm{T} \to \mathrm{T}$, $t \mapsto t + \boldsymbol{t}$. In this respect $\mathbf{0} \in \mathbb{T}$ equals the identity map of T. It is quite obvious now that

$$\mathcal{N}^{\rightarrow} \to \mathcal{T}n(\mathbb{T}), \qquad L \mapsto \mathrm{ti}\,L$$

is a surjective Lie group homomorphism. Its kernel,

$$\mathcal{N}_i := \{L \in \mathcal{N}^{\rightarrow} \mid \mathrm{ti}\, L = 1_{\mathbb{T}}\} = \{L \in \mathcal{N}^{\rightarrow} \mid \tau \circ L = \tau\},$$

is called the *instantaneous Noether group.* It is a nine-dimensional Lie group having the Lie algebra

$$\mathbf{La}(\mathcal{N}_i) = \{H \in \mathrm{Aff}(\mathrm{M}, \mathbf{M}) \mid \boldsymbol{\tau} \circ H = \mathbf{0}, \quad \boldsymbol{H} \cdot \boldsymbol{i} \in \mathrm{A}(\boldsymbol{h})\}.$$

Instantaneous Noether transformations leave every instant invariant.

$\mathcal{T}n(\mathbb{T})$ is not a subgroup of $\mathcal{N}$. For every $\boldsymbol{u} \in V(1)$,

$$\mathcal{T}n(\mathbb{T})_{\boldsymbol{u}} := \{1_{\mathrm{M}} + \boldsymbol{u}\boldsymbol{t} \mid \boldsymbol{t} \in \mathbb{T}\}$$

is a subgroup of the orthochronous Noether group, called the group of $\boldsymbol{u}$-*timelike translations.* The restriction of the homomorphism $L \mapsto \mathrm{ti}\, L$ onto $\mathcal{T}n(\mathbb{T})_{\boldsymbol{u}}$ is a bijection between $\mathcal{T}n(\mathbb{T})_{\boldsymbol{u}}$ and $\mathcal{T}n(\mathbb{T})$.

In other words, given $\boldsymbol{u} \in V(1)$, we can assign to every $\boldsymbol{t} \in \mathbb{T}$ the Noether transformation

$$x \mapsto x + \boldsymbol{u}\boldsymbol{t}$$

called the $\boldsymbol{u}$-timelike translation by $\boldsymbol{t}$.

11.6.6. The Galilean group is not a subgroup of the Noether group. The mapping $\mathcal{N} \to \mathcal{G}$, $L \mapsto \boldsymbol{L}$ is a surjective Lie group homomorphism whose kernel is $\mathcal{T}n(\mathbf{M})$, the translation group of M,

$$\mathcal{T}n(\mathbf{M}) = \{T_{\boldsymbol{a}} \mid \boldsymbol{a} \in \mathbf{M}\} = \{L \in \mathcal{N} \mid \boldsymbol{L} = \mathbf{1}_{\mathbf{M}}\}.$$

As we know, its Lie algebra is $\mathbf{M}$ regarded as the set of constant maps from M into $\mathbf{M}$ (VII.3.3).

For every $o \in \mathrm{M}$,

$$\mathcal{G}_o := \{L \in \mathcal{N} \mid L(o) = o\},$$

called the group of o*-centered Galilean transformations,* is a subgroup of the Noether group and even of the instantaneous Noether group; the restriction of the homomorphism $L \mapsto \boldsymbol{L}$ onto $\mathcal{G}_o$ is a bijection between $\mathcal{G}_o$ and $\mathcal{G}$.

In other words, given $o \in \mathrm{M}$, we can assign to every Galilean transformation $\boldsymbol{L}$ the Noether transformation

$$x \mapsto o + \boldsymbol{L} \cdot (x - o),$$

called the o*-centered Galilean transformation by* $\boldsymbol{L}$.

The subgroup of o*-centered special Galilean transformations*

$$\mathcal{V}_o := \{L \in \mathcal{N}_o \mid \boldsymbol{L} \in \mathcal{V}\}$$

has a special importance.

11.6.7. The three-dimensional orthogonal group is not a subgroup of the Noether group. The mapping $\mathcal{N}^{\rightarrow} \to \mathcal{O}(\boldsymbol{h}),\ L \mapsto \boldsymbol{R}_L$ (where $\boldsymbol{L}\cdot\boldsymbol{i} = \boldsymbol{i}\cdot\boldsymbol{R}_L$) is a surjective Lie group homomorphism having the kernel

$$\mathcal{H} := \{L \in \mathcal{N}^{\rightarrow} \mid \boldsymbol{L}\cdot\boldsymbol{i} = \boldsymbol{i}\} = \{L \in \mathcal{N}^{\rightarrow} \mid \boldsymbol{L} \in \mathcal{V}\}$$

is called the *special Noether group.* Observe that $\mathcal{H}$ is a subgroup of $\mathcal{N}^{+\rightarrow}$.

The special Noether group is a seven-dimensional Lie group having the Lie algebra

$$\{H \in \mathbf{La}(\mathcal{N}) \mid \boldsymbol{H} \in \mathbf{La}(\mathcal{V})\} = \{H \in \mathrm{Aff}(\mathrm{M},\mathbf{M}) \mid \boldsymbol{\tau}\cdot\boldsymbol{H} = 0, \quad \boldsymbol{H}\cdot\boldsymbol{i} = \mathbf{0}\}.$$

For every $\boldsymbol{u} \in V(1)$ and $o \in \mathrm{M}$,

$$\mathcal{O}(\boldsymbol{h})_{\boldsymbol{u},o} := \{L \in \mathcal{N}^{\rightarrow} \mid L(o) = o, \quad \boldsymbol{L}\cdot\boldsymbol{u} = \boldsymbol{u}\},$$

called the group of *o-centered $\boldsymbol{u}$-spacelike orthogonal transformations,* is a subgroup of $\mathcal{N}^{\rightarrow}$ and even of the instantaneous Noether group $\mathcal{N}_i$. The restriction of the homomorphism $\mathcal{N}^{\rightarrow} \to \mathcal{O}(\boldsymbol{h})$ onto $\mathcal{O}(\boldsymbol{h})_{\boldsymbol{u},o}$ is a bijection between $\mathcal{O}(\boldsymbol{h})_{\boldsymbol{u},o}$ and $\mathcal{O}(\boldsymbol{h})$.

In other words, given $(\boldsymbol{u},o) \in V(1)\times \mathrm{M}$, we can assign to every $\boldsymbol{R} \in \mathcal{O}(\boldsymbol{h})$ the Noether transformation

$$x \mapsto o + \boldsymbol{u\tau}\cdot(x-o) + \boldsymbol{R}\cdot\boldsymbol{\pi_u}\cdot(x-o),$$

called the *o-centered $\boldsymbol{u}$-spacelike orthogonal transformation by $\boldsymbol{R}$.*

11.6.8. The Neumann group

$$\mathcal{C} := \{L \in \mathcal{N}_i \mid \boldsymbol{L}\cdot\boldsymbol{i} = \boldsymbol{i}\} = \mathcal{H}\cap\mathcal{N}_i$$

is an important subgroup of the special Noether group. It is a six-dimensional Lie group having the Lie algebra

$$\{H \in \mathbf{La}(\mathcal{N}_i) \mid \boldsymbol{H} \in \mathbf{La}(\mathcal{V})\} = \{H \in \mathrm{Aff}(\mathrm{M},\mathbf{M}) \mid \boldsymbol{\tau}\circ H = 0, \quad \boldsymbol{H}\cdot\boldsymbol{i} = \mathbf{0}\}.$$

Proposition. The Neumann group is a commutative normal subgroup of the Noether group.

Proof. Let K and L be arbitrary Neumann transformations. Since they are instantaneous Noether transformations, for all world points x we have that $L(x)-x$ and $K(x)-x$ are in $\mathbf{S}$. As a consequence, $L(x)-x = \boldsymbol{K}\cdot(L(x)-x) =$

$KL(x) - K(x)$ and similarly, $K(x) - x = LK(x) - L(x)$ from which we conclude that $KL(x) - LK(x) = \mathbf{0}$, i.e. $KL = LK$, the Neumann group is commutative.

Now we have to show that if L is an arbitrary Neumann transformation and G is an arbitrary Noether transformation then $G^{-1}LG$ is a Neumann transformation, too. The range of $\boldsymbol{G} \cdot \boldsymbol{i}$ is in $\mathbf{S}$, hence $\boldsymbol{L} \cdot \boldsymbol{G} \cdot \boldsymbol{i} = \boldsymbol{G} \cdot \boldsymbol{i}$ and so $\boldsymbol{G}^{-1} \cdot \boldsymbol{L} \cdot \boldsymbol{G} \cdot \boldsymbol{i} = \boldsymbol{i}$ which ends the proof.

11.7 The vectorial Noether group

11.7.1. Recall that for an arbitrary world point o, the vectorization of M with origin o, $O_o : \mathrm{M} \to \mathbf{M}$, $x \mapsto x - o$, is an affine bijection.

With the aid of such a vectorization we can 'vectorize' the Noether group as well: if L is a Noether transformation then $O_o \circ L \circ {O_o}^{-1}$ is an affine transformation of $\mathbf{M}$, represented by the matrix (see VI.2.4(*ii*) and Exercise VI.2.5)

$$\begin{pmatrix} 1 & \mathbf{0} \\ L(o) - o & \boldsymbol{L} \end{pmatrix}.$$

The Lie algebra of the Noether group consists of affine maps $H : \mathrm{M} \to \mathbf{M}$ where $\mathbf{M}$ is considered to be a *vector space* (the *sum* of such maps is a part of the Lie algebra structure). Thus the vectorization $H \circ {O_o}^{-1}$ is an affine map $\mathbf{M} \to \mathbf{M}$ where the range is considered to be a vector space. Then it is represented by the matrix (see VI.2.4(*iii*))

$$\begin{pmatrix} 0 & \mathbf{0} \\ H(o) & \boldsymbol{H} \end{pmatrix}.$$

11.7.2. Definition. The *vectorial Noether group* is

$$\left\{ \begin{pmatrix} 1 & \mathbf{0} \\ \boldsymbol{a} & \boldsymbol{L} \end{pmatrix} \,\middle|\, \boldsymbol{a} \in \mathbf{M}, \quad \boldsymbol{L} \in \mathcal{G} \right\}. \quad \blacksquare$$

The vectorial Noether group is a ten-dimensional Lie group, its Lie algebra is the vectorization of the Lie algebra of the Noether group:

$$\left\{ \begin{pmatrix} 0 & \mathbf{0} \\ \boldsymbol{a} & \boldsymbol{H} \end{pmatrix} \,\middle|\, \boldsymbol{a} \in \mathbf{M}, \quad \boldsymbol{H} \in \mathbf{La}(\mathcal{G}) \right\}.$$

An advantage of this matrix representation is that the commutator of two Lie algebra elements can be computed by the difference of their products in different orders.

11.7.3. A vectorization of the Noether group is a Lie group isomorphism between the Noether group and the vectorial Noether group. The following transformation rule shows how the vectorizations depend on the world points serving as origins of the vectorization. Let o and o' be two world points; then

$$T_{o-o'} := O_{o'} \circ O_o{}^{-1} = \begin{pmatrix} 1 & \mathbf{0} \\ o-o' & \mathbf{1}_{\mathrm{M}} \end{pmatrix}$$

and

$$T_{o-o'} \begin{pmatrix} 1 & \mathbf{0} \\ \boldsymbol{a} & \boldsymbol{L} \end{pmatrix} T_{o-o'}{}^{-1} = \begin{pmatrix} 1 & \mathbf{0} \\ \boldsymbol{a} + (\boldsymbol{L} - \mathbf{1}_{\mathrm{M}})(o'-o) & \boldsymbol{L} \end{pmatrix} \qquad (\boldsymbol{a} \in \mathbf{M},\ \boldsymbol{L} \in \mathcal{G}).$$

As concerns the corresponding Lie algebra isomorphisms, we have

$$\begin{pmatrix} 0 & \mathbf{0} \\ \boldsymbol{a} & \boldsymbol{H} \end{pmatrix} T_{o-o'}{}^{-1} = \begin{pmatrix} 0 & \mathbf{0} \\ \boldsymbol{a} + \boldsymbol{H}(o'-o) & \boldsymbol{H} \end{pmatrix} \qquad (\boldsymbol{a} \in \mathbf{M},\ \boldsymbol{H} \in \mathbf{La}(\mathcal{G})).$$

11.8. The split Noether group

11.8.1. With the aid of the splitting corresponding to $\boldsymbol{u} \in V(1)$, we send the transformations of $\mathbf{M}$ into the transformations of $\mathbb{T} \times \mathbf{S}$. Composing a vectorization and a splitting, we convert Noether transformations into affine transformations of $\mathbb{T} \times \mathbf{S}$.

For $o \in \mathrm{M}$ and $\boldsymbol{u} \in V(1)$ put

$$\boldsymbol{\xi}_{\boldsymbol{u},o} := \boldsymbol{\xi}_{\boldsymbol{u}} \circ O_o \colon \mathrm{M} \to \mathbb{T} \times \mathbf{S}, \quad x \mapsto \big(\boldsymbol{\tau} \cdot (x-o),\ \boldsymbol{\pi}_{\boldsymbol{u}} \cdot (x-o)\big).$$

Embedding the affine transformations of $\mathbb{T} \times \mathbf{S}$ into the linear transformations of $\mathbb{R} \times (\mathbb{T} \times \mathbf{S})$ (see VI.2.4(*ii*)) and using the customary matrix representation of such linear maps, we get

$$\boldsymbol{\xi}_{\boldsymbol{u},o} \circ L \circ \boldsymbol{\xi}_{\boldsymbol{u},o}{}^{-1} = \begin{pmatrix} 1 & 0 & \mathbf{0} \\ \boldsymbol{\tau} \cdot \big(L(o)-o\big) & \operatorname{ar} L & \mathbf{0} \\ \boldsymbol{\pi}_{\boldsymbol{u}} \cdot \big(L(o)-o\big) & \boldsymbol{L} \cdot \boldsymbol{u} - (\operatorname{ar} L)\boldsymbol{u} & \boldsymbol{R}_L \end{pmatrix}.$$

The Lie algebra elements of the Noether group are converted into affine maps $\mathbb{T} \times \mathbf{S} \to \mathbb{T} \times \mathbf{S}$ where the range is regarded as a vector space. Then we can represent such maps in a matrix form as well:

$$\boldsymbol{\xi}_{\boldsymbol{u}} \circ H \circ \boldsymbol{\xi}_{\boldsymbol{u},o}{}^{-1} = \begin{pmatrix} 0 & 0 & \mathbf{0} \\ \boldsymbol{\tau} \cdot H(o) & 0 & \mathbf{0} \\ \boldsymbol{\pi}_{\boldsymbol{u}} \cdot H(o) & \boldsymbol{H} \cdot \boldsymbol{u} & \boldsymbol{H} \cdot \boldsymbol{i} \end{pmatrix}.$$

11.8.2. Definition. The *split Noether group* is

$$\left\{ \begin{pmatrix} 1 & 0 & \mathbf{0} \\ t & \pm 1 & \mathbf{0} \\ q & v & \pm R \end{pmatrix} \;\middle|\; t \in \mathbb{T},\ q \in \mathbf{S},\ v \in \frac{\mathbf{S}}{\mathbb{T}},\ R \in \mathcal{SO}(h) \right\}. \quad \blacksquare$$

The split Noether group is a ten-dimensional Lie group having the Lie algebra

$$\left\{ \begin{pmatrix} 0 & 0 & \mathbf{0} \\ t & 0 & \mathbf{0} \\ q & v & A \end{pmatrix} \;\middle|\; t \in \mathbb{T},\quad q \in \mathbf{S},\quad v \in \frac{\mathbf{S}}{\mathbb{T}},\quad A \in \mathrm{A}(h) \right\}.$$

Keep in mind that the group multiplication of split Noether transformations coincides with the usual matrix multiplication and the commutator of Lie algebra elements is the difference of their two products.

11.8.3. Every $u \in V(1)$ and $o \in \mathrm{M}$ establishes a Lie group isomorphism between the Noether group and the split Noether group. Evidently, for different elements of $V(1) \times \mathrm{M}$, the isomorphisms are different. The transformation rule that shows how the isomorphism depends on (u, o) can be obtained by a combination of the transformation rules 11.7.3 and 11.4.2.

Though the Noether group and the split Noether group are isomorphic (they have the same Lie group structure), they are not 'identical': there is no 'canonical' isomorphism between them that we could use to identify them.

The split Noether group is the Noether group of the split nonrelativistic spacetime model $(\mathbb{T} \times \mathbf{S}, \mathbb{T}, \mathbb{L}, \boldsymbol{\tau}, h)$. The spacetime model $(\mathrm{M}, \mathbb{T}, \mathbb{L}, \boldsymbol{\tau}, h)$ and the corresponding split spacetime model are isomorphic, but they cannot be identified, as we pointed out in 1.5.3.

11.8.4. It is a routine to check that the isomorphism established by an arbitrary $(u, o) \in V(1) \times \mathrm{M}$ sends the subgroups of the Noether group listed below on the left-hand side into the subgroups of the split Noether group listed below on the right-hand side:

$\mathcal{T}n(\mathbf{S})$ $$\left\{ \begin{pmatrix} 1 & 0 & \mathbf{0} \\ 0 & 1 & \mathbf{0} \\ q & 0 & 1_{\mathbf{S}} \end{pmatrix} \;\middle|\; q \in \mathbf{S} \right\},$$

$\mathcal{T}n(\mathbf{M})$ $$\left\{ \begin{pmatrix} 1 & 0 & \mathbf{0} \\ t & 1 & \mathbf{0} \\ q & 0 & 1_{\mathbf{S}} \end{pmatrix} \;\middle|\; t \in \mathbb{T},\ q \in \mathbf{S} \right\},$$

$\mathcal{C}$ (Neumann group) $$\left\{ \begin{pmatrix} 1 & 0 & \mathbf{0} \\ 0 & 1 & \mathbf{0} \\ q & v & 1_{\mathbf{S}} \end{pmatrix} \;\middle|\; q \in \mathbf{S},\ v \in \frac{\mathbf{S}}{\mathbb{T}} \right\},$$

$\mathcal{H}$ (special Noether group) $\left\{ \begin{pmatrix} 1 & 0 & 0 \\ t & 1 & 0 \\ \boldsymbol{q} & \boldsymbol{v} & \mathbf{1_S} \end{pmatrix} \;\middle|\; t \in \mathbb{T},\, \boldsymbol{q} \in \mathbf{S},\, \boldsymbol{v} \in \frac{\mathbf{S}}{\mathbb{T}} \right\}$,

$\mathcal{N}_i$ (instantaneous Noether group) $\left\{ \begin{pmatrix} 1 & 0 & 0 \\ 0 & 1 & 0 \\ \boldsymbol{q} & \boldsymbol{v} & \boldsymbol{R} \end{pmatrix} \;\middle|\; \boldsymbol{q} \in \mathbf{S},\, \boldsymbol{v} \in \frac{\mathbf{S}}{\mathbb{T}},\, \boldsymbol{R} \in \mathcal{O}(\boldsymbol{h}) \right\}$.

It is emphasized that the isomorphism established by an arbitrary $(\boldsymbol{u}, o)$ makes a correspondence between the listed subgroups; of course, the correspondences due to different $(\boldsymbol{u}, o)$ and $(\boldsymbol{u}', o')$ are different.

Moreover, the isomorphism established by $(\boldsymbol{u}, o)$ makes correspondences between the following subgroups, too:

$\mathcal{T}n(\mathbb{T})_{\boldsymbol{u}}$ ($\boldsymbol{u}$-timelike translations) $\left\{ \begin{pmatrix} 1 & 0 & 0 \\ t & 1 & 0 \\ \mathbf{0} & \mathbf{0} & \mathbf{1_S} \end{pmatrix} \;\middle|\; t \in \mathbb{T} \right\}$,

$\mathcal{O}(\boldsymbol{h})_{\boldsymbol{u},o}$ (o-centered $\boldsymbol{u}$-spacelike orthogonal transformations) $\left\{ \begin{pmatrix} 1 & 0 & \mathbf{0} \\ 0 & 1 & \mathbf{0} \\ \mathbf{0} & \mathbf{0} & \boldsymbol{R} \end{pmatrix} \;\middle|\; \boldsymbol{R} \in \mathcal{O}(\boldsymbol{h}) \right\}$,

$\mathcal{G}_o$ (o-centered Galilean transformations) $\left\{ \begin{pmatrix} 1 & 0 & \mathbf{0} \\ 0 & \pm 1 & \mathbf{0} \\ \mathbf{0} & \boldsymbol{v} & \pm\boldsymbol{R} \end{pmatrix} \;\middle|\; \boldsymbol{v} \in \frac{\mathbf{S}}{\mathbb{T}},\, \boldsymbol{R} \in \mathcal{SO}(\boldsymbol{h}) \right\}$,

$\mathcal{V}_o$ (o-centered special Galilean transformations) $\left\{ \begin{pmatrix} 1 & 0 & \mathbf{0} \\ 0 & 1 & \mathbf{0} \\ \mathbf{0} & \boldsymbol{v} & \mathbf{1_S} \end{pmatrix} \;\middle|\; \boldsymbol{v} \in \frac{\mathbf{S}}{\mathbb{T}} \right\}$,

and now it is emphasized that the isomorphism established by $(\boldsymbol{u}', o')$, in general, does not make a correspondence between the listed subgroups.

11.8.5. Corresponding to the structure of the split Noether group, the following four subgroups are called its *fundamental subgroups:*

$$\left\{ \begin{pmatrix} 1 & 0 & \mathbf{0} \\ t & 1 & \mathbf{0} \\ \mathbf{0} & \mathbf{0} & \mathbf{1_S} \end{pmatrix} \;\middle|\; t \in \mathbb{T} \right\}, \qquad \left\{ \begin{pmatrix} 1 & 0 & \mathbf{0} \\ 0 & 1 & \mathbf{0} \\ \boldsymbol{q} & \mathbf{0} & \mathbf{1_S} \end{pmatrix} \;\middle|\; \boldsymbol{q} \in \mathbf{S} \right\},$$

$$\left\{ \begin{pmatrix} 1 & 0 & \mathbf{0} \\ 0 & 1 & \mathbf{0} \\ \mathbf{0} & \boldsymbol{v} & \mathbf{1_S} \end{pmatrix} \;\middle|\; \boldsymbol{v} \in \frac{\mathbf{S}}{\mathbb{T}} \right\}, \qquad \left\{ \begin{pmatrix} 1 & 0 & \mathbf{0} \\ 0 & 1 & \mathbf{0} \\ \mathbf{0} & \mathbf{0} & \boldsymbol{R} \end{pmatrix} \;\middle|\; \boldsymbol{R} \in \mathcal{O}(\boldsymbol{h}) \right\}.$$

The isomorphism established by $(\boldsymbol{u}, o) \in V(1) \times \mathrm{M}$ assigns these subgroups to the subgroups $\mathcal{T}n(\mathrm{T})_{\boldsymbol{u}}$, $\mathcal{T}n(\mathbf{S})$, $\mathcal{V}_o$ and $\mathcal{O}(\boldsymbol{h})_{\boldsymbol{u},o}$, respectively.

It is worth repeating the actual form of the corresponding Noether transformations:

$$\mathcal{T}n(\mathbb{T})_{\boldsymbol{u}} : x \mapsto x + \boldsymbol{u}\boldsymbol{t} \qquad (\boldsymbol{t} \in \mathbb{T}),$$

$$\mathcal{T}n(\mathbf{S}) : x \mapsto x + \boldsymbol{q} \qquad (\boldsymbol{q} \in \mathbf{S}),$$

$$\mathcal{V}_o : x \mapsto x + \boldsymbol{v}\boldsymbol{\tau} \cdot (x - o) \qquad \left(\boldsymbol{v} \in \frac{\mathbf{S}}{\mathbb{T}}\right),$$

$$\mathcal{O}(\boldsymbol{h})_{\boldsymbol{u},o} : x \mapsto o + \boldsymbol{u}\boldsymbol{\tau} \cdot (x - o) + \boldsymbol{R} \cdot \boldsymbol{\pi}_{\boldsymbol{u}}(x - o) \qquad (\boldsymbol{R} \in \mathcal{O}(\boldsymbol{h})).$$

11.8.6. Taking a linear bijection $\mathbb{T} \to \mathbb{R}$ and an orthogonal linear bijection $\mathbf{S} \to \mathbb{R}^3$, we can transfer the split Noether group into the following affine transformation group of $\mathbb{R} \times \mathbb{R}^3$,

$$\left\{ \begin{pmatrix} 1 & 0 & \mathbf{0} \\ \eta & \pm 1 & \mathbf{0} \\ \boldsymbol{\xi} & \boldsymbol{\nu} & \pm\boldsymbol{\rho} \end{pmatrix} \middle| \; \eta \in \mathbb{R},\ \boldsymbol{\xi} \in \mathbb{R}^3,\ \boldsymbol{\nu} \in \mathbb{R}^3,\ \boldsymbol{\rho} \in \mathcal{SO}(3) \right\},$$

which we call the *arithmetic Noether group.* This is the Noether group of the arithmetic spacetime model ($\mathcal{O}(3)$ denotes the orthogonal group of $\mathbb{R}^3$ endowed with the usual inner product).

In conventional treatments one considers the arithmetic spacetime model (without an explicit definition) and the arithmetic Noether group which is called there Galilean group. The special form of such transformations yields that one speaks about *the* time inversion ($\eta = 0,\ -1,\ \boldsymbol{\xi} = \mathbf{0},\ \boldsymbol{\nu} = \mathbf{0},\ \boldsymbol{\rho} = \mathbf{0}$), *the* time translations $\eta,\ 1,\ \boldsymbol{\xi} = \mathbf{0},\ \boldsymbol{\nu} = \mathbf{0},\ \boldsymbol{\rho} = \mathbf{0}$), *the* space rotations ($\eta = 0,\ 1,\ \boldsymbol{\xi} = \mathbf{0}$, $\boldsymbol{\nu} = \mathbf{0},\ +\boldsymbol{\rho}$) etc., whereas we know well that such Noether transformations do not exist: there are o-centered $\boldsymbol{u}$-timelike inversions, $\boldsymbol{u}$-timelike translations and o-centered $\boldsymbol{u}$-spacelike rotations etc.

11.9. Exercises

1. Let L be a Noether transformation for which $\boldsymbol{L} = -\mathbf{1}_{\mathrm{M}}$. Then there is a unique $o \in \mathrm{M}$ such that L is the o-centered spacetime inversion.

2. A Noether transformation L is instantaneous, i.e. is in $\mathcal{N}_i$ if and only if all the hyperplanes $t \in \mathrm{T}$ are invariant for L.

3. Prove that for all $o \in \mathrm{M}$,

$$O_o \circ \mathcal{N}_o \circ O_o^{-1} = \left\{ \begin{pmatrix} 1 & \mathbf{0} \\ 0 & \boldsymbol{L} \end{pmatrix} \middle| \; \boldsymbol{L} \in \mathcal{N} \right\}.$$

4. Find $\boldsymbol{\xi}_{u,o} \cdot T_{u,o} \cdot \boldsymbol{\xi}_{u,o}{}^{-1}$ and $\boldsymbol{\xi}_{u,o} \cdot P_{u,o} \cdot \boldsymbol{\xi}_{u,o}{}^{-1}$.

5. Prove that the subgroup generated by $\{T_{u,o} | \boldsymbol{u} \in V(1),\ o \in \mathrm{M}\}$ equals $\{L \in \mathcal{N} \mid \boldsymbol{L} \cdot \boldsymbol{i} = \boldsymbol{i}\}$.

6. For all $\boldsymbol{u} \in V(1),\ o \in \mathrm{M}$ we have

$$(\mathrm{ti}\, T_{u,o})(t) = \tau(o) - (t - \tau(o)) = t - 2(t - \tau(o)) \qquad (t \in \mathrm{T}).$$

7. Prove that the derived Lie algebra of the Noether group, i.e. $[\mathbf{La}(\mathcal{N}), \mathbf{La}(\mathcal{N})]$ equals the Lie algebra of the instantaneous Noether group.

8. Let $L \in \mathcal{T}n(\mathbf{M})$. Then $\boldsymbol{\xi}_{u,o} \cdot L \cdot \boldsymbol{\xi}_{u,o}{}^{-1}$ is the same for all $\boldsymbol{u}$ and o if and only if $L \in \mathcal{T}n(\mathbf{S})$.

9. Take a $\boldsymbol{u} \in V(1)$ and an $o \in \mathrm{M}$. If $\boldsymbol{t} \in \mathbb{T}$, $\boldsymbol{q} \in \mathbf{S}$, $\boldsymbol{v} \in \frac{\mathbf{S}}{\mathbb{T}}$, $\boldsymbol{A} \in \mathrm{A}(\boldsymbol{h})$, then the maps $\mathrm{M} \to \mathbf{M}$

$$H(x) := \begin{cases} (i) & \boldsymbol{ut} \\ (ii) & \boldsymbol{q} \\ (iii) & \boldsymbol{v\tau} \cdot (x - o) \\ (iv) & \boldsymbol{A} \cdot \boldsymbol{\pi_u} \cdot (x - o) \end{cases} \qquad (x \in \mathrm{M})$$

are elements of the Lie algebra of the Noether group. Prove that

$$\mathrm{e}^H(x) = \begin{cases} (i) & x + \boldsymbol{ut} \\ (ii) & x + \boldsymbol{q} \\ (iii) & x + \boldsymbol{v\tau} \cdot (x - o) \\ (iv) & o + \boldsymbol{u\tau} \cdot (x - o) + \mathrm{e}^{\boldsymbol{A}} \cdot \boldsymbol{\pi_u} \cdot (x - o) \end{cases} \qquad (x \in \mathrm{M}).$$

10. Compute the product of two split Noether transformations:

$$\begin{pmatrix} 1 & 0 & \mathbf{0} \\ \boldsymbol{t} & \pm 1 & \mathbf{0} \\ \boldsymbol{q} & \boldsymbol{v} & \pm \boldsymbol{R} \end{pmatrix} \begin{pmatrix} 1 & 0 & \mathbf{0} \\ \boldsymbol{t}' & \pm 1' & \mathbf{0} \\ \boldsymbol{q}' & \boldsymbol{v}' & \pm' \boldsymbol{R}' \end{pmatrix}.$$

11. Let L be a Noether transformation.

If r is a world line function, then $L \circ r \circ (\mathrm{ti}\, L)^{-1}$ is a world line function, too.

If C is a world line, then $L[\mathrm{C}]$ is a world line, too; moreover, if $C = \mathrm{Ran}\, r$, then $L[\mathrm{C}] = \mathrm{Ran}\left(L \circ r \circ (\mathrm{ti}\, L)^{-1}\right)$.

II. SPECIAL RELATIVISTIC SPACETIME MODELS

1. Fundamentals

1.1. Absolute light propagation

1.1.1. In the previous part material points played a fundamental role, light beams were mentioned only here and there. Now we concentrate upon light phenomena.

We experience that a 'pointlike' light beam—let us call it a *light signal*—behaves in some respect similarly to a material point; in particular, it moves relative to observers. There is a noteworthy difference, however: a light signal cannot stand in any space.

This means that the history of a light signal, too, is a curve in spacetime. In the nonrelativistic model such a curve cannot be a world line because a light signal moves relative to all observers. If not a world line, then it must be contained in a simultaneous hyperplane which says that a light signal reflected on a mirror would return to its source exactly when it is emitted, contradicting to our experience.

As a consequence, *absolute simultaneity and the right description of light phenomena exclude each other*, so we have to construct a new spacetime model.

1.1.2. We have the following experience regarding light propagation.

(**L1**) The path of a free light signal is a straight line in the space of an inertial observer.

(**L2**) Every straight line in the space of an inertial observer can be the path of a light signal.

(**L3**) In the space of any observer, a light signal is faster than any material point moving on the same path.

(**L4**) The propagation of a light signal in the space of an observer can be arbitrarily approximated by a motion of a material point.

(**L5**) The round way speed of light signals is the same over all paths.

None of the statements above involves synchronization. The first three ones are evident. The fourth one means that given a start and a goal, for arbitrary (small) time interval t there is a material point starting together with a light signal, arrives at the goal t time later than the light signal.

The last statement is simple and understandable (both Fizeau and Foucault measured round-way light speed), nevertheless it deserves some words. In usual treatments of relativity one accepts the axiom that "the speed of light in vacuum is the same for all inertial observers in every direction". This, however, would make sense only with a synchronization. Nonrelativistic theories work well not mentioning synchronization (we know why), therefore one does not realize that similar axioms are not meaningful without a synchronization; moreover, in usual treatments synchronizations do not appear at all. That is why the light speed becomes mystic there: if both a light signal and you move towards me on the same path then the light signal must be slower to you than to me!

1.1.3. (**L3**) and (**L4**) imply:

Light signals on the same path are equally fast.

Indeed, if a light signal would be faster than another, then a material point approximating conveniently first light signal would be faster than the other.

It follows then:

The history of a light signal is independent of its source.

For an illustration let us consider a lamp on a train and a lamp close to the rail. When they meet (assuming this idealized event) both emit a light signal; the light signals will propagate together both in the space of the Earth and in the space of the train and even in the space of an arbitrary observer.

According to property (**L5**), round-way speed does not depend either on the space point and on the proper time point of the start-goal or on the direction of the path. All these admit us to state:

Light propagation in spacetime is absolute, homogeneous and isotropic.

1.1.4. Knowing that the homogeneous and isotropic round-way speed of light is $c := 2.997\ldots \cdot 10^8 \boldsymbol{m}/\boldsymbol{s}$, we can measure distances by light signals. Let a light signal be emitted in a space point and reflected in another space point; if the proper time interval between starting and returning is $\boldsymbol{t}$, then the distance between the two space points is $\frac{1}{2}c\boldsymbol{t}$.

This means that a distinguished linear isomorphism exists between the measure line of time periods and the measure line of distances: $\mathbb{T} \to \mathbb{L}$, $\boldsymbol{t} \mapsto \frac{1}{2}c\boldsymbol{t}$ which makes it possible to disregard the measure line of distances i.e. to measure distances directly by time periods in such a way that

$$\boldsymbol{m} := 3.336\ldots \cdot 10^{-9}\boldsymbol{s};$$

then the round-way speed of light becomes the real number 1.

This choice makes the formulae of the special relativistic spacetime model much simpler. Of course, even the Planck constant is considered to be the real number 1 and then

$$\boldsymbol{kg} := 8.55\ldots \cdot 10^{50}\frac{1}{\boldsymbol{s}}.$$

1.1.5 According to what has been said in the Introduction, spacetime will be modelled by an affine space.

Then it can be shown – the proof lies outside the purpose of this book – that (**M1**)–(**M3**), (**U**) and (**S4**), and (**L1**)–(**L5**) imply that the absolute proper time progress, absolute light propagation and the Euclidean structure of inertial spaces are described by a Lorentz form.

1.2. The spacetime model

1.2.1. Definition. A *special relativistic spacetime model* is a triplet $(\mathrm{M}, \mathbb{T}, \boldsymbol{g})$, where

— M is spacetime, an oriented four-dimensional real affine space (over the vector space $\mathbf{M}$),

— $\mathbb{T}$ is the measure line of time periods and distances,

— $\boldsymbol{g}\colon \mathbf{M} \times \mathbf{M} \to \mathbb{T} \otimes \mathbb{T}$ is an arrow oriented Lorentz form. ■

Elements of M are called *occurrences* or *world points.* Elements of $\mathbf{M}$ are called *world vectors.*

1.2.2. If $(\mathrm{M}, \mathbb{T}, \boldsymbol{g})$ is a special relativistic spacetime model, then $(\mathbf{M}, \mathbb{T}, \boldsymbol{g})$ is an oriented and arrow-oriented Minkowski vector space. The results and formulae of Section V.4 will be used all over this part. Remember to distinguish between $\boldsymbol{x}^2 := \boldsymbol{x} \cdot \boldsymbol{x}$ and $|\boldsymbol{x}|^2 := |\boldsymbol{x} \cdot \boldsymbol{x}|$; since $\mathbb{T}$ is oriented, the pseudo-length $|\boldsymbol{x}| := \sqrt{|\boldsymbol{x}|^2}$ is meaningful. Moreover, recall

$$\begin{aligned} S &:= \{\boldsymbol{x} \in \mathbf{M} \mid \ \boldsymbol{x}^2 > 0\}, \\ T &:= \{\boldsymbol{x} \in \mathbf{M} \mid \ \boldsymbol{x}^2 < 0\}, \\ L &:= \{\boldsymbol{x} \in \mathbf{M} \mid \ \boldsymbol{x}^2 = 0,\ \boldsymbol{x} \neq 0\}; \end{aligned}$$

the elements of $S_0 := S \cup \{\mathbf{0}\}$, T and L are called *spacelike, timelike* and *lightlike,* respectively.

Furthermore, the arrow orientation indicates the arrow classes $T^{\rightarrow}$ and $L^{\rightarrow}$; for every $\boldsymbol{x} \in T^{\rightarrow}$ and $\boldsymbol{y} \in T^{\rightarrow} \cup L^{\rightarrow}$ we have $\boldsymbol{x} \cdot \boldsymbol{y} < 0$. Then $T^{\leftarrow} := -T^{\rightarrow}$ and $L^{\leftarrow} := -L^{\rightarrow}$ are the other arrow classes and

$$T = T^{\rightarrow} \cup T^{\leftarrow}, \qquad L = L^{\rightarrow} \cup L^{\leftarrow}.$$

$T^{\rightarrow}$ and $L^{\rightarrow}$ are the *future time cone* and the *future light cone*, respectively; their elements are called *future directed.* $T^{\leftarrow}$ and $L^{\leftarrow}$ are the corresponding *past cones* with *past directed* elements.

We often illustrate the world vectors in the plane of the page:

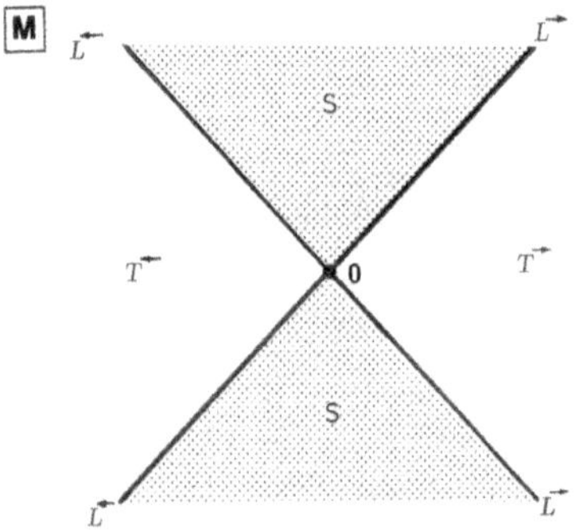

This illustration is based on the following: represent $\mathbb{R} \times \mathbb{R}$ in the plane in the usual way by horizontal and vertical axes, called zeroth and first; draw the sets S, $T^{\rightarrow}$, $L^{\rightarrow}$, etc. corresponding to the Lorentz form

$$((\xi^0, \xi^1), (\eta^0, \eta^1)) \mapsto -\xi^0\eta^0 + \xi^1\eta^1$$

and to the arrow orientation determined by the condition $\xi^0 > 0$; hide the coordinate axes.

We know that T consists of two disjoint open subsets, the two arrow classes which can be well seen in the illustration. On the other hand, S is connected, in spite of the illustration. Keep this slight inaccuracy of the illustration in mind.

1.2.3. Spacetime, too, will be illustrated in the plane of the page. If x is a world point, $x + (T^{\rightarrow} \cup L^{\rightarrow})$ and $x + (T^{\leftarrow} \cup L^{\leftarrow})$ are called the *futurelike* and the *pastlike* part of M, with respect to x.

If $y \in x + (T^{\rightarrow} \cup L^{\rightarrow})$—or, equivalently, $y - x \in (T^{\rightarrow} \cup L^{\rightarrow})$—then we say y is *futurelike with respect to* x (x is *pastlike with respect to* y), or y is *later* than x (x is earlier than y).

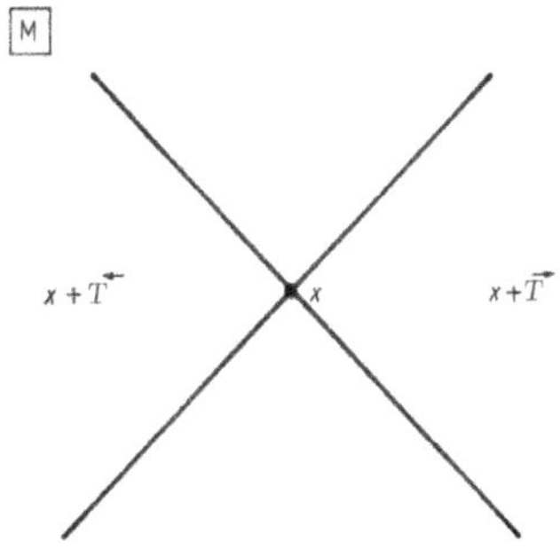

We say that the world points x and y are *spacelike separated, timelike separated, lightlike separated,* if $y-x$ is in S, T, L, respectively.

1.3. Structure of world vectors and covectors

1.3.1. The Euclidean structure of our space is deeply fixed in our mind, therefore we must be careful when dealing with $\mathbf{M}$ which has a nonEuclidean structure; especially when illustrating it in the Euclidean plane of the page. For instance, keep in mind that the centre line of the cone $L^{\rightarrow}$ makes no sense (the centre line would be the set of points that have the same distance from every generatrix of the cone but distance is not meaningful here). The following considerations help us to take in the situation.

Put

$$V(1) := \left\{ \boldsymbol{u} \in \frac{\mathbf{M}}{\mathbb{T}} \;\middle|\; \boldsymbol{u}^2 = -1,\ \boldsymbol{u} \otimes \mathbb{T}^+ \subset T^{\rightarrow} \right\}.$$

We shall see in 2.3.4 that elements of $V(1)$ can be interpreted as *absolute velocity values.*

According to our convention, $V(1)$ is illustrated as follows:

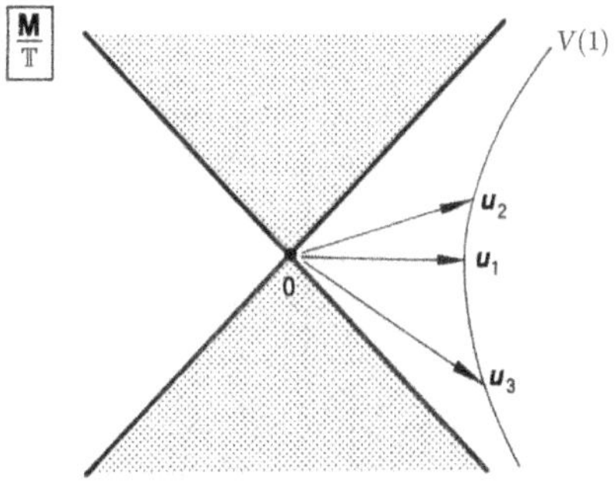

Three elements of $V(1)$ appear in the Figure. Observe that it makes no sense that

— $\boldsymbol{u}_1$ is in the centre line of $T^{\rightarrow}$ (there is no centre line of $T^{\rightarrow}$),
— the angle between $\boldsymbol{u}_1$ and $\boldsymbol{u}_2$ is less than the angle between $\boldsymbol{u}_1$ and $\boldsymbol{u}_3$ (there is no angle between the elements of $V(1)$),
— $\boldsymbol{u}_2$ is longer than $\boldsymbol{u}_1$ (the elements of $V(1)$ have no length).

The reversed Cauchy inequality (see V.4.7) involves the following important and frequently used relation:

$$-\boldsymbol{u}\cdot\boldsymbol{u}' \geq 1$$

for $\boldsymbol{u}, \boldsymbol{u}' \in V(1)$ and equality holds if and only if $\boldsymbol{u} = \boldsymbol{u}'$.

1.3.2. For $\boldsymbol{u} \in V(1)$ put

$$\begin{aligned} \tau_{\boldsymbol{u}} &: \mathbf{M} \to \mathbb{T}, \qquad \boldsymbol{x} \mapsto -\boldsymbol{u}\cdot\boldsymbol{x}, \\ \mathbf{S}_{\boldsymbol{u}} &:= \operatorname{Ker}\tau_{\boldsymbol{u}} = \{\boldsymbol{x} \in \mathbf{M} \mid \boldsymbol{u}\cdot\boldsymbol{x} = 0\}, \\ i_{\boldsymbol{u}} &: \mathbf{S}_{\boldsymbol{u}} \to \mathbf{M}, \qquad \boldsymbol{x} \mapsto \boldsymbol{x}. \end{aligned}$$

Since $\boldsymbol{u}$ is timelike, $\mathbf{S}_{\boldsymbol{u}}$ is a three-dimensional linear subspace consisting of spacelike vectors. According to our convention, $\mathbf{S}_{\boldsymbol{u}}$ is represented by a line that inclines to $L^{\rightarrow}$ with the same angle as $\boldsymbol{u}$:

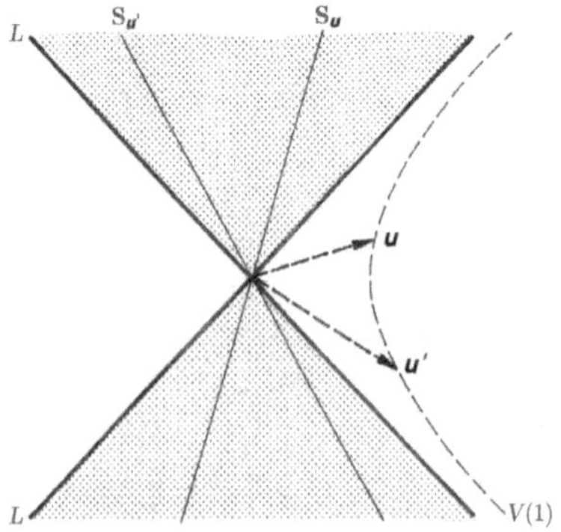

We emphasize that 'inclination to $L^{\rightarrow}$' makes no sense in the structure of the spacetime model; it makes sense only in the rules of the illustration we have chosen.

$\mathbf{S}_{\boldsymbol{u}}$ and $\boldsymbol{u}\otimes\mathbb{T}$ are complementary subspaces in $\mathbf{M}$, thus every vector $\boldsymbol{x}$ can be uniquely decomposed into the sum of components in $\boldsymbol{u}\otimes\mathbb{T}$ and in $\mathbf{S}_{\boldsymbol{u}}$, respectively:

$$\boldsymbol{x} = \boldsymbol{u}(\tau_{\boldsymbol{u}}\cdot\boldsymbol{x}) + \big(\boldsymbol{x} - \boldsymbol{u}(\tau_{\boldsymbol{u}}\cdot\boldsymbol{x})\big) = \boldsymbol{u}(-\boldsymbol{u}\cdot\boldsymbol{x}) + \big(\boldsymbol{x} + \boldsymbol{u}(\boldsymbol{u}\cdot\boldsymbol{x})\big).$$

The linear map

$$\pi_{\boldsymbol{u}} : \mathbf{M} \to \mathbf{S}_{\boldsymbol{u}}, \qquad \boldsymbol{x} \mapsto \boldsymbol{x} + \boldsymbol{u}(\boldsymbol{u} \cdot \boldsymbol{x})$$

is the projection onto $\mathbf{S}_{\boldsymbol{u}}$ along $\boldsymbol{u}$. It is illustrated as follows:

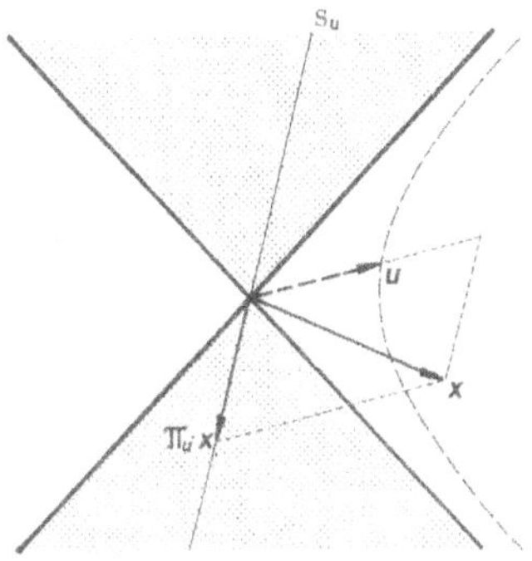

The dashed line is to express that $V(1)$ is in fact a subset of $\frac{\mathbf{M}}{\mathbb{T}}$ and not of $\mathbf{M}$.

1.3.3. For all $\boldsymbol{u} \in V(1)$, the restriction $\boldsymbol{h}_{\boldsymbol{u}}$ of the Lorentz form $\boldsymbol{g}$ onto $\mathbf{S}_{\boldsymbol{u}} \times \mathbf{S}_{\boldsymbol{u}}$ is positive definite. Thus $(\mathbf{S}_{\boldsymbol{u}}, \mathbb{T}, \boldsymbol{h}_{\boldsymbol{u}})$ is a three-dimensional Euclidean vector space.

Accordingly, the pseudo-length of vectors in $\mathbf{S}_{\boldsymbol{u}}$ is in fact a *length* and the *angle* between nonzero vectors in $\mathbf{S}_{\boldsymbol{u}}$ make sense; of course, similar notions for vectors in $\frac{\mathbf{S}_{\boldsymbol{u}}}{\mathbb{A}}$ can be introduced where $\mathbb{A}$ is a measure line. Moreover, all the results obtained in I.1.2.5 can be applied.

It is trivial that every spacelike vector is contained in some $\mathbf{S}_{\boldsymbol{u}}$:

$$S_0 = \bigcup_{\boldsymbol{u} \in V(1)} \mathbf{S}_{\boldsymbol{u}}.$$

Consequently, the pseudo-length of a spacelike vector will be said *length* or *magnitude*. However, we call attention to the fact that this length satisfies the triangle inequality only for two spacelike vectors spanning a spacelike linear subspace (see Exercise V.4.20.2)

1.3.4. The orientation of $\mathbf{M}$ and the arrow orientation of $\boldsymbol{g}$ determine a unique orientation of $\mathbf{S}_{\boldsymbol{u}}$.

Definition. Let $\boldsymbol{u} \in V(1)$. An ordered basis $(\boldsymbol{e}_1, \boldsymbol{e}_2, \boldsymbol{e}_3)$ of $\mathbf{S}_{\boldsymbol{u}}$ is called *positively oriented* if $(\boldsymbol{u}\boldsymbol{t}, \boldsymbol{e}_1, \boldsymbol{e}_2, \boldsymbol{e}_3)$ is a positively oriented basis of $\mathbf{M}$ for some (hence for all) $\boldsymbol{t} \in \mathbb{T}^+$.

1.3.5. Proposition. Let $\boldsymbol{u} \in V(1)$. Then

$$\boldsymbol{\xi}_{\boldsymbol{u}} := (\boldsymbol{\tau}_{\boldsymbol{u}}, \boldsymbol{\pi}_{\boldsymbol{u}}) \colon \mathbf{M} \to \mathbb{T} \times \mathbf{S}_{\boldsymbol{u}}, \quad \boldsymbol{x} \mapsto \big(-\boldsymbol{u}\cdot\boldsymbol{x},\, \boldsymbol{x} + \boldsymbol{u}(\boldsymbol{u}\cdot\boldsymbol{x})\big)$$

is an orientation preserving linear bijection and

$$\boldsymbol{\xi}_{\boldsymbol{u}}^{-1}(t, \boldsymbol{q}) = \boldsymbol{u}t + \boldsymbol{q} \qquad (t \in \mathbb{T},\ \boldsymbol{q} \in \mathbf{S}_{\boldsymbol{u}}). \qquad \blacksquare$$

Keep in mind that $\boldsymbol{x} = \boldsymbol{u}(-\boldsymbol{u}\cdot\boldsymbol{x}) + \boldsymbol{\pi}_{\boldsymbol{u}}\cdot\boldsymbol{x}$ results in the following important formula:

$$\boldsymbol{x}^2 = -(\boldsymbol{u}\cdot\boldsymbol{x})^2 + |\boldsymbol{\pi}_{\boldsymbol{u}}\cdot\boldsymbol{x}|^2 \qquad (\boldsymbol{x} \in \mathbf{M}).$$

1.3.6. Note the striking similarity between the previous formulae and the formulae of the nonrelativistic spacetime model treated in I.1.2. However, behind the resemblance to it there is an important difference: in the nonrelativistic case a single three-dimensional subspace $\mathbf{S}$ appears whereas in the special relativistic case every $\boldsymbol{u} \in V(1)$ indicates its own three-dimensional subspace. Correspondingly, instead of a single $\boldsymbol{\tau}$, now there is a $\boldsymbol{\tau}_{\boldsymbol{u}}$ for all $\boldsymbol{u}$. The range of $\boldsymbol{\xi}_{\boldsymbol{u}}$ is the same set in the nonrelativistic case, whereas it depends on $\boldsymbol{u}$ in the relativistic case.

A further very important difference is that $\mathbf{M}$ and $\mathbf{M}^*$ are different vector spaces in the nonrelativistic case, whereas they are 'nearly the same' in the relativistic case. More precisely, we have the identification (see V.1.3).

$$\frac{\mathbf{M}}{\mathbb{T}\otimes\mathbb{T}} \equiv \mathbf{M}^*,$$

which is established by the Lorentz form $\boldsymbol{g}$.

Of course, we make the identification

$$\frac{\mathbf{S}_{\boldsymbol{u}}}{\mathbb{T}\otimes\mathbb{T}} \equiv \mathbf{S}_{\boldsymbol{u}}^*,$$

too.

According to these identifications $\boldsymbol{u}\otimes\boldsymbol{u}$ is considered a linear map $\mathbf{M} \to \mathbf{M}$, $\boldsymbol{x} \mapsto \boldsymbol{u}(\boldsymbol{u}\cdot\boldsymbol{x})$, so we have

$$\boldsymbol{\pi}_{\boldsymbol{u}} = \mathbf{1}_{\mathbf{M}} + \boldsymbol{u}\otimes\boldsymbol{u} = \mathbf{1}_{\mathbf{M}} - \boldsymbol{u}\otimes\boldsymbol{\tau}_{\boldsymbol{u}},$$

and

$$\boldsymbol{\tau}_{\boldsymbol{u}} \in \mathbb{T}\otimes\mathbf{M}^* \equiv \frac{\mathbf{M}}{\mathbb{T}}, \qquad \boldsymbol{i}_{\boldsymbol{u}} \in \mathbf{M}\otimes\mathbf{S}_{\boldsymbol{u}}^* \equiv \frac{\mathbf{M}\otimes\mathbf{S}_{\boldsymbol{u}}}{\mathbb{T}\otimes\mathbb{T}},$$
$$\boldsymbol{\tau}_{\boldsymbol{u}}^* \in \mathbf{M}^*\otimes\mathbb{T} \equiv \frac{\mathbf{M}}{\mathbb{T}}, \qquad \boldsymbol{i}_{\boldsymbol{u}}^* \in \mathbf{S}_{\boldsymbol{u}}^*\otimes\mathbf{M} \equiv \frac{\mathbf{S}_{\boldsymbol{u}}\otimes\mathbf{M}}{\mathbb{T}\otimes\mathbb{T}},$$

$$\pi_u \in \mathbf{S}_u \otimes \mathbf{M}^* \equiv \frac{\mathbf{S}_u \otimes \mathbf{M}}{\mathbb{T} \otimes \mathbb{T}}.$$

Moreover,

$$\tau_u \cdot i_u = 0, \qquad \pi_u \cdot i_u = 1_{\mathbf{S}_u}$$

and the identifications yield the relations

$$\tau_u \equiv \tau_u^* \equiv -u, \qquad i_u^* \equiv \pi_u.$$

The reader is asked to prove the first formula; as concerns the second one, see the following equalities for $x \in \mathbf{M}$, $q \in \mathbf{S}_u$:

$$(i_u^* \cdot x) \cdot q = x \cdot i_u \cdot q = x \cdot q = (\pi_u \cdot x) \cdot q.$$

1.3.7. For $u, u' \in V(1)$ put

$$v_{u'u} := \frac{u'}{-u' \cdot u} - u.$$

We shall see later that this is the relative velocity of u' with respect to u.

It is an easy task to show that $|v_{u'u}|^2 = |v_{uu'}|^2 = 1 - \frac{1}{(u' \cdot u)^2}$; as a consequence of the reversed Cauchy inequality, $v_{u'u} = 0$ if and only if $u = u'$. Moreover, if $q \in \mathbf{S}_u \cap \mathbf{S}_{u'}$ then $q \cdot v_{u'u} = 0$ which proves the following.

Proposition. $\mathbf{S}_u \cap \mathbf{S}_{u'}$ is a two-dimensional linear subspace if and only if $u \neq u'$ and in this case $v_{u'u} \otimes \mathbb{T}$ $(v_{uu'} \otimes \mathbb{T})$ is a one-dimensional linear subspace of $\mathbf{S}_u$ $(\mathbf{S}_{u'})$, orthogonal to $\mathbf{S}_u \cap \mathbf{S}_{u'}$.

(In other words, $\mathbf{S}_u \cap \mathbf{S}_{u'}$ and $v_{u'u} \otimes \mathbb{T}$ $(v_{uu'} \otimes \mathbb{T})$ are orthogonal complementary subspaces in $\mathbf{S}_u$ $(\mathbf{S}_{u'})$).

1.3.8. For different u and u', $\mathbf{S}_u$ and $\mathbf{S}_{u'}$ are different linear subspaces; however, we can give a distinguished bijection between them which will play a fundamental role concerning observer spaces.

Let $L(u', u)$ be the linear map from $\mathbf{S}_u$ onto $\mathbf{S}_{u'}$ defined in such a way that it leaves the elements of $\mathbf{S}_u \cap \mathbf{S}_{u'}$ invariant and maps the orthogonal complements of this subspace into each other. More precisely,

$$L(u', u) \cdot q := \begin{cases} q & \text{if } q \in \mathbf{S}_u \cap \mathbf{S}_{u'}, \\ -v_{uu'}t & \text{if } q = v_{u'u}t \quad (t \in \mathbb{T}). \end{cases}$$

It is not difficult to see that $L(u', u)$ is an orientation preserving h_u-$h_{u'}$-orthogonal linear bijection between $\mathbf{S}_u$ and $\mathbf{S}_{u'}$. We can extend it to a linear bijection $\mathbf{M} \to \mathbf{M}$ by the requirement

$$L(u', u) \cdot u := u'$$

(recall that the dot product notation allows us to apply linear maps $\mathbf{M} \to \mathbf{M}$ to elements of $\frac{\mathbf{M}}{\mathbb{T}}$).

This linear bijection can be given by a simple formula as follows. Recall that for $\boldsymbol{u}, \boldsymbol{u}' \in V(1)$, $\boldsymbol{u}' \otimes \boldsymbol{u} \in \frac{\mathbf{M}}{\mathbb{T}} \otimes \frac{\mathbf{M}}{\mathbb{T}} \equiv \frac{\mathbf{M}\otimes\mathbf{M}}{\mathbb{T}\otimes\mathbb{T}} \equiv \mathbf{M} \otimes \mathbf{M}^* \equiv \mathrm{Lin}\,(\mathbf{M}, \mathbf{M})$.

Definition. Let $\boldsymbol{u}, \boldsymbol{u}' \in V(1)$. Then

$$\boldsymbol{L}(\boldsymbol{u}', \boldsymbol{u}) := \mathbf{1}_{\mathbf{M}} + \frac{(\boldsymbol{u}' + \boldsymbol{u}) \otimes (\boldsymbol{u}' + \boldsymbol{u})}{1 - \boldsymbol{u}' \cdot \boldsymbol{u}} - 2\boldsymbol{u}' \otimes \boldsymbol{u}$$

is called the *Lorentz boost* from $\boldsymbol{u}$ to $\boldsymbol{u}'$.

Proposition. (*i*) $\boldsymbol{L}(\boldsymbol{u}', \boldsymbol{u})$ is an orientation- and arrow-preserving $\boldsymbol{g}$-orthogonal linear map from $\mathbf{M}$ into $\mathbf{M}$;

(*ii*) $\boldsymbol{L}(\boldsymbol{u}', \boldsymbol{u}) \cdot \boldsymbol{u} = \boldsymbol{u}'$;

(*iii*) $\boldsymbol{L}(\boldsymbol{u}', \boldsymbol{u})$ maps $\mathbf{S}_{\boldsymbol{u}}$ onto $\mathbf{S}_{\boldsymbol{u}'}$, more closely,

— $\boldsymbol{L}(\boldsymbol{u}', \boldsymbol{u}) \cdot \boldsymbol{q} = \boldsymbol{q}$ if $\boldsymbol{q} \in \mathbf{S}_{\boldsymbol{u}} \cap \mathbf{S}_{\boldsymbol{u}'}$,

— $\boldsymbol{L}(\boldsymbol{u}', \boldsymbol{u}) \cdot \boldsymbol{v}_{\boldsymbol{u}'\boldsymbol{u}} = -\boldsymbol{v}_{\boldsymbol{u}\boldsymbol{u}'}$;

(*iv*) $\boldsymbol{L}(\boldsymbol{u}, \boldsymbol{u}) = \boldsymbol{g}$, $\quad \boldsymbol{L}(\boldsymbol{u}', \boldsymbol{u})^{-1} = \boldsymbol{L}(\boldsymbol{u}, \boldsymbol{u}')$

and $\boldsymbol{L}(\boldsymbol{u}', \boldsymbol{u})$ is the unique linear map for which (*i*)–(*iii*) hold.

1.3.9. Since the Lorentz boosts map the corresponding spacelike subspaces onto each other in a 'handsome' manner, we might expect that executing the Lorentz boost from $\boldsymbol{u}$ to $\boldsymbol{u}'$ and then the Lorentz boost from $\boldsymbol{u}'$ to $\boldsymbol{u}''$ we should get the Lorentz boost from $\boldsymbol{u}$ to $\boldsymbol{u}''$; however, this occurs only in some special cases.

Proposition. Let $\boldsymbol{u}$, $\boldsymbol{u}'$, $\boldsymbol{u}''$ be elements of $V(1)$. Then $\boldsymbol{L}(\boldsymbol{u}'', \boldsymbol{u}') \cdot \boldsymbol{L}(\boldsymbol{u}', \boldsymbol{u}) = \boldsymbol{L}(\boldsymbol{u}'', \boldsymbol{u})$ if and only if the three elements of $V(1)$ are coplanar.

Proof. Suppose the equality holds. Then for all $\boldsymbol{q} \in \mathbf{S}_{\boldsymbol{u}} \cap \mathbf{S}_{\boldsymbol{u}''}$

$$\begin{aligned}
\boldsymbol{q} &= \boldsymbol{L}(\boldsymbol{u}'', \boldsymbol{u}') \cdot \boldsymbol{L}(\boldsymbol{u}', \boldsymbol{u}) \cdot \boldsymbol{q} = \\
&= \left(\mathbf{1}_{\mathbf{M}} + \frac{(\boldsymbol{u}'' + \boldsymbol{u}') \otimes (\boldsymbol{u}'' + \boldsymbol{u}')}{1 - \boldsymbol{u}'' \cdot \boldsymbol{u}'} - 2\boldsymbol{u}'' \otimes \boldsymbol{u}'\right) \cdot \left(\boldsymbol{q} + \frac{(\boldsymbol{u}' + \boldsymbol{u})\boldsymbol{u}' \cdot \boldsymbol{q}}{1 - \boldsymbol{u}' \cdot \boldsymbol{u}}\right) = \\
&= \boldsymbol{q} + (\boldsymbol{u}' \cdot \boldsymbol{q}) \left(\frac{\boldsymbol{u}'' + \boldsymbol{u}'}{1 - \boldsymbol{u}'' \cdot \boldsymbol{u}'} + \frac{\boldsymbol{u}' + \boldsymbol{u}}{1 - \boldsymbol{u}' \cdot \boldsymbol{u}} + \right. \\
&\qquad \left. + \frac{(\boldsymbol{u}'' + \boldsymbol{u}')(\boldsymbol{u}'' \cdot \boldsymbol{u}' + \boldsymbol{u}'' \cdot \boldsymbol{u} + \boldsymbol{u}' \cdot \boldsymbol{u} - 1)}{(1 - \boldsymbol{u}'' \cdot \boldsymbol{u}')(1 - \boldsymbol{u}' \cdot \boldsymbol{u})}\right),
\end{aligned}$$

from which we deduce that

— either $\boldsymbol{u}' \cdot \boldsymbol{q} = 0$ for all $\boldsymbol{q} \in \mathbf{S}_{\boldsymbol{u}} \cap \mathbf{S}_{\boldsymbol{u}''}$, implying that $\boldsymbol{u}'$ is in the two-dimensional subspace spanned by $\boldsymbol{u}$ and $\boldsymbol{u}''$, i.e. the three elements of $V(1)$ are coplanar,

— or the last expression in parentheses is zero which implies again that the three elements of $V(1)$ are coplanar. ■

Observe that $\boldsymbol{L}(\boldsymbol{u}'',\boldsymbol{u}')\cdot\boldsymbol{L}(\boldsymbol{u}',\boldsymbol{u})$ maps $\mathbf{S}_{\boldsymbol{u}}$ onto $\mathbf{S}_{\boldsymbol{u}''}$; as a consequence, if it is a Lorentz boost, it must be equal to $\boldsymbol{L}(\boldsymbol{u}'',\boldsymbol{u})$. Thus our result implies that, in general, the product of Lorentz boosts is not a Lorentz boost.

1.4. The arithmetic spacetime model

1.4.1. Let us take the Minkowski vector space $(\mathbb{R}^{1+3},\mathbb{R},\boldsymbol{G})$ treated in V.4.19 and endowed with the standard orientation and arrow orientation. Considering $\mathbb{R}^{1+3}$ to be an affine space, we easily find that $(\mathbb{R}^{1+3},\mathbb{R},\boldsymbol{G})$ is a special relativistic spacetime model which we call the *arithmetic special relativistic spacetime model.*

Similarly to the nonrelativistic case, the arithmetic spacetime model has the peculiar property that the same object, $\mathbb{R}^{1+3}$, represents the affine space of world points and the vector space of world vectors (and even the vector space of covectors). We follow our nonrelativistic convention that the world points will be denoted by Greek letters, whereas world vectors (and covectors) will be denoted by Latin letters.

We find it convenient to write the elements of the *affine space* $\mathbb{R}^{1+3}$ in the form (ξ^i); the elements of the vector space $\mathbb{R}^{1+3}$ in the form $(x^i)=(x^0,\boldsymbol{x})$, and the elements of $(\mathbb{R}^{1+3})^*$ in the form $(k_i)=(k_0,\boldsymbol{k})$.

Recall that the identification $(\mathbb{R}^{1+3})^*\equiv\mathbb{R}^{1+3}$ established by $\boldsymbol{G}$ gives

$$x_0=-x^0,\qquad x_\alpha=x^\alpha\qquad(\alpha=1,2,3).$$

Correspondingly, the dot product of (x^i) and (y^i) equals

$$x^iy_i,$$

where the Einstein summation convention is applied: a summation is carried out from 0 to 3 for identical subscripts and superscripts.

1.4.2. In the arithmetic spacetime model

$$V(1)=\left\{(u^i)\in\mathbb{R}^{1+3}|u^iu_i=-1,\quad u^0>0\right\}.$$

Here, too, we find a misleading feature of this spacetime model: $V(1)$ seems to have a distinguished, simplest element, namely $(1,\mathbf{0})$. $\boldsymbol{\pi}_{(1,\mathbf{0})}$ is the canonical projection $\mathbb{R}^{1+3}\to\{0\}\times\mathbb{R}^3$.

For an arbitrary element (u^i) of $V(1)$ we can define

$$v^\alpha:=\frac{u^\alpha}{u^0}\qquad(\alpha=1,2,3),\qquad\boldsymbol{v}:=(v^1,v^2,v^3)\in\mathbb{R}^3;$$

then with the usual norm $|\ |$ on $\mathbb{R}^3$ we have $|\boldsymbol{v}| < 1$ and

$$u^0 = \frac{1}{\sqrt{1 - |\boldsymbol{v}|^2}}$$

and

$$(u^i) = \frac{1}{\sqrt{1 - |\boldsymbol{v}|^2}}(1, \boldsymbol{v}). \tag{$*$}$$

We easily find that $\boldsymbol{v}$ is exactly the relative velocity of (u^i) with respect to the 'basic velocity value' $(1, \mathbf{0})$ (see 1.3.7).

1.4.3. It is then obvious that

$$\mathbf{S}_{(u^i)} = \left\{(x^i) \in \mathbb{R}^{1+3} \,|\, x^0 = \boldsymbol{x} \cdot \boldsymbol{v}\right\}.$$

Unlike the nonrelativistic case, $\boldsymbol{\pi}_{(u^i)}$ for a general (u^i) in $V(1)$ is an uneasy object because it maps onto a three-dimensional linear subspace in $\mathbb{R}^{1+3}$ which is different from $\{0\} \times \mathbb{R}^3$. Thus the values of $\boldsymbol{\pi}_{(u^i)}$ cannot be given directly by triplets of real numbers. However, as it is known, in textbooks one usually deals with triplets (and quartets) of real numbers. We can achieve this by always referring to the space of the 'basic velocity value' with the aid of the corresponding Lorentz boost, i.e. instead of $\boldsymbol{\pi}_{(u^i)}$ taking $\boldsymbol{L}\big((1, \mathbf{0}), (u^i)\big) \cdot \boldsymbol{\pi}_{(u^i)}$, whose range is $\{0\} \times \mathbb{R}^3$.

1.4.4. The Lorentz boost from (u'^i) to (u^i) is given by the matrix

$$L^i_k := \delta^i_k + \frac{(u^i + u'^i)(u_k + u'_k)}{1 - u^j u'_j} - 2u^i u'_k.$$

If (u'^i) is the 'basic velocity value', then it becomes

$$\begin{pmatrix} u^0 & u^1 & u^2 & u^3 \\ u^1 & 1 + \dfrac{(u^1)^2}{1 + u^0} & \dfrac{u^1 u^2}{1 + u^0} & \dfrac{u^1 u^3}{1 + u^0} \\ u^2 & \dfrac{u^1 u^2}{1 + u^0} & 1 + \dfrac{(u^2)^2}{1 + u^0} & \dfrac{u^2 u^3}{1 + u^0} \\ u^3 & \dfrac{u^1 u^3}{1 + u^0} & \dfrac{u^2 u^3}{1 + u^0} & 1 + \dfrac{(u^3)^2}{1 + u^0} \end{pmatrix}.$$

Using formula $(*)$ in 1.4.2 and the notation

$$\kappa := \frac{1}{\sqrt{1 - |v|^2}},$$

we find that

$$L\big((1,\mathbf{0}),(u^i)\big) = \kappa \begin{pmatrix} 1 & v^1 & v^2 & v^3 \\ v^1 & \frac{1}{\kappa}+\frac{\kappa}{1+\kappa}(v^1)^2 & \frac{\kappa}{1+\kappa}v^1v^2 & \frac{\kappa}{1+\kappa}v^1v^3 \\ v^2 & \frac{\kappa}{1+\kappa}v^1v^2 & \frac{1}{\kappa}+\frac{\kappa}{1+\kappa}(v^2)^2 & \frac{\kappa}{1+\kappa}v^2v^3 \\ v^3 & \frac{\kappa}{1+\kappa}v^1v^3 & \frac{\kappa}{1+\kappa}v^2v^3 & \frac{1}{\kappa}+\frac{\kappa}{1+\kappa}(v^3)^2 \end{pmatrix}.$$

This shows what a complicated form $L\big((1,\mathbf{0}),(u^i)\big)\cdot\boldsymbol{\pi}_{(u^i)}$ has; later (see 7.1.4) we give it in detail.

1.4.5. The previous matrix is the usual 'Lorentz transformation'. Most frequently one considers the special case $v^2 = v^3 = 0$, $v := v^1$; then $\kappa = \frac{1}{\sqrt{1-v^2}}$ and $\frac{\kappa^2 v^2}{1+\kappa} = \kappa - 1$, thus the previous matrix reduces to

$$\kappa \begin{pmatrix} 1 & v & 0 & 0 \\ v & 1 & 0 & 0 \\ 0 & 0 & 1/\kappa & 0 \\ 0 & 0 & 0 & 1/\kappa \end{pmatrix}.$$

1.5. Classification of physical quantities

1.5.1. We introduce notions similar to those in the nonrelativistic spacetime model. Let $\mathbb{A}$ be a measure line. Then the elements of

$$\begin{aligned} &\mathbb{A} && \text{are called} && \textit{scalars of type } \mathbb{A}, \\ &\mathbb{A}\otimes\mathbf{M} && \text{are called} && \textit{vectors of type } \mathbb{A}, \\ &\frac{\mathbf{M}}{\mathbb{A}} && \text{are called} && \textit{vectors of cotype } \mathbb{A}, \\ &\mathbb{A}\otimes(\mathbf{M}\otimes\mathbf{M}) && \text{are called} && \textit{tensors of type } \mathbb{A}, \\ &\frac{\mathbf{M}\otimes\mathbf{M}}{\mathbb{A}} && \text{are called} && \textit{tensors of cotype } \mathbb{A}. \end{aligned}$$

Covectors of type $\mathbb{A}$, etc. are defined similarly with $\mathbf{M}^*$ instead of $\mathbf{M}$.

In particular, the elements of $\mathbf{M}\otimes\mathbf{M}$ and $\mathbf{M}^*\otimes\mathbf{M}^*$ are called *tensors* and *cotensors*, respectively; the elements of $\mathbf{M}\otimes\mathbf{M}^*$ and $\mathbf{M}^*\otimes\mathbf{M}$ are *mixed tensors.*

A very important feature of the special relativistic spacetime model is that covectors can be identified with vectors of cotype $\mathbb{T}\otimes\mathbb{T}$. As a consequence, e.g. a covector of type $\mathbb{A}$ is identified with a vector of type $\frac{\mathbb{A}}{\mathbb{T}\otimes\mathbb{T}}$.

1.5.2. According to our convention, the dot product between vectors (covectors) of different types makes sense. For instance, for $\boldsymbol{u} \in V(1) \subset \frac{\mathbf{M}}{\mathbb{T}}$ and for

$$z \in \mathbb{A} \otimes \mathbf{M} \quad \text{we have} \quad u \cdot z \in \mathbb{T} \otimes \mathbb{A}, \quad z^2 \in (\mathbb{A} \otimes \mathbb{A}) \otimes (\mathbb{T} \otimes \mathbb{T}),$$
$$w \in \frac{\mathbf{M}}{\mathbb{A}} \quad \text{we have} \quad u \cdot w \in \frac{\mathbb{T}}{\mathbb{A}}, \quad w^2 \in \frac{\mathbb{T} \otimes \mathbb{T}}{\mathbb{A} \otimes \mathbb{A}}.$$

In particular, $z^2 \in \mathbb{R}$ for $z \in \frac{\mathbf{M}}{\mathbb{T}}$.

Since $(\mathbb{A} \otimes \mathbb{A}) \otimes (\mathbb{T} \otimes \mathbb{T}) \equiv (\mathbb{A} \otimes \mathbb{T}) \otimes (\mathbb{A} \otimes \mathbb{T})$ has a natural orientation, we can speak of its positive and negative elements. Thus a vector z of type $\mathbb{A}$ is called

$$\textit{spacelike} \text{ if } z^2 > \mathbf{0} \text{ or } z = \mathbf{0},$$
$$\textit{timelike} \text{ if } z^2 < \mathbf{0},$$
$$\textit{lightlike} \text{ if } z^2 = \mathbf{0},\ z \neq \mathbf{0}.$$

It can be easily shown that z is spacelike if and only if $z \in \mathbb{A} \otimes S_0$, etc. Moreover, a measure line $\mathbb{A}$ is oriented, hence $\mathbb{A}^+$ makes sense. Consequently, we say that a timelike (lightlike) vector z of type $\mathbb{A}$ is *future directed* if $z \in \mathbb{A}^+ \otimes T^{\rightarrow}$ $(z \in \mathbb{A}^+ \otimes L^{\rightarrow})$.

1.6. Comparison of spacetime models

1.6.1. Definition. The special relativistic spacetime model $(\mathrm{M}, \mathbb{T}, g)$ is *isomorphic* to the special relativistic spacetime model $(\mathrm{M}', \mathbb{T}', g')$ if there are

(*i*) an orientation- and arrow-preserving affine bijection $F\colon \mathrm{M} \to \mathrm{M}'$, over the linear bijection $\boldsymbol{F}$,

(*ii*) an orientation preserving linear bijection $\boldsymbol{Z}\colon \mathbb{T} \to \mathbb{T}'$

such that

$$g' \circ (\boldsymbol{F} \times \boldsymbol{F}) = (\boldsymbol{Z} \otimes \boldsymbol{Z}) \circ g.$$

The pair $(F, \boldsymbol{Z})$ is an *isomorphism* between the two spacetime models.

If the two models coincide, an isomorphism is called an *automorphism.* An automorphism of the form $(F, \mathbf{1}_{\mathbb{T}})$ is *strict.* ■

Three diagrams illustrate the isomorphism:

$$\begin{array}{ccccccc}
\mathrm{M} & \qquad & \mathbb{T} & \qquad & \mathbf{M} \times \mathbf{M} & \xrightarrow{g} & \mathbb{T} \otimes \mathbb{T} \\
{\scriptstyle F}\big\downarrow & & \big\downarrow{\scriptstyle \boldsymbol{Z}} & & {\scriptstyle \boldsymbol{F} \times \boldsymbol{F}}\big\downarrow & & \big\downarrow{\scriptstyle \boldsymbol{Z} \otimes \boldsymbol{Z}}. \\
\mathrm{M}' & & \mathbb{T}' & & \mathbf{M}' \times \mathbf{M}' & \underset{g'}{\longrightarrow} & \mathbb{T}' \otimes \mathbb{T}'
\end{array}$$

The definition is quite natural and simple, needs no comment.

1.6.2. Proposition. The special relativistic spacetime model $(\mathrm{M}, \mathbb{T}, g)$ is isomorphic to the arithmetic spacetime model.

Proof. Take
(*i*) a positive element $\boldsymbol{s}$ of $\mathbb{T}$,
(*ii*) a positively oriented $\boldsymbol{g}$-orthogonal basis $(\boldsymbol{e}_0, \boldsymbol{e}_1, \boldsymbol{e}_2, \boldsymbol{e}_3)$, normed to $\boldsymbol{s}$, of $\mathbf{M}$, for which $\boldsymbol{e}_0$ is future directed,
(*iii*) an element o of M.
Then

$$F : \mathrm{M} \to \mathbb{R}^4, \qquad x \mapsto \left(\frac{\boldsymbol{e}_k \cdot (x - o)}{\boldsymbol{e}_k^2} \,\middle|\, k = 0, 1, 2, 3 \right),$$

$$\boldsymbol{Z} : \mathbb{T} \to \mathbb{R}, \qquad \boldsymbol{t} \mapsto \frac{\boldsymbol{t}}{\boldsymbol{s}}$$

is an isomorphism. ■

This isomorphism has the inverse

$$\mathbb{R}^4 \to \mathrm{M}, \qquad \xi \mapsto o + \sum_{k=0}^{3} \xi^k \boldsymbol{e}_k,$$

$$\mathbb{R} \to \mathbb{T}, \qquad \alpha \mapsto \alpha \boldsymbol{s}.$$

1.6.3. An important consequence of the previous result is that *any two special relativistic spacetime models are isomorphic,* i.e. are of the same kind. The special relativistic spacetime model as a mathematical structure is unique. This means that there is a unique 'special relativistic physics'.

Note: the special relativistic spacetime models are of the same kind but, in general, are not identical. They are isomorphic, but, in general, there is no 'canonical' isomorphism between them, we cannot identify them by a distinguished isomorphism. The situation is the same as what we encountered for nonrelativistic spacetime models.

Since all special relativistic spacetime models are isomorphic, we can use an arbitrary one for investigation and application. However, an actual model can have additional structures. For instance, in the arithmetic spacetime model, spacetime is a vector space, $V(1)$ has a distinguished element. This model tempts us to multiply world points by real numbers (though this has no physical meaning and that is why it is not meaningful in the abstract spacetime model), to speak about time and space, consider spacetime as the Cartesian product of time and space (whereas neither time nor space exists absolutely), etc.

To avoid such confusions, we should keep away from similar specially constructed models for theoretical investigations and applications of the special relativistic spacetime model. However, for solving special problems, for executing some particular calculations, we can choose a convenient concrete model, like in the nonrelativistic case.

1.6.4. Present day physics uses tacitly the arithmetic spacetime model. One represents time points by real numbers, space points by triplets of real numbers. To obtain such representations, one chooses a unit for time periods, an initial time point, a distance unit, an initial space point and an orthogonal spatial basis whose elements have unit length.

However, all the previous notions in usual circumstances have merely a heuristic sense. The isomorphism established in 1.6.2 will give these notions a mathematically precise meaning. We shall see later that $\boldsymbol{s}$ is the time unit (and the distance unit), $\boldsymbol{e}_0$ characterizes an observer which produces its own time and space, the spacelike vectors $\boldsymbol{e}_1$, $\boldsymbol{e}_2$, $\boldsymbol{e}_3$ correspond to the spatial basis, o includes the initial time point and space point in some way.

1.7. The $\boldsymbol{u}$-split spacetime model

1.7.1. The arithmetic spacetime model is useful for solving particular problems, for executing practical calculations. Moreover, at present, one usually expounds theories, too, in the frame of the arithmetic spacetime model, so we have to translate every notion into the arithmetic language. As in the nonrelativistic case, it is convenient to introduce an 'intermediate' spacetime model between the abstract and the arithmetic ones.

1.7.2. Let $(\mathrm{M}, \mathbb{T}, \boldsymbol{g})$ be a special relativistic spacetime model and use the notations introduced in this chapter. Take a $\boldsymbol{u} \in V(1)$ and define the Lorentz form

$$\boldsymbol{g}_{\boldsymbol{u}} \colon (\mathbb{T} \times \mathbf{S}_{\boldsymbol{u}}) \times (\mathbb{T} \times \mathbf{S}_{\boldsymbol{u}}) \to \mathbb{T} \otimes \mathbb{T}, \qquad ((\boldsymbol{t}', \boldsymbol{q}'), (\boldsymbol{t}, \boldsymbol{q})) \mapsto -\boldsymbol{t}'\boldsymbol{t} + \boldsymbol{q}' \cdot \boldsymbol{q}.$$

Put

$$\begin{aligned} S &:= \{(\boldsymbol{t}, \boldsymbol{q}) \mid |\boldsymbol{q}| > |\boldsymbol{t}|\}, \\ T &:= \{(\boldsymbol{t}, \boldsymbol{q}) \mid |\boldsymbol{q}| < |\boldsymbol{t}|g\}, \\ L &:= \{(\boldsymbol{t}, \boldsymbol{q}) \mid |\boldsymbol{q}| = |\boldsymbol{t}| \neq \mathbf{0}\}. \end{aligned}$$

Endow $\mathbb{T} \times \mathbf{S}_{\boldsymbol{u}}$ with the product orientation and $\boldsymbol{g}_{\boldsymbol{u}}$ with the arrow orientation determined by

$$T^{\rightarrow} := \{(\boldsymbol{t}, \boldsymbol{q}) \in T \mid \boldsymbol{t} > \mathbf{0}\}.$$

Then $(\mathbb{T} \times \mathbf{S}_{\boldsymbol{u}}, \mathbb{T}, \boldsymbol{g}_{\boldsymbol{u}})$ is a special relativistic spacetime model, called the $\boldsymbol{u}$-*split special relativistic spacetime model.*

It is quite obvious that for all $o \in \mathrm{M}$,

$$\begin{aligned} &\mathrm{M} \to \mathbb{T} \times \mathbf{S}_{\boldsymbol{u}}, \qquad && x \mapsto \boldsymbol{\xi}_{\boldsymbol{u}} \cdot (x - o), \\ &\mathbb{T} \to \mathbb{T}, && \boldsymbol{t} \mapsto \boldsymbol{t} \end{aligned}$$

is an isomorphism between the two special relativistic spacetime models.

1.7.3. In the $\boldsymbol{u}$-split model

$$V(1)=\left\{(\alpha,\boldsymbol{w})\in\mathbb{R}\times\frac{\mathbf{S}_{\boldsymbol{u}}}{\mathbb{T}}\;\middle|\;-\alpha^2+|\boldsymbol{w}|^2=-1,\ \alpha>0\right\}=$$
$$=\left\{\frac{1}{\sqrt{1-|\boldsymbol{v}|^2}}(1,\boldsymbol{v})\;\middle|\;\boldsymbol{v}\in\frac{\mathbf{S}_{\boldsymbol{u}}}{\mathbb{T}},\ |\boldsymbol{v}|<1\right\}.$$

There is a simplest element in it: $(1,\mathbf{0})$.

1.8. Exercises

1. To be later (futurelike) is a transitive relation on M: if y is futurelike with respect to x and z is futurelike with respect to y then z is futurelike with respect to x.

2. $V(1)$ is a three-dimensional submanifold of M; its tangent space at $\boldsymbol{u}$ is $\frac{\mathbf{S}_{\boldsymbol{u}}}{\mathbb{T}}$ (see Exc.VI.4.14.3). For every $\boldsymbol{u}\in V(1)$,

$$\frac{\mathbf{S}_{\boldsymbol{u}}}{\mathbb{T}}\to V(1),\qquad \boldsymbol{w}\mapsto\boldsymbol{u}\sqrt{1+|\boldsymbol{w}|^2}+\boldsymbol{w},$$

and

$$\left\{\boldsymbol{v}\in\frac{\mathbf{S}_{\boldsymbol{u}}}{\mathbb{T}}\;\middle|\;|\boldsymbol{v}|<1\right\}\to V(1),\qquad \boldsymbol{v}\mapsto\frac{\boldsymbol{u}+\boldsymbol{v}}{\sqrt{1-|\boldsymbol{v}|^2}}$$

are global parametrizations of $V(1)$ having inverses

$$\boldsymbol{u}'\mapsto\boldsymbol{\pi}_{\boldsymbol{u}}\cdot\boldsymbol{u}'=\boldsymbol{u}'+(\boldsymbol{u}\cdot\boldsymbol{u}')\boldsymbol{u}=\frac{\boldsymbol{v}_{\boldsymbol{u}'\boldsymbol{u}}}{\sqrt{1-|\boldsymbol{v}_{\boldsymbol{u}'\boldsymbol{u}}|^2}}$$

and

$$\boldsymbol{u}'\mapsto\frac{\boldsymbol{\pi}_{\boldsymbol{u}}\cdot\boldsymbol{u}'}{-\boldsymbol{u}\cdot\boldsymbol{u}'}=\boldsymbol{v}_{\boldsymbol{u}'\boldsymbol{u}},$$

respectively.

3. Prove that for all $\boldsymbol{u}\in V(1)$

$$\mathbb{R}\times\left\{\boldsymbol{n}\in\frac{\mathbf{S}_{\boldsymbol{u}}}{\mathbb{T}}\;\middle|\;|\boldsymbol{n}|=1\right\}\to V(1),\quad(\alpha,\boldsymbol{n})\mapsto\frac{\boldsymbol{u}+\boldsymbol{n}\tanh\alpha}{\sqrt{1-\tanh^2\alpha}}=\boldsymbol{u}\cosh\alpha+\boldsymbol{n}\sinh\alpha$$

is a smooth map which is a bijection between $\mathbb{R}_0^+ \times \left\{ \boldsymbol{n} \in \frac{\mathbf{S}_{\boldsymbol{u}}}{\mathbb{T}} \mid |\boldsymbol{n}| = 1 \right\}$ and $V(1)$, having the inverse

$$\boldsymbol{u}' \mapsto \left(\operatorname{arcosh}(-\boldsymbol{u} \cdot \boldsymbol{u}'), \ \frac{\boldsymbol{\pi}_{\boldsymbol{u}} \cdot \boldsymbol{u}'}{\sqrt{(\boldsymbol{u} \cdot \boldsymbol{u}')^2 - 1}} \right).$$

4. Let $\boldsymbol{u} \in V(1)$, $\boldsymbol{n} \in \frac{\mathbf{S}_{\boldsymbol{u}}}{\mathbb{T}}$, $|\boldsymbol{n}| = 1$. Take $\alpha, \beta \in \mathbb{R}$ and put

$$\begin{aligned} \boldsymbol{u}' &:= \boldsymbol{u}\cosh\alpha + \boldsymbol{n}\sinh\alpha, \quad \boldsymbol{n}' := \boldsymbol{L}(\boldsymbol{u}', \boldsymbol{u}) \cdot \boldsymbol{n} = \boldsymbol{u}\sinh\alpha + \boldsymbol{n}\cosh\alpha \\ \boldsymbol{u}'' &:= \boldsymbol{u}'\cosh\beta + \boldsymbol{n}'\sinh\beta = \boldsymbol{u}\cosh(\alpha+\beta) + \boldsymbol{n}\sinh(\alpha+\beta), \\ \boldsymbol{u}''' &:= \boldsymbol{u}\cosh\beta + \boldsymbol{n}\sinh\beta. \end{aligned}$$

Prove that

$$\boldsymbol{L}(\boldsymbol{u}'', \boldsymbol{u}') = \boldsymbol{L}(\boldsymbol{u}''', \boldsymbol{u}).$$

(Hint: $\boldsymbol{L}(\boldsymbol{u}'', \boldsymbol{u}') \cdot \boldsymbol{u} = \boldsymbol{u}'''$ and $\mathbf{S}_{\boldsymbol{u}} \cap \mathbf{S}_{\boldsymbol{u}'''} = \mathbf{S}_{\boldsymbol{u}'} \cap \mathbf{S}_{\boldsymbol{u}''}$.)

5. Use the notations of the preceding exercise and prove that

$$\boldsymbol{L}(\boldsymbol{u}''', \boldsymbol{u}) \cdot \boldsymbol{L}(\boldsymbol{u}', \boldsymbol{u}) = \boldsymbol{L}(\boldsymbol{u}'', \boldsymbol{u})$$

i.e.

$$\begin{aligned} &\boldsymbol{L}(\boldsymbol{u}\cosh\beta + \boldsymbol{n}\sinh\beta, \boldsymbol{u}) \cdot \boldsymbol{L}(\boldsymbol{u}\cosh\alpha + \boldsymbol{n}\sinh\alpha, \boldsymbol{u}) = \\ &\qquad = \boldsymbol{L}\big(\boldsymbol{u}\cosh(\alpha+\beta) + \boldsymbol{n}\sinh(\alpha+\beta), \boldsymbol{u}\big). \end{aligned}$$

6. Let $\boldsymbol{u} \in V(1)$, $\boldsymbol{m}, \boldsymbol{n} \in \frac{\mathbf{S}_{\boldsymbol{u}}}{\mathbb{T}}$, $|\boldsymbol{m}| = |\boldsymbol{n}| = 1$, $\boldsymbol{m} \cdot \boldsymbol{n} = 0$. Take an $0 \neq \alpha \in \mathbb{R}$ and put

$$\boldsymbol{u}' := \boldsymbol{u}\cosh\alpha + \boldsymbol{n}\sinh\alpha, \qquad \boldsymbol{u}'' := \boldsymbol{u}\cosh\alpha + \boldsymbol{m}\sinh\alpha.$$

Then $\boldsymbol{n}' := \boldsymbol{L}(\boldsymbol{u}', \boldsymbol{u}) \cdot \boldsymbol{n} = \boldsymbol{u}\sinh\alpha + \boldsymbol{n}\cosh\alpha$ and $\boldsymbol{L}(\boldsymbol{u}'', \boldsymbol{u}) \cdot \boldsymbol{n} = \boldsymbol{n}$. Prove that $\boldsymbol{L}(\boldsymbol{u}'', \boldsymbol{u}') \cdot \boldsymbol{n}'$ is not parallel to $\boldsymbol{n}$.

7. Let $\boldsymbol{u}, \boldsymbol{u}' \in V(1)$. Then $\boldsymbol{u}' \otimes \mathbb{T}$ and $\mathbf{S}_{\boldsymbol{u}}$ are complementary subspaces. The *projection onto* $\mathbf{S}_{\boldsymbol{u}}$ *along* $\boldsymbol{u}' \otimes \mathbb{T}$ is the linear map

$$\boldsymbol{P}_{\boldsymbol{u}\boldsymbol{u}'} := \mathbf{1}_{\mathrm{M}} + \frac{\boldsymbol{u}' \otimes \boldsymbol{u}}{-\boldsymbol{u}' \cdot \boldsymbol{u}} : \mathrm{M} \to \mathrm{M}, \qquad x \mapsto x + \boldsymbol{u}' \frac{\boldsymbol{u} \cdot x}{-\boldsymbol{u}' \cdot \boldsymbol{u}}.$$

Prove that

(*i*) the restriction of $\boldsymbol{P}_{\boldsymbol{u}\boldsymbol{u}'}$ onto $\mathbf{S}_{\boldsymbol{u}'}$ is a bijection between $\mathbf{S}_{\boldsymbol{u}'}$ and $\mathbf{S}_{\boldsymbol{u}}$;

(*ii*) the restriction of $\boldsymbol{P}_{\boldsymbol{u}\boldsymbol{u}'}$ onto $\mathbf{S}_{\boldsymbol{u}} \cap \mathbf{S}_{\boldsymbol{u}'}$ is the identity;

(*iii*) $\boldsymbol{P}_{\boldsymbol{u}\boldsymbol{u}'} \cdot \boldsymbol{v}_{\boldsymbol{u}\boldsymbol{u}'} = \sqrt{1 - |\boldsymbol{v}_{\boldsymbol{u}\boldsymbol{u}'}|^2}\, \boldsymbol{v}_{\boldsymbol{u}'\boldsymbol{u}}$.

2. World lines

2.1. History of a masspoint: world line

2.1.1. As in the nonrelativistic spacetime model, the history of a masspoint will be described by a curve in the special relativistic spacetime model as well. However, it is not obvious here, what kind of curves can be allowed.

Our heuristic considerations regarding the affine structure of spacetime imply that the history of a free masspoint has to be described by a straight line. We can easily demonstrate that such a straight line must be directed by a timelike vector. Indeed, it cannot be lightlike because this would mean that there is a light signal resting with respect to the masspoint. Suppose that the straight line is directed by a spacelike vector, choose two different points on the line and draw the corresponding future light cones: the cones intersect each other. As a consequence, two light signals emitted successively by the masspoint would meet which contradicts our experience.

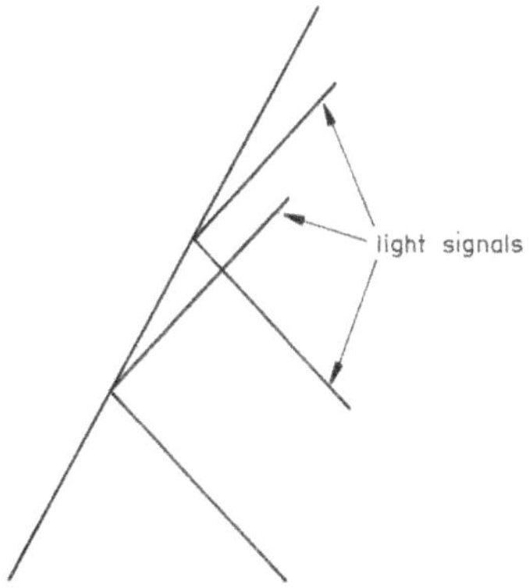

A simple generalization—in accordance with I.2.2—yields that the existence of a masspoint must be described by a curve whose tangent vectors are timelike.

We call attention to the fact that up to now we have spoken about light signals and masspoint histories in a heuristic sense. The following definition gives these notions a precise meaning in the spacetime model.

2.1.2. Definition. 1. A straight line segment in M, directed by a lightlike vector, is called a *light signal.*

2. A *world line* is a connected piecewise twice differentiable curve in M whose tangent vectors are timelike.

Proposition. Let C be a world line. Then $y - x$ is timelike for every $x, y \in C$, $x \neq y$. In other words,

$$C \setminus \{x\} \subset x + T \qquad (x \in C).$$

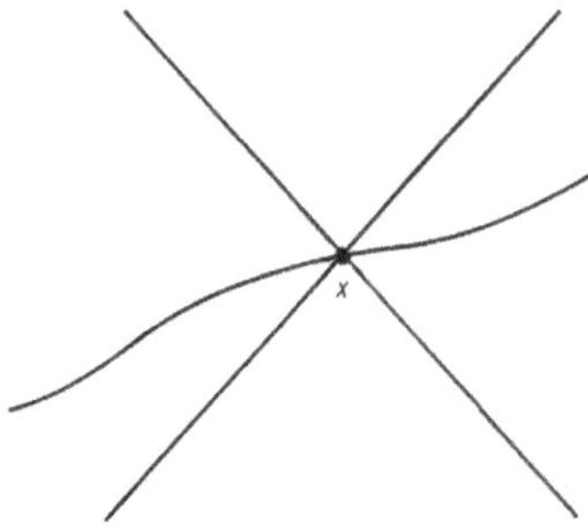

Proof. Suppose the statement is not true: there is an $x \in C$ such that $C \setminus \{x\}$ is not contained in $x + T$. Let $p\colon \mathbb{R} \rightarrowtail M$ be a parametrization of C, $p(0) = x$. Then there is a $0 \neq \alpha \in \operatorname{Dom} p$ such that $p(\alpha) - p(0)$ is not timelike. For the sake of definiteness we can assume $\alpha > 0$. Then

$$a := \inf\{\alpha \in \operatorname{Dom} p |\ \alpha > 0,\ p(\alpha) - p(0) \notin T\} > 0.$$

Indeed, if this infimum were zero then there would be a sequence $\alpha_n > 0$ $(n \in \mathbb{N})$ such that $\lim_{n\to\infty} \alpha_n = 0$ and $\frac{p(\alpha_n)-p(0)}{\alpha_n} \notin T$ for all n implying $\dot{p}(0) = \lim_{n\to\infty} \frac{p(\alpha_n)-p(0)}{\alpha_n} \notin T$ because the set of timelike vectors is open (the complement of T is closed). Because of the same reason, $p(a) - p(0) = \lim_{n\to\infty} \big(p(\alpha_n) - p(0)\big) \notin T$.

Thus $p(\alpha) - p(0)$ is timelike for $0 < \alpha < a$ and $p(a) - p(0)$ is not timelike. Since p is continuous, $p(a) - p(0)$ must be in the closure of T, i.e. it is lightlike: $\big(p(a) - p(0)\big)^2 = 0$.

Lagrange's mean value theorem, applied to the function $[0, a] \to \mathbb{T} \otimes \mathbb{T}$, $\alpha \mapsto \big(p(\alpha) - p(0)\big)^2$ ensures the existence of a $c \in]0, a[$ such that $2\big(p(c) - p(0)\big) \cdot \dot{p}(c) = 0$. Since $\dot{p}(c)$ is timelike, this means that $p(c) - p(0)$ is spacelike, a contradiction.

2.1.3. The previous result and the arrow orientation (which gives rise to the relation 'later', see 1.2.3) allow us to define an order—an orientation—on a world line as follows.

Proposition. Let $p\colon \mathbb{R} \rightarrowtail M$ be a parametrization of the world line C. Then one of the following two possibilities occurs:

(*i*) $\alpha < \beta$ if and only if $p(\alpha)$ is earlier than $p(\beta)$,

(*ii*) $\alpha < \beta$ if and only if $p(\alpha)$ is later than $p(\beta)$

for all $\alpha, \beta \in \mathrm{Dom}\, p$.

Proof. $\dot{p}$ is a continuous function having values in T and defined on an interval, thus its range is connected which means that the range of $\dot{p}$ is contained either in $T^{\rightarrow}$ or in $T^{\leftarrow}$.

(*i*) Suppose $\mathrm{Ran}\,\dot{p} \subset T^{\rightarrow}$ and select an arbitrary α from the domain of p. Then $\{\beta \in \mathrm{Dom}\, p \mid \alpha < \beta\} \to T$, $\beta \mapsto \frac{p(\beta)-p(\alpha)}{\beta-\alpha}$ is a continuous function defined on an interval, hence its range is contained in $T^{\rightarrow}$ or in $T^{\leftarrow}$. Since $\lim\limits_{\beta\to\alpha} \frac{p(\beta)-p(\alpha)}{\beta-\alpha} = \dot{p}(\alpha) \in T^{\rightarrow}$ and $T^{\rightarrow}$ is open, we conclude that $\frac{p(\beta)-p(\alpha)}{\beta-\alpha} \in T^{\rightarrow}$, which implies that $p(\beta) - p(\alpha)$ is in $T^{\rightarrow}$, i.e. $p(\alpha)$ is earlier than $p(\beta)$ for all $\alpha < \beta$.

(*ii*) Similar considerations yield the desired result if $\mathrm{Ran}\,\dot{p} \subset T^{\leftarrow}$.

Definition. A parametrization p of a world line is called *progressive* (*regressive*) if $\alpha < \beta$ implies that $p(\alpha)$ is earlier (later) than $p(\beta)$ for all $\alpha, \beta \in \mathrm{Dom}\, p$.

A world line is considered oriented by progressive parametrizations. ■

The reader easily verifies that the orientation is correctly defined: if p and q are progressive parametrizations of a world line, then $p^{-1} \circ q \colon \mathbb{R} \rightarrowtail \mathbb{R}$ is strictly monotone increasing.

Note that the proposition holds and the definition can be applied also in case of parametrizations that are defined on an oriented one-dimensional affine space.

2.1.4. If x and y are different points of a world line then they are timelike separated.

Conversely, if x and y are timelike separated world points then there is a world line C such that $x, y \in \mathrm{C}$. Indeed, the straight line passing through x and y is such a world line. Note the important fact that there are many world lines containing x and y:

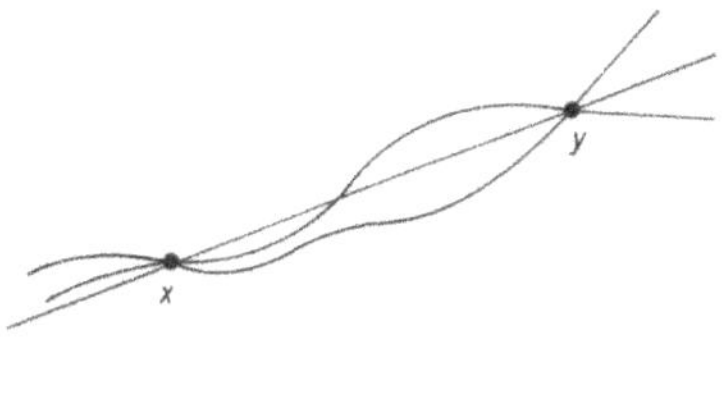

2.2. Proper time of world lines

2.2.1. The Lorentz form $\boldsymbol{g}$ includes the description of proper time progresses, too. Let the occurrence y be later than x and recall the pseudolength $|y-x| := \sqrt{-\boldsymbol{g}(y-x, y-x)}$; then $\boldsymbol{u} := \frac{y-x}{|y-x|}$ is an absolute velocity value. The inertial proper time between x and y is $\boldsymbol{\tau_u}(y-x) = |y-x|$ (which can be explained by 1.3.2. and the similar formulae in the nonrelativistic case).

Take now a 'world line' consisting of two consecutive nonparallel straight line segments (according to our present definition, such a line is not a world line because it is not differentiable at one point, that is why we put the quotation mark; we use such broken world lines for our heuristic consideration and later we permit them by a precise definition, too). Let z be the breaking point, let x be earlier than z, z earlier than y. Then we measure the time passed between x and y along the broken world line by the sum of the time passed along the straight line segments: $|z-x| + |y-z|$.

The generalization to a broken world line consisting of several straight line segments is trivial.

Let now C be an arbitrary world line, $x, y \in \mathrm{C}$, x is earlier than y. It is intuitively clear that we can approximate the time passed between x and y along C by the time passed along broken lines approximating C.

Take a progressive parametrization p of the world line C. Then an approximation of the time passed between x and y along C has the form

$$\sum_{k=1}^{n} |p(\alpha_{k+1}) - p(\alpha_k)|$$

which 'nearly equals'

$$\sum_{k=1}^{n} |\dot{p}(\alpha_k)|(\alpha_{k+1} - \alpha_k).$$

We recognize an integral approximating sum. This suggests the following definition (the reader is asked to study Section VI.7).

Definition. Let x and y be timelike separated world points or $x = y$. If C is a world line passing through x and y (i.e. $x, y \in \mathrm{C}$) then

$$\mathbf{t}_{\mathrm{C}}(x, y) := \int_x^y |d\mathrm{C}|$$

is called the *time passed between x and y along C.*

The time passed between x and y along a straight line is called the *inertial time between x and y* and is denoted by $\mathbf{t}(x, y)$. ■

We emphasize that the integral formula for the time passing along a world line is a *definition* and not a statement.

Evidently,

$$\mathbf{t}(x,y)=\begin{cases}|y-x| & \text{if } x \text{ is earlier than } y,\\ -|y-x| & \text{if } y \text{ is earlier than } x.\end{cases}$$

2.2.2. The time passed between two world points along different world lines can be different. The longest time passes along the inertial world line:

Proposition. Let x be a world point earlier than the world point y. If C is a world line containing x and y then

$$\mathbf{t}_{\mathrm{C}}(x,y)\le \mathbf{t}(x,y)$$

and equality holds if and only if C is a straight line segment between x and y.

Proof. Let $z\in \mathrm{C}$ be a world point which is earlier than y and later than x. Then the reversed triangle inequality (V.4.10) results in $\mathbf{t}(x,z)+\mathbf{t}(z,y)\le \mathbf{t}(x,y)$, where equality holds if and only if z is on the straight line passing through x and y. As a consequence, the time passed between x and y along a broken line (defined to be the sum of times passed along the corresponding straight line segments) is smaller than the inertial time between x and y. The definition of $\mathbf{t}_{\mathrm{C}}(x,y)$ as an integral involves that $\mathbf{t}_{\mathrm{C}}(x,y)$ can be obtained as the infimum of times passed between x and y along broken lines.

2.2.3. We call attention to the fact that in our customary illustration the same time period passed along different inertial world lines is represented, in general, by segments of different lengths.

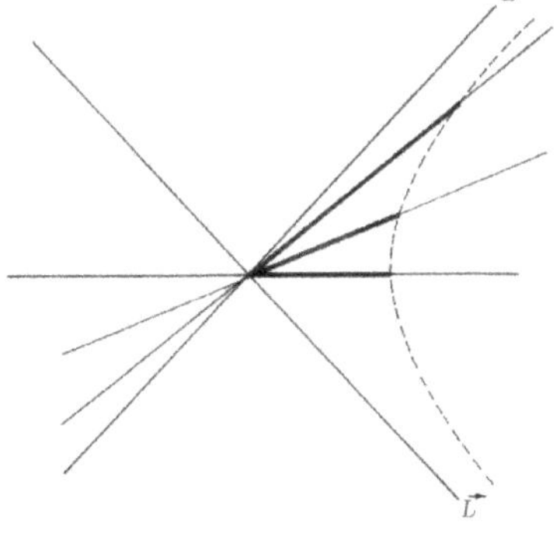

The same length corresponds to the same time period on two inertial world lines if and only if the two illustrating straight lines have the same inclination to the two lines of $L^{\rightarrow}$:

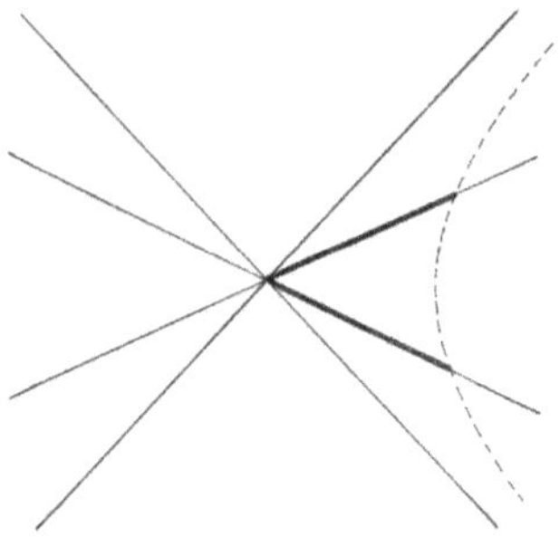

2.3. World line functions

2.3.1. Definition. Let C be a world line, $x_o \in \mathrm{C}$. Then the mapping

$$\mathrm{C} \to \mathbb{T}, \qquad x \mapsto \mathbf{t}_{\mathrm{C}}(x_o, x)$$

is called the *proper time* of C starting from x_o. ■

Since every tangent vector $\boldsymbol{x} \neq \mathbf{0}$ of the world line C is timelike i.e. $|\boldsymbol{x}| \neq \mathbf{0}$, according to Proposition VI.7.5, the inverse of the proper time,

$$r : \mathbb{T} \rightarrowtail \mathrm{M}$$

defined by

$$r\big(\mathbf{t}_{\mathrm{C}}(x_o, x)\big) = x \qquad (x \in \mathrm{C})$$

and having the property

$$\mathbf{t}_{\mathrm{C}}\big(x_o, r(\boldsymbol{t})\big) = \boldsymbol{t} \qquad (\boldsymbol{t} \in \mathrm{Dom}\, r)$$

is a progressive parametrization of C, called the *proper time parametrization* of C starting from x_o. We know that for all $\boldsymbol{t} \in \mathrm{Dom}\, r$

$$\dot{r}(\boldsymbol{t}) \in \frac{\mathbf{M}}{\mathbb{T}}, \qquad \dot{r}(\boldsymbol{t}) \text{ is future directed timelike,}$$

moreover, Proposition VI.7.5 implies

$$|\dot{r}(\boldsymbol{t})| = 1;$$

all these mean that

$$\dot{r}(\boldsymbol{t}) \in V(1) \qquad (\boldsymbol{t} \in \operatorname{Dom} r).$$

2.3.2. According to the previous considerations, if C is a world line then there is a parametrization $r\colon \mathbb{T} \rightarrowtail \mathrm{M}$ of C (i.e. r is defined on an interval, is twice differentiable, its range is C) such that $\dot{r}(\boldsymbol{t}) \in V(1)$ for all $\boldsymbol{t} \in \operatorname{Dom} r$.

From the properties of integration on curves we derive that

$$\mathrm{t}_{\mathrm{C}}(x, y) = \int_{r^{-1}(x)}^{r^{-1}(y)} |\dot{r}(\boldsymbol{t})| d\boldsymbol{t} = r^{-1}(y) - r^{-1}(x).$$

As a consequence, if r_1 and r_2 are parametrizations with the above property then there is a $\boldsymbol{t}_o \in \mathbb{T}$ such that $\operatorname{Dom} r_2 = \boldsymbol{t}_o + \operatorname{Dom} r_1$ and $r_2(\mathbf{t}) = r_1(\boldsymbol{t} - \boldsymbol{t}_o)$ $(\boldsymbol{t} \in \operatorname{Dom} r_2)$.

Indeed, choosing an element x_o of C and putting $\boldsymbol{t}_o := r_2^{-1}(x_o) - r_1^{-1}(x_o)$ we get $r_1^{-1}(x) = r_2^{-1}(x) - \boldsymbol{t}_o$ which gives the desired result with the notation $\boldsymbol{t} := r_2^{-1}(x)$.

2.3.3. Our results suggest how to introduce the notion of world line functions which allows us to admit piecewise differentiability as in the nonrelativistic case.

Definition. A function $r\colon \mathbb{T} \rightarrowtail \mathrm{M}$ is called a *world line function* if

(*i*) $\operatorname{Dom} r$ is an interval,

(*ii*) r is piecewise twice continuously differentiable,

(*iii*) $\dot{r}(\boldsymbol{t})$ is in $V(1)$ for all $\boldsymbol{t} \in \operatorname{Dom} r$ where r is differentiable.

A subset C of M is a *world line* if it is the range of a world line function.

The world line function r and the world line $\operatorname{Ran} r$ is *global* if $\operatorname{Dom} r = \mathbb{T}$. ■

2.3.4. If r is a world line function then differentiating the constant mapping $\boldsymbol{t} \mapsto \dot{r}(\boldsymbol{t}) \cdot \dot{r}(\boldsymbol{t}) = -1$ defined on the differentiable pieces of $\operatorname{Dom} r$ we get that

$$\dot{r}(\boldsymbol{t}) \cdot \ddot{r}(\boldsymbol{t}) = 0, \qquad \text{i.e.} \qquad \ddot{r}(\boldsymbol{t}) \in \frac{\mathbf{S}_{\dot{r}(\boldsymbol{t})}}{\mathbb{T} \otimes \mathbb{T}},$$

and the same is true for right and left derivatives where r is not differentiable.

The functions $\dot{r}\colon \mathbb{T} \rightarrowtail V(1)$ and $\ddot{r}\colon \mathbb{T} \rightarrowtail \frac{\mathbf{M}}{\mathbb{T} \otimes \mathbb{T}}$ can be interpreted as the *absolute velocity* and the *absolute acceleration* of the material point whose history is described by r.

That is why we call the elements of $V(1)$ *absolute velocity values* and the spacelike elements in $\frac{\mathbf{M}}{\mathbb{T} \otimes \mathbb{T}}$ *absolute acceleration values.*

2.3.5. Recall that $V(1)$ is a three-dimensional smooth submanifold of $\frac{\mathbb{M}}{\mathbb{T}}$. The elements of $\left\{\boldsymbol{v} \in \frac{\mathbb{M}}{\mathbb{T}} \mid 0 < \boldsymbol{v}^2 < 1\right\}$ will be called *relative velocity values*; later we shall see the motivation of this name.

Note the following important facts.

(*i*) Absolute velocity values are future directed timelike vectors of cotype $\mathbb{T}$. They do not form either a vector space or an affine space. The pseudo-length of every velocity value is 1. There is no zero velocity value. Velocity values have no angles between themselves.

(*ii*) Relative velocity values are spacelike vectors of cotype $\mathbb{T}$. They do not form a vector space. The *magnitude* of a relative velocity value (see 1.3.3) is a real number less than 1. A relative velocity can be smaller than another; there is a *zero* relative velocity value. If $\boldsymbol{u} \in V(1)$ then $\left\{\boldsymbol{v} \in \frac{\mathbf{S}_{\boldsymbol{u}}}{\mathbb{T}} \mid |\boldsymbol{v}| < 1\right\}$ is an open ball in a three-dimensional Euclidean vector space and it consists of relative velocity values. The *angle* between such relative velocities makes sense.

(*iii*) Absolute acceleration values are the spacelike vectors of cotype $\mathbb{T} \otimes \mathbb{T}$. The *magnitude* of an acceleration value is meaningful, it is an element of $\frac{\mathbb{R}}{\mathbb{T}}$. An acceleration value can be smaller than another; there is a *zero* acceleration value. If $\boldsymbol{u} \in V(1)$, then $\frac{\mathbf{S}_{\boldsymbol{u}}}{\mathbb{T} \otimes \mathbb{T}}$ is a three-dimensional Euclidean vector space consisting of acceleration values. The *angle* between such acceleration values makes sense.

The absence of magnitudes of absolute velocity values means that 'quickness' makes no absolute sense; it is not meaningful that a material object exists more quickly than another. An absolute velocity value characterizes somehow the *tendency* of the history of a material point. Masspoints can move slowly or quickly *relative to each other.*

2.4. Classification of world lines

2.4.1. We would like to classify the world lines as we did it in the nonrelativistic case. The notion of an inertial world line is straightforward. However, uniformly accelerated world lines and twist-free world lines give us some trouble.

If we copied the nonrelativistic definition, i.e. we required that the acceleration of a world line function r be constant, $\ddot{r} = \boldsymbol{a}$, where $\boldsymbol{a}$ is a spacelike element of $\frac{\mathbb{M}}{\mathbb{T} \otimes \mathbb{T}}$, then there would be a $\boldsymbol{c} \in V(1)$ such that $\dot{r}(\boldsymbol{t}) = \boldsymbol{c} + \boldsymbol{a}\boldsymbol{t}$ ($\boldsymbol{t} \in \operatorname{Dom} r$). Since $\dot{r}$ and $\ddot{r}$ are $\boldsymbol{g}$-orthogonal, $\boldsymbol{c}+\boldsymbol{a}\boldsymbol{t}$ and $\boldsymbol{a}$ are $\boldsymbol{g}$-orthogonal: $\boldsymbol{c} \cdot \boldsymbol{a} + |\boldsymbol{a}|^2 \boldsymbol{t} = \mathbf{0}$ for all $\boldsymbol{t} \in \operatorname{Dom} r$ which implies $\boldsymbol{a} = \mathbf{0}$. There would be no uniformly accelerated world lines except the inertial ones.

The problem lies in the fact that the momentary absolute acceleration values of a world line belong to the subspace $\boldsymbol{g}$-orthogonal to the corresponding absolute velocity value; if the velocity value changes then the corresponding subspace changes as well: changing velocity involves changing acceleration.

Nevertheless, we do not have to give up the notion of uniform acceleration. We have established a natural mapping between two subspaces $\boldsymbol{g}$-orthogonal to two absolute velocity values: the corresponding Lorentz boost (see 1.3.8). Then we may require that the world line function r is uniformly accelerated if $\ddot{r}(\boldsymbol{s})$ is mapped into $\ddot{r}(\boldsymbol{t})$ by the Lorentz boost from $\dot{r}(\boldsymbol{s})$ to $\dot{r}(\boldsymbol{t})$.

A similar requirement for $\frac{\ddot{r}}{|\ddot{r}|}$ leads us to twist-free world line functions.

2.4.2. Definition. A twice continuously differentiable world line function r and the corresponding world line is called

(*i*) *inertial* if $\ddot{r} = 0$,

(*ii*) *uniformly accelerated* if $\boldsymbol{L}(\dot{r}(\boldsymbol{t}), \dot{r}(\boldsymbol{s})) \cdot \ddot{r}(\boldsymbol{s}) = \ddot{r}(\boldsymbol{t})$ for all $\boldsymbol{t}, \boldsymbol{s} \in \mathrm{Dom}\, r$,

(*iii*) *twist-free* if $|\ddot{r}(\boldsymbol{t})| \boldsymbol{L}(\dot{r}(\boldsymbol{s}), \dot{r}(\boldsymbol{t})) \cdot \ddot{r}(\boldsymbol{s}) = |\ddot{r}(\boldsymbol{s})| \ddot{r}(\boldsymbol{t})$ for all $\boldsymbol{t}, \boldsymbol{s} \in \mathrm{Dom}\, r$. ■

It is quite evident that a twice continuously differentiable world line function r is inertial if and only if there are an $x_o \in \mathrm{M}$ and a $\boldsymbol{u}_o \in V(1)$ such that

$$r(\boldsymbol{t}) = x_o + \boldsymbol{u}_o \boldsymbol{t} \qquad (\boldsymbol{t} \in \mathrm{Dom}\, r).$$

2.4.3. Let r be a twice continuously differentiable world line function and put

$$\boldsymbol{u} := \dot{r} \colon \mathbb{T} \rightarrowtail V(1).$$

If r is uniformly accelerated, then, by definition,

$$\dot{\boldsymbol{u}}(\boldsymbol{s}) - \frac{\big(\boldsymbol{u}(\boldsymbol{t}) + \boldsymbol{u}(\boldsymbol{s})\big)\big(\boldsymbol{u}(\boldsymbol{t}) \cdot \dot{\boldsymbol{u}}(\boldsymbol{s})\big)}{1 - \boldsymbol{u}(\boldsymbol{t}) \cdot \boldsymbol{u}(\boldsymbol{s})} = \dot{\boldsymbol{u}}(\boldsymbol{t}) \qquad (\boldsymbol{t}, \boldsymbol{s} \in \mathrm{Dom}\, r). \qquad (*)$$

Fix an $\boldsymbol{s} \in \mathrm{Dom}\, r$, put $\boldsymbol{u}_o := \boldsymbol{u}(\boldsymbol{s}) \in V(1)$ and $\boldsymbol{a}_o := \dot{\boldsymbol{u}}(\boldsymbol{s}) \in \frac{\mathbf{S}_{\boldsymbol{u}_o}}{\mathbb{T} \otimes \mathbb{T}}$ to obtain the following first-order differential equation for $\boldsymbol{u}$:

$$\dot{\boldsymbol{u}} = \boldsymbol{a}_o + \frac{(\boldsymbol{u} + \boldsymbol{u}_o)(\boldsymbol{u} \cdot \boldsymbol{a}_o)}{1 - \boldsymbol{u} \cdot \boldsymbol{u}_o}.$$

Unfortunately, it is rather complicated.

Another differential equation can be derived, too, by using $\boldsymbol{u}(\boldsymbol{s}) \cdot \dot{\boldsymbol{u}}(\boldsymbol{s}) = 0$ and observing that $|\ddot{r}| = |\dot{\boldsymbol{u}}| =: \alpha$ is constant (the Lorentz boosts are $\boldsymbol{g}$-orthogonal maps). We obtain the equality

$$\dot{\boldsymbol{u}}(\boldsymbol{s}) - \dot{\boldsymbol{u}}(\boldsymbol{t}) = \frac{\big(\boldsymbol{u}(\boldsymbol{t}) + \boldsymbol{u}(\boldsymbol{s})\big)\big(\boldsymbol{u}(\boldsymbol{t}) - \boldsymbol{u}(\boldsymbol{s})\big) \cdot \dot{\boldsymbol{u}}(\boldsymbol{s})}{1 - \boldsymbol{u}(\boldsymbol{t}) \cdot \boldsymbol{u}(\boldsymbol{s})}$$

from $(*)$; dividing it by $\boldsymbol{s} - \boldsymbol{t}$ and letting $\boldsymbol{s}$ tend to $\boldsymbol{t}$ we get the extremely simple second-order differential equation

$$\ddot{\boldsymbol{u}} = \alpha^2 \boldsymbol{u}$$

whose general solution has the form

$$u(t) = u_o \cosh \alpha t + \frac{a_o}{\alpha} \sinh \alpha t \qquad (t \in \mathbb{T}), \qquad (**)$$

where $u_o \in V(1)$, $a_o \in \frac{\mathbf{S}_{u_o}}{\mathbb{T}\otimes\mathbb{T}}$, $|a_o| = \alpha$.

Equality $(**)$ has been derived from $(*)$. It is not hard to see that $t \mapsto u(t)$ defined by $(**)$ satisfies $(*)$, i.e. $(*)$ and $(**)$ are equivalent.

Finally, a simple integration results in the following.

Proposition. function r is uniformly accelerated if and only if there are an $x_o \in \mathrm{M}$, a $u_o \in V(1)$ and an $a_o \in \frac{\mathbf{S}_{u_o}}{\mathbb{T}\otimes\mathbb{T}}$ such that

$$r(t) = x_o + u_o \frac{\sinh |a_o| t}{|a_o|} + a_o \frac{\cosh |a_o| t - 1}{|a_o|^2} \qquad (t \in \mathrm{Dom}\, r).$$

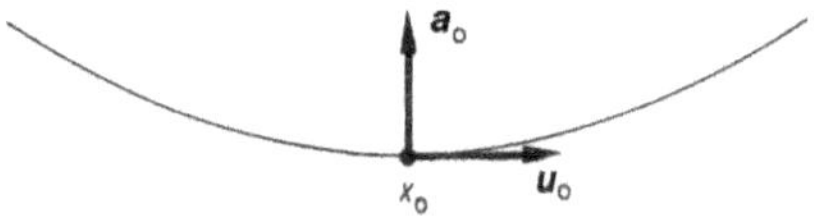

2.4.4. If the twice differentiable world line function r is twist-free, then there are $u_o \in V(1)$, $n_o \in \frac{\mathbf{S}_{u_o}}{\mathbb{T}}$, $|n_o| = 1$ such that for $u := \dot{r}$ the following differential equation holds:

$$\dot{u} = |\dot{u}| \left(n_o + \frac{(u + u_o)(u \cdot n_o)}{1 - u \cdot u_o} \right).$$

The method applied to uniformly accelerated world line functions to derive another differential equation works here as well. The reader is asked to perform the calculations to obtain

$$u|\dot{u}|^4 = \ddot{u}|\dot{u}|^2 - \dot{u}(\dot{u} \cdot \ddot{u}),$$

or

$$u|\dot{u}|^2 = \left(\mathbf{1}_{\mathrm{M}} - \frac{\dot{u} \otimes \dot{u}}{|\dot{u}|^2} \right) \cdot \ddot{u},$$

provided that $\dot{u}$ is nowhere zero.

2.5. World horizons

2.5.1. The light signals starting from a world point x are in $x + L^{\to}$, masspoints existing in x continue their existence in $x + T^{\to}$: every phenomenon occurring in x can influence only the occurrences in $x + (T^{\to} \cup L^{\to})$, the futurelike part of spacetime with respect to x.

Conversely, only the occurrences in $x + (T^{\leftarrow} \cup L^{\leftarrow})$ can influence an occurrence in x.

Consider a world line C. If $x + (T^{\to} \cup L^{\to})$ does not intersect C, then an occurrence in x cannot influence the masspoint whose history is described by C; in other words, the masspoint cannot have information about the occurrence in x. That is why we call

$$\{x \in \mathrm{M} \mid \mathrm{C} \cap (x + (T^{\to} \cup L^{\to})) = \emptyset\} = \{x \in \mathrm{M} \mid (\mathrm{C} - x) \cap (T^{\to} \cup L^{\to}) = \emptyset\}$$

the *indifferent region* of spacetime with respect to C.

It can be shown that it is a closed set (Exercise 2.7.3) whose boundary is called the *world horizon* of the world line C.

Obviously the indifferent region is void if and only if the world horizon is void.

2.5.2. Consider a world line function r. Then a world point x is not indifferent to the corresponding world line if and only if there is a $\boldsymbol{t} \in \mathrm{Dom}\, r$ such that $r(\boldsymbol{t}) - x \in T^{\to} \cup L^{\to}$ i.e.

$$\big(r(\boldsymbol{t}) - x\big)^2 \leq \mathbf{0}$$

and

$$\boldsymbol{u} \cdot \big(r(\boldsymbol{t}) - x\big) < \mathbf{0}$$

for an arbitrary $\boldsymbol{u} \in V(1)$.

2.5.3. The world horizon of an inertial world line is empty.

Indeed, take the inertial world line $x_o + \boldsymbol{u}_o \otimes \mathbb{T}$, an arbitrary world point x and look for $\boldsymbol{t} \in \mathbb{T}$ satisfying

$$(x_o + \boldsymbol{u}_o\boldsymbol{t} - x)^2 \leq \mathbf{0},$$
$$\boldsymbol{u}_o \cdot (x_o + \boldsymbol{u}_o\boldsymbol{t} - x) < \mathbf{0}.$$

Since $(x_o - x)^2 = |\boldsymbol{\pi}_{\boldsymbol{u}_o} \cdot (x_o - x)|^2 - |\boldsymbol{u}_o \cdot (x_o - x)|^2$, the inequalities can be written in the form

$$|\boldsymbol{\pi}_{\boldsymbol{u}_o} \cdot (x_o - x)|^2 - |\boldsymbol{t} - \boldsymbol{u}_o \cdot (x_o - x)|^2 \leq \mathbf{0},$$
$$\boldsymbol{t} - \boldsymbol{u}_o \cdot (x_o - x) > \mathbf{0};$$

they are satisfied for every

$$t > u_o \cdot (x_o - x) + |\pi_{u_o} \cdot (x_o - x)|.$$

2.5.4. The indifferent region of spacetime with respect to the uniformly accelerated global world line described by

$$t \mapsto x_o + u_o \frac{\sinh |a_o| t}{|a_o|} + a_o \frac{\cosh |a_o| t - 1}{|a_o|^2} \qquad (t \in \mathbb{T})$$

is

$$\{x \in \mathrm{M} | (|a_o| u_o + a_o) \cdot (x_o - x) \geq 1\}. \qquad (*)$$

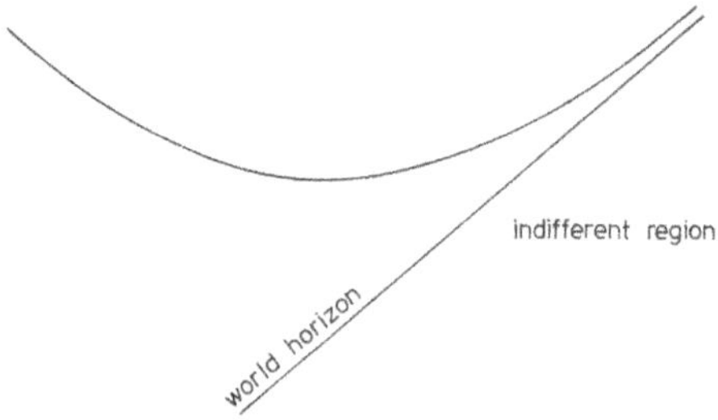

Indeed, according to 2.5.2, the world point x is not indifferent if and only if there is a t for which

$$x^2 - \frac{\sinh^2 \alpha t}{\alpha^2} + \frac{(\cosh \alpha t - 1)^2}{\alpha^2} + 2u_o \cdot x \frac{\sinh \alpha t}{\alpha} + 2a_o \cdot x \frac{\cosh \alpha t - 1}{\alpha^2} \leq 0,$$

$$u_o \cdot x - \frac{\sinh \alpha t}{\alpha} < 0$$

where

$$x := x_o - x, \qquad \alpha := |a_o|.$$

The second inequality holds if t is large enough.
The first inequality can be written in the form

$$x^2 + 2u_o \cdot x \frac{\sinh \alpha t - \cosh \alpha t + 1}{\alpha} + 2(a_o \cdot x + \alpha u_o \cdot x - 1) \frac{\cosh \alpha t - 1}{\alpha^2} \leq 0.$$

Since $\sinh \alpha t - \cosh \alpha t$ tends to zero and $\cosh \alpha t$ tends to plus infinity as t tends to plus infinity, we see that if $(a_o \cdot x + \alpha u \cdot x - 1) < 0$ then both inequalities hold for $\boldsymbol{t}$ large enough, i.e. an x out of the set $(*)$ is not indifferent with respect to the world line.

Take now an x in the set $(*)$ such that $\boldsymbol{u}_o \cdot \boldsymbol{x} = \mathbf{0}$. Then $\boldsymbol{x}$ is a nonzero spacelike vector, thus $\boldsymbol{x}^2 > \mathbf{0}$ and we see that the previous inequality does not hold for any $\boldsymbol{t}$ because $\cosh \alpha \boldsymbol{t} - 1 \geq 0$: x is indifferent with respect to the world line.

To end the proof, note that $\alpha \boldsymbol{u}_o + \boldsymbol{a}_o$ is a lightlike vector, hence x is indifferent if and only if $x + \lambda(\alpha \boldsymbol{u}_o + \boldsymbol{a}_o)$ is indifferent for some $\lambda \in \mathbb{T}$. Choose λ in such a way that $\boldsymbol{u}_o \cdot \big(x_o - x - \lambda(\alpha \boldsymbol{u}_o + \boldsymbol{a}_o)\big) = \mathbf{0}$.

2.6. Newtonian equation

2.6.1. Recall that now the measure line of masses is $\mathbb{T}^* = \frac{\mathbb{R}}{\mathbb{T}}$. Since acceleration values are elements of $\frac{\mathbf{M}}{\mathbb{T}\otimes\mathbb{T}}$ and 'the product of mass and acceleration equals the force', the force values are elements of $\mathbb{T}^* \otimes \frac{\mathbf{M}}{\mathbb{T}\otimes\mathbb{T}} \equiv \frac{\mathbf{M}}{\mathbb{T}\otimes\mathbb{T}\otimes\mathbb{T}} \equiv \frac{\mathbf{M}^*}{\mathbb{T}}$; moreover, we take into account that the momentary acceleration value of a mass-point is $\boldsymbol{g}$-orthogonal to the corresponding velocity value. In what follows, from the dynamical point of view, it is convenient to denote absolute velocity values by $\dot{x}$.

Thus we accept that a *force field* is a differentiable mapping

$$\boldsymbol{f}\colon \mathrm{M} \times V(1) \rightarrowtail \frac{\mathbf{M}^*}{\mathbb{T}}$$

such that

$$\dot{x} \cdot \boldsymbol{f}(x, \dot{x}) = \mathbf{0} \qquad \big((x, \dot{x}) \in \operatorname{Dom} \boldsymbol{f}\big).$$

The history of the material point with mass m under the action of the force field $\boldsymbol{f}$ is described by the Newtonian equation

$$m\ddot{x} = \boldsymbol{f}(x, \dot{x})$$

i.e. the world line function modelling the history is a solution of this differential equation.

2.6.2. Some of the most important force fields in special relativity, too, can be derived from potentials; e.g. the electromagnetic field. However, the gravitational field cannot be described by a potential; this problem will be discussed later (Chapter III).

A *potential* is a twice differentiable mapping

$$\boldsymbol{K}\colon \mathrm{M} \rightarrowtail \mathbf{M}^*$$

(in other words, a potential is a twice differentiable covector field).

The *field strength* corresponding to $\boldsymbol{K}$ is $\mathrm{D} \wedge \boldsymbol{K}\colon \mathrm{M} \rightarrowtail \mathbf{M}^* \wedge \mathbf{M}^*$ (the antisymmetric or exterior derivative of $\boldsymbol{K}$, see VI.3.6).

The force field $\boldsymbol{f}$ has a potential (is derived from a potential) if
— there is an open subset $\mathrm{O} \subset \mathrm{M}$ such that $\mathrm{Dom}\,\boldsymbol{f} = \mathrm{O} \times V(1)$,
— there is a potential $\boldsymbol{K}$ defined on O such that

$$\boldsymbol{f}(x, \dot{x}) = \boldsymbol{F}(x) \cdot \dot{x} \qquad (x \in \mathrm{O}, \dot{x} \in V(1))$$

where $\boldsymbol{F} := \mathrm{D} \wedge \boldsymbol{K}$.

It is worth mentioning: $\boldsymbol{F}(x)$ is antisymmetric, hence $\dot{x} \cdot \boldsymbol{F}(x) \cdot \dot{x} = \mathbf{0}$, as it must be for a force field.

2.6.3. In the nonrelativistic spacetime model a force field can be independent of either of its variables, in particular, it can be a constant map. In the present case, on the contrary, a nonzero force field cannot be independent of velocity, in particular, it cannot be a constant map.

We could try to define a constant force field in such a way that the corresponding Lorentz boosts map its values into each other, i.e. $\boldsymbol{f}$ would be constant if $\boldsymbol{L}(\dot{x}', \dot{x}) \cdot \boldsymbol{f}(x, \dot{x}) = \boldsymbol{f}(x, \dot{x}')$ for all possible x, $\dot{x}$ and $\dot{x}'$. However, such a nonzero field cannot exist (Exercise 2.7.5): *there is no nonzero special relativistic constant force field*!

2.7. Exercises

1. Prove that the uniformly accelerated world line function given in 2.4.3 satisfies

$$r(\boldsymbol{t}) = x_o + \boldsymbol{u}_o \boldsymbol{t} + \frac{\boldsymbol{a}_o}{2} \boldsymbol{t}^2 + \mathrm{ordo}(\boldsymbol{t}^3).$$

2. Let $\boldsymbol{u}_o \in V(1)$, $\boldsymbol{a}_o \in \frac{\mathbf{S}_{\boldsymbol{u}_o}}{\mathbb{T} \otimes \mathbb{T}}$ and $\beta\colon \mathbb{T} \rightarrowtail \mathbb{T}$ a continuously differentiable function defined on an interval. Demonstrate that the world line function r for which

$$\dot{r} = \boldsymbol{u}_o \sqrt{\beta^2 + 1} + \boldsymbol{a}_o \beta$$

holds is twist-free.

3. The indifferent region of spacetime with respect to the world line C has the complement

$$\bigcup_{z \in C} \{z + (T^{\leftarrow} \cup L^{\leftarrow})\}.$$

Using $L^{\rightarrow} + T^{\rightarrow} = T^{\rightarrow}$ show that it equals

$$\bigcup_{z \in C} \{z + T^{\leftarrow}\}$$

which, being a union of open sets, is open. Consequently, the indifferent part of spacetime with respect to C is closed.

4. Let r be a global world line function and put $\boldsymbol{u} := \dot{r}$. Prove that the world horizon of the corresponding world line is empty if one of the following conditions holds:

(i) there exist $\lim_{t\to\infty} \boldsymbol{u}(t)$,

(ii) $\boldsymbol{u}$ is periodic, i.e. there is a $\boldsymbol{t}_o > \boldsymbol{0}$ such that $\boldsymbol{u}(\boldsymbol{t}+\boldsymbol{t}_o) = \boldsymbol{u}(\boldsymbol{t})$ for all $\boldsymbol{t} \in \mathbb{T}$.

(Hint: (i) $V(1)$ is closed, hence the limit belongs to it. (ii) Put $\boldsymbol{z}_o := \int_0^{\boldsymbol{t}_o} \boldsymbol{u}(\boldsymbol{t})d\boldsymbol{t}$, $\boldsymbol{u}_o := \frac{\boldsymbol{z}_o}{|\boldsymbol{z}_o|}$ and consider the inertial world line $r(\boldsymbol{t}_o) + \boldsymbol{u}_o \otimes \mathbb{T}$.)

5. Let $\boldsymbol{s}\colon V(1) \to \mathbf{M}$ be a function such that

$$\boldsymbol{u}\cdot\boldsymbol{s}(\boldsymbol{u}) = 0 \quad\text{and}\quad \boldsymbol{s}(\boldsymbol{u}') := \boldsymbol{L}(\boldsymbol{u}',\boldsymbol{u})\cdot\boldsymbol{s}(\boldsymbol{u}) \qquad (\boldsymbol{u}',\boldsymbol{u}\in V(1)).$$

Prove that $\boldsymbol{s} = \boldsymbol{0}$. (Hint: $\boldsymbol{L}(\boldsymbol{u}'',\boldsymbol{u}')\cdot\boldsymbol{L}(\boldsymbol{u}',\boldsymbol{u})\cdot\boldsymbol{s}(\boldsymbol{u}) = \boldsymbol{L}(\boldsymbol{u}'',\boldsymbol{u})\cdot\boldsymbol{s}(\boldsymbol{u})$ must hold; applying Proposition 1.3.9 find appropriate $\boldsymbol{u}''$ and $\boldsymbol{u}'$ for a fixed $\boldsymbol{u}$ in such a way that the equality fails.)

3. Observers and synchronizations

3.1. The notions of an observer and its space

3.1.1. In most of the textbooks one says that special relativity concerns only inertial observers, treating of noninertial observers is possible only in the framework of general relativity. We emphasize that this is not true.

The difference between special relativity and general relativity does not lie in observers which is evident from our point of view: spacetime models are defined without the notion of observers; on the contrary, observers are defined by means of spacetime models.

Noninertial observers are right objects in the special relativistic spacetime model. Inertial observers and noninertial observers differ only in the level of mathematical tools they require. Inertial observers remain in the nice and simple framework of affine spaces while the deep treatment of noninertial observers needs the same mathematical tools as the treatment of general relativistic spacetime models: the theory of pseudo-Riemannian manifolds.

We can repeat word by word what we said in I.3.1.1 to motivate the following definition.

Definition. An *observer* is a smooth map $\boldsymbol{U}\colon \mathrm{M} \rightarrowtail V(1)$ whose domain is connected.

If $\mathrm{Dom}\,\boldsymbol{U} = \mathrm{M}$, the observer is *global.*

The observer is called *inertial* if it is a constant map. ■

$V(1)$ is a subset of $\frac{\mathrm{M}}{\mathbb{T}}$; differentiability (smoothness) of a map from M into $V(1)$ means differentiability (smoothness) of the map from M into $\frac{\mathrm{M}}{\mathbb{T}}$.

3.1.2. Let $\boldsymbol{U}$ be an observer. Integral curves of the differential equation

$$(x\colon\ \mathbb{T} \rightarrowtail \mathrm{M})? \qquad \dot{x} = \boldsymbol{U}(x)$$

are evidently world lines.

The maximal integral curves of this differential equation will be called *$\boldsymbol{U}$-lines.*

As in the nonrelativistic case,

$$\boldsymbol{A}_U := \mathrm{D}\boldsymbol{U} \cdot \boldsymbol{U}\colon\ \mathrm{M} \rightarrowtail \frac{\mathbf{M}}{\mathbb{T} \otimes \mathbb{T}}$$

is the *acceleration field* corresponding to $\boldsymbol{U}$.

3.1.3. Again we can repeat the arguments confirming that the space of an observer is the set of its maximal integral curves.

Definition. Let $\boldsymbol{U}$ be an observer. Then S_U, the set of maximal integral curves of $\boldsymbol{U}$, is the *space* of the observer $\boldsymbol{U}$ or the *$\boldsymbol{U}$-space.* ■

Again a maximal integral curve of $\boldsymbol{U}$ is called a $\boldsymbol{U}$-line if considered to be a subset of M and it is called a $\boldsymbol{U}$-space point if considered to be an element of S_U.

$C_U(x)$ will stand for the (unique) $\boldsymbol{U}$-line passing through x; we say that $C_U(x)$ is the $\boldsymbol{U}$-space point that x is *incident* with.

It can be shown that, in general, the $\boldsymbol{U}$-space can be endowed with a smooth structure in a natural way, thus limits, differentiability, etc. will make sense. However, in this book we avoid the general theory of smooth manifolds, that is why, in general, we do not deal with the structure of observer spaces. Later the spaces of some special observers, important from the point of view of applications, will be treated in detail.

Because of the absence of absolute synchronization, here we cannot classify observers as in I.3.2.1.

3.2. The notions of a synchronization and its time

3.2.1. Now synchronizations play a nontrivial role, in contrast to the nonrelativistic case. Of course, we can repeat the arguments in I.3.3.1 that a synchronization is an equivalence relation having the property that different occurrences of any world line and light signal cannot be simultaneous.

Thus, x and y cannot be simultaneous if $y - x$ is timelike or lightlike.

Definition. A *synchronization* $\mathcal{S}$ is a smooth equivalence relation defined on a connected open subset of spacetime such that every equivalence class is a three-dimensional submanifold whose every tangent space is transversal to all timelike and lightlike vectors. The set $\mathrm{T}_{\mathcal{S}}$ of the equivalence classes is the *time* of the synchronization $\mathcal{S}$ or the $\mathcal{S}$*-time. An everywhere defined synchronization is called global.* ■

An equivalence class of a synchronization is called a *world surface*; in particular, that of $\mathcal{S}$ is a $\mathcal{S}$-surface if considered to be a subset of M and is called a $\mathcal{S}$-time point or $\mathcal{S}$-instant if considered to be an element of $\mathrm{T}_{\mathcal{S}}$. According to the definition, the tangent spaces of $\mathcal{S}$-surfaces must be spacelike.

A $\mathcal{S}$-surface consists of $\mathcal{S}$*-simultaneous* occurrences. $\tau_{\mathcal{S}}(x)$ will stand for the (unique) $\mathcal{S}$-surface containing x.

The smoothness of a synchronization $\mathcal{S}$, given on $\mathrm{G} \subset \mathrm{M}$, means the following: to every $x \in \mathrm{G}$ there is a unique $\boldsymbol{U}_{\mathcal{S}}(x) \in V(1)$ such that the tangent space of the equivalence class (world surface) at x equals $\mathbf{S}_{\boldsymbol{U}_{\mathcal{S}}(x)}$; then $\boldsymbol{U}_{\mathcal{S}} : \mathrm{M} \rightarrowtail V(1) \subset \frac{\mathrm{M}}{\mathrm{T}}$ is required to be smooth.

3.2.2. We can prove, similarly to the corresponding assertion for world lines, that if F is a world surface and $x \in \mathrm{F}$, then $\mathrm{F} \setminus \{x\} - x \subset S$ consists of spacelike vectors.

Consequently, if F is a world surface and C is a world line then $\mathrm{C} \cap \mathrm{F}$ is either void or contains a single element which we shall denote by $\mathrm{C} \star \mathrm{F}$.

Definition. Let $\mathcal{S}$ be a synchronization, $t, s \in \mathrm{T}_{\mathcal{S}}$. We say that s is *later* than t (t is *earlier* than s) if there are $x \in t$ and $y \in s$ such that y is later than x. ■

It cannot occur that both of t and s are earlier than the other. Indeed, let s, t and x, y be as in the definition. Then for all $y' \in s$, $x' \in t$ we have $y' - x' = (y' - y) + (y - x) + (x - x')$. Because of the properties of world surfaces, $y' - y$ and $x - x'$ are spacelike vectors. Thus, in view of Exercise V.4.22.2, if $y - x \in T^{\rightarrow}$ (t is earlier than s) then $y' - x' \notin T^{\leftarrow}$ (s is not earlier than t).

We easily find that 'later' is an ordering (a reflexive, antisymmetric and transitive relation) on $\mathrm{T}_{\mathcal{S}}$. However, it need not be total: there can be t and s in $\mathrm{T}_{\mathcal{S}}$ such that neither of them is later than the other:

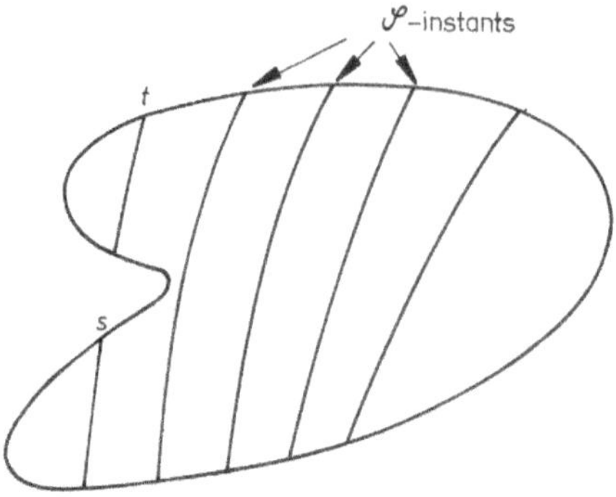

We say that the synchronization $\mathcal{S}$ is *well posed* if the relation 'later' on $\mathrm{T}_{\mathcal{S}}$ is a total ordering.

It can be shown that every world point x_o in the domain of $\mathcal{S}$ has a neighbourhood such that the restriction of $\mathcal{S}$ onto this neighbourhood is well posed. If C_o is the $\boldsymbol{U}_{\mathcal{S}}$-line passing through x_o, then $\{x \in \mathrm{M} | \mathrm{C}_o \cap \tau_{\mathcal{S}}(x) \neq \emptyset\}$ is such a neighbourhood.

3.2.3. Definition. A *reference frame* is a pair $(\mathcal{S}, \boldsymbol{U})$ where $\mathcal{S}$ is a synchronization and $\boldsymbol{U}$ is an observer having a common domain. ■

A reference frame $(\mathcal{S}, \boldsymbol{U})$ *splits* a part of spacetime (its domain) into $\mathcal{S}$-time and $\boldsymbol{U}$-space:

$$\xi_{\mathcal{S},\boldsymbol{U}}\colon \mathrm{M} \rightarrowtail \mathrm{T}_{\mathcal{S}} \times \mathrm{S}_{\boldsymbol{U}}, \qquad x \mapsto \big(\tau_{\mathcal{S}}(x), C_{\boldsymbol{U}}(x)\big).$$

3.2.4. It is worth mentioning that a synchronization determines an observer uniquely but the contrary does not hold. For an observer $\boldsymbol{U}$, in general, there are no world surfaces whose tangent spaces at every x is $\mathbf{S}_{\boldsymbol{U}(x)}$ (see 6.7.6.).

3.3. Global inertial observers and their spaces

3.3.1. Let us consider a global inertial observer; we shall refer to it by its constant absolute velocity value $\boldsymbol{u}$.

The observer space $\mathrm{S}_{\boldsymbol{u}}$ is the set of straight lines directed by $\boldsymbol{u}$; more closely,

$$C_{\boldsymbol{u}}(x) = x + \boldsymbol{u} \otimes \mathbb{T} := \{x + \boldsymbol{u}\boldsymbol{t} \mid \boldsymbol{t} \in \mathbb{T}\}$$

which turns to be an affine space over $\mathbf{S}_{\boldsymbol{u}}$, similarly to I.4.2.

Proposition. $S_{\boldsymbol{u}}$, the space of the inertial observer $\boldsymbol{u}$, endowed with the subtraction

$$q' - q := \boldsymbol{\pi_u} \cdot (x' - x) \qquad (x' \in q', x \in q)$$

is an affine space over $\mathbf{S}_{\boldsymbol{u}}$. ∎

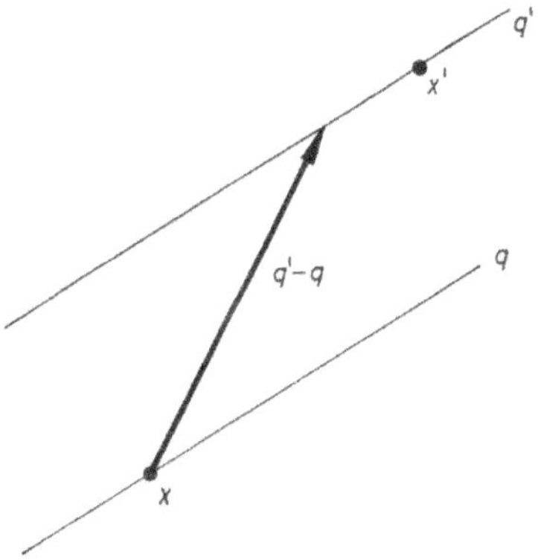

Proof. For the proof we have to show only that the subtraction is well defined: if $y' \in q'$ and $y \in q$, too, then $\boldsymbol{\pi_u} \cdot (x' - x) = \boldsymbol{\pi_u} \cdot (y' - y)$, which is evident. ∎

Recall that the restriction of the Lorentz form to $\mathbf{S}_{\boldsymbol{u}}$, denoted by $\boldsymbol{h_u}$, is positive definite: $(\mathbf{S}_{\boldsymbol{u}}, \mathbb{T}, \boldsymbol{h_u})$ is a Euclidean vector space. Thus we can say that the space of a global inertial observer is a three-dimensional oriented Euclidean affine space.

3.3.2. In the nonrelativistic model the spaces of different inertial observers are different affine spaces over the same vector space, whereas here the spaces of different inertial observers are **different affine spaces** over **different vector spaces**. This very important fact gets lost in the usual treatments where all inertial spaces are represented in the same way by triplets of numbers.

3.4. Inertial reference frames

3.4.1.

Definition. A synchronization is called *affine* if its world surfaces are parallel affine hyperlanes in M. ∎

We have immediately:

Proposition. A synchronization $\mathcal{S}$ is affine if and only there is a $\boldsymbol{u}_s \in V(1)$ such that the $\mathcal{S}$-surfaces are hyperplanes directed by $\mathbf{S}_{\boldsymbol{u}_s}$. ∎

Such a synchronization is called the synchronization due to $\boldsymbol{u}_s$ or $\boldsymbol{u}_s$-synchronization and $\mathrm{T}_{\boldsymbol{u}_s}$ will denote the corresponding time (the set of hyperplanes directed by $\mathbf{S}_{\boldsymbol{u}_s}$).

3.4.2. Observers and synchronizations, in general, are complicated objects in the relativistic spacetime model. Exceptions are inertial observers and affine synchronizations.

Definition. An *inertial reference frame* consists of an affine synchronization and a global inertial observer. ∎

Let $\boldsymbol{u}$ be a global inertial observer and let us take a global $\boldsymbol{u}_s$-synchronization; the corresponding inertial reference frame will be denoted by $(\boldsymbol{u}_s, \boldsymbol{u})$.

In this reference frame the observer measure time periods between $\boldsymbol{u}_s$-instants by the proper time passed in the $\boldsymbol{u}$-space points.

Let t and s be two $\boldsymbol{u}_s$-instants, and let y and x be occurrences in t and s, respectively. If $\boldsymbol{t}$ is the time period measured by $\boldsymbol{u}$ between s and t then $\boldsymbol{ut} + (x - y) \in \mathbf{S}_{\boldsymbol{u}_s}$ i.e. $\boldsymbol{u}_s \cdot (\boldsymbol{ut} + (x - y)) = \mathbf{0}$. Then we have

Proposition $\mathrm{T}_{\boldsymbol{u}_s}$ endowed with the subtraction

$$(t - s)_{\boldsymbol{u}} := \frac{-\boldsymbol{u}_s \cdot (y - x)}{-\boldsymbol{u}_s \cdot \boldsymbol{u}} \qquad (t, s \in \mathrm{T}_{\boldsymbol{u}_s},\ y \in t,\ x \in s)$$

is an affine space over $\mathbb{T}$. ∎

Proof. For the proof we have to show only that the subtraction is well defined i.e. does not depend on the choice of the occurrences in t and s which is evident. ∎

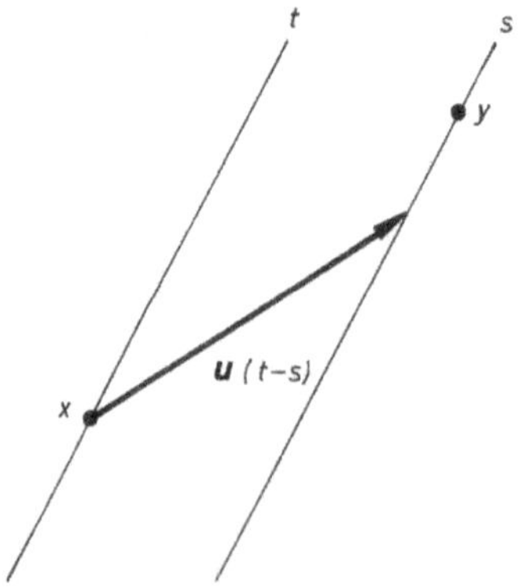

3.5. Standard inertial frames

3.5.1.

Definition. A *standard inertial frame* is an inertial reference frame in which the synchronization and the observer are given by the same absolute velocity value. ■

We shall refer to a standard inertial frame by the corresponding absolute velocity value. Note that the same letter $\boldsymbol{u}$ can denote an observer, a synchronization and a standard inertial frame; the actual role of $\boldsymbol{u}$ will always be specified, thus, hopefully, we avoid misunderstandings.

3.5.2. In practice, synchronization is established nowadays by radio signals (i.e. light signals). A chronometer ticks in the studio and at a determined tick – let us call it 12 o'clock – emits a radio signal. Hearing the signal, we set our clock. Of course, in principle, we do not set the clock to 12 because 'some time has elapsed between emission and reception'. The elapsed time is calculated from the distance covered by the radio signal and from the velocity of light. But this is a vicious circle: the velocity of light makes sense only with a given synchronization. The right setting is that *we establish a synchronization in which one-way light speed equals the round-way speed.*

It will be found that such a synchronization cannot be correct on the Earth, being a uniformly rotating observer (see Exercise 6.9.11.). In the following we investigate how this method works for an inertial observer.

3.5.3. Let us consider a global inertial observer $\boldsymbol{u}$. A light signal is emitted in a $\boldsymbol{u}$-space point q and, reflected in some other $\boldsymbol{u}$-space point, it returns. Let the occurrences of emission, reflection and return be x_e, y and x_r, respectively. Then there is an occurrence x and a time period $\boldsymbol{t}$ such that $x_e = x - \boldsymbol{ut}$, $x_r = x + \boldsymbol{ut}$. Thus, it takes $2\boldsymbol{t}$ for the light signal to travel back and forth; the one-way speed equals the two-way speed if x and y are defined to be simultaneous.

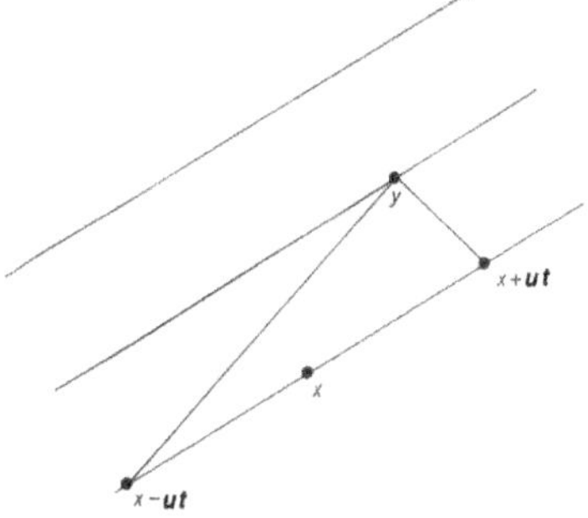

Then $y-(x-\boldsymbol{u}t)$ and $y-(x+\boldsymbol{u}t)$ are lightlike vectors:

$$\big(y-(x-\boldsymbol{u}t)\big)^2=\mathbf{0}, \qquad \big(y-(x+\boldsymbol{u}t)\big)^2=\mathbf{0},$$
$$(y-x)^2+2(y-x)\cdot\boldsymbol{u}t-t^2=\mathbf{0}, \qquad (y-x)^2-2(y-x)\cdot\boldsymbol{u}t-t^2=\mathbf{0},$$

which give

$$y-x\in\mathbf{S}_{\boldsymbol{u}}; \quad \text{in other words,} \quad y\in x+\mathbf{S}_{\boldsymbol{u}}.$$

We have obtained the affine synchronization due to $\boldsymbol{u}$. Thus we have:

Proposition. The one-way speed of light equals the round-way speed only in standard inertial frames. ■

3.5.4. Note that the vector defined between the $\boldsymbol{u}$-space points q' and q in 3.3.1. is the vector between arbitrary $\boldsymbol{u}$-simultaneous occurrences of q' and q. Indeed, if $x'\in q'$ and $x\in q$ such that $x'-x\in\mathbf{S}_{\boldsymbol{u}}$ then $\boldsymbol{\pi}_{\boldsymbol{u}}\cdot(x'-x)=x'-x$.

3.6. Standard splitting of spacetime

3.6.1. Let us take a standard inertial frame $\boldsymbol{u}$.

This inertial frame assigns to every world point x the $\boldsymbol{u}$-time point $\tau_{\boldsymbol{u}}(x)$, the set of world points $\boldsymbol{u}$-simultaneous with x:

$$\tau_{\boldsymbol{u}}(x)=x+\mathbf{S}_{\boldsymbol{u}},$$

as well as the $\boldsymbol{u}$-space point $C_{\boldsymbol{u}}(x)$ that x is incident with:

$$C_{\boldsymbol{u}}(x)=x+\boldsymbol{u}\otimes\mathbb{T}.$$

It is worth listing the following relations regarding the affine structures of $\mathrm{T}_{\boldsymbol{u}}$ and of $\mathrm{S}_{\boldsymbol{u}}$ as well as the mappings $\tau_{\boldsymbol{u}}:\mathrm{M}\to\mathrm{T}_{\boldsymbol{u}}$ and $C_{\boldsymbol{u}}:\mathrm{M}\to\mathbf{S}_{\boldsymbol{u}}$:

(*i*) $(y+\mathbf{S}_{\boldsymbol{u}})-(x+\mathbf{S}_{\boldsymbol{u}})=-\boldsymbol{u}\cdot(y-x)$ $\quad (x,y\in\mathrm{M})$,

(*ii*) $(x+\boldsymbol{x}+\mathbf{S}_{\boldsymbol{u}})=(x+\mathbf{S}_{\boldsymbol{u}})-\boldsymbol{u}\cdot\boldsymbol{x}$ $\quad (x\in\mathrm{M},\ \boldsymbol{x}\in\mathbf{M})$,

(*iii*) $x+\mathbf{S}_{\boldsymbol{u}}=y+\mathbf{S}_{\boldsymbol{u}}$ if and only if $y-x$ is $\boldsymbol{g}$-orthogonal to $\boldsymbol{u}$,

and

(*iv*) $(x'+\boldsymbol{u}\otimes\mathbb{T})-(x+\boldsymbol{u}\otimes\mathbb{T})=\boldsymbol{\pi}_{\boldsymbol{u}}\cdot(x'-x)$ $\quad (x',x\in\mathrm{M})$,

(*v*) $(x+\boldsymbol{x})+\boldsymbol{u}\otimes\mathbb{T}=(x+\boldsymbol{u}\otimes\mathbb{T})+\boldsymbol{\pi}_{\boldsymbol{u}}\cdot\boldsymbol{x}$ $\quad (x\in\mathrm{M},\ \boldsymbol{x}\in\mathbf{M})$,

(*vi*) $x+\boldsymbol{u}\otimes\mathbb{T}=x'+\boldsymbol{u}\otimes\mathbb{T}$ if and only if $x'-x$ is parallel to $\boldsymbol{u}$;

moreover,

(*vii*) $(y+\mathbf{S}_{\boldsymbol{u}})\cap(x+\boldsymbol{u}\otimes\mathbb{T})=\big\{x+\boldsymbol{u}\big(-\boldsymbol{u}\cdot(y-x)\big)\big\}$ $\quad (x,y\in\mathrm{M})$

or, in another form,

$$(y+\mathbf{S}_{\boldsymbol{u}})\star(x+\boldsymbol{u}\otimes\mathbb{T})=x+\boldsymbol{u}\big(-\boldsymbol{u}\cdot(y-x)\big)$$

3.6.2. It is trivial by the previous formulae (i) and (iv) that

$$\tau_{\boldsymbol{u}}\colon \mathrm{M} \to \mathrm{T}_{\boldsymbol{u}}, \qquad x \mapsto x + \mathbf{S}_{\boldsymbol{u}}$$

is an affine map over $\boldsymbol{\tau}_{\boldsymbol{u}} = -\boldsymbol{u}$ and

$$C_{\boldsymbol{u}}\colon \mathrm{M} \to \mathrm{S}_{\boldsymbol{u}}, \qquad x \mapsto x + \boldsymbol{u} \otimes \mathbb{T}$$

is an affine map over $\boldsymbol{\pi}_{\boldsymbol{u}}$.

Definition.

$$\xi_{\boldsymbol{u}} := (\tau_{\boldsymbol{u}}, C_{\boldsymbol{u}})\colon \mathrm{M} \to \mathrm{T}_{\boldsymbol{u}} \times \mathrm{S}_{\boldsymbol{u}}, \qquad x \mapsto (x + \mathbf{S}_{\boldsymbol{u}},\, x + \boldsymbol{u} \otimes \mathbb{T})$$

is the *splitting of spacetime according to the standard inertial frame* $\boldsymbol{u}$.

Proposition. The splitting $\xi_{\boldsymbol{u}}$ is an orientation preserving affine bijection over the linear map $\boldsymbol{\xi}_{\boldsymbol{u}} = (\boldsymbol{\tau}_{\boldsymbol{u}}, \boldsymbol{\pi}_{\boldsymbol{u}})$ (cf. 1.3.5) and

$$\xi_{\boldsymbol{u}}^{-1}(t, q) = q \star t \qquad (t \in \mathrm{T}_{\boldsymbol{u}},\ q \in \mathrm{S}_{\boldsymbol{u}}).$$

3.6.3. We can simplify a number of formulae and calculations by choosing a $\boldsymbol{u}$-time point t_o and a $\boldsymbol{u}$-space point q_o and vectorizing $\boldsymbol{u}$-time and $\boldsymbol{U}$-space:

$$\mathrm{T}_{\boldsymbol{u}} \to \mathbb{T}, \qquad t \mapsto t - t_o,$$
$$\mathrm{S}_{\boldsymbol{u}} \to \mathbf{S}_{\boldsymbol{u}}, \qquad q \mapsto q - q_o.$$

Choosing t_o and q_o is equivalent to choosing a 'spacetime reference origin' $o \in \mathrm{M}$: $\{o\} \in q_o \cap t_o$, $\tau_{\boldsymbol{u}}(o) = t_o$, $C_{\boldsymbol{u}}(o) = q_o$.

The pair $(\boldsymbol{u}, o)$ is called a *standard inertial frame with reference origin.* We can establish the *vectorized splitting* of spacetime due to $(\boldsymbol{u}, o)$:

$$\begin{aligned}\xi_{\boldsymbol{u},o}\colon \mathrm{M} \to \mathbb{T} \times \mathbf{S}_{\boldsymbol{u}}, \quad x \mapsto \big(\tau_{\boldsymbol{u}}(x) - \tau_{\boldsymbol{u}}(o),\, C_{\boldsymbol{u}}(x) - C_{\boldsymbol{u}}(o)\big) &= \\ = \big(-\boldsymbol{u} \cdot (x - o),\, \boldsymbol{\pi}_{\boldsymbol{u}} \cdot (x - o)\big)&.\end{aligned}$$

Thus, if O_o denotes the vectorization of M with origin o then

$$\xi_{\boldsymbol{u},o} = \boldsymbol{\xi}_{\boldsymbol{u}} \circ O_o.$$

3.7. Exercise

Choose the zero in $\mathbb{R}^{1+3}$ to be a reference origin for the standard reference frame of the 'basic observer' $(1, \mathbf{0})$ in the arithmetic spacetime model. Then the vectorized splitting of spacetime is the identity map of $\mathbb{R}^{1+3}$.

4. Kinematics

4.1. Motions relative to a standard inertial frame

4.1.1. Consider a standard inertial frame $\boldsymbol{u}$.

Take a world line function r.

Then the function $\tau_{\boldsymbol{u}} \circ r \colon \mathbb{T} \rightarrowtail \mathrm{T}_{\boldsymbol{u}}$ assigns $\boldsymbol{u}$-time points to proper time points of r. This function is piecewise twice differentiable and its derivative

$$(\tau_{\boldsymbol{u}} \circ r)^{\cdot} = \boldsymbol{\tau_u} \cdot \dot{r} = -\boldsymbol{u} \cdot \dot{r}$$

is everywhere positive (see 1.3.1). Consequently, $\tau_{\boldsymbol{u}} \circ r$ is strictly monotone increasing, has a monotone increasing inverse

$$z_{\boldsymbol{u}} := (\tau_{\boldsymbol{u}} \circ r)^{-1} \colon \mathrm{T}_{\boldsymbol{u}} \rightarrowtail \mathbb{T}$$

which gives the proper time points of r corresponding to $\boldsymbol{u}$-time points; moreover, its derivative comes from the inverse of the derivative of $\tau_{\boldsymbol{u}} \circ r$:

$$\dot{z}_{\boldsymbol{u}}(t) = \frac{1}{-\boldsymbol{u} \cdot \dot{r}(z_{\boldsymbol{u}}(t))} \qquad (t \in \mathrm{Dom}\, z_{\boldsymbol{u}}).$$

4.1.2. The history of a material point is described by a world line function r. A standard inertial frame $\boldsymbol{u}$ perceives this history as a motion described by a function $r_{\boldsymbol{u}}$ assigning to $\boldsymbol{u}$-time points the $\boldsymbol{u}$-space points where the material point is at that $\boldsymbol{u}$-time point.

To establish this function, select a $\boldsymbol{u}$-time point t; find the corresponding proper time point $z_{\boldsymbol{u}}(t)$ and the spacetime position $r(z_{\boldsymbol{u}}(t))$ of the material point; look for the $\boldsymbol{u}$-space point $C_{\boldsymbol{u}}\big(r(z_{\boldsymbol{u}}(t))\big)$ that the world point in question is incident with.

Definition.

$$r_{\boldsymbol{u}} \colon \mathrm{T}_{\boldsymbol{u}} \rightarrowtail \mathrm{S}_{\boldsymbol{u}}, \qquad t \mapsto C_{\boldsymbol{u}}\big(r(z_{\boldsymbol{u}}(t))\big) = r(z_{\boldsymbol{u}}(t)) + \boldsymbol{u} \otimes \mathbb{T}$$

is called the *motion relative to* $\boldsymbol{u}$, or the $\boldsymbol{u}$*-motion*, corresponding to the world line function r.

4.1.3. The question arises whether the history, i.e. the world line function, can be regained from the motion. Later a positive answer will be given (Section 4.4).

4.1.4. Some formulae and calculations become simpler if we use a vectorization of $\boldsymbol{u}$-time and $\boldsymbol{u}$-space, i.e. we introduce a reference origin o (see 3.4.3.).

Then $\tau_{\boldsymbol{u}} \circ r - \tau_{\boldsymbol{u}}(o) = -\boldsymbol{u} \cdot (r - o) \colon \mathbb{T} \rightarrowtail \mathbb{T}$ is differentiable, its derivative equals the derivative of $\tau_{\boldsymbol{u}} \circ r$, hence it is strictly monotone increasing, its inverse

$$z_{\boldsymbol{u},o} := \big(-\boldsymbol{u} \cdot (r - o)\big)^{-1} \colon \mathbb{T} \rightarrowtail \mathbb{T}$$

is monotone increasing as well and

$$\dot{z}_{\boldsymbol{u},o}(\boldsymbol{t}) = \frac{1}{-\boldsymbol{u} \cdot \dot{r}(z_{\boldsymbol{u},o}(\boldsymbol{t}))} \qquad (\boldsymbol{t} \in \operatorname{Dom} z_{\boldsymbol{u},o}).$$

The *motion relative to* $(\boldsymbol{u}, o)$ is

$$r_{\boldsymbol{u},o} \colon \mathbb{T} \rightarrowtail \mathbf{S}_{\boldsymbol{u}}, \qquad \boldsymbol{t} \mapsto r_{\boldsymbol{u}}(\boldsymbol{t}) - C_{\boldsymbol{u}}(o) = \boldsymbol{\pi}_{\boldsymbol{u}} \cdot \big(r(z_{\boldsymbol{u},o}(t)) - o\big).$$

4.2. Relative velocities

4.2.1. Proposition. Let $\boldsymbol{u}$ be a standard inertial frame and let r be a differentiable world line function; then $r_{\boldsymbol{u}}$ is differentiable and

$$\dot{r}_{\boldsymbol{u}} = \left(\frac{\dot{r}}{-\boldsymbol{u} \cdot \dot{r}} - \boldsymbol{u}\right) \circ z_{\boldsymbol{u}}.$$

Proof. Recalling that $C_{\boldsymbol{u}} \colon \mathrm{M} \to \mathrm{S}_{\boldsymbol{u}}$ is an affine map over $\boldsymbol{\pi}_{\boldsymbol{u}}$, we obtain

$$\dot{r}_{\boldsymbol{u}}(t) = \frac{\mathrm{d}}{\mathrm{d}t} C_{\boldsymbol{u}}\big(r\big(z_{\boldsymbol{u}}(t)\big)\big) = \boldsymbol{\pi}_{\boldsymbol{u}} \cdot \dot{r}\big(z_{\boldsymbol{u}}(t)\big)\dot{z}_{\boldsymbol{u}}(t);$$

then taking into account the formula in 4.1.1 for the derivative of $z_{\boldsymbol{u}}$, we easily find the desired equality. ■

It is evident that, choosing a reference origin o, we have

$$\dot{r}_{\boldsymbol{u},o} = \left(\frac{\dot{r}}{-\boldsymbol{u} \cdot \dot{r}} - \boldsymbol{u}\right) \circ z_{\boldsymbol{u},o}.$$

4.2.2. Since $r_{\boldsymbol{u}}$ describes the motion, relative to the standard inertial frame $\boldsymbol{u}$, of a material point, $\dot{r}_{\boldsymbol{u}}$ is the relative velocity function of the material point. This suggests the following definition.

Definition. Let $\boldsymbol{u}$ and $\boldsymbol{u}'$ be elements of $V(1)$. Then

$$\boldsymbol{v}_{\boldsymbol{u}'\boldsymbol{u}} := \frac{\boldsymbol{u}'}{-\boldsymbol{u} \cdot \boldsymbol{u}'} - \boldsymbol{u}$$

is called the *standard relative velocity of* u' *with respect to* u.

Proposition. For all $u, u' \in V(1)$
(*i*) $v_{u'u}$ is in $\frac{\mathbf{S}_u}{\mathbb{T}}$,
(*ii*) $v_{u'u} = -v_{uu'}$ if and only if $u = u'$,
(*iii*) $|v_{u'u}|^2 = |v_{uu'}|^2 = 1 - \frac{1}{(u\cdot u')^2} < 1$.

Proof. (*i*) is trivial, (*iii*) is demonstrated by a simple calculation. To show (*ii*) suppose

$$\frac{u'}{-u\cdot u'} - u = \frac{u}{-u'\cdot u} - u';$$

multiply the equality by u to have

$$0 = \frac{-1}{-u'\cdot u} - u'\cdot u, \qquad (u'\cdot u)^2 = 1.$$

According to the reversed Cauchy inequality (see 1.3.1) this is equivalent to $u = u'$. ■

Earlier we obtained that $\mathbf{S}_u$ and $\mathbf{S}_{u'}$ are different if and only if $u \neq u'$ and in this case $v_{u'u} \otimes \mathbb{T}$ ($v_{uu'} \otimes \mathbb{T}$) is a one-dimensional linear subspace in $\mathbf{S}_u$ ($\mathbf{S}_{u'}$), orthogonal to $\mathbf{S}_u \cap \mathbf{S}_{u'}$ (see 1.3.7) which offers an alternative proof of (*ii*).

4.2.3. Let us take now two standard inertial frames with constant velocity values u and u'. Then $v_{u'u}$ and $v_{uu'}$ are the relative velocities of the observers with respect to each other. Then (*iii*) of the previous proposition implies that $v_{u'u} = \mathbf{0}$ if and only if $u = u'$. Moreover, (*ii*) says that *in contradistinction to the nonrelativistic case and to our habitual 'evidence', the relative velocity of* u' *with respect to* u *is not the opposite of the relative velocity of* u *with respect to* u', *except the trivial case* $u = u'$.

It is worth emphasizing this fact because in most of the textbooks one takes it for granted that $v_{u'u}$ and $-v_{uu'}$ are equal: 'if an observer moves with velocity v relative to another then the second observer moves with velocity $-v$ relative to the first one'. This comes from the fact that vectors are given there by components with respect to convenient bases and then the components of $v_{u'u}$ and $v_{uu'}$ become opposite to each other.

The reason of nonequality of $v_{u'u}$ and $-v_{uu'}$ is that the spaces of *different* inertial observers are affine spaces over *different* vector spaces.

However, we have a nice relation between the two vector spaces in question: the Lorentz boost from u to u' maps $\mathbf{S}_u$ onto $\mathbf{S}_{u'}$ in a natural way and maps $v_{u'u}$ into $-v_{uu'}$.

Having the equality

$$L(u', u)\cdot v_{u'u} = -v_{uu'}$$

(see 1.3.8), we already know how to choose bases in $\mathbf{S}_{\boldsymbol{u}}$ and in $\mathbf{S}_{\boldsymbol{u}'}$ to get the mentioned usual relation between the components of relative velocities: take an arbitrary ordered basis $(\boldsymbol{e}_1, \boldsymbol{e}_2, \boldsymbol{e}_3)$ in $\mathbf{S}_{\boldsymbol{u}}$ and the basis $\big(\boldsymbol{e}'_i := \boldsymbol{L}(\boldsymbol{u}', \boldsymbol{u}) \cdot \boldsymbol{e}_i \,\big|\, i = 1, 2, 3\big)$ in $\mathbf{S}_{\boldsymbol{u}'}$. Now

$$\text{if} \qquad \boldsymbol{v}_{\boldsymbol{u}'\boldsymbol{u}} = \sum_{i=1}^{3} v^i \boldsymbol{e}_i, \qquad \text{then} \qquad \boldsymbol{v}_{\boldsymbol{u}\boldsymbol{u}'} = \sum_{i=1}^{3} (-v^i) \boldsymbol{e}'_i.$$

4.2.4. We often shall use the equalities

$$-\boldsymbol{u} \cdot \boldsymbol{u}' = \frac{1}{\sqrt{1 - |\boldsymbol{v}_{\boldsymbol{u}'\boldsymbol{u}}|^2}}$$

and

$$\boldsymbol{u}' = \frac{\boldsymbol{u} + \boldsymbol{v}_{\boldsymbol{u}'\boldsymbol{u}}}{\sqrt{1 - |\boldsymbol{v}_{\boldsymbol{u}'\boldsymbol{u}}|^2}}$$

deriving from 4.2.2 (iii) and (i) and from the definition of $\boldsymbol{v}_{\boldsymbol{u}'\boldsymbol{u}}$.

4.2.5. The relative velocities in the nonrelativistic spacetime model form a Euclidean vector space. Here the relative velocities with respect to a fixed $\boldsymbol{u} \in V(1)$ form the unit open ball in the Euclidean vector space $\frac{\mathbf{S}_{\boldsymbol{u}}}{\mathbb{T}}$:

$$B_{\boldsymbol{u}} := \left\{ \boldsymbol{v} \in \frac{\mathbf{S}_{\boldsymbol{u}}}{\mathbb{T}} \,\middle|\, |\boldsymbol{v}|^2 < 1 \right\}.$$

The set of all relative velocities is $\bigcup_{\boldsymbol{u} \in V(1)} B_{\boldsymbol{u}}$, a complicated subset of $\frac{\mathbf{M}}{\mathbb{T}}$, which does not admit a vector space structure.

4.3. Addition of relative velocities

4.3.1. As a consequence of the structure of relative velocites, 'addition of relative velocities' is not a vector addition, i.e. if $\boldsymbol{u}$, $\boldsymbol{u}'$, $\boldsymbol{u}''$ are different elements of $V(1)$ then—in contradistinction to the nonrelativistic case—we have

$$\boldsymbol{v}_{\boldsymbol{u}''\boldsymbol{u}} \neq \boldsymbol{v}_{\boldsymbol{u}''\boldsymbol{u}'} + \boldsymbol{v}_{\boldsymbol{u}'\boldsymbol{u}}.$$

The left-hand side is an element of $\frac{\mathbf{S}_{\boldsymbol{u}}}{\mathbb{T}}$; the right-hand side is the sum of elements in $\frac{\mathbf{S}_{\boldsymbol{u}'}}{\mathbb{T}}$ and in $\frac{\mathbf{S}_{\boldsymbol{u}}}{\mathbb{T}}$ which indicates that they cannot be, in general, equal.

We might think that the convenient Lorentz boost helps us, i.e. we get an equality if $\boldsymbol{v}_{\boldsymbol{u}''\boldsymbol{u}'}$ is replaced with $\boldsymbol{L}(\boldsymbol{u},\boldsymbol{u}')\cdot\boldsymbol{v}_{\boldsymbol{u}''\boldsymbol{u}'}$; however,

$$\boldsymbol{v}_{\boldsymbol{u}''\boldsymbol{u}} \neq \boldsymbol{L}(\boldsymbol{u},\boldsymbol{u}')\cdot\boldsymbol{v}_{\boldsymbol{u}''\boldsymbol{u}'} + \boldsymbol{v}_{\boldsymbol{u}'\boldsymbol{u}}$$

because the length of the vector on the right-hand side can be greater than 1.

4.3.2. Nevertheless, the relative velocity on the left hand side of the above inequality is a linear combination of the relative velocities on the right hand side. For the sake of brevity, let us introduce the notations

$$\boldsymbol{v}'' := \boldsymbol{v}_{\boldsymbol{u}''\boldsymbol{u}}, \quad \boldsymbol{v}' := \boldsymbol{L}(\boldsymbol{u},\boldsymbol{u}')\cdot\boldsymbol{v}_{\boldsymbol{u}''\boldsymbol{u}'}, \quad \boldsymbol{v} := \boldsymbol{v}_{\boldsymbol{u}'\boldsymbol{u}},$$

$$\alpha := -\boldsymbol{u}'\cdot\boldsymbol{u} = \frac{1}{\sqrt{1-|\boldsymbol{v}|^2}}, \quad \beta := -\boldsymbol{u}''\cdot\boldsymbol{u}' = \frac{1}{\sqrt{1-|\boldsymbol{v}'|^2}},$$

$$\gamma := -\boldsymbol{u}''\cdot\boldsymbol{u} = \frac{1}{\sqrt{1-|\boldsymbol{v}''|^2}} = \alpha\beta(1+\boldsymbol{v}\cdot\boldsymbol{v}').$$

Then we obtain by straightforward calculations that

$$\boldsymbol{v}'' = \frac{\beta}{\gamma}\boldsymbol{v} + \frac{\alpha(\beta+\gamma)}{\gamma(1+\alpha)}\boldsymbol{v}'.$$

4.3.3. If $\boldsymbol{v}'$ is parallel to $\boldsymbol{v}$, we get the most frequently cited Einstein formula

$$\boldsymbol{v}'' = \frac{\boldsymbol{v}+\boldsymbol{v}'}{1+|\boldsymbol{v}|\,|\boldsymbol{v}'|}.$$

4.4. History regained from motion

4.4.1. Given a motion relative to a standard inertial frame $\boldsymbol{u}$, i.e. a piecewise twice differentiable function $m\colon \mathrm{T}_{\boldsymbol{u}} \rightarrowtail \mathrm{S}_{\boldsymbol{u}}$, can we determine the corresponding world line function r such that $m = r_{\boldsymbol{u}}$?

Since $r_{\boldsymbol{u}} = C_{\boldsymbol{u}} \circ r \circ z_{\boldsymbol{u}}$ and $\tau_{\boldsymbol{u}} \circ r$ is the inverse of $z_{\boldsymbol{u}}$, we have

$$(1_{\mathrm{T}_{\boldsymbol{u}}}, r_{\boldsymbol{u}}) = (\tau_{\boldsymbol{u}} \circ r \circ z_{\boldsymbol{u}},\, C_{\boldsymbol{u}} \circ r \circ z_{\boldsymbol{u}}) = \xi_{\boldsymbol{u}} \circ r \circ z_{\boldsymbol{u}}.$$

Consequently, given the motion m,

$$r := \xi_{\boldsymbol{u}}^{-1} \circ (1_{\mathrm{T}_{\boldsymbol{u}}}, m) \circ z_{\boldsymbol{u}}^{-1}$$

will be the corresponding world line function.

Similarly, if the vectorized motion $\boldsymbol{m}\colon \mathbb{T} \rightarrowtail \mathbf{S}_{\boldsymbol{u}}$ is known, then

$$r := \xi_{\boldsymbol{u},o}^{-1} \circ (\mathbf{1}_{\mathbb{T}}, \boldsymbol{m}) \circ z_{\boldsymbol{u},o}^{-1}$$

is the required world line function which can be given by a simple formula:

$$t \mapsto o + \boldsymbol{m}\left(z_{\boldsymbol{u},o}^{-1}(t)\right) + \boldsymbol{u} z_{\boldsymbol{u},o}(t).$$

4.4.2. The previous formulae are not satisfactory yet because $z_{\boldsymbol{u}}$ and $z_{\boldsymbol{u},o}$ are defined by $\boldsymbol{u}$ and $(\boldsymbol{u}, o)$ together with the world line function r to be found; we have to determine them—or their inverse—from $\boldsymbol{u}$, $(\boldsymbol{u}, o)$ and the motion m or $\boldsymbol{m}$.

Equalities in 4.1.1 and 4.1.4 result in

$$\left(z_{\boldsymbol{u}}^{-1}\right)^{\cdot} = \frac{1}{\sqrt{1 - |\dot{m}|^2}}, \qquad \left(z_{\boldsymbol{u},o}^{-1}\right)^{\cdot} = \frac{1}{\sqrt{1 - |\dot{\boldsymbol{m}}|^2}}.$$

$\dot{m}$ and $\dot{\boldsymbol{m}}$ are given functions, hence $z_{\boldsymbol{u}}^{-1}$ and $z_{\boldsymbol{u},o}^{-1}$ can be obtained by a simple integration.

4.4.3. Let us consider the 'basic reference frame' $(1, \mathbf{0})$ with reference origin $(0, \mathbf{0})$ in the arithmetic spacetime model (see Exercise 3.5). A motion is given by a function $\boldsymbol{m}\colon \mathbb{R} \rightarrowtail \mathbb{R}^3$. Let $h\colon \mathbb{R} \rightarrowtail \mathbb{R}$ be a primitive function of $\frac{1}{\sqrt{1-|\dot{\boldsymbol{m}}|^2}}$. Then $\mathbb{R} \rightarrowtail \mathbb{R} \times \mathbb{R}^3$, $t \mapsto \left(h(t), \boldsymbol{m}(h(t))\right)$ is the world line function regained from the motion $\boldsymbol{m}$.

4.5. Relative accelerations

Let r be a world line function and let $\boldsymbol{u}$ be a standard inertial frame. Then $r_{\boldsymbol{u}}$ is twice differentiable and a differentiation of the equality in 4.2.1 yields

$$\ddot{r}_{\boldsymbol{u}} = \left(\frac{1}{(\boldsymbol{u} \cdot \dot{r})^2} \left(\mathbf{1}_{\mathrm{M}} + \frac{\dot{r} \otimes \boldsymbol{u}}{-\boldsymbol{u} \cdot \dot{r}} \right) \cdot \ddot{r} \right) \circ z_{\boldsymbol{u}}.$$

If a reference origin o is chosen as well, $\ddot{r}_{\boldsymbol{u},o}$ is given by a similar formula, with $z_{\boldsymbol{u},o}$ instead of $z_{\boldsymbol{u}}$.

We see that, in contradistinction to the nonrelativistic case, the relative acceleration does not equal the absolute one. Of course, the relative acceleration takes values in $\frac{\mathbf{S}_{\boldsymbol{u}}}{\mathbb{T} \otimes \mathbb{T}}$, the absolute acceleration takes values in $\frac{\mathbf{S}_{\dot{r}(t)}}{\mathbb{T} \otimes \mathbb{T}}$.

4.6. Some particular motions

4.6.1. Take the inertial world line function

$$r(\mathbf{t}) = x_o + \boldsymbol{u}_o\mathbf{t} \qquad (\mathbf{t} \in \mathbb{T}) \quad (*).$$

Consider a standard inertial frame $\boldsymbol{u}$.
Then

$$\dot{z}_{\boldsymbol{u}} = \frac{1}{-\boldsymbol{u}\cdot\boldsymbol{u}_o}$$

(see 4.1.1) from which we get immediately

$$z_{\boldsymbol{u}}(t) = \frac{t - t_o}{-\boldsymbol{u}\cdot\boldsymbol{u}_o}$$

for some $t_o \in \mathrm{T}_{\boldsymbol{u}}$ Consequently (see 3.4.1(v)),

$$r_{\boldsymbol{u}}(t) = \left(x_o + \boldsymbol{u}_o\frac{t-t_o}{-\boldsymbol{u}\cdot\boldsymbol{u}_o}\right) + \boldsymbol{u}\otimes\mathbb{T} = \left(x_o + \boldsymbol{u}\otimes\mathbb{T}\right) + \frac{\boldsymbol{\pi}_{\boldsymbol{u}}\cdot\boldsymbol{u}_o}{-\boldsymbol{u}\cdot\boldsymbol{u}_o}(t - t_o) =$$
$$= q_{x_o} + \boldsymbol{v}_{\boldsymbol{u}_o\boldsymbol{u}}(t - t_o) \qquad (t \in \mathrm{T}_{\boldsymbol{u}}),$$

where $q_{x_o} := x_o + \boldsymbol{u}\otimes\mathbb{T}$ is the $\boldsymbol{u}$-space point that x_o is incident with.

This is a uniform and rectilinear motion.

Conversely, suppose that we are given a uniform and rectilinear motion relative to the inertial frame, i.e. there are a $q \in \mathrm{S}_{\boldsymbol{u}}$, a $t_o \in \mathrm{T}_{\boldsymbol{u}}$ and a $\boldsymbol{v} \in \frac{\mathbf{S}_{\boldsymbol{u}}}{\mathbb{T}}$ such that

$$r_{\boldsymbol{u}}(t) = q + \boldsymbol{v}(t - t_o) \qquad (t \in \mathrm{T}_{\boldsymbol{u}}).$$

Then letting x_o denote the unique world point in the intersection of q and t_o and putting $\boldsymbol{u}_o := \frac{\boldsymbol{u}+\boldsymbol{v}}{\sqrt{1-|\boldsymbol{v}|^2}}$, the world line function of form ($*$) gives rise to the given motion.

4.6.2. Take the uniformly accelerated world line function

$$r(\mathbf{t}) = x_o + \boldsymbol{u}_o\frac{\sinh\alpha\mathbf{t}}{\alpha} + \boldsymbol{a}_o\frac{\cosh\alpha\mathbf{t} - 1}{\alpha^2} \qquad (\mathbf{t} \in \mathbb{T})$$

where $\alpha := |\boldsymbol{a}_o|$; then

$$\dot{r}(\mathbf{t}) = \boldsymbol{u}_o\cosh\alpha\mathbf{t} + \boldsymbol{a}_o\frac{\sinh\alpha\mathbf{t}}{\alpha}.$$

Consider a standard inertial frame $\boldsymbol{u}$. The formulae will be more tractable by choosing a reference origin o. Then

$$z_{\boldsymbol{u},o}^{-1}(\mathbf{t}) = -\boldsymbol{u}\cdot(r(\mathbf{t}) - o) =$$
$$= -\boldsymbol{u}\cdot(x_o - o) - \boldsymbol{u}\cdot\boldsymbol{u}_o\frac{\sinh\alpha\mathbf{t}}{\alpha} - \boldsymbol{u}\cdot\boldsymbol{a}_o\frac{\cosh\alpha\mathbf{t} - 1}{\alpha^2} \qquad (\mathbf{t} \in \mathbb{T}).$$

Let us consider the special case $\boldsymbol{u}\cdot\boldsymbol{a}_o = 0$; then

$$z_{\boldsymbol{u},o}(\boldsymbol{t}) = \frac{\operatorname{arsinh}\alpha\frac{\boldsymbol{t}-\boldsymbol{t}_o}{-\boldsymbol{u}\cdot\boldsymbol{u}_o}}{\alpha} \qquad (\boldsymbol{t}\in\mathbb{T}),$$

where $\boldsymbol{t}_o := -\boldsymbol{u}\cdot(x_o - o)$. Thus

$$\begin{aligned} r_{\boldsymbol{u},o}(\boldsymbol{t}) &= \boldsymbol{\pi}_{\boldsymbol{u}}\cdot\left(x_o - o + \boldsymbol{u}_o\frac{\boldsymbol{t}-\boldsymbol{t}_o}{-\boldsymbol{u}\cdot\boldsymbol{u}_o} + \boldsymbol{a}_o\frac{\sqrt{1+\alpha^2\frac{(\boldsymbol{t}-\boldsymbol{t}_o)^2}{(\boldsymbol{u}\cdot\boldsymbol{u}_o)^2}}-1}{\alpha^2}\right) = \\ &= \boldsymbol{q}_o + \boldsymbol{v}_{\boldsymbol{u}_o\boldsymbol{u}}(\boldsymbol{t}-\boldsymbol{t}_o) + \boldsymbol{b}_o\frac{\sqrt{1+\beta^2(\boldsymbol{t}-\boldsymbol{t}_o)^2}-1}{\beta^2} \qquad (\boldsymbol{t}\in\mathbb{T}), \end{aligned}$$

where

$$\boldsymbol{q}_o := \boldsymbol{\pi}_{\boldsymbol{u}}\cdot(x_o - o), \qquad \boldsymbol{b}_o := \boldsymbol{a}_o\left(1-|\boldsymbol{v}_{\boldsymbol{u}_o\boldsymbol{u}}|^2\right), \qquad \beta^2 := \alpha^2\left(1-|\boldsymbol{v}_{\boldsymbol{u}_o\boldsymbol{u}}|^2\right).$$

4.7. Standard speed of light

4.7.1. We wish to determine the motion of a light signal with respect to a standard inertial frame. The procedure will be similar to that in Sections 4.1 and 4.2.

We introduce the notation

$$V(0) := \left\{\boldsymbol{w}\in\frac{\mathbf{M}}{\mathbb{T}}\;\middle|\; \boldsymbol{w}^2 = 0,\ \boldsymbol{w}\otimes\mathbb{T}^+\subset L^{\to}\right\}.$$

The elements of $V(0)$ are future directed lightlike vectors of cotype $\mathbb{T}$. Though the notation is similar to $V(1)$, observe a significant difference: if two elements in $V(1)$ are parallel then they are equal; on the other hand, if $\boldsymbol{w}$ is in $V(0)$ then $\alpha\boldsymbol{w}$, too, is in $V(0)$ for all $\alpha\in\mathbb{R}^+$.

Let $\boldsymbol{u}$ be a standard inertial frame. Let us consider a light signal F, i.e. a straight line directed by a vector in $V(0)$. The motion of the light signal with respect to the observer is described by

$$f_{\boldsymbol{u}}\colon \mathrm{T}_{\boldsymbol{u}}\to\mathrm{S}_{\boldsymbol{u}}, \qquad t\mapsto(\mathrm{F}\star t)+\boldsymbol{u}\otimes\mathbb{T}$$

where $\mathrm{F}\star t$ denotes the unique element in the intersection of the straight line F and the hyperplane t.

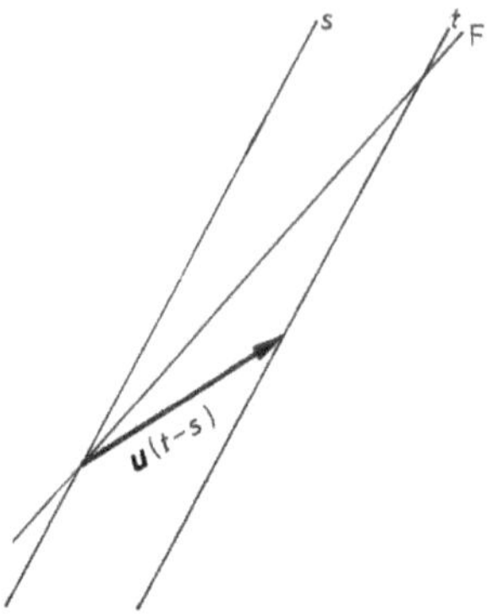

If $t, s \in \mathrm{T}_{\boldsymbol{u}}$, then $t - s = -\boldsymbol{u} \cdot (\mathrm{F} \star t - \mathrm{F} \star s)$ by the definition of the affine structure of $\mathrm{T}_{\boldsymbol{u}}$ (see 3.4.2); then we easily find that

$$\mathrm{F} \star t - \mathrm{F} \star s = \boldsymbol{w} \frac{t - s}{-\boldsymbol{u} \cdot \boldsymbol{w}},$$

hence

$$f_{\boldsymbol{u}}(t) - f_{\boldsymbol{u}}(s) = \boldsymbol{\pi}_{\boldsymbol{u}} \cdot (\mathrm{F} \star t - \mathrm{F} \star s) = \left(\frac{\boldsymbol{w}}{-\boldsymbol{u} \cdot \boldsymbol{w}} - \boldsymbol{u} \right) (t - s).$$

Thus the light signal moves uniformly on a straight line relative to the observer.

4.7.2. Definition. Let $\boldsymbol{w} \in V(0)$ and $\boldsymbol{u} \in V(1)$. Then

$$\boldsymbol{v}_{\boldsymbol{wu}} := \frac{\boldsymbol{w}}{-\boldsymbol{u} \cdot \boldsymbol{w}} - \boldsymbol{u}$$

is the *standard relative velocity of* $\boldsymbol{w}$ *with respect to* $\boldsymbol{u}$.

Proposition. $\boldsymbol{v}_{\boldsymbol{wu}}$ is an element of $\frac{\mathrm{S}_{\boldsymbol{u}}}{\mathbb{T}}$ and

$$|\boldsymbol{v}_{\boldsymbol{wu}}| = 1. \qquad \blacksquare$$

Observe that given an arbitrary $\boldsymbol{u}$, the lightlike vectors $\boldsymbol{w}$ and $\alpha\boldsymbol{w}$ have the same relative velocities with respect to $\boldsymbol{u}$.

We do not define the relative velocity of $\boldsymbol{u}$ with respect to $\boldsymbol{w}$.

According to the previous proposition, the magnitude of the standard relative velocity is the same number, namely 1, for every light signal and every inertial observer.

Thus we obtained explicitly that the given synchronization procedure of inertial observers by light signals does result in the one-way speed of light being equal the two-way speed.

4.7.3. Recall that relative velocity values have magnitude but two relative velocity values need not have an angle between themselves; however, relative velocities with respect to the same element of $V(1)$ do form an angle.

Now we look for the relation between certain angles formed by relative velocity values. The physical situation is similar to that in I.6.2.3. A car is going on a straight road and it is raining. The raindrops hit the road and the car at different angles relative to the direction of the road. What is the relation between the two angles? Now we can treat another question, too, considering instead of raindrops light beams (continuous sequences of light signals) arriving from the sun.

Let $\boldsymbol{u}$ and $\boldsymbol{u}'$ be different elements of $V(1)$ (representing the absolute velocity values of the road and of the car, respectively). If $\boldsymbol{w}$ is an element of $V(1) \cup V(0)$ (representing the absolute velocity value of the raindrops or the absolute direction of the light beam), $\boldsymbol{w} \neq \boldsymbol{u}$, $\boldsymbol{w} \neq \boldsymbol{u}'$, then

$$\theta(\boldsymbol{w}) := \arccos \frac{\boldsymbol{v_{wu}} \cdot \boldsymbol{v_{u'u}}}{|\boldsymbol{v_{wu}}|\, |\boldsymbol{v_{u'u}}|}, \qquad \theta'(\boldsymbol{w}) := \arccos \frac{\boldsymbol{v_{wu'}} \cdot (-\boldsymbol{v_{uu'}})}{|\boldsymbol{v_{wu'}}|\, |\boldsymbol{v_{uu'}}|}$$

are the angles formed by the relative velocity values in question. A simple calculation verifies that

$$\cos\theta(\boldsymbol{w}) = \frac{\frac{|\boldsymbol{v_{wu'}}|}{|\boldsymbol{v_{wu}}|}\cos\theta'(\boldsymbol{w}) + \frac{|\boldsymbol{v_{u'u}}|}{|\boldsymbol{v_{wu}}|}}{1 + |\boldsymbol{v_{wu'}}|\, |\boldsymbol{v_{uu'}}| \cos\theta'(\boldsymbol{w})}.$$

If $\boldsymbol{w} \in V(0)$ then $|\boldsymbol{v_{wu}}| = |\boldsymbol{v_{wu'}}| = 1$ and

$$\cos\theta(\boldsymbol{w}) = \frac{\cos\theta'(\boldsymbol{w}) + |\boldsymbol{v_{uu'}}|}{1 + |\boldsymbol{v_{uu'}}| \cos\theta'(\boldsymbol{w})}.$$

This formula is known as the *aberration of light*: two different inertial observers see the same light beam under different angles with respect to their relative velocities; the angles are related by the above formula.

4.8. Motions relative to a nonstandard inertial reference frame

4.8.1. Consider an inertial reference frame $(\boldsymbol{u}_s, \boldsymbol{u})$. Take an inertial world line C directed by $\boldsymbol{u}' \in V(1)$. The corresponding history of a material point is described by the reference frame as a motion $\mathrm{T}_{\boldsymbol{u}_s} \to \mathrm{S}_{\boldsymbol{u}}$.

Let t, s be $\mathrm{T}_{\boldsymbol{u}_s}$-instants. The material point at s and at t meets the $\boldsymbol{u}$-space points q and p; let x and y be the occurrences of meetings, respectively. Thus, the velocity of this motion is

$$\boldsymbol{v}_{\boldsymbol{u}'(\boldsymbol{u}_s,\boldsymbol{u})} = \frac{p-q}{(t-s)_{\boldsymbol{u}}} = \frac{\boldsymbol{\pi}\cdot(y-x)}{-\boldsymbol{u}_s\cdot(y-x)}(-\boldsymbol{u}_s\cdot\boldsymbol{u}) = \frac{\boldsymbol{\pi}\cdot\boldsymbol{u}'}{-\boldsymbol{u}_s\cdot\boldsymbol{u}'}(-\boldsymbol{u}_s\cdot\boldsymbol{u}) =$$
$$= \boldsymbol{v}_{\boldsymbol{u}'\boldsymbol{u}}\frac{(-\boldsymbol{u}\cdot\boldsymbol{u}')(-\boldsymbol{u}_s\cdot\boldsymbol{u})}{-\boldsymbol{u}_s\cdot\boldsymbol{u}'}$$

where, of course, $\boldsymbol{v}_{\boldsymbol{u}'\boldsymbol{u}}$ is the standard velocity of $\boldsymbol{u}'$ with respect to $\boldsymbol{u}$.

The motion of an inertial material point with respect an arbitrary inertial reference frame is rectilinear and uniform.

4.8.2. The same formula holds for a light signal directed by $\boldsymbol{w}$:

$$\boldsymbol{v}_{\boldsymbol{w}(\boldsymbol{u}_s,\boldsymbol{u})} = \boldsymbol{v}_{\boldsymbol{w}\boldsymbol{u}}\frac{(-\boldsymbol{u}\cdot\boldsymbol{w})(-\boldsymbol{u}_s\cdot\boldsymbol{u})}{-\boldsymbol{u}_s\cdot\boldsymbol{w}}.$$

To make our result more apparent, we write

$$\boldsymbol{u}_s = \frac{\boldsymbol{u}+v_s\boldsymbol{n}_s}{\sqrt{1-v_s^2}}, \qquad \boldsymbol{w} = \boldsymbol{u}+\boldsymbol{n}_w$$

where $\boldsymbol{n}_s$ and $\boldsymbol{n}_w$ are unit vectors of cotype $\mathbb{T}$ in the $\boldsymbol{u}$-space, indicating the 'direction of the synchronization' and the direction of the path of the light signal, and v_s is a read number describing how much the inertial reference frame $(\boldsymbol{u}_s, \boldsymbol{u})$ deviates from a the standard one. Then we have

$$\boldsymbol{v}_{\boldsymbol{w}(\boldsymbol{u}_s,\boldsymbol{u})} = \boldsymbol{v}_{\boldsymbol{w}\boldsymbol{u}}\frac{1-v_s\boldsymbol{n}_s\cdot\boldsymbol{n}_w}{\sqrt{1-v_s^2}}.$$

The one-way light speed is different in different directions (except the standard case $v_s = 0$); the least value is $\frac{1}{1+v_s}$, the greatest one is $\frac{1}{1-v_s}$.

4.9. Exercises

1. Treat the motion, relative to inertial reference frames, corresponding to an arbitrary world line function r.

2. Estimate the one-way light speed from Budapest to London and from London to Budapest (distance: 1450 km) if the synchronization is established in such a way that the standard 12 o'clock in London and the standard 12,0005 o'clock in Budapest are simultaneous.

5. Some comparisons between different spaces and times

5.1. Physically equal vectors in different spaces

5.1.1. It is an important fact that the spaces of different global inertial observers are affine spaces over different vector spaces. Thus it has no meaning, in general, that a straight line segment (a vector) in the space of an inertial observer coincides with a straight line segment (with a vector) in the space of another inertial observer.

Let us consider two different global inertial observers with constant velocity values $\boldsymbol{u}$ and $\boldsymbol{u}'$, respectively. Their spaces are affine spaces over $\mathbf{S}_{\boldsymbol{u}}$ and $\mathbf{S}_{\boldsymbol{u}'}$, respectively. We know that $\mathbf{S}_{\boldsymbol{u}} \cap \mathbf{S}_{\boldsymbol{u}'}$ is a two-dimensional subspace, orthogonal to $\boldsymbol{v}_{\boldsymbol{u}'\boldsymbol{u}}$ and to $\boldsymbol{v}_{\boldsymbol{u}\boldsymbol{u}'}$.

If a vector between two points in the $\boldsymbol{u}$-space lies in $\mathbf{S}_{\boldsymbol{u}} \cap \mathbf{S}_{\boldsymbol{u}'}$ then we can find two points in the $\boldsymbol{u}'$-space having the same vector connecting them. We have troubles only with vector outside this two-dimensional subspace.

To relate other vectors, too, we start from the rational agreement that 'if you move with respect to me in some direction in my space then I move with respect to you in the opposite direction in your space'. This suggests that the mathematically different vectors $\lambda \boldsymbol{v}_{\boldsymbol{u}'\boldsymbol{u}}$ in $\mathbf{S}_{\boldsymbol{u}}$ and $-\lambda \boldsymbol{v}_{\boldsymbol{u}\boldsymbol{u}'}$ in $\mathbf{S}_{\boldsymbol{u}'}$ could be considered the 'same' from a physical point of view for all $\lambda \in \mathbb{T}$. More generally, every vector in $\mathbf{S}_{\boldsymbol{u}}$ has the form

$$\lambda \boldsymbol{v}_{\boldsymbol{u}'\boldsymbol{u}} + \boldsymbol{q} \qquad (\lambda \in \mathbb{T},\ \boldsymbol{q} \in \mathbf{S}_{\boldsymbol{u}} \cap \mathbf{S}_{\boldsymbol{u}'})$$

and every vector in $\mathbf{S}_{\boldsymbol{u}'}$ has the form

$$\lambda' \boldsymbol{v}_{\boldsymbol{u}\boldsymbol{u}'} + \boldsymbol{q}' \qquad (\lambda' \in \mathbb{T},\ \boldsymbol{q}' \in \mathbf{S}_{\boldsymbol{u}} \cap \mathbf{S}_{\boldsymbol{u}'}).$$

The observers agree that two such vectors are considered to be the 'same' if and only if

$$\lambda' = -\lambda, \qquad \boldsymbol{q}' = \boldsymbol{q}.$$

We have a nice tool to express this agreement: the Lorentz boost.

Definition. The vectors $\boldsymbol{q}'$ in $\mathbf{S}_{\boldsymbol{u}'}$ and $\boldsymbol{q}$ in $\mathbf{S}_{\boldsymbol{u}}$ are called *physically equal* if and only if

$$\boldsymbol{L}(\boldsymbol{u}', \boldsymbol{u}) \cdot \boldsymbol{q} = \boldsymbol{q}'. \qquad \blacksquare$$

We emphasize that the equality of vectors in different observer spaces is not obvious; here we agreed to define it conveniently.

5.1.2. To be physically equal in different observer spaces, according to our convention, is a symmetric relation, but *is not a transitive relation.*

Indeed, if $\boldsymbol{q}' = \boldsymbol{L}(\boldsymbol{u}', \boldsymbol{u}) \cdot \boldsymbol{q}$ then $\boldsymbol{q} = \boldsymbol{L}(\boldsymbol{u}, \boldsymbol{u}') \cdot \boldsymbol{q}'$: the relation is symmetric.

However, if $\boldsymbol{u}$, $\boldsymbol{u}'$ and $\boldsymbol{u}''$ are not coplanar, then there are $\boldsymbol{q}$, $\boldsymbol{q}'$ and $\boldsymbol{q}''$ in such a way that

$$\boldsymbol{q}' = \boldsymbol{L}(\boldsymbol{u}', \boldsymbol{u}) \cdot \boldsymbol{q}, \qquad \boldsymbol{q}'' = \boldsymbol{L}(\boldsymbol{u}'', \boldsymbol{u}') \cdot \boldsymbol{q}',$$
$$\boldsymbol{q}'' \text{ is not parallel to } \boldsymbol{L}(\boldsymbol{u}'', \boldsymbol{u}) \cdot \boldsymbol{q}$$

(Exercise 1.9.6); $\boldsymbol{q}'$ is physically equal to $\boldsymbol{q}$, $\boldsymbol{q}''$ is physically equal to $\boldsymbol{q}'$, but $\boldsymbol{q}''$ is not physically equal to $\boldsymbol{q}$: the relation is not transitive.

In particular, 'if a straight line in your space is parallel to a straight line in my space and a line in his space is parallel to your line then his line need not be parallel to mine'.

This is a rather embarrassing situation but there is no escape. The truth of the common sense that the standard relative velocity of an observer with respect to another is physically equivalent to the opposite of the other standard relative velocity and the transitivity of parallelism exclude each other.

5.2. How to perceive spaces of other reference frames?

5.2.1. In 5.1.1 an agreement is settled what equality—in particular, parallelism—of vectors in different observer spaces means.

Now the question arises how a straight line segment in the space of an inertial observer is perceived by another observer. The question and the answer are formulated correctly as follows (cf. I.7.1.2). It is most important that the perception does not depend only on the observer: the synchronization is involved, too.

Let $\boldsymbol{u}_o$ and $\boldsymbol{u}$ be standard inertial frames. Let H_o be a subset (a geometrical figure) in the $\boldsymbol{u}_o$-space (this is independent of any synchronization). The corresponding figure *perceived by the reference frame* $\boldsymbol{u}$ at the $\boldsymbol{u}$-instant t—called the *trace* of H_o at t in $\mathbf{S}_{\boldsymbol{u}}$—is the set of $\boldsymbol{u}$-space points that coincide at t with the points of H_o:

$$\{q \star t + \boldsymbol{u} \otimes \mathbb{T} \mid q \in \mathrm{H}_o\}$$

where $q \star t$ denotes the unique world point in the intersection of the line q and the hyperplane t.

Introducing the map

$$P_t\colon \mathrm{S}_{\boldsymbol{u}_o} \to \mathrm{S}_{\boldsymbol{u}}, \qquad q \mapsto q \star t + \boldsymbol{u} \otimes \mathbb{T}$$

we see that the trace of H_o at t equals $P_t[\mathrm{H}_o]$. It is quite easy to see (recall the definition of subtraction in the observer spaces) that

$$P_t(q_2) - P_t(q_1) = q_2 \star t - q_1 \star t = q_2 - q_1 + \boldsymbol{u}_o \frac{\boldsymbol{u} \cdot (q_2 - q_1)}{-\boldsymbol{u}_o \cdot \boldsymbol{u}} =$$
$$= \boldsymbol{P}_{\boldsymbol{u}\boldsymbol{u}_o} \cdot (q_2 - q_1),$$

where $\boldsymbol{P}_{\boldsymbol{u}\boldsymbol{u}_o}$ is the projection onto $\mathbf{S}_{\boldsymbol{u}}$ along $\boldsymbol{u}_o \otimes \mathbb{T}$ (see Exercise 1.9.7).

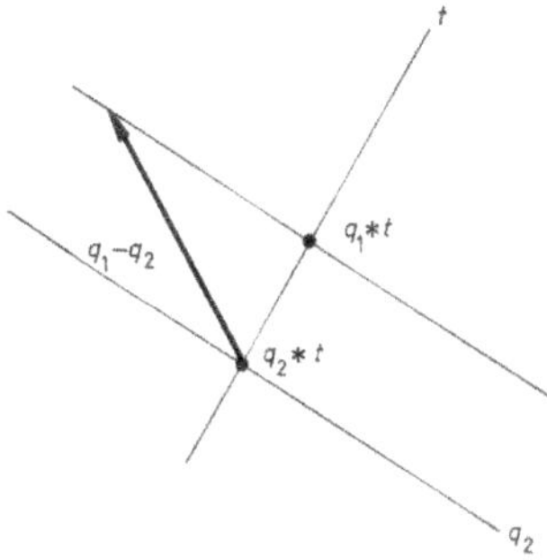

Since the restriction of $\boldsymbol{P}_{\boldsymbol{u}\boldsymbol{u}_o}$ onto $\mathbf{S}_{\boldsymbol{u}_o}$, denoted by $\boldsymbol{A}_{\boldsymbol{u}\boldsymbol{u}_o}$, is a linear bijection between $\mathbf{S}_{\boldsymbol{u}_o}$ and $\mathbf{S}_{\boldsymbol{u}}$, P_t is an affine bijection over $\boldsymbol{A}_{\boldsymbol{u}\boldsymbol{u}_o}$.

5.2.2. We can easily find that

$$\boldsymbol{A}_{\boldsymbol{u}\boldsymbol{u}_o} \cdot \boldsymbol{q} = \boldsymbol{q} \quad \text{if} \quad \boldsymbol{q} \in \mathbf{S}_{\boldsymbol{u}} \cap \mathbf{S}_{\boldsymbol{u}_o} \quad \text{i.e. if } \boldsymbol{q} \text{ is orthogonal to } \boldsymbol{v}_{\boldsymbol{u}\boldsymbol{u}_o},$$

$$\boldsymbol{A}_{\boldsymbol{u}\boldsymbol{u}_o} \cdot \boldsymbol{v}_{\boldsymbol{u}\boldsymbol{u}_o} = -\sqrt{1 - |\boldsymbol{v}_{\boldsymbol{u}\boldsymbol{u}_o}|^2}\, \boldsymbol{v}_{\boldsymbol{u}_o\boldsymbol{u}}.$$

The linear bijection $\boldsymbol{A}_{\boldsymbol{u}\boldsymbol{u}_o}$ resembles the restriction onto $\mathbf{S}_{\boldsymbol{u}_o}$ of the Lorentz boost $\boldsymbol{L}(\boldsymbol{u}, \boldsymbol{u}_o)$; an essential difference is that it maps $\boldsymbol{v}_{\boldsymbol{u}\boldsymbol{u}_o}$ into $-\boldsymbol{v}_{\boldsymbol{u}_o\boldsymbol{u}}$ *multiplied by a real number less than* 1. Consequently, $\boldsymbol{A}_{\boldsymbol{u}\boldsymbol{u}_o}$ *is not an orthogonal map*; it does not preserve either lengths or angles which is illustrated as follows:

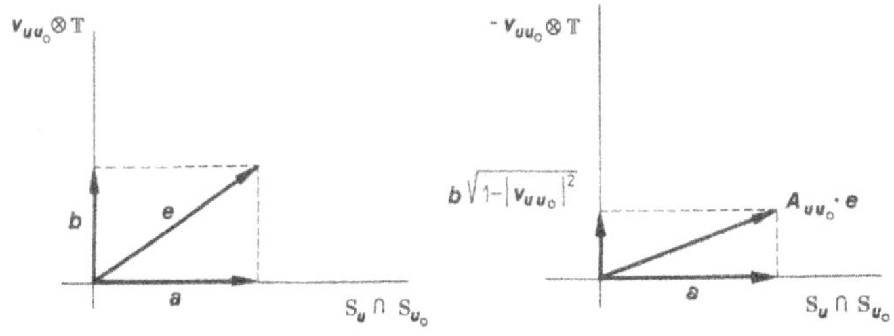

5.2.3. Every figure in the $\boldsymbol{u}_o$-space is of the form $q_o + \mathbf{H}_o$, where $q_o \in \mathrm{S}_{\boldsymbol{u}_o}$ and $\mathbf{H}_o \subset \mathbf{S}_{\boldsymbol{u}_o}$; then $P_t[q_o + \mathbf{H}_o] = P_t(q_o) + \boldsymbol{A}_{\boldsymbol{u}\boldsymbol{u}_o}[\mathbf{H}_o]$. Consequently, the perceived figure and the original one are not congruent, in general.

If L_o is a straight line segment in the $\boldsymbol{u}_o$-space, then its trace is a straight line segment, too. However, the perceived segment and the original one are not parallel, in general: if L_o is directed by the vector $\boldsymbol{e}_o$, $L_o = q_o + \mathbb{R}\boldsymbol{e}_o$, then $P_t[L_o]$ is directed by $\boldsymbol{A}_{\boldsymbol{u}\boldsymbol{u}_o} \cdot \boldsymbol{e}_o$.

$\boldsymbol{A}_{\boldsymbol{u}\boldsymbol{u}_o} \cdot \boldsymbol{e}_o$ is parallel to $\boldsymbol{e}_o$, by definition, if there is a real number λ such that

$$\boldsymbol{A}_{\boldsymbol{u}\boldsymbol{u}_o} \cdot \boldsymbol{e}_o = \lambda \boldsymbol{L}(\boldsymbol{u}, \boldsymbol{u}_o) \cdot \boldsymbol{e}_o$$

which occurs if and only if $\boldsymbol{e}_o$ is in $\mathbf{S}_{\boldsymbol{u}} \cap \mathbf{S}_{\boldsymbol{u}_o}$ or in $\boldsymbol{v}_{\boldsymbol{u}\boldsymbol{u}_o} \otimes \mathbb{T}$.

Thus a straight line segment L_o in the $\boldsymbol{u}_o$-space is perceived by $\boldsymbol{u}$ to be parallel to L_o if and only if L_o is

— either orthogonal to $\boldsymbol{v}_{\boldsymbol{u}\boldsymbol{u}_o}$

— or parallel to $\boldsymbol{v}_{\boldsymbol{u}\boldsymbol{u}_o}$.

5.2.4. Let L_1 and L_2 be crossing straight lines in the $\boldsymbol{u}_o$-space. Then $\boldsymbol{u}$ perceives at every $\boldsymbol{u}$-instant that they are crossing straight lines. However, the angle formed by L_1 and L_2 and the angle formed by the perceived straight lines differ, in general.

Let L_1 and L_2 be directed by $\boldsymbol{e}_1$ and $\boldsymbol{e}_2$, respectively. If θ_o denotes the angle formed by L_1 and L_2 then

$$\cos\theta_o = \frac{\boldsymbol{e}_1 \cdot \boldsymbol{e}_2}{|\boldsymbol{e}_1|\,|\boldsymbol{e}_2|}.$$

For the angle θ perceived by $\boldsymbol{u}$ we have

$$\cos\theta = \frac{(\boldsymbol{A}_{\boldsymbol{u}\boldsymbol{u}_o} \cdot \boldsymbol{e}_1) \cdot (\boldsymbol{A}_{\boldsymbol{u}\boldsymbol{u}_o} \cdot \boldsymbol{e}_2)}{|\boldsymbol{A}_{\boldsymbol{u}\boldsymbol{u}_o} \cdot \boldsymbol{e}_1|\,|\boldsymbol{A}_{\boldsymbol{u}\boldsymbol{u}_o} \cdot \boldsymbol{e}_2|} = \frac{\cos\theta_o - \alpha_1\alpha_2}{\sqrt{1-\alpha_1^2}\sqrt{1-\alpha_2^2}},$$

where

$$\alpha_1 := \frac{\boldsymbol{u} \cdot \boldsymbol{e}_1}{-(\boldsymbol{u}_o \cdot \boldsymbol{u})|\boldsymbol{e}_1|} = \boldsymbol{v}_{\boldsymbol{u}\boldsymbol{u}_o} \cdot \frac{\boldsymbol{e}_1}{|\boldsymbol{e}_1|},$$

$$\alpha_2 := \frac{\boldsymbol{u} \cdot \boldsymbol{e}_2}{-(\boldsymbol{u}_o \cdot \boldsymbol{u})|\boldsymbol{e}_2|} = \boldsymbol{v}_{\boldsymbol{u}\boldsymbol{u}_o} \cdot \frac{\boldsymbol{e}_2}{|\boldsymbol{e}_2|}.$$

Thus θ and θ_o are equal if and only if $\alpha_1 = \alpha_2 = 0$, i.e. if and only if both $\boldsymbol{e}_1$ and $\boldsymbol{e}_2$ are orthogonal to the relative velocity $\boldsymbol{v}_{\boldsymbol{u}\boldsymbol{u}_o}$.

5.3. Lorentz contraction

5.3.1. A straight line segment orthogonal to the relative velocity $\boldsymbol{v}_{\boldsymbol{u}\boldsymbol{u}_o}$ in the $\boldsymbol{u}_o$-space is perceived by $\boldsymbol{u}$ as a straight line segment parallel to the original one and having the same length.

A straight line segment parallel to the relative velocity $\boldsymbol{v}_{\boldsymbol{u}\boldsymbol{u}_o}$ in the $\boldsymbol{u}_o$-space is perceived by $\boldsymbol{u}$ as a shorter straight line segment parallel to the original one. This is the famous *Lorentz contraction* which will be detailed as follows.

A straight line segment in the $\boldsymbol{u}_o$-space can be represented by one of its extremities and the vector between its extremities. Since parallel segments are perceived in a similar way, we can consider only the vector $\boldsymbol{e}_o \in \mathbf{S}_{\boldsymbol{u}_o}$ between the extremities.

The perception of $\boldsymbol{e}_o$ by $\boldsymbol{u}$ yields $\boldsymbol{e} := \boldsymbol{A}_{\boldsymbol{u}\boldsymbol{u}_o} \cdot \boldsymbol{e}_o$. A simple calculation shows that

$$|\boldsymbol{e}|^2 = |\boldsymbol{e}_o|^2 - \frac{(\boldsymbol{u} \cdot \boldsymbol{e}_o)^2}{(\boldsymbol{u} \cdot \boldsymbol{u}_o)^2} = |\boldsymbol{e}_o|^2 - (\boldsymbol{v}_{\boldsymbol{u}\boldsymbol{u}_o} \cdot \boldsymbol{e}_o)^2.$$

The perceived length, in general, is smaller than the proper one.

More closely, the perceived length equals the original one if and only if the segment is orthogonal to the relative velocity; otherwise the perceived length is smaller than the original one. The perceived length is the smallest if the segment is parallel to the relative velocity:

$$|\boldsymbol{e}| = |\boldsymbol{e}_o|\sqrt{1 - |\boldsymbol{v}_{\boldsymbol{u}\boldsymbol{u}_o}|^2} \qquad \text{if } \boldsymbol{e}_o \text{ is parallel to } \boldsymbol{v}_{\boldsymbol{u}\boldsymbol{u}_o}.$$

5.3.2. One often says that the travelling length is smaller than the proper (or rest) length: 'a moving rod is contracted, becomes shorter'.

Then one continues: let us imagine two rods having the same proper length and resting in the spaces of different observers; both observers will perceive the *other* rod to be shorter than its own one.

A number of paradoxes can arise from this situation: 'I say that your rod is shorter then mine, you say that my rod is shorter than yours; which of us is right?'

The paradox stems from the fact that usually one does not speak about synchronizations. And the answer is: both of us are right if both of us apply our different standard synchronizations. It is most important that the trace, as it is defined, depends on a synchronization. The observer $\boldsymbol{u}$ is not obliged to use its own standard synchronization; if it uses the $\boldsymbol{u}_o$-synchronization then there is no contraction.

We emphasize that the Lorentz contraction formula does not state any real physical contraction at all.

5.3.3. Suppose you do not believe that the contraction is illusory and you want determine experimentally which of us is right. The experiment seems extremely simple: you catch my rod (which is moving relative to you) and having stopped it you put it close to your one and then you will see which of them is shorter.

We consider an ideal case: you seize the moving rod all at once so that it stops instantaneously.

Let us translate the situation into our mathematical language. My rod is described by a line segment in the u_o-space:

$$L_o = q_o + [0,1]e_o$$

and e_o is taken to be parallel to v_{uu_o}.

Then the rod has the length (the proper length) $d_o := |e_o|$ with respect to u_o and the length (the travelling length) $d := \sqrt{1-|v_{uu_o}|^2}d_o$ perceived by u.

At a u-instant t the history of each point of the rod will be changed into an inertial history with the velocity value u; then you get

$$L = \{q \star t + u \otimes \mathbb{T} \mid q \in L_o\} = q + [0,1]e,$$

where

$$q := q_o \star t + u \otimes \mathbb{T}, \qquad e := A(u, u_o) \cdot e_o.$$

The segment (the seized rod) L is in the u-space and has the length $|e| = d$, the length of L_o perceived by u.

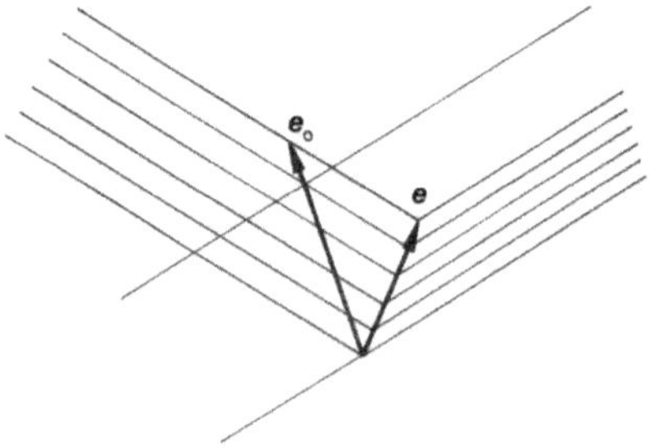

5.3.4. You can relax: you showed that my rod is 'really' shorter than yours. But then you think that I can execute a similar experiment to show that your rod is 'really' shorter than mine. Again the same disturbing situation.

To solve the seeming contradiction, note that in your experiment your rod continues to exist without any effect on it, while my rod is affected by your seizure, and in my experiment your rod is affected. The seizure means a physical change in the rod which causes contraction.

Let us analyze the problem more thoroughly.

(*i*) The rod resting in the u_o-space moves relative to the standard inertial frame u which perceives that the length of the rod is d. At a u-instant t the observer u seizes the rod, stops it, and discovers that this rod has the same length d. According to u, *the rod did not change length during the seizure,* in other words, the observer u sees the rod as *rigid* and this is well understandable from its point of view because the rod is stopped *at an instant* with respect to $\boldsymbol{u}$, i.e. every point of the rod stops *simultaneously* with respect to the standard synchronization of $\boldsymbol{u}$.

(*ii*) The rod rests in the $\boldsymbol{u}_o$-space. As $\boldsymbol{u}$ seizes the rod, the observer $\boldsymbol{u}_o$ sees that the rod begins to move but not instantaneously with respect to $\boldsymbol{u}_o$: the points of the rod begin the movement at different $\boldsymbol{u}_o$-instants! First the backward extremity (from the point of view of the relative velocity of $\boldsymbol{u}$ with respect to $\boldsymbol{u}_o$) starts and then successively the other points, at last the forward extremity. Evidently, $\boldsymbol{u}_o$ sees the rod is not rigid, it contracts during the *time interval* of seizure.

(*iii*) The rod experiences that it moves relative to $\boldsymbol{u}$ which begins to stop it in such a way that first the forward extremity (from the point of view of the relative velocity of the rod (i.e. of $\boldsymbol{u}_o$) with respect to $\boldsymbol{u}$) stops and then successively the other points and at last the backward extremity. The rod experiences contraction during the procedure of seizure.

We have examined three standpoints. Two of them concern inertial frames and the third concerns a noninertial one.

5.3.5. The ideal case that every point of the rod changes its velocity abruptly at a $\boldsymbol{u}$-instant can be replaced by the more realistic one that every point of the rod changes its velocity from $\boldsymbol{u}_o$ to $\boldsymbol{u}$ during a $\boldsymbol{u}$-time interval, as the following Figure shows.

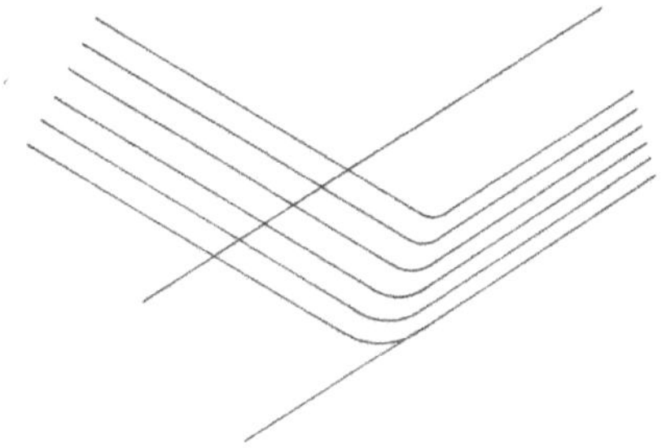

5.3.6. We emphasize again that 'stopping instantaneously' in the previous explanation refers to a $\boldsymbol{u}$-instant.

The reader is asked to analyze the problem that the rod is seized by $\boldsymbol{u}$ instantaneously with respect to a $\boldsymbol{u}_o$-instant (i.e. every point of the rod is stopped by $\boldsymbol{u}$ simultaneously with respect to the standard synchronization of $\boldsymbol{u}_o$).

5.4. The tunnel paradox

5.4.1. Consider a train and a tunnel such that the proper length of the train is greater than the proper length of the tunnel. Let the travelling train enter the tunnel.

The observer resting with respect to the tunnel perceives Lorentz contraction on the train, thus it sees that, if the velocity of the train is high enough, the train is entirely in the tunnel during a time interval.

On the contrary, the observer resting with respect to the train perceives Lorentz contraction on the tunnel, thus it experiences that the train is never entirely in the tunnel.

Which of them is right? We know that both. However, it seems to be a very strange situation because the observer resting with respect to the tunnel says that 'when the train is entirely inside the tunnel, I close both gates of the tunnel, thus I confine the train in the tunnel, I am right and the observer in the train is wrong'.

5.4.2. On the basis of our previous examination we can remove the paradox easily.

In the assertion above 'when' means that the gates become closed *simultaneously with respect to the tunnel.*

From the point of view of the train the gates will not be closed simultaneously: first the forward gate closes and later the backward gate. *When* the forward gate closes, the forepart of the train is in the tunnel (the back part is still outside), *when* the backward gate closes, the back part of the train is in the tunnel (the forepart is already outside).

'I confined the train in the tunnel' means that the closed gates hinder the train from leaving the tunnel. But how do they do this? The train is moving; it must be stopped to be confined definitely in the tunnel: some apparatus in the tunnel brakes the train or the train hits against the front gate which is so strongly closed that it stops the train. In any case, as we have seen, stopping means a real contraction of the train, consequently, it finds room in the tunnel; however, then the train *ceases to be inertial* in all its existence, that is why the assertion 'I am never entirely in the tunnel' (true for an inertial train) will be false.

5.5. No measuring rods

5.5.1. A physical observer makes measurements in his space: measures the distance between two points, the length of a line, etc. In practice such measurements are based on measuring rods: one takes a rod, carries it to the figure to be measured, puts it consecutively on convenient places … One supposes that during all this procedure the rod is *absolutely rigid:* it remains a straight line segment and its length does not change.

This assumption is justified nonrelativistically. But we have seen previously that the rod when seized to be carried elsewhere changes its shape depending on how it is moved: in the special relativistic spacetime model the absolute rigid rod is not a meaningful notion.

5.5.2. Spacetime measurements in the nonrelativistic case are based on chronometers (showing the absolute time) and measuring rods (that are absolutely rigid).

Spacetime measurements in the special relativistic case are based on chronometeres (showing their proper times) and light signals.

5.6. Time dilation

5.6.1. Recall that given an absolute velocity value $\boldsymbol{u}$, $\boldsymbol{u}$-time is an affine space over $\mathbb{T}$. Since $\mathbb{T}$ is oriented, later and earlier makes sense between $\boldsymbol{u}$-instants: t is earlier than s (s is later than t) if $s - t$ is positive.

A unique $\boldsymbol{u}$-instant $\tau_{\boldsymbol{u}}(x)$ is assigned to every world point x. Consequently, we can decide which of two arbitrary world points is later according to $\boldsymbol{u}$.

The $\boldsymbol{u}$-time interval between the world points x and y is

$$\mathbf{t}_{\boldsymbol{u}}(x,y) := \tau_{\boldsymbol{u}}(y) - \tau_{\boldsymbol{u}}(x) = -\boldsymbol{u} \cdot (y - x).$$

The world point y is *later* than the world point x (x is *earlier* than y) according to $\boldsymbol{u}$ if the $\boldsymbol{u}$-time interval between x and y is positive.

Neither of x and y is later according to $\boldsymbol{u}$ if and only if they are simultaneous according to $\boldsymbol{u}$.

If y is futurelike with respect to x (i.e. if they are lightlike or timelike separated) then y is later than x according to all inertial observers.

5.6.2. Fix two different world points x and y on an inertial world line with absuolute velocity $\boldsymbol{u}'$. Then $\mathbf{t}_{\boldsymbol{u}'}(x,y) = |y - x|$ is the inertial time between x and y, and $y - x = \boldsymbol{u}'\mathbf{t}_{\boldsymbol{u}'}(x,y)$.

Consequently, we have the following formula for the $\boldsymbol{u}$-time interval between x and y:

$$\mathbf{t}_{\boldsymbol{u}}(x,y) = -(\boldsymbol{u} \cdot \boldsymbol{u}')\mathbf{t}_{\boldsymbol{u}'}(x,y) = \frac{\mathbf{t}_{\boldsymbol{u}'}(x,y)}{\sqrt{1 - |\boldsymbol{v}_{\boldsymbol{u}\boldsymbol{u}'}|^2}}.$$

$\mathbf{t}_{\boldsymbol{u}}(x,y) = \mathbf{t}_{\boldsymbol{u}'}(x,y)$ if and only if $\boldsymbol{u} = \boldsymbol{u}'$. In any other cases $\mathbf{t}_{\boldsymbol{u}}(x,y)$ is greater than $\mathbf{t}'_{\boldsymbol{u}}(x,y)$.

5.6.3. Let us illustrate our result as follows.

Let us consider an inertial (pointlike) chronometer. Let x and y be the occurrences between which the chronometer measures one hour. An inertial observer, moving relative to the chronometer, and using its own standard synchronization perceives more than one hour between the occurrences. This is the famous *time dilation.*

The chronometer and the observer move with respect to each other. In usual formulations one considers that the observer is at rest and the chronometer is moving and one says that 'a moving clock works more slowly than a clock at rest'.

Then one continues: let us imagine two clocks resting in the spaces of different observers; both of them will perceive the *other* clock working more slowly.

A number of paradoxes can arise from this situation: 'I say that your clock works more slowly, you say that my clock works more slowly; which of us is right?'

The paradox comes from the fact that usually one does not take the different synchronizations entering in the problem into account. The observer $\boldsymbol{u}$ is not obliged to use its own standard synchronization; if it uses the $\boldsymbol{u}_o$-synchronization, then there is no time dilation.

We emphasize that the time dilation formula does not state any real physical dilation of time at all.

5.7. The twin paradox

5.7.1. Let us consider two twins, Alice and Bob. Both are launched in separate missiles. Alice says that Bob is moving relative to her, hence his time passes more slowly and she observes that when she is forty then he is only twenty. On the other hand, Bob says that Alice is moving relative to him and he observes that when he is forty then she is only twenty. Which of them is right?

Keeping in mind that only illusory and no physical time dilations are in question, we can be convinced that both are right. How can it be possible?

The paradox is based on our everyday concept of absolute time, i.e. that 'when' has an absolute meaning. However, the first 'when' means simultaneity with respect to Alice and the second 'when' means simultaneity with respect to Bob; we know well that these simultaneities are different.

5.7.2. Suppose the twins do not believe that time dilation is illusory and they want an experimental test: let them meet and then a simple inspection will determine which of them is older.

However, the time dilation formula concerns *inertial frames.* It is excusable that both missiles are considered to be inertial. But if they remain inertial then the twins never meet. If the twins meet, then at least one of them *ceases to be inertial.* Anyhow, the mutually equivalent situation of the twins breaks. It will not be true that both are right saying 'when I am forty then my twin is only twenty'.

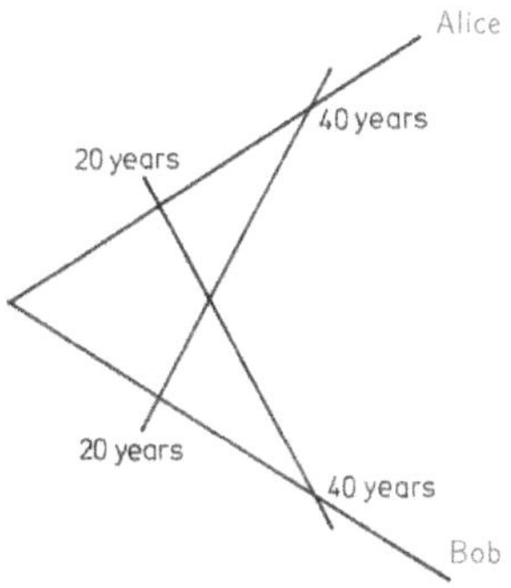

Let the twins meet. Both existed somehow between the two occurrences, the departure and the arrival, and their proper times passed during their existence. The times elapsed depend on their existence and need not be equal. It can happen that Alice is older than Bob; e.g. if Alice remains inertial (the inertial time between two world points is always greater than a time passed on a noninertial world line, see 2.2.3). This difference of proper times is an absolute fact and has nothing to do with the illusory time dilation.

It is important to distinguish between illusory time dilation concerning two standard inertial frames and actually different times passed along two world lines between two world points.

5.8. Experiments concerning time

5.8.1. Cosmic rays produce unstable particles called muons in the ionosphere. These particles have a well-defined average lifetime T. Some of those muons reach the Earth. Detecting the magnitude v of their velocity in the space of the Earth with respect to the standard synchronization and knowing the height d of the place where they are created, we can calculate the time of their travel (uniform and rectilinear motion seems a good approximation). It turns out that the time of travel d/v exceeds the lifetime T of muons. Thus it seems to the Earth that the moving muons live more than their lifetime.

The muon in question exists inertially thus it 'feels' the inertial time t_o of its travel which is less than its lifetime; the Earth (considered as a standard inertial frame) perceives a longer time (time dilation):

$$\frac{d}{v} = \frac{t_o}{\sqrt{1 - v^2}} > T > t_o.$$

It is interesting that someone can argue in another way, too. The muon perceives the distance $d\sqrt{1 - v^2}$ between its birth place and the Earth (Lorentz contraction), hence it travels for $t_o = \frac{d\sqrt{1-v^2}}{v}$.

5.8.2. Let us suppose that, simultaneously with the muon in the ionosphere (muon I), a muon is produced and remains resting on the Earth (muon E). According to what has been said, muon E decays before muon I arrives at the earth: muon E 'sees' that time passes more slowly for muon I.

Of course, muon I 'sees' as well that time passes more slowly for muon E. Then one could suspect a contradiction (paradox): according to muon I, muon E would be alive at the end of the travel of muon I.

There is no contradiction: *simultaneously* in the present context means simultaneously according to the earth, i.e. to muon E, and muon I is produced simultaneously according to muon E. Then, according to muon I, the other muon is born earlier; consequently, muon I, though sees time passing more slowly for muon E, will observe that the life of muon E ends before muon I meets the earth.

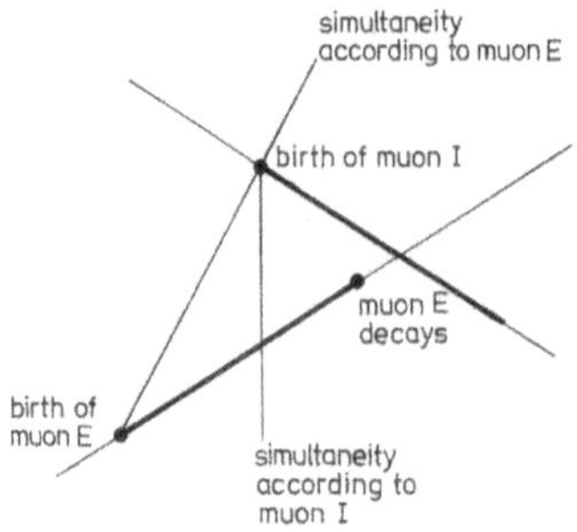

5.8.3. Experiments with unstable particles revolved in an accelerator show the physical fact that different times can pass between two world points along different world lines, as it will be explained.

Suppose two muons are produced 'at the same time and at the same place' i.e. a single world point corresponds to their birth. One one of them (muon R) remains resting beside the accelerator, the other (muon A) is constrained to revolve in the accelerator. The muons meet several times. Then muon R decays, but muon A continues to revolve and meets again the void place of muon R; we see (resting with muon R) as if muon A had a longer life time. Nevertheless, both muons have the same proper lifetime T.

The world line of muon R is inertial while muon A has a noninertial world line; the two world lines intersect each other several times. Different times t_R and t_A pass along the different world lines of the muons between their two successive meetings. Inertial time is always greater than a noninertial: $t_A < t_R$. That is why there can be a natural number n such that $nt_A < T < nt_R$, i.e. muon R does not last until the n-th meeting but muon A survives it.

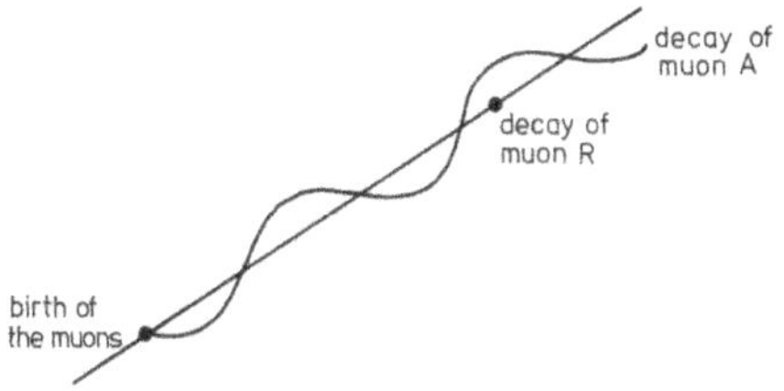

5.9. Exercises

1. Prove that the addition formula 4.3.2 of relative velocities remains valid for $\boldsymbol{u}, \boldsymbol{u}' \in V(1)$, $\boldsymbol{u}'' \in V(0)$.

2. Take the motion treated in 4.6.2. Demonstrate that

$$\lim_{t\to\infty} \dot{r}_{\boldsymbol{u},o}(\boldsymbol{t}) = \boldsymbol{v}_{\boldsymbol{u}\boldsymbol{u}_o} + \frac{\boldsymbol{b}_o}{\beta}, \qquad \lim_{t\to\infty} |\dot{r}_{\boldsymbol{u},o}(\boldsymbol{t})| = 1.$$

3. Consider the uniformly accelerated world line treated in 4.6.2. Try to describe the corresponding motion relative to an inertial observer with constant velocity value $\boldsymbol{u}$ which is not $\boldsymbol{g}$-orthogonal to $\boldsymbol{a}_o$.

4. Let x and y be different world points simultaneous with respect to an affine synchronization (x: a plane lands in London at 12:00; y: a train leaves Paris at 12:00). Then there is an affine synchronization according to which x is later than y and there is an affine synchronization according to which x is earlier than y.

5. We have a clock that can measure a proper time period of $10^{-8}\boldsymbol{s}$. At which relative velocity magnitude can we perceive a time dilation in a minute? (Keep in mind that $1 \equiv 2.9979\ldots \cdot 10^8 \boldsymbol{m}/\boldsymbol{s}$.)

6. Let $\boldsymbol{u} \in V(1)$, $\boldsymbol{v} \in \frac{\mathbf{S}_{\boldsymbol{u}}}{\mathbb{T}}$, $|\boldsymbol{v}| < 1$. Let $\boldsymbol{t}_o \in \mathbb{T}^+$. Consider the world line function r that passes through the world point x_o and

$$(i) \qquad \dot{r}(\boldsymbol{t}) = \frac{\boldsymbol{u}+\boldsymbol{v}\sin(\boldsymbol{t}/\boldsymbol{t}_o)}{\sqrt{1-|\boldsymbol{v}|^2\sin^2(\boldsymbol{t}/\boldsymbol{t}_o)}} \qquad (\boldsymbol{t} \in \mathbb{T});$$

$$(ii) \qquad \dot{r}(\boldsymbol{t}) = \boldsymbol{u}\sqrt{1+|\boldsymbol{v}|^2\sin^2(\boldsymbol{t}/\boldsymbol{t}_o)} + \boldsymbol{v}\sin(\boldsymbol{t}/\boldsymbol{t}_o) \qquad (\boldsymbol{t} \in \mathbb{T}).$$

Prove that the inertial world line $x_o + \boldsymbol{u} \otimes \mathbb{T}$ and the world line $\operatorname{Ran} r$ intersect each other in $x_o + 2\pi n \boldsymbol{t}_o$ for all integers n.

Evidently, $\boldsymbol{t}_o$ is the time passed along $\operatorname{Ran} r$ between two consecutive intersections. Estimate the time passed between two consecutive intersections along the inertial world line.

6. Some special noninertial observers*

6.1. General reference frames

6.1.1. As said, in general, the treatment of noninertial observers in the relativistic spacetime model requires the theory of pseudo-Riemannian manifolds.

Fortunately, to describe some special and important aspects of noninertial special relativistic observers, we can avoid the theory of manifolds; nevertheless, we shall meet some complications.

First of all, together with observers, we always must deal with synchronizations, too. The first question is how a convenient synchronization can be assigned to an observer. The answer is immediate for an inertial observer: by light signals. Is a similar synchronization procedure satisfactory for a noninertial observer?

Let us take two space points q and q' of a noninertial observer $\boldsymbol{U}$. A light signal starting at the world point x^- incident with q meets q' at y; the reflected light signal meets q at x^+. Then the world point x incident with q would be considered simultaneous with y if the proper time passed between x^- and x equals the proper time passed between x and x^+.

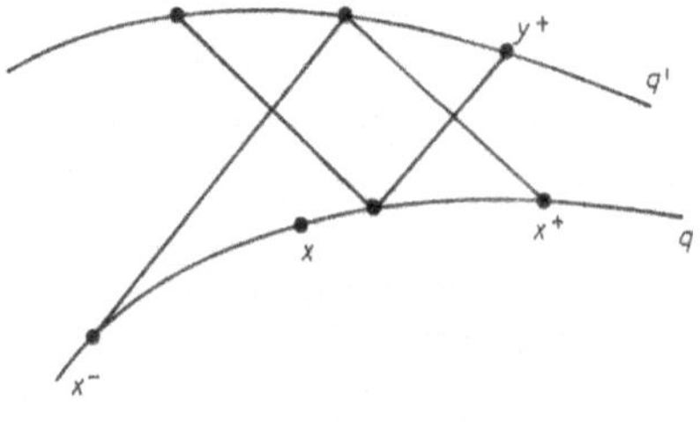

Unfortunately, the synchronization defined by light signals starting from q' does not necessarily coincide with the synchronization defined by light signals starting from q (see Exercise 6.9.12).

Let us accept that synchronization defined by light signals works well 'infinitesimally'. This means that the $\boldsymbol{U}$-line passing through the world point x, in a neighbourhood of x, can be approximated by a straight line directed by $\boldsymbol{U}(x)$. Thus we can say that in a neighbourhood of x the world points approximately simultaneous with x according to $\boldsymbol{U}$ are the elements of $x + \mathbf{S}_{\boldsymbol{U}(x)}$. The smaller the neighbourhood, the better the approximation we get. A clear reasoning leads us then to the idea that world points simultaneous with each other according to $\boldsymbol{U}$ would constitute a hypersurface whose tangent space at every x equals $\mathbf{S}_{\boldsymbol{U}(x)}$. Such a definition of synchronization does not depend on the $\boldsymbol{U}$-space point ($\boldsymbol{U}$-line) from which light signals start. However, it may happen that there is no such hypersurface at all (see 6.7.6)! And even if such hypersurfaces exist, it may happen that the proper times passed between two such hypersurfaces along different $\boldsymbol{U}$-lines are different (see 6.6.5), thus the synchronization is not satisfactory in all respects.

Definition. An observer $\boldsymbol{U}$ is *regular* if there is a (necessary unique) synchronization $\mathcal{S}_{\boldsymbol{U}}$ such that the tangent space of $\tau_{\mathcal{S}_{\boldsymbol{U}}}(x)$ at x equals $\mathbf{S}_{\boldsymbol{U}(x)}$.

6.1.2. In general, there is no *natural* synchronization with respect to a noninertial observer; consequently, there is no natural time of such an observer. Nevertheless, of course, a noninertial observer can choose some sort of *artificial* synchronization (e.g. chooses one of its space points and makes the synchronization by light signals relative to this space point; on the Earth one makes such a synchronization relative to Greenwich).

6.2. Distances in observer spaces

6.2.1. How distances are measured in an observer space? Let $\boldsymbol{U}$ be an observer and suppose a synchronization $\mathcal{S}$ is given on the domain of $\boldsymbol{U}$. We would like to determine the distance between two $\boldsymbol{U}$-space points q and q' at an $\mathcal{S}$-instant t.

First we make the following heuristic considerations. Let us put $x := q \star t$ and suppose q' is 'close' to q. According to the 'infinitesimal' synchronization which is reasonable from the point of view of the observer, $y' := q' \star (x + \mathbf{S}_{\boldsymbol{U}(x)})$ is the world point on q' that is approximately simultaneous with x in a natural way. Then $\boldsymbol{d} := |y' - x|$ is the approximate value of the distance to be determined.

The world point $x' := q' \star t$ is simultaneous with x according to $\mathcal{S}$. Since $y' \approx x' + \boldsymbol{U}(x')\frac{\boldsymbol{U}(x)\cdot(x'-x)}{-\boldsymbol{U}(x)\cdot\boldsymbol{U}(x')} \approx x' + \boldsymbol{U}(x)\big(\boldsymbol{U}(x)\cdot(x'-x)\big)$, we see that $\boldsymbol{d} \approx |\boldsymbol{\pi}_{\boldsymbol{U}(x)} \cdot (x' - x)|$.

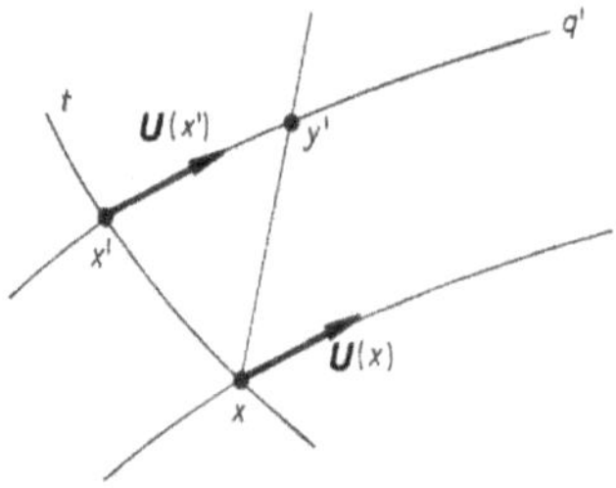

We have got a formula for 'infinitesimal' distances from which we can define the length of a curve in a natural way by an integration. The distance between two observer space points will be defined to be the least length of curves connecting the space points.

Before going further, the reader is asked to study Section VI.7.

6.2.2. Definition. Let $\boldsymbol{U}$ be an observer. A subset L of the observer space $\mathrm{S}_{\boldsymbol{U}}$ is called a *curve* if there is a synchronization $\mathcal{S}$ on $\mathrm{Dom}\,\boldsymbol{U}$ such that $L_t := \{q \star t \mid q \in L,\ q \cap t \neq \emptyset\}$ is either void or a curve in M for all $t \in \mathrm{T}_{\mathcal{S}}$. ■

Note that in fact L_t is contained in the hypersurface t.

We say that the curve L *connects* the $\boldsymbol{U}$-space points q_1 and q_2 if L_t connects $q_1 \star t$ and $q_2 \star t$ for all $t \in \mathrm{T}_{\mathcal{S}}$, provided that $q_1 \cap t$ and $q_2 \cap t$ are not void.

6.2.3. Definition. Let L be a curve in $\mathrm{S}_{\boldsymbol{U}}$. Then

$$\ell_t(L) := \ell_{\boldsymbol{U}}(L_t) := \int_{L_t} |\boldsymbol{\pi}_{\boldsymbol{U}(\cdot)} \mathrm{d}L_t|$$

is called the *length* of L at the $\mathcal{S}$-time point t.

The *distance* between the $\boldsymbol{U}$-space points q and q' at the $\mathcal{S}$-time point t is

$$\boldsymbol{d}_t(q, q') := \inf\{\ell_t(L) \mid L \text{ is a curve connecting } q \text{ and } q'\}. \quad ■$$

It is worth describing explicitly that if p_t is a parametrization of L_t then

$$\ell_t(L) = \int_{\mathrm{Dom}\,p_t} |\boldsymbol{\pi}_{\boldsymbol{U}(p_t(a))} \cdot \dot{p}_t(a)| \mathrm{d}a =$$

$$= \int_{\mathrm{Dom}\,p_t} \sqrt{|\dot{p}_t(a)|^2 + \big(\boldsymbol{U}(p_t(a)) \cdot \dot{p}_t(a)\big)^2} \mathrm{d}a.$$

Note the special case when $\boldsymbol{U}$ is regular and the $\boldsymbol{U}$-synchronization is taken; then $\boldsymbol{\pi}_{\boldsymbol{U}(x)} \cdot \boldsymbol{x} = \boldsymbol{x}$ if $x \in t$ and $\boldsymbol{x}$ is a tangent vector of t at x. Consequently,

$$\ell_t(L) = \int_{L_t} |\mathrm{d}L_t| = \int_{\mathrm{Dom}\,p_t} |\dot{p}_t(a)| \mathrm{d}a \qquad \text{for a regular observer.}$$

In particular, if $\boldsymbol{U}$ is inertial and $\boldsymbol{U}$-time is used then, for all t, $\boldsymbol{d}_t(q,q')$ equals the distance $|q'-q|$ defined earlier.

Keep in mind the following important remark: suppose the $\mathcal{S}$-instant t is a hyperplane; then there is a unique $\boldsymbol{u}_o \in V(1)$ such that t is directed by $\mathbf{S}_{\boldsymbol{u}_o}$. The distance between the $\boldsymbol{U}$-space points at t does not equal, in general, the distance perceived by the standard inertial frame $\boldsymbol{u}_o$. Recall, e.g. the case that $\boldsymbol{U}$ is an inertial observer with the velocity value $\boldsymbol{u}$ (see Section 5.3).

6.2.4. Definition. The observer $\boldsymbol{U}$ is called *rigid* if there is a synchronization $\mathcal{S}$ on $\mathrm{Dom}\,\boldsymbol{U}$ in such a way that if

— L is an arbitrary curve in S_U,

— $t, t' \in \mathrm{T}_{\mathcal{S}}$ and $q \cap t \neq \emptyset$, $q \cap t' \neq \emptyset$ for all $q \in L$,

then $\ell_t(L) = \ell_{t'}(L)$. ■

Note that rigidity of observers is a highly complicated notion in the special relativistic spacetime model, in contradistinction to the nonrelativistic case.

6.2.5. The following assertions can be proved by means of the tools of smooth manifolds.

(*i*) Our definition of a curve in S_U involves a synchronization; nevertheless, it does not depend on synchronization: if there is a synchronization with the required conditions then these conditions are satisfied for all other synchronizations as well.

(*ii*) The distance between two $\boldsymbol{U}$-space points at an $\mathcal{S}$-instant is defined by an infimum; this infimum is in fact a minimum, i.e. for each $\mathcal{S}$-instant t there is a curve connecting the points whose length at t equals the distance between the $\boldsymbol{U}$-space points at t.

(*iii*) Our definition of rigidity involves a synchronization; nevertheless, it does not depend on synchronization: if there is a synchronization with respect to which the observer is rigid, then the observer is rigid with respect to all other synchronization as well.

6.3. A method of finding the observer space

6.3.1. To find the space of an observer, i.e. the $\boldsymbol{U}$-lines, we have to find the solutions of the differential equation

$$(x\colon\ \mathbb{T} \rightarrowtail \mathrm{M})? \qquad \dot{x} = \boldsymbol{U}(x).$$

A frequently applicable method is to transform the differential equation by

$$\xi_{\boldsymbol{u}_o,o}\colon\ \mathrm{M} \to \mathbb{T} \times \mathbf{S}_{\boldsymbol{u}_o}, \qquad x \mapsto \big(-\boldsymbol{u}_o \cdot (x-o),\, \boldsymbol{\pi}_{\boldsymbol{u}_o} \cdot (x-o)\big)$$

according to VI.6.3, where $\boldsymbol{u}_o$ is a suitably chosen element of $V(1)$.

The transformed differential equation will have the form

$$((\boldsymbol{t},\boldsymbol{q})\colon\ \mathbb{T}\rightarrowtail\mathbb{T}\times\mathbf{S}_{\boldsymbol{u}_o})?\qquad (\boldsymbol{t},\boldsymbol{q})^{\cdot}=\left(-\boldsymbol{u}_o\cdot\boldsymbol{U}(o+\boldsymbol{u}_o\boldsymbol{t}+\boldsymbol{q}),\boldsymbol{\pi}_{\boldsymbol{u}_o}\cdot\boldsymbol{U}(o+\boldsymbol{u}_o\boldsymbol{t}+\boldsymbol{q})\right),$$

i.e.

$$(\boldsymbol{t}\colon\ \mathbb{T}\rightarrowtail\mathbb{T})?\qquad \dot{\boldsymbol{t}}=-\boldsymbol{u}_o\cdot\boldsymbol{U}(o+\boldsymbol{u}_o\boldsymbol{t}+\boldsymbol{q}),$$

$$(\boldsymbol{q}\colon\ \mathbb{T}\rightarrowtail\mathbf{S}_{\boldsymbol{u}_o})?\qquad \dot{\boldsymbol{q}}=\boldsymbol{\pi}_{\boldsymbol{u}_o}\cdot\boldsymbol{U}(o+\boldsymbol{u}_o\boldsymbol{t}+\boldsymbol{q}).$$

Let $\boldsymbol{s}\mapsto\boldsymbol{t}(\boldsymbol{s})$ and $\boldsymbol{s}\mapsto\boldsymbol{q}(\boldsymbol{s})$ denote the solutions of these differential equations with the initial conditions $\boldsymbol{t}(\boldsymbol{0})=\boldsymbol{0}$, $\boldsymbol{q}(\boldsymbol{0})=\boldsymbol{q}_o$, where $\boldsymbol{q}_o$ is an arbitrary element in $\mathbf{S}_{\boldsymbol{u}_o}$ such that $o+\boldsymbol{q}_o$ is in the domain of $\boldsymbol{U}$. Then

$$\boldsymbol{s}\mapsto o+\boldsymbol{u}_o\boldsymbol{t}(\boldsymbol{s})+\boldsymbol{q}(\boldsymbol{s})$$

is the world line function giving the $\boldsymbol{U}$-line passing through $o+\boldsymbol{q}_o$.

It is worth using more precise notations: let x be an element of $(o+\mathbf{S}_{\boldsymbol{u}_o})\cap(\operatorname{Dom}\boldsymbol{U})$; then $\boldsymbol{s}\mapsto\boldsymbol{t}_x(\boldsymbol{s})$ and $\boldsymbol{s}\mapsto\boldsymbol{q}_x(\boldsymbol{s})$ will denote the solutions of the differential equations with the initial conditions $\boldsymbol{t}_x(\boldsymbol{0})=\boldsymbol{0}$, $\boldsymbol{q}_x(\boldsymbol{0})=x-o$. Then

$$\mathbb{T}\rightarrowtail\mathrm{M},\qquad \boldsymbol{s}\mapsto r_x(\boldsymbol{s})=o+\boldsymbol{u}_o\boldsymbol{t}_x(\boldsymbol{s})+\boldsymbol{q}_x(\boldsymbol{s})$$

is the world line function giving the $\boldsymbol{U}$-space point that x is incident with.

6.3.2. Consider the $\boldsymbol{u}_o$-synchronization. Then, according to 4.1.1, $\boldsymbol{s}\mapsto\boldsymbol{t}_x(\boldsymbol{s})$ gives $\boldsymbol{u}_o$-time as a function of the proper time of the $\boldsymbol{U}$-line passing through x; in other words, $\boldsymbol{t}_x(\boldsymbol{s})$ is the $\boldsymbol{u}_o$-time passed between $t_o:=o+\mathbf{S}_{\boldsymbol{u}_o}$ and $t_x(\boldsymbol{s}):=r_x(\boldsymbol{s})+\mathbf{S}_{\boldsymbol{u}_o}$: $\boldsymbol{t}_x(\boldsymbol{s})=t_x(\boldsymbol{s})-t_o$.

This function is strictly monotone increasing; its inverse, denoted by $\mathbb{T}\rightarrowtail\mathbb{T}$, $\boldsymbol{t}\mapsto\boldsymbol{s}_x(\boldsymbol{t})$, gives the proper time between the $\boldsymbol{u}_o$-instants t_o and $t_o+\boldsymbol{t}$ passed in the $\boldsymbol{U}$-space point that x is incident with.

6.4. Uniformly accelerated observer I

6.4.1. In the special relativistic spacetime model the definition of a uniformly accelerated observer is not so straightforward as in the nonrelativistic case. We know that here the acceleration of a uniformly accelerated world line function is not constant, thus a uniformly accelerated observer will not be an observer with constant acceleration field. Anyhow, we wish to find an observer whose lines are uniformly accelerated.

Omitting the thorny way of searching, let us take an observer satisfying the requirements and study its properties.

Let $o \in \mathrm{M}$, $\boldsymbol{u}_o \in V(1)$ and $\mathbf{0} \neq \boldsymbol{a}_o \in \frac{\mathbf{S}_{\boldsymbol{u}_o}}{\mathbb{T}\otimes\mathbb{T}}$ and define the global observer

$$\boldsymbol{U}(x) := \boldsymbol{u}_o\sqrt{1 + |\boldsymbol{a}_o|^2\big(\boldsymbol{u}_o\cdot(x-o)\big)^2} - \boldsymbol{a}_o\big(\boldsymbol{u}_o\cdot(x-o)\big) \qquad (x \in \mathrm{M}).$$

Note that $\boldsymbol{u}_o = \boldsymbol{U}(o)$.
The observer has the acceleration field

$$\boldsymbol{A}_U(x) = \boldsymbol{a}_o\sqrt{1 + |\boldsymbol{a}_o|^2\big(\boldsymbol{u}_o\cdot(x-o)\big)^2} - \boldsymbol{u}_o|\boldsymbol{a}_o|^2\big(\boldsymbol{u}_o\cdot(x-o)\big) \qquad (x \in \mathrm{M}),$$

thus $\boldsymbol{a}_o = \boldsymbol{A}_U(o)$.
It is trivial that

$$\boldsymbol{U}(x+\boldsymbol{q}) = \boldsymbol{U}(x), \qquad \boldsymbol{A}_U(x+\boldsymbol{q}) = \boldsymbol{A}_U(x) \qquad (x \in \mathrm{M}, \boldsymbol{q} \in \mathbf{S}_{\boldsymbol{u}_o}),$$

i.e. $\boldsymbol{U}$ and $\boldsymbol{A}_U$ are constant on the hyperplanes directed by $\mathbf{S}_{\boldsymbol{u}_o}$.
As a consequence, the translation of a $\boldsymbol{U}$-line by a vector in $\mathbf{S}_{\boldsymbol{u}_o}$ is a $\boldsymbol{U}$-line, too.

6.4.2. Transforming the differential equation of the observer according to 6.3, we get

$$\dot{\boldsymbol{t}} = \sqrt{1 + |\boldsymbol{a}_o|^2\boldsymbol{t}^2},$$
$$\dot{\boldsymbol{q}} = \boldsymbol{a}_o\boldsymbol{t}.$$

The first equation, with the initial value $\boldsymbol{t}(\mathbf{0}) = \mathbf{0}$, has the solution

$$\boldsymbol{t}(\boldsymbol{s}) = \frac{\sinh|\boldsymbol{a}_o|\boldsymbol{s}}{|\boldsymbol{a}_o|} \qquad (\boldsymbol{s} \in \mathbb{T}).$$

Then the second equation becomes very simple and we find its solutions in the form

$$\boldsymbol{q}(\boldsymbol{s}) = \boldsymbol{a}_o\frac{\cosh|\boldsymbol{a}_o|\boldsymbol{s} - 1}{|\boldsymbol{a}_o|^2} + \boldsymbol{q}_o \qquad (\boldsymbol{s} \in \mathbb{T}).$$

Hence we obtain that the $\boldsymbol{U}$-line passing through $x \in o + \mathbf{S}_{\boldsymbol{u}_o}$ is given by the world line function

$$r_x(\boldsymbol{s}) = x + \boldsymbol{u}_o\frac{\sinh|\boldsymbol{a}_o|\boldsymbol{s}}{|\boldsymbol{a}_o|} + \boldsymbol{a}_o\frac{\cosh|\boldsymbol{a}_o|\boldsymbol{s} - 1}{|\boldsymbol{a}_o|^2} \qquad (\boldsymbol{s} \in \mathbb{T}).$$

It is not hard to see that every $\boldsymbol{U}$-line meets the hyperplane $o + \mathbf{S}_{\boldsymbol{u}_o}$, hence every $\boldsymbol{U}$-line can be given by such a world line function: all $\boldsymbol{U}$-lines are uniformly accelerated.

6.4.3. Because $\boldsymbol{U}$ and $\boldsymbol{A}_{\boldsymbol{U}}$ are constant on the hyperplanes directed by $\mathbf{S}_{\boldsymbol{u}_o}$, all $\boldsymbol{U}$-lines have the same velocity and the same acceleration on the hyperplanes directed by $\mathbf{S}_{\boldsymbol{u}_o}$.

Using the notations of 6.4, we see that

$$s_x(\boldsymbol{t}) = \frac{\operatorname{arsinh}|\boldsymbol{a}_o|\boldsymbol{t}}{|\boldsymbol{a}_o|} =: \boldsymbol{s}(\boldsymbol{t}) \qquad (\boldsymbol{t} \in \mathbb{T})$$

for all x in $o + \mathbf{S}_{\boldsymbol{u}_o}$. Thus, given two $\boldsymbol{u}_o$-instants, the same time passes along all $\boldsymbol{U}$-lines between them.

Because of these properties of $\boldsymbol{U}$-lines it seems suitable to associate with $\boldsymbol{U}$ the $\boldsymbol{u}_o$-synchronization to form a reference frame.

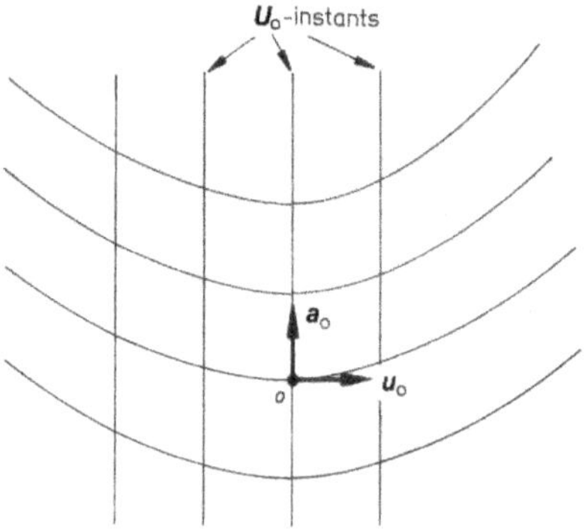

6.4.4. Let us examine whether this observer is regular.

Evidently, $o + \mathbf{S}_{\boldsymbol{u}_o}$ is a world surface $\boldsymbol{g}$-orthogonal to $\boldsymbol{U}$.

Introduce the notation

$$h(x) :=$$

$$:= \boldsymbol{a}_o \cdot (x - o) - \sqrt{1 + |\boldsymbol{a}_o|^2 (\boldsymbol{u}_o \cdot (x - o))^2} + \operatorname{artanh}\sqrt{1 + |\boldsymbol{a}_o|^2 (\boldsymbol{u}_o \cdot (x - o))^2}$$

for $x \in \mathrm{M}$, $x \notin o + \mathbf{S}_{\boldsymbol{u}_o}$ and put for $\lambda \in \mathbb{R}$

$$t_\lambda := \begin{cases} \{x \in \mathrm{M} \mid \ h(x) = \ln\lambda, \ -\boldsymbol{u}_o \cdot (x - o) > \mathbf{0}\} & \text{if } \lambda > 0 \\ o + \mathbf{S}_{\boldsymbol{u}_o} & \text{if } \lambda = 0 \\ \{x \in \mathrm{M} \mid h(x) = \ln(-\lambda), \ -\boldsymbol{u}_o \cdot (x - o) < 0\} & \text{if } \lambda < 0. \end{cases}$$

Evidently, h is a differentiable function outside $o + \mathbf{S}_{\boldsymbol{u}_o}$ and

$$\mathrm{D}h(x) = \boldsymbol{a}_o + \frac{\boldsymbol{u}_o\sqrt{1 + |\boldsymbol{a}_o|^2 (\boldsymbol{u}_o \cdot (x - o))^2}}{-\boldsymbol{u}_o \cdot (x - o)} = \frac{\boldsymbol{U}(x)}{-\boldsymbol{u}_o \cdot (x - o)}.$$

As a consequence, for all $\lambda \in \mathbb{R}$, t_λ is a three-dimensional submanifold whose tangent space at x equals $\mathrm{Ker}\,\mathrm{D}h(x) = \mathbf{S}_{U(x)}$. This means that t_λ-s are $\boldsymbol{U}$-surfaces, $\boldsymbol{U}$ is regular and

$$\mathrm{T}_U = \{t_\lambda \mid \lambda \in \mathbb{R}\}.$$

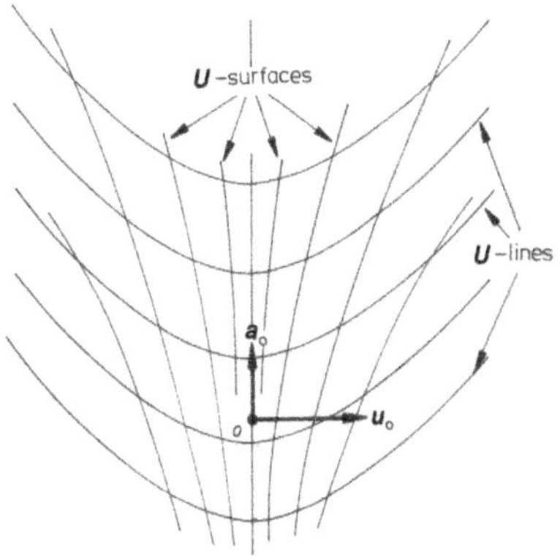

6.4.5. If q_1 and q_2 are $\boldsymbol{U}$-lines, then $q_2 \star t - q_1 \star t$ is the same for all $\boldsymbol{u}_o$-instants t. In other words, the vector and the distance perceived by the standard inertial frame $\boldsymbol{u}_o$ between two $\boldsymbol{U}$-space points is the same for all $\boldsymbol{u}_o$-instants. We can say that $\boldsymbol{u}_o$ perceives $\boldsymbol{U}$ to be rigid and rotation-free. Is $\boldsymbol{U}$ rigid and rotation-free?

We have not defined when an observer is rotation-free, thus we can answer only the question regarding rigidity as defined in 6.3.4.

This observer $\boldsymbol{U}$ is not rigid. Let us take a $\boldsymbol{u}_o$-instant t. For all $x \in t$ we have $-\boldsymbol{u}_o \cdot (x - o) = t - t_o$ (where $t_o := o + \mathbf{S}_{\boldsymbol{u}_o}$), thus

$$\boldsymbol{U}(x) = \boldsymbol{a}_o(t - t_o) + \boldsymbol{u}_o\sqrt{1 + |\boldsymbol{a}_o|^2(t - t_o)^2} =: \boldsymbol{u}_t \qquad (x \in t,\ t \in \mathrm{T}_{\boldsymbol{u}_o})$$

($\boldsymbol{U}$ is constant on the $\boldsymbol{u}_o$-instants).

The formula in 6.4.3 gives us the proper time $\boldsymbol{s}(\boldsymbol{t})$ passed in every $\boldsymbol{U}$-space point between the $\boldsymbol{u}_o$-instants t_o and $t := t_o + \boldsymbol{t}$:

$$\boldsymbol{t} = \frac{\sinh |\boldsymbol{a}_o| \boldsymbol{s}(\boldsymbol{t})}{|\boldsymbol{a}_o|}.$$

Let L_o be a curve in $o + \mathbf{S}_{\boldsymbol{u}_o}$; then the set of $\boldsymbol{U}$-space points that meet L_o, $L := \{q \in \mathrm{S}_U \mid q \cap L_o \neq \emptyset\}$ is a curve in the observer space. Indeed, if p_o is a parametrization of L_o, then

$$p_t := p_o + \boldsymbol{u}_o \boldsymbol{t} + \frac{\boldsymbol{a}_o}{|\boldsymbol{a}_o|^2}\left(\sqrt{1 + |\boldsymbol{a}_o|^2 \boldsymbol{t}^2} - 1\right)$$

is a parametrization of L_t (see 6.3.2) for $t = t_o + \boldsymbol{t} \in T_{\boldsymbol{u}_o}$.

Then

$$\dot{p}_t = \dot{p}_o$$

and

$$\boldsymbol{U}\big(p_t(a)\big) \cdot \dot{p}_t(a) = -\boldsymbol{a}_o \cdot \dot{p}_o(a)\boldsymbol{t} \qquad (a \in \operatorname{Dom} p_o);$$

consequently,

$$|\dot{p}_t|^2 + \big((\boldsymbol{U} \circ p_t) \cdot \dot{p}_t\big)^2 = |\dot{p}_o|^2 + \big(\boldsymbol{a}_o \cdot \dot{p}_o\big)^2 \boldsymbol{t}^2,$$

which shows that the length of curves depends on the $\boldsymbol{u}_o$-time points: the observer is not rigid.

6.4.6. The length of curves in $\boldsymbol{U}$-space, consequently the distance between $\boldsymbol{U}$-space points, in general, decreases prior to t_o and increases after t_o, as $\boldsymbol{u}_o$-time passes. This is well understandable from a heuristic point of view. Though we defined Lorentz contraction between two inertial observers, we can say e.g. that after t_o the space points of $\boldsymbol{U}$ move faster and faster with respect to $\boldsymbol{u}_o$, thus their distances seem more and more contracted with respect to the standard inertial grame $\boldsymbol{u}_o$; that is, their distances must increase continually in order that the distances perceivedd by $\boldsymbol{u}_o$ be constant.

6.5. Uniformly accelerated observer II

6.5.1. Let $o \in M$, $\boldsymbol{u}_o \in V(1)$ and $\boldsymbol{0} \neq \boldsymbol{a}_o \in \frac{\mathbf{S}_{\boldsymbol{u}_o}}{\mathbb{T}\otimes\mathbb{T}}$, put

$$\boldsymbol{B}(x) := \big(\boldsymbol{a}_o \cdot (x - o)\big)\boldsymbol{u}_o - \big(\boldsymbol{u}_o \cdot (x - o)\big)\boldsymbol{a}_o$$

for $x \in M$ and define the nonglobal observer by

$$\operatorname{Dom}\boldsymbol{U} := \{x \in \mathbf{M} \mid \boldsymbol{B}(x) \text{ is future directed timelike}\}$$
$$\boldsymbol{U}(x) := \frac{\boldsymbol{B}(x)}{|\boldsymbol{B}(x)|} \qquad (x \in \operatorname{Dom}\boldsymbol{U}).$$

Note that $\boldsymbol{B}(x)$ is future directed timelike if and only if

$$\boldsymbol{0} > \big(\boldsymbol{B}(x)\big)^2 = -\big(\boldsymbol{a}_o \cdot (x - o)\big)^2 + \big(\boldsymbol{u}_o \cdot (x - o)\big)^2 |\boldsymbol{a}_o|^2,$$
$$\boldsymbol{0} > \boldsymbol{u}_o \cdot \boldsymbol{B}(x) = -\boldsymbol{a}_o \cdot (x - o).$$

Then we find that a world point x for which $x - o$ lies in the plane generated by $\boldsymbol{u}_o$ and $\boldsymbol{a}_o$ is in the domain of $\boldsymbol{U}$ if and only if $x - o$ is spacelike and $\boldsymbol{a}_o \cdot (x - o) > \boldsymbol{0}$.

6.5.2. If $\boldsymbol{q}$ is a world vector $\boldsymbol{g}$-orthogonal to both $\boldsymbol{u}_o$ and $\boldsymbol{a}_o$ then

$$\operatorname{Dom}\boldsymbol{U} + \boldsymbol{q} = \operatorname{Dom}\boldsymbol{U}$$

and

$$\boldsymbol{U}(x+\boldsymbol{q}) = \boldsymbol{U}(x) \qquad (x \in \operatorname{Dom}\boldsymbol{U},\ \boldsymbol{q} \in \mathbf{S}_{\boldsymbol{u}_o},\ \boldsymbol{a}_o \cdot \boldsymbol{q} = 0).$$

The observer has the acceleration field

$$\boldsymbol{A}_{\boldsymbol{U}}(x) = \frac{\big(\boldsymbol{a}_o \cdot (x-o)\big)\boldsymbol{a}_o - |\boldsymbol{a}_o|^2 \big(\boldsymbol{u}_o \cdot (x-o)\big)\boldsymbol{u}_o}{|\boldsymbol{B}(x)|^2} \qquad (x \in \operatorname{Dom}\boldsymbol{U}).$$

Then we easily find that

$\boldsymbol{U}(x) = \boldsymbol{u}_o$	if and only if $x-o$ is in $\mathbf{S}_{\boldsymbol{u}_o}$,
$\boldsymbol{A}_{\boldsymbol{U}}(x) \neq \boldsymbol{a}_o$	for all $x \in \operatorname{Dom}\boldsymbol{U}$,
$\boldsymbol{A}_{\boldsymbol{U}}(x) = \frac{\boldsymbol{a}_o}{\boldsymbol{a}_o \cdot (x-o)}$	if and only if $x-o$ is in $\mathbf{S}_{\boldsymbol{u}_o}$.

6.5.3. Let us introduce the notation

$$\boldsymbol{n}_o := \frac{\boldsymbol{a}_o}{|\boldsymbol{a}_o|}.$$

If $\lambda \in \mathbb{R}$ then

$$\boldsymbol{u}_\lambda := \frac{\boldsymbol{u}_o + (\tanh\lambda)\boldsymbol{n}_o}{\sqrt{1-(\tanh\lambda)^2}} = \boldsymbol{u}_o \cosh\lambda + \boldsymbol{n}_o \sinh\lambda$$

is in $V(1)$ and we easily find that

$$\boldsymbol{U}(x) = \boldsymbol{u}_\lambda \qquad (x \in \operatorname{Dom}\boldsymbol{U},\ x-o \in \mathbf{S}_{\boldsymbol{u}_\lambda}).$$

Thus $t_\lambda := \big(o + \mathbf{S}_{\boldsymbol{u}_\lambda}\big) \cap (\operatorname{Dom}\boldsymbol{U})$ is a $\boldsymbol{U}$-surface. To every $x \in \operatorname{Dom}\boldsymbol{U}$ there is such a $\boldsymbol{U}$-surface containing x, given by

$$\lambda_x := \operatorname{artanh}\left(\frac{-\boldsymbol{u}_o \cdot (x-o)}{\boldsymbol{n}_o \cdot (x-o)}\right).$$

This means that $\boldsymbol{U}$ is regular, and

$$\mathrm{T}_{\boldsymbol{U}} := \{t_\lambda \mid \lambda \in \mathbb{R}\}.$$

6.5.4. To find the U-lines we use the method outlined in 6.3.
Transforming the differential equation $\dot{x} = U(x)$ by $\xi_{u_o,o}$ we get

$$\dot{t} = \frac{n_o \cdot q}{\sqrt{(n_o \cdot q)^2 - t^2}}, \qquad (*)$$

$$\dot{q} = \frac{n_o t}{\sqrt{(n_o \cdot q)^2 - t^2}}. \qquad (**)$$

Equation $(**)$ implies $n_o \cdot \dot{q} = \frac{t}{\sqrt{(n_o \cdot q)^2 - t^2}}$ which, together with equation $(*)$, results in

$$(n_o \cdot q)(n_o \cdot \dot{q}) = t\dot{t}$$

implying

$$(n_o \cdot q)^2 - t^2 = \text{const} =: \frac{1}{\alpha^2}.$$

Then differentiating equation $(*)$ we obtain

$$\ddot{t} = \alpha^2 t$$

from which—taking the initial values $t(0) = 0$, $\dot{t}(0) = 1$—we infer

$$t(s) = \frac{\sinh \alpha s}{\alpha}.$$

As a consequence, equation $(**)$ takes an extremely simple form, and we find its solutions easily:

$$q(s) = n_o \frac{\cosh \alpha s - 1}{\alpha} + q_o.$$

Hence we obtain that the U-line passing through $x \in \left(o + \mathbf{S}_{u_o}\right) \cap (\operatorname{Dom} U)$ is given by the world line function

$$r_x(s) = x + u_o \frac{\sinh |a_x| s}{|a_x|} + a_x \frac{\cosh |a_x| s - 1}{|a_x|^2} \qquad (s \in \mathbb{T})$$

where

$$a_x := \frac{a_o}{a_o \cdot (x - o)} = \frac{n_o}{n_o \cdot (x - o)}.$$

It is not hard to see that every U-line meets the hyperplane $o + \mathbf{S}_{u_o}$, hence every U-line can be given by such a world line function; all U-lines are uniformly accelerated.

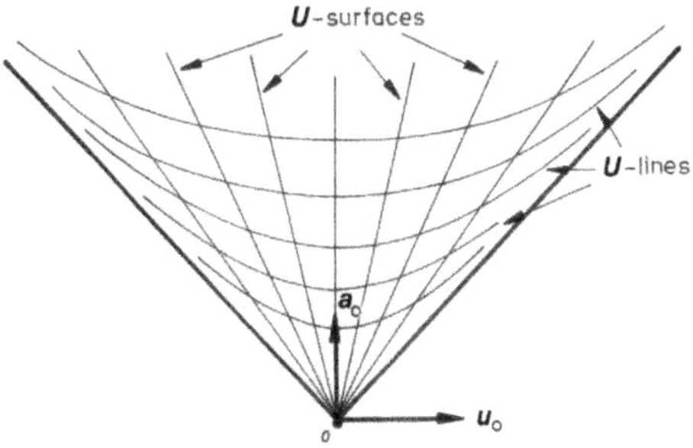

6.5.5. The present observer $\boldsymbol{U}$ serves as an example to show that the observer is regular, but different times pass in different $\boldsymbol{U}$-space points (along different $\boldsymbol{U}$-lines) between two $\boldsymbol{U}$-instants.

Let us consider the $\boldsymbol{U}$-line passing through $x \in o + \mathbf{S}_{\boldsymbol{u}_o}$, described by the world line function r_x given previously; a simple calculation yields that $r_x(\boldsymbol{s})$ is in the $\boldsymbol{U}$-surface t_λ if and only if $\boldsymbol{s} = \frac{\lambda}{|\boldsymbol{a}_x|}$. In other words,

$$\boldsymbol{s}_x(\lambda) := \frac{\lambda}{|\boldsymbol{a}_x|} = \big(\boldsymbol{n}_o \cdot (x - o)\big)\lambda$$

which clearly depends on x, is the time passed between the $\boldsymbol{U}$-time points t_0 and t_λ in the $\boldsymbol{U}$-space point that x is incident with.

6.5.6. Now we shall show that this observer is rigid.

Let L_o be a curve in $o + \mathbf{S}_{\boldsymbol{u}_o}$; then the set of $\boldsymbol{U}$-space points that meet L_o, $L := \{q \in \mathrm{S}_{\boldsymbol{U}} | \, q \cap L_o \neq \emptyset\}$ is a curve in the observer space. Indeed, if p_o is a parametrization of L_o, then, according to the previous result on proper times,

$$p_{t_\lambda} := p_o + \big(\boldsymbol{n}_o \cdot (p_o - o)\big)\big(\boldsymbol{u}_o \sinh\lambda + \boldsymbol{n}_o(\cosh\lambda - 1)\big)$$

is a parametrization of L_{t_λ} for all $t_\lambda \in \mathrm{T}_{\boldsymbol{U}}$.

Then

$$\dot{p}_{t_\lambda} = \dot{p}_o + (\boldsymbol{n}_o \cdot \dot{p}_o)\big(\boldsymbol{u}_o \sinh\lambda + \boldsymbol{n}_o(\cosh\lambda - 1)\big)$$

and

$$|\dot{p}_{t_\lambda}| = |\dot{p}_o| \qquad (t_\lambda \in \mathrm{T}_{\boldsymbol{U}}).$$

Since $\boldsymbol{U}$ is regular and $\boldsymbol{U}$-time is considered, $\boldsymbol{U}(p_{t_\lambda}(a)) \cdot \dot{p}_{t_\lambda}(a) = \mathbf{0}$ for all $\lambda \in \mathbb{R}$ and $a \in \mathrm{Dom}\, p_o = \mathrm{Dom}\, p_{t_\lambda}$, this means that $\ell_{t_\lambda}(L) = \ell_o(L)$ for all $t_\lambda \in \mathrm{T}_{\boldsymbol{U}}$. It is not hard to see that every curve in the observer space can be obtained from a curve in $o + \mathbf{S}_{u_o}$ by the previous method; consequently, the observer is rigid.

6.5.7. Two uniformly accelerated observers have been treated. Neither of them possesses all the good properties of the uniformly accelerated observer in the nonrelativistic spacetime model. It is an open question whether we can find a special relativistic observer $\boldsymbol{U}$ such that

(*i*) all $\boldsymbol{U}$-lines are uniformly accelerated,

(*ii*) $\boldsymbol{U}$ and $\boldsymbol{A}_U$ are constant on each instant (world surface) of a synchronization,

(*iii*) $\boldsymbol{U}$ is rigid.

The observer in 6.4 does not satisfy (*iii*); the observer in 6.5 does not satisfy (*ii*).

6.6. Uniformly rotating observer I

6.6.1. In defining the uniformly rotating observer we encounter problems similar to those in the previous section and, in the same manner, we find two possibilities but neither of them possesses all the good properties of the nonrelativistic uniformly rotating observer.

Let $o \in \mathrm{M}$, $\boldsymbol{u}_o \in V(1)$ and let $\Omega\colon \mathbf{S}_{u_o} \to \frac{\mathbf{S}_{u_o}}{\mathbb{T}}$ be a nonzero antisymmetric linear map and define the global observer

$$\boldsymbol{U}(x) := \Omega \cdot \boldsymbol{\pi}_{u_o} \cdot (x - o) + \boldsymbol{u}_o \sqrt{1 + |\Omega \cdot \boldsymbol{\pi}_{u_o} \cdot (x - o)|^2} \qquad (x \in \mathrm{M}).$$

Note that

$$\boldsymbol{u}_o = \boldsymbol{U}(o),$$

and

$$\boldsymbol{U}(x + \boldsymbol{q}) = \boldsymbol{U}(x) \qquad (x \in \mathrm{M},\ \boldsymbol{q} \in \mathrm{Ker}\,\Omega).$$

The observer has the acceleration field

$$\boldsymbol{A}_U(x) = \Omega \cdot \Omega \cdot \boldsymbol{\pi}_{u_o} \cdot (x - o) \qquad (x \in \mathrm{M}).$$

6.6.2. To find the $\boldsymbol{U}$-lines we apply the well-proved method: transforming the differential equation $\dot{x} = \boldsymbol{U}(x)$ by $\boldsymbol{\xi}_{u_o,o}$ we get

$$\dot{t} = \sqrt{1 + |\Omega \cdot \boldsymbol{q}|^2},$$
$$\dot{\boldsymbol{q}} = \Omega \cdot \boldsymbol{q}.$$

The second equation can be solved immediately:

$$q(s) = e^{s\Omega} \cdot q_o \qquad (s \in \mathbb{T}).$$

Then the first equation becomes $\dot{t} = \sqrt{1 + |\Omega \cdot q_o|^2}$ having the solution—with the initial value $t(0) = 0$—

$$t(s) = s\sqrt{1 + |\Omega \cdot q_o|^2} \qquad (s \in \mathbb{T}).$$

Thus the U-line passing through $x \in o + \mathbf{S}_{u_o}$ (the U-space point that x is incident with) is given by the world line function

$$r_x(s) = o + u_o s\sqrt{1 + |\Omega \cdot (x - o)|^2} + e^{s\Omega} \cdot (x - o) \qquad (s \in \mathbb{T}).$$

It is not hard to see that every U-line meets the hyperplane $o + \mathbf{S}_{u_o}$, thus every U-line is of this form.

6.6.3. Note that the U-line passing through $o + e$, where e is in $\operatorname{Ker}\Omega$, is a straight line directed by u_o; then the set of U-space points

$$\{o + e + u_o \otimes \mathbb{T} \mid e \in \operatorname{Ker}\Omega\}$$

can be interpreted as the *axis of rotation.*

If x is in $o + \mathbf{S}_{u_o}$, then $x - o$ can be decomposed into a sum $e_x + q_x$ where e_x is in $\operatorname{Ker}\Omega$ and q_x is orthogonal to $\operatorname{Ker}\Omega$. Then the U-line above can be written in the form

$$r_x(s) = o + e_x + u_o s\sqrt{1 + \omega^2 |q_x|^2} + e^{s\Omega} \cdot q_x, \qquad (*)$$

where ω is the magnitude of Ω (see Exercise V.3.21.1).

Hence all the U-lines are composed of an inertial line (with a proper time 'accelerated' relative to the proper time of the points of the axis) and a uniform rotation.

Let us consider the u_o-synchronization..

Put $t_o := o + \mathbf{S}_{u_o}$. Then

$$s_x(t) = \frac{t}{\sqrt{1 + |\Omega \cdot (x - o)|^2}} = \frac{t}{\sqrt{1 + |\Omega \cdot q_x|^2}}$$

time passes between the u_o-instants t_o and $t_o + t$ in the U-space point that x is incident with.

The distance perceived by the standard inertial frame u_o at the u_o-instant $t_o + t$ between the U-space point that x is incident with and $o + e_x + u_o \otimes \mathbb{T}$ (the axis of rotation) equals

$$|r_x(s_x(t)) - (o + e_x + u_o t)| = |q_x|,$$

which is independent of t.

6.6.4. Because of the term $s \mapsto e^{s\Omega} \cdot (x - o)$ in $(*)$ we can state that the time period T of rotation is the same for all U-space points (out of the axis of rotation), concerning their proper times: $T = \frac{2\pi}{\omega}$.

On the other hand, concerning u_o-time, the time period of rotation of a U-space point having the u_o-distance $d > 0$ from the axis of rotation equals $T_o(d) := \frac{2\pi}{\omega}\sqrt{1 + \omega^2 d^2}$; it increases from $\frac{2\pi}{\omega}$ to infinity as d increases from zero to infinity.

The following Figure illustrates the situation. Two U-line segments are represented; the proper time passed along both segments equals $\frac{2\pi}{\omega}$.

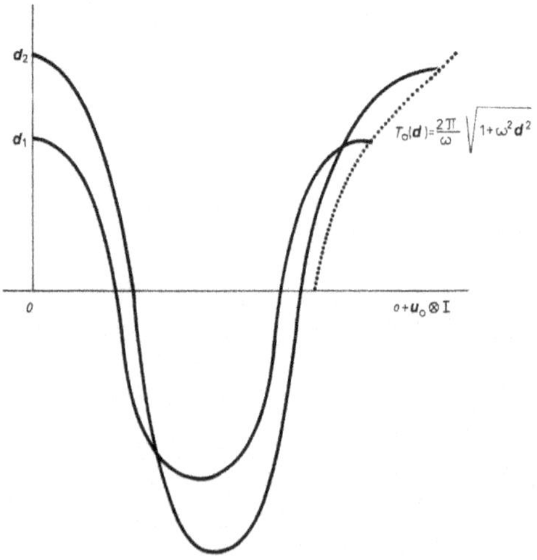

Another figure shows the plane in the u_o-space, orthogonal to $\mathrm{Ker}\,\Omega$, and illustrates the angles of rotation of U-space points during a u_o-time interval $\frac{2\pi}{\omega}$.

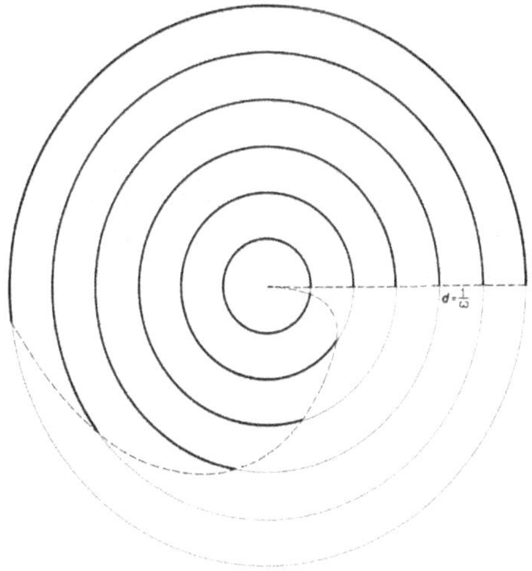

6.6.5. This observer is not rigid.

Let L_o be a curve in $o + \mathbf{S}_{\boldsymbol{u}_o}$; then the set of $\boldsymbol{U}$-space points that meet L_o, $L := \{q \in \mathrm{S}_U \mid q \cap L_o \neq \emptyset\}$ is a curve in the observer space. Indeed, if p_o is a parametrization of L_o, then

$$p_t(a) := o + \boldsymbol{u}_o\boldsymbol{t} + \exp\left(\frac{\boldsymbol{t}\Omega}{\sqrt{1 + |\Omega \cdot (p_o(a) - o)|^2}}\right) \cdot (p_o(a) - o)$$
$$(a \in \mathrm{Dom}\, p_o)$$

is a parametrization of L_t for $t = t_o + \boldsymbol{t} \in \mathrm{T}_{\boldsymbol{u}_o}$. Then

$$\dot{p}_t = \exp\left(\frac{\boldsymbol{t}\Omega}{\sqrt{1 + |\Omega \cdot (p_o - o)|^2}}\right) \cdot \left(\frac{\boldsymbol{t}(\Omega \cdot p_o) \cdot (\Omega \cdot \dot{p}_o)}{\left(1 + |\Omega \cdot (p_o - o)|^2\right)^{3/2}} \cdot \Omega \cdot (p_o - o) + \dot{p}_o\right).$$

We easily find that

$$\boldsymbol{U} \circ p_t = \exp\left(\frac{\boldsymbol{t}\Omega}{\sqrt{1 + |\Omega \cdot (p_o - o)|^2}}\right) \cdot \Omega \cdot (p_o - o) + \boldsymbol{u}_o\sqrt{1 + |\Omega \cdot (p_o - o)|^2}.$$

Then, using

$$(\mathrm{e}^{\alpha\Omega} \cdot \boldsymbol{q}_1) \cdot (\mathrm{e}^{\alpha\Omega} \cdot \boldsymbol{q}_2) = \boldsymbol{q}_1 \cdot \boldsymbol{q}_2$$

for all $q_1, q_2 \in \mathbf{S}_{u_o}$ and $\alpha \in \mathbb{R}$, the reader can demonstrate without difficulty that $|\dot{p}_t|^2 + ((U \circ p_t) \cdot \dot{p}_t)^2$ depends on t: the observer is not rigid.

6.6.6. This observer is not regular. It is easy to show that there is no world surface g-orthogonal to U and passing through o.

Suppose such a world surface F exists. Then F has $\mathbf{S}_{u_o}$ as its tangent space at o.

For all $q \in \mathbf{S}_{u_o}$,

$$f(a) := o + aq \qquad (a \in \mathbb{R})$$

is a function such that $f(0) = o$ and

$$(U \circ f) \cdot \dot{f}(a) = a^2 \left(\Omega \cdot q + u_o \sqrt{1 + a^2 |\Omega \cdot q|^2} \right) \cdot q = \mathbf{0}.$$

The curve (in fact a straight line) $\mathrm{Ran} f$ passes through $o \in \mathrm{F}$ and all of its tangent vectors are g-orthogonal to the corresponding values of U which would imply that $\mathrm{Ran} f \subset \mathrm{F}$. Since q is arbitrary in $\mathbf{S}_{u_o}$, this means that $o + \mathbf{S}_{u_o} = \mathrm{F}$; in particular, every tangent space of F equals $\mathbf{S}_{u_o}$. However, if $x \in o + \mathbf{S}_{u_o} = \mathrm{F}$ and $x - o$ is not in $\mathrm{Ker}\,\Omega$ then $U(x) \neq u_o$; thus the tangent space of F at x is not g-orthogonal to $U(x)$: a contradiction.

6.7. Uniformly rotating observer II

6.7.1. Let $o \in \mathrm{M}$, $u_o \in V(1)$ and $\Omega : \mathbf{S}_{u_o} \to \frac{\mathbf{S}_{u_o}}{\mathbb{T}}$ be a nonzero antisymmetric linear map and define the nonglobal observer

$$\mathrm{Dom}\, U := \left\{ x \in \mathrm{M} \mid |\Omega \cdot \pi_{u_o} \cdot (x - o)|^2 < 1 \right\},$$

$$U(x) := \frac{u_o + \Omega \cdot \pi_{u_o} \cdot (x - o)}{\sqrt{1 - |\Omega \cdot \pi_{u_o} \cdot (x - o)|^2}} \qquad (x \in \mathrm{Dom}\, U).$$

If q is in $\mathrm{Ker}\,\Omega$, then

$$\mathrm{Dom}\, U + q = \mathrm{Dom}\, U$$

and

$$U(x + q) = U(x) \qquad (x \in \mathrm{Dom}\, U,\ q \in \mathrm{Ker}\,\Omega).$$

The observer has the acceleration field

$$A_U(x) = \frac{\Omega \cdot \Omega \cdot \pi_{u_o} \cdot (x - o)}{1 - |\Omega \cdot \pi_{u_o} \cdot (x - o)|^2} \qquad (x \in \mathrm{Dom}\, U).$$

6.7.2. To find the $\boldsymbol{U}$-lines, we again use the known transformation and we obtain

$$\dot{\boldsymbol{t}} = \frac{1}{\sqrt{1 - |\Omega \cdot \boldsymbol{q}|^2}},$$
$$\dot{\boldsymbol{q}} = \frac{\Omega \cdot \boldsymbol{q}}{\sqrt{1 - |\Omega \cdot \boldsymbol{q}|^2}}.$$

Now we apply a new trick: 'dividing' the second equation by the first one we get a very simple differential equation which has the following correct meaning. Consider the initial conditions

$$\boldsymbol{t}(\mathbf{0}) = \mathbf{0}, \qquad \boldsymbol{q}(\mathbf{0}) = x - o,$$

where $x \in \left(o + \mathbf{S}_{\boldsymbol{u}_o}\right) \cap (\mathrm{Dom}\, \boldsymbol{U})$. The formula for the derivative of inverse function results in—with the notations of 6.4—

$$\frac{\mathrm{d}\boldsymbol{s}_x(t)}{\mathrm{d}\boldsymbol{t}} = \sqrt{1 - |\Omega \cdot \boldsymbol{q}(s_x(t))|^2}. \qquad (*)$$

Then introducing the function $\boldsymbol{t} \mapsto \mathsf{q}(\boldsymbol{t}) := \boldsymbol{q}(\boldsymbol{s}_x(\boldsymbol{t}))$ we get the differential equation

$$\frac{\mathrm{d}\mathsf{q}(\boldsymbol{t})}{\mathrm{d}\boldsymbol{t}} = \dot{\boldsymbol{q}}(\boldsymbol{s}_x(\boldsymbol{t}))\frac{\mathrm{d}\boldsymbol{s}_x(\boldsymbol{t}))}{\mathrm{d}\boldsymbol{t}} = \Omega \cdot \mathsf{q}(\boldsymbol{t})$$

which has the solution

$$\mathsf{q}(\boldsymbol{t}) = \mathrm{e}^{\boldsymbol{t}\Omega} \cdot (x - o) \qquad (\boldsymbol{t} \in \mathbb{T}).$$

Consequently $|\Omega \cdot \boldsymbol{q}(\boldsymbol{s}_x(\boldsymbol{t}))| = |\Omega \cdot (x - o)|$, thus equation $(*)$ becomes trivial having the solution—with the initial condition $\boldsymbol{s}_x(\mathbf{0}) = \mathbf{0}$—

$$\boldsymbol{s}_x(\boldsymbol{t}) = \boldsymbol{t}\sqrt{1 - |\Omega \cdot (x - o)|^2}.$$

Finally we obtain

$$\boldsymbol{t}_x(\boldsymbol{s}) = \frac{\boldsymbol{s}}{\sqrt{1 - |\Omega \cdot (x - o)|^2}},$$
$$\boldsymbol{q}_x(\boldsymbol{s}) = \exp\left(\frac{\boldsymbol{s}}{\sqrt{1 - |\Omega \cdot (x - o)|^2}}\right) \cdot (x - o)$$

from which we regain the world line function giving the $\boldsymbol{U}$-line passing through $x \in o + \mathbf{S}_{\boldsymbol{u}_o}$:

$$r_x(\boldsymbol{s}) = o + \boldsymbol{u}_o \frac{\boldsymbol{s}}{\sqrt{1 - |\Omega \cdot (x - o)|^2}} + \exp\left(\frac{\boldsymbol{s}\Omega}{\sqrt{1 - |\Omega \cdot (x - o)|^2}}\right) \cdot (x - o)$$
$$(\boldsymbol{s} \in \mathbb{T}).$$

It is not hard to see that every U-line meets the hyperplane $o+\mathbf{S}_{u_o}$, thus every U-line is of this form.

6.7.3. Note that the U-line passing through $o+e$, where e is in $\mathrm{Ker}\,\Omega$, is a straight line directed by u_o; then the set of U-space points

$$\{o+e+u_o\otimes\mathbb{T} \mid e\in\mathrm{Ker}\,\Omega\}$$

is interpreted as the *axis of rotation.*

If x is in $o+\mathbf{S}_{u_o}$, then $x-o$ can be decomposed into a sum e_x+q_x, where e_x is in $\mathrm{Ker}\,\Omega$ and q_x is orthogonal to $\mathrm{Ker}\,\Omega$. Then the above given world line function can be written in the form

$$r_x(s)=o+e_x+u_o\frac{s}{\sqrt{1-\omega^2|q_x|^2}}+\exp\left(\frac{s\Omega}{\sqrt{1-\omega^2|q_x|^2}}\right)\cdot q_x, \qquad (**)$$

where ω is the magnitude of Ω.

Hence all the U-lines are composed of an inertial line (with a proper time 'accelerated' relative to the proper time of the points of the axis) and a uniform rotation.

Let us consider the u_o-synchronization. Put $t_o:=o+\mathbf{S}_{u_o}$. Then

$$s_x(t)=t\sqrt{1-|\Omega\cdot(x-o)|^2}$$

is the time passed between the u_o-instants t_o and t_o+t in the U-space point that x is incident with.

The distance perceived by the standard inertial frame u_o at the u_o-instant t_o+t between the U-space point that x is incident with and $o+e_x+u_o\otimes\mathbb{T}$ (the axis of rotation) equals

$$\left|r_x(s_x(t))-(o+e_x+u_ot)\right|=|q_x|$$

which is independent of t.

6.7.4. Because of the term $s\mapsto e^{t_x(s)\Omega}\cdot(x-o)$ in $(**)$ we can state that the time period T_o of rotation is the same for all U-space points (out of the axis of rotation), concerning the u_o-time: $T_o=\frac{2\pi}{\omega}$.

On the other hand, concerning the proper times of U-space points, the time period of rotation of a U-space point having the u_o-distance $0<d<\frac{1}{\omega}$ from the axis of rotation equals $T(d):=\frac{2\pi}{\omega}\sqrt{1-\omega^2d^2}$; it decreases from $\frac{2\pi}{\omega}$ to zero as d increases from zero to $\frac{1}{\omega}$.

The following figure illustrates the situation. Two $\boldsymbol{U}$-line segments are represented; the proper time passed along both segments equals $\frac{2\pi}{\omega}$.

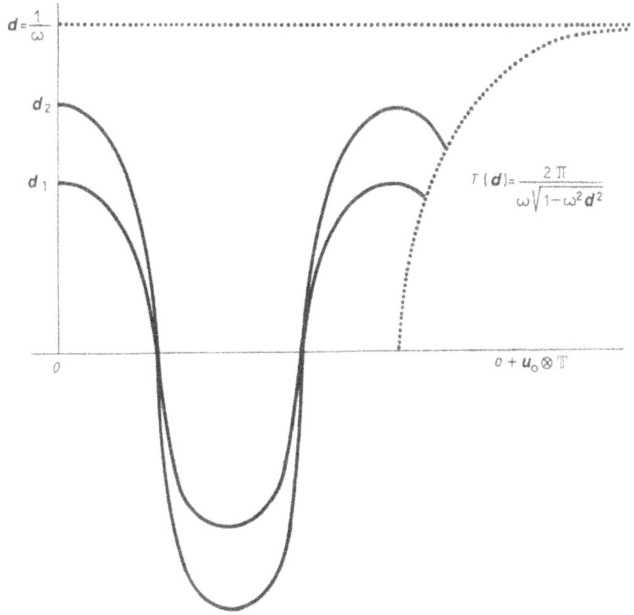

6.7.5. This observer is rigid.

Let L_o be a curve in $o + \mathbf{S}_{\boldsymbol{u}_o}$; then the set of $\boldsymbol{U}$-space points that meet L_o, $L := \{q \in \mathbf{S}_{\boldsymbol{U}} | \, q \cap L_o \neq \emptyset\}$ is a curve in the observer space. Indeed, if p_o is a parametrization of L_o then

$$p_t := o + \boldsymbol{u}_o \boldsymbol{t} + \mathrm{e}^{\boldsymbol{t}\Omega} \cdot (p_o - o)$$

is a parametrization of L_t for $t = t_o + \boldsymbol{t} \in \mathrm{T}_{\boldsymbol{u}_o}$. Then

$$\dot{p}_t = \mathrm{e}^{\boldsymbol{t}\Omega} \cdot \dot{p}_o$$

and we easily find that

$$\boldsymbol{U} \circ p_t = \frac{\boldsymbol{u}_o + \Omega \cdot \mathrm{e}^{\boldsymbol{t}\Omega} \cdot (p_o - o)}{\sqrt{1 - |\Omega \cdot (p_o - o)|^2}}$$

and

$$(\boldsymbol{U} \circ p_t) \cdot \dot{p}_t = \frac{(\Omega \cdot (p_o - o)) \cdot \dot{p}_o}{\sqrt{1 - |\Omega \cdot (p_o - o)|^2}}.$$

Consequently,

$$|\dot{p}_t|^2 + ((\boldsymbol{U} \circ p_t) \cdot \dot{p}_t)^2 = |\dot{p}_o|^2 + \frac{(\dot{p}_o \cdot \Omega \cdot (p_o - o))^2}{1 - |\Omega \cdot (p_o - o)|^2}$$

is independent of t: the observer is rigid.

6.7.6. This observer furnishes a good instance that the laws of Euclidean geometry do not hold necessarily in the space of a rigid noninertial observer.

Since the observer is rigid, all the lengths in $\boldsymbol{U}$-space can be calculated by curves in $(o + \mathbf{S}_{\boldsymbol{u}_o}) \cap (\mathrm{Dom}\, \boldsymbol{U})$ which can be reduced to curves in

$$\mathbf{S}_{\boldsymbol{u}_o} \cap (\mathrm{Dom}\, \boldsymbol{U} - o) =$$
$$= \mathrm{Ker}\,\Omega + \{\boldsymbol{q} \in \mathbf{S}_{\boldsymbol{u}_o} |\ \boldsymbol{q} \text{ is orthogonal to } \mathrm{Ker}\,\Omega,\ |\boldsymbol{q}| < \frac{1}{\omega}\} =: \mathbf{S}_\Omega.$$

If L_o is a curve in $(o + \mathbf{S}_{\boldsymbol{u}_o}) \cap (\mathrm{Dom}\, \boldsymbol{U})$ then $\mathbf{L} := L_o - o$ is a curve in $\mathbf{S}_\Omega$; if p_o is a parametrization of L_o then $\boldsymbol{p} := p_o - o$ is a parametrization of $\mathbf{L}$.

$\mathbf{S}_\Omega$ is a subset of the Euclidean vector space $\mathbf{S}_{\boldsymbol{u}_o}$ in which distances and curve lengths have a well-defined meaning; however, now curves in $\mathbf{S}_\Omega$ will represent curves in $\boldsymbol{U}$-space and their lengths will be calculated in this sense. Thus, to avoid misunderstanding, we shall say $\boldsymbol{U}$-length and $\boldsymbol{U}$-distance, indicating it in notations, too.

A curve $\mathbf{L}$ in $\mathbf{S}_\Omega$ has the $\boldsymbol{U}$-length

$$\ell_U(\mathbf{L}) = \int_{\mathrm{Dom}\, \boldsymbol{p}} \sqrt{|\dot{\boldsymbol{p}}(a)|^2 + \frac{(\dot{\boldsymbol{p}}(a) \cdot \Omega \cdot \boldsymbol{p}(a))^2}{1 - |\Omega \cdot \boldsymbol{p}(a)|^2}}\, da. \qquad (***)$$

Take arbitrary elements $\boldsymbol{x}$ and $\boldsymbol{y}$ in $\mathbf{S}_\Omega$. Then we easily find for the straight line segment connecting $\boldsymbol{x}$ and $\boldsymbol{y}$, $]\boldsymbol{x}, \boldsymbol{y}[:= \{\boldsymbol{x} + a(\boldsymbol{y} - \boldsymbol{x}) \mid 0 < a < 1\}$ that

$$\ell_U(]\boldsymbol{x}, \boldsymbol{y}[) \geq |\boldsymbol{y} - \boldsymbol{x}|$$

and equality holds if and only if $\boldsymbol{x} \cdot \Omega \cdot \boldsymbol{y} = \mathbf{0}$ which is equivalent to the fact that the straight line passing through $\boldsymbol{x}$ and $\boldsymbol{y}$ meets the kernel of Ω.

Suppose the straight line passing through $\boldsymbol{x}$ and $\boldsymbol{y}$ meets $\mathrm{Ker}\,\Omega$ and $\mathbf{L}$ is a broken line connecting $\boldsymbol{x}$ and $\boldsymbol{y}$; then the previous inequality implies

$$\ell_U(\mathbf{L}) \geq |\boldsymbol{y} - \boldsymbol{x}| = \ell_U(]\boldsymbol{x}, \boldsymbol{y}[).$$

As a consequence, the inequality above will be valid for an arbitrary $\mathbf{L}$ connecting $\boldsymbol{x}$ and $\boldsymbol{y}$ because $\ell(\mathbf{L})$ is obtained as the supremum of U-lengths of broken lines approximating the curve $\mathbf{L}$. Since the U-distance $d_U(\boldsymbol{x},\boldsymbol{y})$ between $\boldsymbol{x}$ and $\boldsymbol{y}$ is the infimum of curve lengths connecting $\boldsymbol{x}$ and $\boldsymbol{y}$ we see that

$$d_U(\boldsymbol{x},\boldsymbol{y}) = |\boldsymbol{y}-\boldsymbol{x}|$$

if the straight line passing through $\boldsymbol{x}$ and $\boldsymbol{y}$ intersects $\operatorname{Ker}\Omega$.

Let $\boldsymbol{d}$ be an element of $\mathbb{T}$, $\mathbf{0} < \boldsymbol{d} < \frac{1}{\omega}$, and put

$$\mathrm{C}_{\boldsymbol{d}} := \{\boldsymbol{q} \in \mathbf{S}_\Omega |\ \boldsymbol{q} \text{ is orthogonal to } \operatorname{Ker}\Omega,\ |\boldsymbol{q}| = \boldsymbol{d}\}.$$

Evidently, if $\boldsymbol{q} \in \mathrm{C}_{\boldsymbol{d}}$ then $-\boldsymbol{q} \in \mathrm{C}_{\boldsymbol{d}}$ as well. Moreover, according to our previous result, the U-distance between $\boldsymbol{q}$ and $-\boldsymbol{q}$ equals $|\boldsymbol{q}-(-\boldsymbol{q})| = 2\boldsymbol{d}$.

This means that $\mathrm{C}_{\boldsymbol{d}}$ represents a circle of radius $\boldsymbol{d}$ in the observer space. Let us calculate the circumference of this circle.

Choosing the parametrization

$$\boldsymbol{p}\colon\ ]-\pi,\pi] \to \mathrm{C}_{\boldsymbol{d}}, \qquad a \mapsto \exp\left(a\frac{\Omega}{\omega}\right)\cdot \boldsymbol{q}_o$$

where $\boldsymbol{q}_o$ is an arbitrarily fixed element of $\mathrm{C}_{\boldsymbol{d}}$, we find

$$\dot{\boldsymbol{p}} = \frac{\Omega}{\omega}\cdot\boldsymbol{p}, \qquad |\dot{\boldsymbol{p}}|^2 = \boldsymbol{d}^2,$$
$$|\Omega\cdot\boldsymbol{p}|^2 = \omega^2\boldsymbol{d}^2, \qquad (\dot{\boldsymbol{p}}\cdot\Omega\cdot\boldsymbol{p})^2 = \omega^2\boldsymbol{d}^4.$$

Applying formula $(***)$ we obtain

$$\ell_U(\mathrm{C}_{\boldsymbol{d}}) = \frac{2\pi\boldsymbol{d}}{\sqrt{1-\omega^2\boldsymbol{d}^2}}.$$

The circumference of the circle of radius $\boldsymbol{d}$ is longer than $2\pi\boldsymbol{d}$.

6.8. Exercises

1. Take an inertial observer $\boldsymbol{u}$ and an affine synchronization due to $\boldsymbol{u}_o \neq \boldsymbol{u}$. Demonstrate that the distances in U-space calculated at $\mathrm{T}_{\boldsymbol{u}_o}$-instants according to definition 6.3.3 equal the distances defined earlier in U-space.

2. Take a synchronization $\mathcal{S}$ whose time-points are subsets of (not necessarily parallel) affine subsspaces. Supposs that the observer U (a velocity field) is constant on the $\mathcal{S}$-time points. Then for every $\mathcal{S}$-time point there is a $\boldsymbol{u}_t \in V(1)$ such that $U(x) = \boldsymbol{u}_t$ for all x in t. Suppose t is directed by $\mathbf{S}$. Then

$$|\boldsymbol{\pi}_{\boldsymbol{u}_t}\cdot(\boldsymbol{q}_1+\boldsymbol{q}_2)| \le |\boldsymbol{\pi}_{\boldsymbol{u}_t}\cdot\boldsymbol{q}_1| + |\boldsymbol{\pi}_{\boldsymbol{u}_t}\cdot\boldsymbol{q}_2|$$

for all $\boldsymbol{q}_1, \boldsymbol{q}_2 \in \mathbf{S}$. As a consequence, straight lines realize the distance between the points of t, thus

$$\boldsymbol{d}_t(q_1, q_2) = |\boldsymbol{\pi}_{\boldsymbol{u}_t}(q_2 \star t - q_1 \star t)| \qquad (q_1, q_2 \in \mathrm{S}_U).$$

3. Let $\boldsymbol{U}$ be the uniformly accelerated observer treated in 6.5. Then

$$\boldsymbol{v}_{\boldsymbol{U}(x)\boldsymbol{u}_o} = \frac{\boldsymbol{a}_o\big((-\boldsymbol{u}_o \cdot (x-o))\big)}{\sqrt{1 + |\boldsymbol{a}_o|^2\big(\boldsymbol{u}_o \cdot (x-o)\big)^2}} \qquad (x \in \mathrm{M}).$$

4. Let $\boldsymbol{U}$ be as before. Verify that every $\boldsymbol{U}$-line is obtained from a chosen one by a translation with a vector in $\mathbf{S}_{\boldsymbol{u}_o}$. In other words, S_U endowed with the subtraction

$$q' - q := x' - x \qquad (x' \in q,\ x \in q,\ x' - x \in \mathbf{S}_{\boldsymbol{u}_o})$$

is an affine space over $\mathbf{S}_{\boldsymbol{u}_o}$.

5. Let $\boldsymbol{U}$ be the uniformly accelerated observer treated in 6.6. Then the $\boldsymbol{U}$-line passing through $x \in o + \mathbf{S}_{\boldsymbol{u}_o}$ intersects t_λ if and only if $\boldsymbol{a}_o \cdot (x-o) < \ln|\lambda|$.

6. Show that

$$\mathrm{Dom}\,\boldsymbol{U} = \{o + \alpha\boldsymbol{u}_o + \beta\boldsymbol{a}_o + \boldsymbol{q} \mid \beta > \mathbf{0},\ \beta^2|\boldsymbol{a}_o|^2 > \alpha^2,\ \boldsymbol{u}_o \cdot \boldsymbol{q} = \mathbf{0},\ \boldsymbol{a}_o \cdot \boldsymbol{q} = \mathbf{0}\}$$

for the uniformly accelerated observer treated in 6.6.

7. Let $\boldsymbol{U}$ be as before. Then

$$\boldsymbol{v}_{\boldsymbol{U}(x)\boldsymbol{u}_o} = \frac{\boldsymbol{a}_o\big(-\boldsymbol{u}_o \cdot (x-o)\big)}{\boldsymbol{a}_o \cdot (x-o)} \qquad (x \in \mathrm{Dom}\,\boldsymbol{U}).$$

8. Show that the distance perceived by the standard inertial frame $\boldsymbol{u}_o$ between the space points of the uniformly accelerated observer treated in 6.6. is not constant in $\boldsymbol{u}_o$-time. Give an explanation similar to that in 6.5.6.

9. Verify that

$$\mathrm{Dom}\,\boldsymbol{U} = o + \boldsymbol{u}_o \otimes \mathbb{T} + \mathrm{Ker}\,\Omega + \left\{\boldsymbol{q} \in \mathbf{S}_{\boldsymbol{u}_o} \mid \boldsymbol{q} \text{ is orthogonal to } \mathrm{Ker}\,\Omega,\ |\boldsymbol{q}| < \frac{1}{\omega}\right\}$$

for the uniformly rotating observer treated in 6.8.

10. Demonstrate that

$$\boldsymbol{v}_{\boldsymbol{U}(x)\boldsymbol{u}_o} = \frac{\Omega \cdot \boldsymbol{\pi}_{\boldsymbol{u}_o} \cdot (x-o)}{\sqrt{1 + |\Omega \cdot \boldsymbol{\pi}_{\boldsymbol{u}_o} \cdot (x-o)|^2}} \qquad (x \in \mathrm{M})$$

and

$$v_{U(x)u_o} = \Omega \cdot \pi_{u_o} \cdot (x - o) \qquad (x \in \operatorname{Dom} U)$$

where U is the uniformly rotating observer treated in 6.7. and 6.8, respectively.

11. The uniformly rotating observer treated in 6.8. is not regular.

12. Let o be a world point and consider the observer

$$U(x) := \frac{x - o}{|x - o|} \qquad (x \in o + T^{\to}).$$

Prove that

$$\mathrm{S}_U = \{o + u \otimes \mathbb{T}^+ \mid u \in V(1)\}.$$

U is regular and

$$\{V(1)t \mid t \in \mathbb{T}^+\}$$

is the set of U-surfaces (U-instants).

Show that if this observer defined synchronization like an inertial observer (light signals and mirrors, see 3.2.) then simultaneity would depend on the U-space point of the light source.

13. Let $o \in \mathrm{M}$, $u_o \in V(1)$, $\gamma \in \mathbb{T}^*$ and define the observer

$$\operatorname{Dom} U := \{x \in \mathrm{M} \mid \gamma^2 |\pi_{u_o} \cdot (x - o)|^2 < 1\},$$

$$U(x) := \frac{u_o + \gamma \pi_{u_o} \cdot (x - o)}{\sqrt{1 - \gamma^2 |\pi_{u_o} \cdot (x - o)|^2}} \qquad (x \in \operatorname{Dom} U).$$

Applying the method given in 6.4. find that the U-line passing through $x \in (o + \mathbf{S}_{u_o}) \cap (\operatorname{Dom} U)$ is given by the world line function

$$r_x(s) = o + u_o t_x(s) + e^{\gamma t_x(s)}(x - o)$$

where $s \mapsto t_x(s)$ is the solution of the differential equation

$$(t \colon \mathbb{T} \rightarrowtail \mathbb{T})? \qquad \dot{t} = \frac{1}{\sqrt{1 - \gamma^2 |x - o|^2 e^{2\gamma t}}}$$

with the initial condition $t(0) = 0$.

14. Let $o \in \mathrm{M}$, $u_o \in V(1)$, $\gamma \in \mathbb{T}^*$ and define the observer

$$U(x) := u_o \sqrt{1 + \gamma^2 |\pi_{u_o} \cdot (x - o)|^2} + \gamma \pi_{u_o}(x - o) \qquad (x \in \mathrm{M}).$$

Applying the method given in 6.4 find that the U-line passing through $x \in o + \mathbf{S}_{u_o}$ is given by the world line function

$$r_x(s) = o + u_o t_x(s) + e^{\gamma s}(x - o)$$

where $s \mapsto t_x(s)$ is the function for which $t_x(0) = 0$ holds and has the derivative $s \mapsto \sqrt{1 + \gamma^2 |x - o|^2 e^{2\gamma s}}$.

15. Compare the observers of the previous two exercises with the nonrelativistic observer in Exercise I.5.4.9.

7. Vector splittings

7.1. Splitting of vectors

7.1.1. For $u \in V(1)$ we have already defined

$$\tau_u : \mathbf{M} \to \mathbb{T}, \qquad x \mapsto -u \cdot x$$

and

$$\pi_u : \mathbf{M} \to \mathbf{S}_u, \qquad x \mapsto x - (\tau_u \cdot x)u = x + (u \cdot x)u$$

i.e. with the usual identifications,

$$\tau_u = -u, \qquad \pi_u = \mathbf{1}_{\mathbf{M}} + u \otimes u$$

(see 1.3.2) and the linear bijection $\xi_u := (\tau_u, \pi_u) : \mathbf{M} \to \mathbb{T} \times \mathbf{S}_u$ having the inverse

$$(t, q) \mapsto ut + q$$

(see 1.3.5).

Definition. $\tau_u \cdot x = -u \cdot x$ and $\pi_u \cdot x$ are called the u*-timelike component* and the u*-spacelike component* of the vector x. $(-u \cdot x, \pi_u \cdot x)$ is the u-split form of x. $\xi_u = (\tau_u, \pi_u)$ is the *splitting* of $\mathbf{M}$ corresponding to u, or the u*-splitting* of $\mathbf{M}$. ■

Note that

$$x \cdot y = -(u \cdot x)(u \cdot y) + (\pi_u \cdot x) \cdot (\pi_u \cdot y),$$

in particular,

$$x^2 = -(u \cdot x)^2 + |\pi_u \cdot x|^2$$

for all $x, y \in \mathbf{M}$. In other words,

$$\text{if} \quad \xi_u \cdot x = (t, q) \quad \text{then} \quad x^2 = -t^2 + |q|^2.$$

7.1.2. If $\mathbb{A}$ is a measure line, $\mathbb{A} \otimes \mathbf{M}$ $\left(\frac{\mathbf{M}}{\mathbb{A}}\right)$ is split into $(\mathbb{A} \otimes \mathbb{T}) \times (\mathbb{A} \otimes \mathbf{S}_u)$ $\left(\frac{\mathbb{T}}{\mathbb{A}} \times \frac{\mathbf{S}_u}{\mathbb{A}}\right)$ by ξ_u; thus the u-timelike component and the u-spacelike component of a vector of type $\mathbb{A}$ (cotype $\mathbb{A}$) are in $\mathbb{A}\otimes\mathbb{T}$ $\left(\frac{\mathbb{T}}{\mathbb{A}}\right)$ and in $\mathbb{A}\otimes\mathbf{S}_u$ $\left(\frac{\mathbf{S}_u}{\mathbb{A}}\right)$, respectively.

In particular, ξ_u splits $\frac{\mathbf{M}}{\mathbb{T}}$ into $\mathbb{R}\times\frac{\mathbf{S}_u}{\mathbb{T}}$ and for all $u'\in V(1)$

$$\xi_u\cdot u' = \left(-u\cdot u',\, u' + (u\cdot u')u\right) = \frac{1}{\sqrt{1-|v_{u'u}|^2}}(1, v_{u'u}).$$

7.1.3. In contradistinction to the nonrelativistic case, here not only the u-spacelike component but also the u-timelike component of vectors depend on u. The transformation rule that shows how the u-components of a vector vary with u, is much more complicated here than in the nonrelativistic case.

Proposition. Let $u, u'\in V(1)$. Then for all $(t,q)\in\mathbb{T}\times\mathbf{S}_u$ we have

$$\begin{aligned}(\xi_{u'}\cdot\xi_u^{-1})\cdot(t,q) &= ((-u'\cdot u)t - u'\cdot q,\ (u+(u'\cdot u)u')t + q + (u'\cdot q)u') \\ &= \Bigg(\frac{1}{\sqrt{1-|v_{u'u}|^2}}(t - v_{u'u}\cdot q), \\ &\frac{1}{\sqrt{1-|v_{u'u}|^2}}\left(v_{uu'}t - \frac{v_{uu'} + v_{u'u}\sqrt{1-|v_{u'u}|^2}}{|v_{u'u}|^2}(v_{u'u}\cdot q)\right) + q\Bigg).\end{aligned}$$

Proof. The first equality is quite simple. The second one is derived with the aid of the formulae in 4.3.2 and the relation $u'\cdot q = -(u'\cdot u)\left(\frac{u'}{-u'\cdot u} - u\right)\cdot q$ which is true because $u\cdot q = 0$. ■

Note that both $v_{u'u}$ and $v_{uu'}$ appear in that formula.

7.1.4. The previous formula is not a good transformation rule: we want to compare the u'-components of a vector with its u-components (t,q). However, the u'-components and the u-components are in different spaces: (t,q) is in $\mathbb{T}\times\mathbf{S}_u$ and $(\xi_{u'}\cdot\xi_u^{-1})\cdot(t,q)$ is in $\mathbb{T}\times\mathbf{S}_{u'}$, they cannot be compared directly. To obtain a convenient formula, we have to relate $\mathbf{S}_{u'}$ and $\mathbf{S}_u$; we have agreed that such a relation is established by the corresponding Lorentz boost. Thus, leaving invariant the first component, we shall transform the second component of $(\xi_{u'}\cdot\xi_u^{-1})\cdot(t,q)$ by $L(u,u')$.

Definition. Let $u, u'\in V(1)$. Then

$$\xi_{u'u} := \left(\mathbf{1}_{\mathbb{T}}\times L(u,u')|_{\mathbf{S}_{u'}}\right)\cdot(\xi_{u'}\cdot\xi_u^{-1})$$

is called the *vector transformation rule* from u-splitting into u'-splitting. ∎

Proposition. For all $(t, q) \in \mathbb{T} \times \mathbf{S}_u$ we have

$$\xi_{u'u} \cdot (t, q) = \left(\frac{1}{\sqrt{1 - |v_{u'u}|^2}} (t - v_{u'u} \cdot q), \right.$$

$$\left. \frac{1}{\sqrt{1 - |v_{u'u}|^2}} \left[-v_{u'u} \left(t - \frac{1 - \sqrt{1 - |v_{u'u}|^2}}{|v_{u'u}|^2} (v_{u'u} \cdot q) \right) \right] + q \right). \quad ∎$$

In connection with this formula we mention the following frequently useful relation:

$$\frac{1 - \sqrt{1 - |v_{u'u}|^2}}{|v_{u'u}|^2} = \frac{1}{1 + \sqrt{1 - |v_{u'u}|^2}}.$$

7.1.5. The previous formula is a bit fearsome. We can make it more apparent decomposing q into a sum of vectors parallel and orthogonal to $v_{u'u}$:

$$(t, q) = (t, q_{\parallel}) + (0, q_{\perp})$$

where $q_{\parallel}$ is parallel to $v_{u'u}$, i.e. there is a $\lambda \in \mathbb{T}$ such that $q_{\parallel} = \lambda v_{u'u}$ and $q_{\perp}$ is orthogonal to $v_{u'u}$, i.e. $v_{u'u} \cdot q_{\perp} = 0$.

Then we easily find that

$$\xi_{u'u} \cdot (0, q_{\perp}) = (0, q_{\perp}),$$

$$\xi_{u'u} \cdot (t, q_{\parallel}) = \frac{1}{\sqrt{1 - |v_{u'u}|^2}} \left(t - v_{u'u} \cdot q_{\parallel}, -v_{u'u} t + q_{\parallel} \right).$$

7.1.6. The last formula—in a slightly different form—appears in the literature as the formula of Lorentz transformation. To get the usual form we put $v := v_{u'u}$; let (t, q) denote the u-components of a vector and let (t', q') denote its u'-components mapped by the Lorentz boost $L(u, u')$ into $\mathbb{T} \times \mathbf{S}_u$; then supposing q is parallel to v we have

$$t' = \frac{1}{\sqrt{1 - |v|^2}} (t - v \cdot q), \qquad q' = \frac{1}{\sqrt{1 - |v|^2}} (-vt + q).$$

This (or its equivalent in the arithmetic spacetime model) is the usual 'Lorentz transformation' formula.

We emphasize that $\boldsymbol{q}'$ *is not* the $\boldsymbol{u}'$-spacelike component of the vector having the $\boldsymbol{u}$-components $(\boldsymbol{t},\boldsymbol{q})$; it is the Lorentz-boosted $\boldsymbol{u}'$-spacelike component.

Lorentz transformations (see Section 9) are transformations of vectors, i.e. mappings from $\mathbf{M}$ into $\mathbf{M}$; the transformation rule is a mapping from $\mathbb{T}\times\mathbf{S}_{\boldsymbol{u}}$ into $\mathbb{T}\times\mathbf{S}_{\boldsymbol{u}}$. Transformation rules and Lorentz transformations are different mathematical objects. Of course, there is some connection between them. We easily find that

$$\xi_{u'u} = \xi_u \cdot \boldsymbol{L}(\boldsymbol{u},\boldsymbol{u}') \cdot \xi_u^{-1}$$

where $\boldsymbol{L}(\boldsymbol{u},\boldsymbol{u}')$ is the Lorentz boost from $\boldsymbol{u}'$ into $\boldsymbol{u}$.

In the split spacetime model $\mathbf{M}$ and $\mathbb{T}\times\mathbf{S}_{\boldsymbol{u}}$ coincide: *the special structure of the split spacetime model (and the arithmetic spacetime model) involves the possibility of confusing transformation rules with Lorentz transformations.*

7.1.7. Using a matrix form of the linear maps $\mathbb{T}\times\mathbf{S}_{\boldsymbol{u}} \to \mathbb{T}\times\mathbf{S}_{\boldsymbol{u}}$ (see IV.3.7) we can write

$$\xi_{u'u} = \kappa(\boldsymbol{v}_{u'u}) \begin{pmatrix} 1 & -\boldsymbol{v}_{u'u} \\ -\boldsymbol{v}_{u'u} & \boldsymbol{D}(\boldsymbol{v}_{u'u}) \end{pmatrix},$$

where

$$\kappa(\boldsymbol{v}) := \frac{1}{\sqrt{1-|\boldsymbol{v}|^2}},$$

$$\boldsymbol{D}(\boldsymbol{v}) := \frac{1}{\kappa(\boldsymbol{v})}\left(\mathbf{1}_{\mathbf{S}_u} + \frac{\kappa(\boldsymbol{v})^2}{\kappa(\boldsymbol{v})+1}\boldsymbol{v}\otimes\boldsymbol{v}\right)$$

for $\boldsymbol{v} \in \frac{\mathbf{S}_u}{\mathbb{T}}$, $|\boldsymbol{v}| < 1$.

7.2. Splitting of covectors

7.2.1. For $\boldsymbol{u} \in V(1)$, $\mathbf{M}^*$ is split by the transpose of the inverse of ξ_u:

$$\eta_u := \left(\xi_u^{-1}\right)^* : \ \mathbf{M}^* \to \left(\mathbb{T}\times\mathbf{S}_{\boldsymbol{u}}\right)^* \equiv \mathbb{T}^*\times\mathbf{S}_{\boldsymbol{u}}^*.$$

Then for all $\boldsymbol{k} \in \mathbf{M}^*$, $(\boldsymbol{t},\boldsymbol{q}) \in \mathbf{S}_{\boldsymbol{u}}$ we have

$$(\eta_u \cdot \boldsymbol{k})\cdot(\boldsymbol{t},\boldsymbol{q}) = \boldsymbol{k}\cdot\xi_u^{-1}\cdot(\boldsymbol{t},\boldsymbol{q}) = \boldsymbol{k}\cdot(\boldsymbol{u}\boldsymbol{t}+\boldsymbol{q}) = (\boldsymbol{k}\cdot\boldsymbol{u})\boldsymbol{t} + \boldsymbol{k}\cdot\boldsymbol{q}.$$

Of course, instead of $\boldsymbol{k}$ in $\boldsymbol{k}\cdot\boldsymbol{q}$ we can write $\boldsymbol{k}|_{\mathbf{S}_u} = \boldsymbol{i}_u^*\cdot\boldsymbol{k} = \boldsymbol{k}\cdot\boldsymbol{i}_u \in \mathbf{S}_{\boldsymbol{u}}^*$. Then we can state that

$$\eta_u\cdot\boldsymbol{k} = (\boldsymbol{k}\cdot\boldsymbol{u},\, \boldsymbol{k}\cdot\boldsymbol{i}_u) = (\boldsymbol{u}\cdot\boldsymbol{k}, \boldsymbol{i}_u^*\cdot\boldsymbol{k}) \qquad (\boldsymbol{k}\in\mathbf{M}^*).$$

This form is suitable for a comparison with the nonrelativistic case. However, we can get a form more convenient from the point of view of applications. Applying the usual identifications we have $\boldsymbol{i}_{\boldsymbol{u}}^* = \boldsymbol{\pi}_{\boldsymbol{u}}$ (see 1.3.6), thus

$$\boldsymbol{\eta}_{\boldsymbol{u}} \cdot \boldsymbol{k} = (\boldsymbol{u} \cdot \boldsymbol{k},\ \boldsymbol{\pi}_{\boldsymbol{u}} \cdot \boldsymbol{k}) \qquad (\boldsymbol{k} \in \mathbf{M}^*).$$

Recall the identification $\mathbf{M}^* \equiv \frac{\mathbf{M}}{\mathbb{T}\otimes\mathbb{T}}$ which implies that $\boldsymbol{k}$ can be split as a vector of cotype $\mathbb{T}\otimes\mathbb{T}$, too:

$$\boldsymbol{\xi}_{\boldsymbol{u}} \cdot \boldsymbol{k} = (-\boldsymbol{u} \cdot \boldsymbol{k}, \boldsymbol{\pi}_{\boldsymbol{u}} \cdot \boldsymbol{k}) \qquad (\boldsymbol{k} \in \mathbf{M}^*).$$

The two splittings are nearly the same. In the literature (in a somewhat different setting) the split form of $\boldsymbol{k} \in \mathbf{M}^*$ by $\boldsymbol{\eta}_{\boldsymbol{u}}$ and $\boldsymbol{\xi}_{\boldsymbol{u}}$ are called the *covariant* and the *contravariant* components of $\boldsymbol{k}$, respectively.

Of course, in view of $\mathbf{M} \equiv \mathbb{T}\otimes\mathbb{T}\otimes\mathbf{M}^*$, also the elements of $\mathbf{M}$ can be split by $\boldsymbol{\eta}_{\boldsymbol{u}}$: a vector, too, has covariant and contravariant components.

Introducing the notation

$$\boldsymbol{j}_{\boldsymbol{u}}\colon\ \mathbb{T}\times\mathbf{S}_{\boldsymbol{u}} \to \mathbb{T}\times\mathbf{S}_{\boldsymbol{u}}, \qquad (\boldsymbol{t},\boldsymbol{q}) \mapsto (-\boldsymbol{t},\boldsymbol{q})$$

we have (with the usual identifications)

$$\boldsymbol{\eta}_{\boldsymbol{u}} = \boldsymbol{j}_{\boldsymbol{u}} \cdot \boldsymbol{\xi}_{\boldsymbol{u}}.$$

Note that $\boldsymbol{\eta}_{\boldsymbol{u}}^{-1} = \boldsymbol{\xi}_{\boldsymbol{u}}^{-1} \cdot \boldsymbol{j}_{\boldsymbol{u}}$, i.e.

$$\boldsymbol{\eta}_{\boldsymbol{u}}^{-1} \cdot (\boldsymbol{e},\boldsymbol{p}) = -\boldsymbol{e}\boldsymbol{u} + \boldsymbol{p} \qquad (\boldsymbol{e} \in \mathbb{T}^*, \boldsymbol{p} \in \mathbf{S}_{\boldsymbol{u}}^*).$$

7.2.2. The *covector transformation rule* is defined to be

$$\boldsymbol{\eta}_{\boldsymbol{u}'\boldsymbol{u}} := \big(\mathbf{1}_{\mathbb{T}} \times \boldsymbol{L}(\boldsymbol{u},\boldsymbol{u}')|_{\mathbf{S}_{\boldsymbol{u}}}\big) \cdot \boldsymbol{\eta}_{\boldsymbol{u}'} \cdot \boldsymbol{\eta}_{\boldsymbol{u}}^{-1}.$$

It can be easily deduced from the vector transformation rule that, apart from a negative sign, they are the same. Indeed,

$$\big(\mathbf{1}_{\mathbb{T}} \times \boldsymbol{L}(\boldsymbol{u},\boldsymbol{u}')|_{\mathbf{S}_{\boldsymbol{u}'}}\big)\boldsymbol{j}_{\boldsymbol{u}'} = \boldsymbol{j}_{\boldsymbol{u}} \cdot \big(\mathbf{1}_{\mathbb{T}} \times \boldsymbol{L}(\boldsymbol{u},\boldsymbol{u}')|_{\mathbf{S}_{\boldsymbol{u}'}}\big),$$

thus

$$\boldsymbol{\eta}_{\boldsymbol{u}'\boldsymbol{u}} = \boldsymbol{j}_{\boldsymbol{u}} \cdot \boldsymbol{\xi}_{\boldsymbol{u}'\boldsymbol{u}} \cdot \boldsymbol{j}_{\boldsymbol{u}}.$$

Consequently, if $(\boldsymbol{e},\boldsymbol{p}) \in \mathbb{T}^* \times \mathbf{S}_{\boldsymbol{u}}^*$ and $\boldsymbol{p}$ is parallel to $\boldsymbol{v}_{\boldsymbol{u}'\boldsymbol{u}}$ then

$$\boldsymbol{\eta}_{\boldsymbol{u}'\boldsymbol{u}} \cdot (\boldsymbol{e},\boldsymbol{p}) = \frac{1}{\sqrt{1-|\boldsymbol{v}_{\boldsymbol{u}'\boldsymbol{u}}|^2}}\big(\boldsymbol{e} + \boldsymbol{v}_{\boldsymbol{u}'\boldsymbol{u}} \cdot \boldsymbol{p},\ \boldsymbol{v}_{\boldsymbol{u}'\boldsymbol{u}}\boldsymbol{e} + \boldsymbol{p}\big).$$

7.2.3. It is worth mentioning that $\mathbf{S}_{\boldsymbol{u}}^*$ can be considered to be a linear subspace of $\mathbf{M}^*$, since $\mathbf{S}_{\boldsymbol{u}}^* \equiv \frac{\mathbf{S}_{\boldsymbol{u}}}{\mathbb{T}\otimes\mathbb{T}} \subset \frac{\mathbf{M}}{\mathbb{T}\otimes\mathbb{T}} \equiv \mathbf{M}^*$ and

$$\mathbf{S}_{\boldsymbol{u}}^* = \{\boldsymbol{k} \in \mathbf{M}^* \mid \boldsymbol{k}\cdot\boldsymbol{u} = 0\},$$

in other words, $\mathbf{S}_{\boldsymbol{u}}^*$ is the annihilator of $\boldsymbol{u}\otimes\mathbb{T}$.

In the nonrelativistic case $\mathbf{S}^*$ is not a linear subspace of $\mathbf{M}^*$. For all $\boldsymbol{u} \in V(1)$ there is a linear subspace $\mathbf{S}^*\cdot\boldsymbol{\pi}_{\boldsymbol{u}}$ of $\mathbf{M}^*$, the annihilator of $\boldsymbol{u}\otimes\mathbb{T}$, but it is not the dual of any linear subspace in $\mathbf{M}$.

Observe that the special relativistic vector transformation rule which is nearly the same as the covector transformation rule resembles a combination of the nonrelativistic vector and covector transformation rules.

We emphasize that in the special relativistic case *there is no absolute spacelike vector and there is no absolute timelike covector,* in contradistinction to the nonrelativistic case.

7.3. Splitting of vector fields

7.3.1. In applications vector fields $\mathrm{M} \rightarrowtail \mathbf{M}$ and covector fields $\mathrm{M} \rightarrowtail \mathbf{M}^*$ appear frequently. Evidently, a covector field can be considered a vector field of cotype $\mathbb{T}\otimes\mathbb{T}$. Their splitting according to standard inertial frames can be treated analogously to the nonrelativistic case (see I.8.5).

The *half* $\boldsymbol{u}$*-split form* of the covector field $\boldsymbol{K}$ according to the standard inertial frame $\boldsymbol{u}$ is

$$(-V_{\boldsymbol{u}}, \boldsymbol{A}_{\boldsymbol{u}}) := \boldsymbol{\eta}_{\boldsymbol{u}}\cdot\boldsymbol{K} \colon \mathrm{M} \rightarrowtail \mathbb{T}^*\times\mathbf{S}_{\boldsymbol{u}}^*, \qquad x \mapsto= \big(\boldsymbol{u}\cdot\boldsymbol{K}(x),\, \boldsymbol{\pi}_{\boldsymbol{u}}\cdot\boldsymbol{K}(x)\big).$$

The *completely* $\boldsymbol{u}$*-split form* of $\boldsymbol{K}$ is

$$\boldsymbol{\eta}_{\boldsymbol{u}}\cdot\boldsymbol{K}\circ\xi_{\boldsymbol{u}}^{-1} \colon \mathrm{T}\times\mathrm{S}_{\boldsymbol{u}} \rightarrowtail \mathbb{T}^*\times\mathbf{S}^*, \qquad (t,q) \mapsto \big(\boldsymbol{u}\cdot\boldsymbol{K}(q\star t),\, \boldsymbol{\pi}_{\boldsymbol{u}}\cdot\boldsymbol{K}(q\star t)\big)$$

where $q\star t := \xi_{\boldsymbol{u}}^{-1}(t,q)$.

7.3.2. Potentials are covector fields. We can introduce the scalar potential and the vector potential according to an observer by the previous split forms. Regarding the transformation rule concerning scalar potentials and vector potentials we can repeat essentially what we said in I.8.5.3; of course, the transformation rule will be significantly more complicated.

An important difference between the nonrelativistic spacetime model and the special relativistic one is that here *there are no absolute scalar potentials* because there are no absolute timelike covectors. This forecasts that the description of gravitation in the relativistic case will differ significantly from its description in the nonrelativistic case where absolute scalar potentials are used.

7.3.3. In contradistinction to the nonrelativistic case, force fields are split differently according to different observers.

Let us take a force field $\boldsymbol{f}\colon M \times V(1) \rightarrowtail \frac{\mathbf{M}^*}{\mathbb{T}}$. Because of the property $\boldsymbol{f}(x,\dot{x}) \cdot \dot{x} = \mathbf{0}$ for all $(x,\dot{x}) \in \operatorname{Dom}\boldsymbol{f}$, the $\boldsymbol{u}$-spacelike component and the $\boldsymbol{u}$-timelike component of $\boldsymbol{f}$ are not independent. Using the formula in 7.1.1 we get

$$0 = \boldsymbol{f}(x,\dot{x}) \cdot \dot{x} = -\big(\boldsymbol{u} \cdot \boldsymbol{f}(x,\dot{x})\big)(\boldsymbol{u} \cdot \dot{x}) + \big(\boldsymbol{\pi}_{\boldsymbol{u}} \cdot f(x,\dot{x})\big) \cdot (\boldsymbol{\pi}_{\boldsymbol{u}} \cdot \dot{x}),$$

which yields

$$-\boldsymbol{u} \cdot \boldsymbol{f}(x,\dot{x}) = \big(\boldsymbol{\pi}_{\boldsymbol{u}} \cdot \boldsymbol{f}(x,\dot{x})\big) \cdot \boldsymbol{v}_{\dot{x}\boldsymbol{u}}.$$

7.3.4. Splittings of vector fields according to rigid observers in the nonrelativistic case can be treated in the mathematical framework of affine spaces. However, splittings according to general observers require the theory of manifolds.

In the special relativistic case splittings according to noninertial observers can be treated only in the framework of manifolds and they do not appear here.

7.4. Exercises

1. Show that $\boldsymbol{\pi}_{\boldsymbol{u}} \cdot \boldsymbol{x} = (\boldsymbol{u} \wedge \boldsymbol{x}) \cdot \boldsymbol{u}$ for all $\boldsymbol{u} \in V(1)$, $x \in \mathbf{M}$.
2. Take the arithmetic spacetime model. Give the completely split form of the vector field

$$(\xi^0, \boldsymbol{\xi}) \mapsto (\xi^1 + \xi^2, \cos(\xi^0 - \xi^3), 0, 0)$$

according to the global inertial observer with the velocity value $\frac{1}{\sqrt{1-v^2}}(1, v, 0, 0)$.

8. Tensor splittings

8.1. Splitting of tensors

8.1.1. The various tensors—elements of $\mathbf{M}\otimes\mathbf{M}$, $\mathbf{M}\otimes\mathbf{M}^*$, etc.—are split according to $\boldsymbol{u} \in V(1)$ by the maps $\boldsymbol{\xi}_{\boldsymbol{u}}\otimes\boldsymbol{\xi}_{\boldsymbol{u}}$, $\boldsymbol{\xi}_{\boldsymbol{u}}\otimes\boldsymbol{\eta}_{\boldsymbol{u}}$, etc. as in the nonrelativistic case. However, now it suffices to deal with $\boldsymbol{\xi}_{\boldsymbol{u}} \otimes \boldsymbol{\xi}_{\boldsymbol{u}}$ because the identification $\mathbf{M}^* \equiv \frac{\mathbf{M}}{\mathbb{T}\otimes\mathbb{T}}$ and $\boldsymbol{\eta}_{\boldsymbol{u}} = \boldsymbol{j}_{\boldsymbol{u}} \cdot \boldsymbol{\xi}_{\boldsymbol{u}}$ (see 7.2.1) allow us to derive the other splittings from this one.

With the usual identifications we have

$$\boldsymbol{\xi}_{\boldsymbol{u}}\otimes\boldsymbol{\xi}_{\boldsymbol{u}}\colon\ \mathbf{M}\otimes\mathbf{M} \to (\mathbb{T}\times\mathbf{S}_{\boldsymbol{u}})\otimes(\mathbb{T}\times\mathbf{S}_{\boldsymbol{u}}) \equiv (\mathbb{T}\otimes\mathbb{T})\times(\mathbb{T}\otimes\mathbf{S}_{\boldsymbol{u}})\times(\mathbf{S}_{\boldsymbol{u}}\otimes\mathbb{T})\times(\mathbf{S}_{\boldsymbol{u}}\otimes\mathbf{S}_{\boldsymbol{u}}),$$

and for $\boldsymbol{T} \in \mathbf{M} \otimes \mathbf{M}$:

$$(\boldsymbol{\xi}_u \otimes \boldsymbol{\xi}_u) \cdot \boldsymbol{T} = \boldsymbol{\xi}_u \cdot \boldsymbol{T} \cdot \boldsymbol{\xi}_u^* = \boldsymbol{\xi}_u \cdot \boldsymbol{T} \cdot \boldsymbol{\eta}_u^{-1} = \begin{pmatrix} \boldsymbol{u} \cdot \boldsymbol{T} \cdot \boldsymbol{u} & -\boldsymbol{u} \cdot \boldsymbol{T} \cdot \boldsymbol{\pi}_u^* \\ -\boldsymbol{\pi}_u \cdot \boldsymbol{T} \cdot \boldsymbol{u} & \boldsymbol{\pi}_u \cdot \boldsymbol{T} \cdot \boldsymbol{\pi}_u^* \end{pmatrix} =$$

$$= \begin{pmatrix} \boldsymbol{u} \cdot \boldsymbol{T} \cdot \boldsymbol{u} & -\boldsymbol{u} \cdot \boldsymbol{T} - \boldsymbol{u}(\boldsymbol{u} \cdot \boldsymbol{T} \cdot \boldsymbol{u}) \\ -\boldsymbol{T} \cdot \boldsymbol{u} - \boldsymbol{u}(\boldsymbol{u} \cdot \boldsymbol{T} \cdot \boldsymbol{u}) & \boldsymbol{T} + \boldsymbol{u} \otimes (\boldsymbol{u} \cdot \boldsymbol{T}) + (\boldsymbol{T} \cdot \boldsymbol{u}) \otimes \boldsymbol{u} + \boldsymbol{u} \otimes \boldsymbol{u}(\boldsymbol{u} \cdot \boldsymbol{T} \cdot \boldsymbol{u}) \end{pmatrix},$$

for $\boldsymbol{L} \in \mathbf{M} \otimes \mathbf{M}^*$:

$$(\boldsymbol{\xi}_u \otimes \boldsymbol{\eta}_u) \cdot \boldsymbol{L} = \boldsymbol{\xi}_u \cdot \boldsymbol{L} \cdot \boldsymbol{\eta}_u^* = \boldsymbol{\xi}_u \cdot \boldsymbol{L} \cdot \boldsymbol{\xi}_u^{-1} = \begin{pmatrix} -\boldsymbol{u} \cdot \boldsymbol{L} \cdot \boldsymbol{u} & -\boldsymbol{u} \cdot \boldsymbol{L} \cdot \boldsymbol{\pi}_u^* \\ \boldsymbol{\pi}_u \cdot \boldsymbol{L} \cdot \boldsymbol{u} & \boldsymbol{\pi}_u \cdot \boldsymbol{L} \cdot \boldsymbol{\pi}_u^* \end{pmatrix},$$

for $\boldsymbol{P} \in \mathbf{M}^* \otimes \mathbf{M}$:

$$(\boldsymbol{\eta}_u \otimes \boldsymbol{\xi}_u) \cdot \boldsymbol{P} = \boldsymbol{\eta}_u \cdot \boldsymbol{P} \cdot \boldsymbol{\xi}_u^* = \boldsymbol{\eta}_u \cdot \boldsymbol{P} \cdot \boldsymbol{\eta}_u^{-1} = \begin{pmatrix} -\boldsymbol{u} \cdot \boldsymbol{P} \cdot \boldsymbol{u} & \boldsymbol{u} \cdot \boldsymbol{P} \cdot \boldsymbol{\pi}_u^* \\ -\boldsymbol{\pi}_u \cdot \boldsymbol{P} \cdot \boldsymbol{u} & \boldsymbol{\pi}_u \cdot \boldsymbol{P} \cdot \boldsymbol{\pi}_u^* \end{pmatrix},$$

for $\boldsymbol{F} \in \mathbf{M}^* \otimes \mathbf{M}^*$:

$$(\boldsymbol{\eta}_u \otimes \boldsymbol{\eta}_u) \cdot \boldsymbol{F} = \boldsymbol{\eta}_u \cdot \boldsymbol{F} \cdot \boldsymbol{\eta}_u^* = \boldsymbol{\eta}_u \cdot \boldsymbol{F} \cdot \boldsymbol{\xi}_u^{-1} = \begin{pmatrix} \boldsymbol{u} \cdot \boldsymbol{F} \cdot \boldsymbol{u} & \boldsymbol{u} \cdot \boldsymbol{F} \cdot \boldsymbol{\pi}_u^* \\ \boldsymbol{\pi}_u \cdot \boldsymbol{F} \cdot \boldsymbol{u} & \boldsymbol{\pi}_u \cdot \boldsymbol{F} \cdot \boldsymbol{\pi}_u^* \end{pmatrix}.$$

8.1.2. The splittings corresponding to different velocity values $\boldsymbol{u}$ and $\boldsymbol{u}'$ are different. The tensor transformation rule that shows how the splittings depend on velocity values is rather complicated, much more complicated than in the nonrelativistic case. We shall study it only for antisymmetric tensors.

8.2. Splitting of antisymmetric tensors

8.2.1. If $\boldsymbol{T}$ is an antisymmetric tensor, i.e. $\boldsymbol{T} \in \mathbf{M} \wedge \mathbf{M}$, then $\boldsymbol{u} \cdot \boldsymbol{T} \cdot \boldsymbol{u} = 0$, $\boldsymbol{u} \cdot \boldsymbol{T} \cdot \boldsymbol{\pi}_u^* = -(\boldsymbol{\pi}_u \cdot \boldsymbol{T} \cdot \boldsymbol{u})^*$ and $\boldsymbol{\pi}_u \cdot \boldsymbol{T} \cdot \boldsymbol{\pi}_u^* \in \mathbf{S}_u \wedge \mathbf{S}_u$ which means (of course) that the $\boldsymbol{u}$-split form of $\boldsymbol{T}$ is antisymmetric as well. Thus $\boldsymbol{u}$-splitting maps the elements of $\mathbf{M} \wedge \mathbf{M}$ into elements of form

$$\begin{pmatrix} 0 & -\boldsymbol{a} \\ \boldsymbol{a} & \boldsymbol{A} \end{pmatrix} \equiv (\boldsymbol{a}, \boldsymbol{A})$$

where $\boldsymbol{a} \in \mathbf{S}_u \otimes \mathbb{T} \equiv \mathbb{T} \otimes \mathbf{S}_u$, $\boldsymbol{A} \in \mathbf{S}_u \wedge \mathbf{S}_u$.

The corresponding formula in 8.1.1 gives for $\boldsymbol{T} \in \mathbf{M} \wedge \mathbf{M}$

$$\boldsymbol{\xi}_u \cdot \boldsymbol{T} \cdot \boldsymbol{\xi}_u^* = \big(-\boldsymbol{T} \cdot \boldsymbol{u}, \boldsymbol{T} + (\boldsymbol{T} \cdot \boldsymbol{u}) \wedge \boldsymbol{u}\big).$$

Definition. $-\boldsymbol{T}\cdot\boldsymbol{u}$ and $\boldsymbol{T}+(\boldsymbol{T}\cdot\boldsymbol{u})\wedge\boldsymbol{u}$ are called the $\boldsymbol{u}$*-timelike component* and the $\boldsymbol{u}$*-spacelike component* of the antisymmetric tensor $\boldsymbol{T}$.

8.2.2. The following *transformation rule* shows how splittings depend on velocity values.

Proposition. Let $\boldsymbol{u},\boldsymbol{u}'\in V(1)$. Then

$$\xi_{\boldsymbol{u}'\boldsymbol{u}}\cdot(\boldsymbol{a},\boldsymbol{A})\cdot\xi^*_{\boldsymbol{u}'\boldsymbol{u}}=$$

$$=\left(\frac{1}{\sqrt{1-|\boldsymbol{v}_{\boldsymbol{u}'\boldsymbol{u}}|^2}}\left(\boldsymbol{a}+\boldsymbol{v}_{\boldsymbol{u}'\boldsymbol{u}}\frac{1-\sqrt{1-|\boldsymbol{v}_{\boldsymbol{u}'\boldsymbol{u}}|^2}}{|\boldsymbol{v}_{\boldsymbol{u}'\boldsymbol{u}}|^2}(\boldsymbol{v}_{\boldsymbol{u}'\boldsymbol{u}}\cdot\boldsymbol{a})+\boldsymbol{A}\cdot\boldsymbol{v}_{\boldsymbol{u}'\boldsymbol{u}}\right),\right.$$

$$\left.\frac{1}{\sqrt{1-|\boldsymbol{v}_{\boldsymbol{u}'\boldsymbol{u}}|^2}}\left(-\boldsymbol{a}-\boldsymbol{A}\cdot\boldsymbol{v}_{\boldsymbol{u}'\boldsymbol{u}}\frac{1-\sqrt{1-|\boldsymbol{v}_{\boldsymbol{u}'\boldsymbol{u}}|^2}}{|\boldsymbol{v}_{\boldsymbol{u}'\boldsymbol{u}}|^2}\right)\wedge\boldsymbol{v}_{\boldsymbol{u}'\boldsymbol{u}}+\boldsymbol{A}\right).$$

Proof. Using the matrix forms we have

$$\xi_{\boldsymbol{u}'\boldsymbol{u}}\cdot(\boldsymbol{a},\boldsymbol{A})\cdot\xi^*_{\boldsymbol{u}'\boldsymbol{u}}=$$

$$=\kappa(\boldsymbol{v}_{\boldsymbol{u}'\boldsymbol{u}})^2\begin{pmatrix}1 & -\boldsymbol{v}_{\boldsymbol{u}'\boldsymbol{u}}\\ -\boldsymbol{v}_{\boldsymbol{u}'\boldsymbol{u}} & \boldsymbol{D}(\boldsymbol{v}_{\boldsymbol{u}'\boldsymbol{u}})\end{pmatrix}\begin{pmatrix}0 & -\boldsymbol{a}\\ \boldsymbol{a} & \boldsymbol{A}\end{pmatrix}\begin{pmatrix}1 & -\boldsymbol{v}_{\boldsymbol{u}'\boldsymbol{u}}\\ -\boldsymbol{v}_{\boldsymbol{u}'\boldsymbol{u}} & \boldsymbol{D}(\boldsymbol{v}_{\boldsymbol{u}'\boldsymbol{u}})\end{pmatrix},$$

from which we can get the desired formula.

8.2.3. The previous fearsome formula becomes nicer if we write $(\boldsymbol{a},\boldsymbol{A})$ as the sum of components parallel and orthogonal to the relative velocity:

$$\boldsymbol{a}=\boldsymbol{a}_{\|}+\boldsymbol{a}_{\perp},\qquad \boldsymbol{A}=\boldsymbol{A}_{\|}+\boldsymbol{A}_{\perp}$$

where $\boldsymbol{a}_{\|}$ is parallel to $\boldsymbol{v}_{\boldsymbol{u}'\boldsymbol{u}}$, $\boldsymbol{a}_{\perp}$ is orthogonal to $\boldsymbol{v}_{\boldsymbol{u}'\boldsymbol{u}}$, and the kernel of $\boldsymbol{A}_{\|}$ is parallel to $\boldsymbol{v}_{\boldsymbol{u}'\boldsymbol{u}}$, i.e. $\boldsymbol{A}_{\|}\cdot\boldsymbol{v}_{\boldsymbol{u}'\boldsymbol{u}}=\boldsymbol{0}$ and the kernel of $\boldsymbol{A}_{\perp}$ is orthogonal to $\boldsymbol{v}_{\boldsymbol{u}'\boldsymbol{u}}$, i.e. $(\boldsymbol{A}_{\perp}\cdot\boldsymbol{v}_{\boldsymbol{u}'\boldsymbol{u}})\wedge\boldsymbol{v}_{\boldsymbol{u}'\boldsymbol{u}}=-|\boldsymbol{v}_{\boldsymbol{u}'\boldsymbol{u}}|^2\boldsymbol{A}_{\perp}$ (see Exercise V.3.21.1). Then we easily find

$$\xi_{\boldsymbol{u}'\boldsymbol{u}}\cdot(\boldsymbol{a}_{\|},\boldsymbol{A}_{\|})\cdot\xi^*_{\boldsymbol{u}'\boldsymbol{u}}=(\boldsymbol{a}_{\|},\boldsymbol{A}_{\|}),$$

$$\xi_{\boldsymbol{u}'\boldsymbol{u}}\cdot(\boldsymbol{a}_{\perp},\boldsymbol{A}_{\perp})\cdot\xi^*_{\boldsymbol{u}'\boldsymbol{u}}=\frac{1}{\sqrt{1-|\boldsymbol{v}_{\boldsymbol{u}'\boldsymbol{u}}|^2}}(\boldsymbol{a}_{\perp}+\boldsymbol{A}_{\perp}\cdot\boldsymbol{v}_{\boldsymbol{u}'\boldsymbol{u}},\ -\boldsymbol{a}_{\perp}\wedge\boldsymbol{v}_{\boldsymbol{u}'\boldsymbol{u}}+\boldsymbol{A}_{\perp}).$$

8.2.4. The splitting and the transformation rule of antisymmetric cotensors i.e. elements of $\mathbf{M}^*\wedge\mathbf{M}^*$ are the same, apart from a negative sign. The details are left to the reader.

It is interesting that here, in contradistinction to the nonrelativistic case, $\mathrm{M}\wedge\mathrm{M}^*$ makes sense because of the identification $\mathrm{M}^* \equiv \frac{\mathrm{M}}{\mathbb{T}\otimes\mathbb{T}}$. The mixed tensor $\boldsymbol{H}\in\mathrm{M}\wedge\mathrm{M}^*$ has the $\boldsymbol{u}$-split form

$$\boldsymbol{\xi}_{\boldsymbol{u}}\cdot\boldsymbol{H}\cdot\boldsymbol{\xi}_{\boldsymbol{u}}^{-1} = \begin{pmatrix} 0 & \boldsymbol{H}\cdot\boldsymbol{u} \\ \boldsymbol{H}\cdot\boldsymbol{u} & \boldsymbol{H}+(\boldsymbol{H}\cdot\boldsymbol{u})\wedge\boldsymbol{u}\end{pmatrix}$$

which, as a matrix, is not antisymmetric. It need not be antisymmetric, because the symmmetric or antisymmetric properties of matrices refer to these properties of linear maps regarding duals without any identifications (see IV.1.5 and V.4.19).

8.3. Splitting of tensor fields

8.3.1. The splitting of various tensor fields according to inertial observers can be treated analogously to the nonrelativistic case.

The antisymmetric cotensor field $\boldsymbol{F}$ has the *half split form* according to the standard inertial frame $\boldsymbol{u}$

$$\begin{aligned}(\boldsymbol{E}_{\boldsymbol{u}},\boldsymbol{B}_{\boldsymbol{u}}) := \boldsymbol{\eta}_{\boldsymbol{u}}\cdot\boldsymbol{F}\cdot\boldsymbol{\eta}_{\boldsymbol{u}}^*\colon\ \mathrm{M} &\rightarrowtail (\mathbf{S}_{\boldsymbol{u}}^*\otimes\mathbb{T}^*)\times(\mathbf{S}_{\boldsymbol{u}}^*\wedge\mathbf{S}_{\boldsymbol{u}}^*),\\ x &\mapsto \big(\boldsymbol{F}(x)\cdot\boldsymbol{u},\ \boldsymbol{F}(x)+\big(\boldsymbol{F}(x)\cdot\boldsymbol{u}\big)\wedge\boldsymbol{u}\big)\end{aligned}$$

and the *completely split form*

$$\begin{aligned}\boldsymbol{\eta}_{\boldsymbol{u}}\cdot\boldsymbol{F}\cdot\boldsymbol{\eta}_{\boldsymbol{u}}^*\circ\xi_{\boldsymbol{u}}^{-1}:\mathrm{T}_{\boldsymbol{u}}\times\mathrm{S}_{\boldsymbol{u}} &\rightarrowtail (\mathbf{S}^*\otimes\mathbb{T}^*)\times(\mathbf{S}^*\wedge\mathbf{S}^*),\\ (t,q) &\mapsto \big(\boldsymbol{E}_{\boldsymbol{u}}(q\star t),\boldsymbol{B}_{\boldsymbol{u}}(q\star t)\big).\end{aligned}$$

8.3.2. The electromagnetic field is described by an antisymmetric cotensor field $\boldsymbol{F}$ which is the exterior derivative of a potential $\boldsymbol{K}$, $\boldsymbol{F}=\mathrm{D}\wedge\boldsymbol{K}$. The electric field and the magnetic field relative to the inertial observer $\boldsymbol{U}$ are the corresponding components of the completely split form of $\boldsymbol{F}$.

The relation between the completely split form $(-V_{\boldsymbol{u}}^c,\boldsymbol{A}_{\boldsymbol{u}}^c)$ of $\boldsymbol{K}$ and the completely split form $\big(\boldsymbol{E}_{\boldsymbol{u}}^c,\boldsymbol{B}_{\boldsymbol{u}}^c\big)$ of $\boldsymbol{F}$ is exactly the same as in the nonrelativistic case:

$$\boldsymbol{E}_{\boldsymbol{u}}^c = -\partial_0\boldsymbol{A}_{\boldsymbol{u}}^c - \nabla V_{\boldsymbol{u}}^c, \qquad \boldsymbol{B}_{\boldsymbol{u}}^c = \nabla\wedge\boldsymbol{A}_{\boldsymbol{u}}^c.$$

Since the force field defined by the potential $\boldsymbol{K}$ equals

$$\boldsymbol{f}(x,\dot{x}) = \boldsymbol{F}(x)\cdot\dot{x} \qquad\qquad \big(x\in\mathrm{Dom}\boldsymbol{K},\ \dot{x}\in V(1)\big),$$

where $\boldsymbol{F}:=\mathrm{D}\wedge\boldsymbol{K}$, we can state again that a masspoint in the world point x having the velocity value $\boldsymbol{u}'$ 'feels' only the $\boldsymbol{u}'$-timelike component of the field; a masspoint always 'feels' the electric field according to its instantaneous velocity value.

Because of the more complicated transformation rule in the special relativistic case the Lorentz force is expressed by the $\boldsymbol{U}$-electric field and the $\boldsymbol{U}$-magnetic field more complicatedly than in the nonrelativistic case.

8.4. Exercises

1. Let $\boldsymbol{x} \in \mathrm{M}$, $\boldsymbol{T} \in \mathrm{M} \otimes \mathrm{M}$ and $\boldsymbol{L} \in \mathrm{M} \otimes \mathrm{M}^*$. Give the $\boldsymbol{u}$-split form of $\boldsymbol{T} \cdot \boldsymbol{x}$ and $\boldsymbol{L} \cdot \boldsymbol{x}$ using the $\boldsymbol{u}$-split form of $\boldsymbol{x}$, $\boldsymbol{T}$ and $\boldsymbol{L}$.

2. Let $\boldsymbol{T}$ and $\boldsymbol{L}$ as before. Give the $\boldsymbol{u}$-split form of $\boldsymbol{T} \cdot \boldsymbol{L}$ and $\boldsymbol{L} \cdot \boldsymbol{T}$ using the $\boldsymbol{u}$-split form of $\boldsymbol{T}$ and $\boldsymbol{L}$.

3. Recall the nondegenarate bilinear form (see V.4.15)

$$(\mathrm{M} \wedge \mathrm{M}^*) \times (\mathrm{M} \wedge \mathrm{M}^*) \to \mathbb{R}, \qquad (\boldsymbol{F}, \boldsymbol{H}) \mapsto \boldsymbol{F} \bullet \boldsymbol{H} := -\frac{1}{2} \operatorname{Tr} \boldsymbol{F} \cdot \boldsymbol{H}.$$

Express $\boldsymbol{F} \cdot \boldsymbol{H}$ using the $\boldsymbol{u}$-timelike and the $\boldsymbol{u}$-spacelike components of $\boldsymbol{F}$ and $\boldsymbol{H}$.

9. Reference systems

9.1. The notion of a reference system

9.1.1. We can repeat word by word what we said in I. 7.1.1 with the single exception that instead of (absolute) time now we have to consider an (artificial) time derived from a synchronization.

Recall that a reference frame $(\mathcal{S}, \boldsymbol{U})$ i.e. a synchronization $\mathcal{S}$ together with an observer $\boldsymbol{U}$ establishes the splitting $\xi_{\mathcal{S},U} = (\tau_{\mathcal{S}}, C_U)\colon \mathrm{M} \rightarrowtail \mathrm{T}_{\mathcal{S}} \times \mathrm{S}_U$.

Definition. A *reference system* is a quartet $(\mathcal{S}, \boldsymbol{U}, T_{\mathcal{S}}, S_U)$ where

(*i*) $\mathcal{S}$ is a synchronization,

(*ii*) $\boldsymbol{U}$ is an observer,

(*iii*) $T_{\mathcal{S}}\colon \mathrm{T}_{\mathcal{S}} \rightarrowtail \mathbb{R}$ is a strictly monotone increasing mapping,

(*iv*) $S_U\colon \mathrm{S}_U \rightarrowtail \mathbb{R}^3$ is a mapping

such that $(T_{\mathcal{S}} \times S_U) \circ \xi_{\mathcal{S},U} = (T_{\mathcal{S}} \circ \tau_{\mathcal{S}},\ S_U \circ C_U)\colon \mathrm{M} \rightarrowtail \mathbb{R} \times \mathbb{R}^3$ is an orientation preserving coordinatization. ■

We call $T_{\mathcal{S}}$ and S_U the *coordinatization* of $\mathcal{S}$-time and $\boldsymbol{U}$-space, respectively, in spite of the fact that we introduced the notion of coordinatization only for affine spaces and, in general, neither $\mathrm{T}_{\mathcal{S}}$ nor S_U is an affine space. (We mention that in any case $\mathrm{T}_{\mathcal{S}}$ and S_U can be endowed with a smooth structure and in the framework of smooth structures $T_{\mathcal{S}}$ and S_U do become a coordinatization.)

Note that condition (iii) involves that $T_\mathcal{S}$ is defined on a subset of $\mathrm{T}_\mathcal{S}$ where the ordering 'later' is total; consequently, the coordinatization of spacetime is defined on a subset of $\mathrm{Dom}\,\boldsymbol{U}$ where the synchronization is well posed.

9.1.2. Definition. Let us consider a coordinatization $K: \mathrm{M} \rightarrowtail \mathbb{R} \times \mathbb{R}^3$. As usual, we number the coordinates of $\mathbb{R} \times \mathbb{R}^3$ from zero to three. Accordingly, we find convenient to use the notation $K = (\kappa^0, \boldsymbol{\kappa}): \mathrm{M} \rightarrowtail \mathbb{R} \times \mathbb{R}^3$ for the coordinatizations of spacetime. Then the equality

$$\mathrm{D}\boldsymbol{\kappa}(x) \cdot \partial_0 K^{-1}\big(K(x)\big) = \mathbf{0}$$

well-known and used in the nonrelativistic case will hold now as well, since its deduction rests only on the affine structure of M.

We say that a coordinatization K is *referencelike* if there is a reference system $(\mathcal{S}, \boldsymbol{U}, T_\mathcal{S}, S_U)$ such that $K = (T_\mathcal{S} \times S_U) \circ \xi_{\mathcal{S},U}$. In that case

$$\kappa^0 = T_\mathcal{S} \circ \tau_\mathcal{S}, \qquad \boldsymbol{\kappa} = S_U \circ C_U.$$

9.1.3. Proposition. A coordinatization $K = (\kappa^0, \boldsymbol{\kappa}): \mathrm{M} \rightarrowtail \mathbb{R} \times \mathbb{R}^3$ corresponds to a reference system if and only if

(i) K is orientation preserving,

(ii) $\partial_0 K^{-1}\big(K(x)\big)$ is a future directed timelike vector,

(iii) $-\big(\mathrm{D}\kappa^0\big)(x)$ is a future directed timelike vector

for all $x \in \mathrm{Dom}\,K$.

Then

$$\boldsymbol{U}(x) = \frac{\partial_0 K^{-1}\big(K(x)\big)}{\big|\partial_0 K^{-1}\big(K(x)\big)\big|} \qquad (x \in \mathrm{Dom}\,K), \tag{1}$$

is the corresponding observer and the corresponding synchronization $\mathcal{S}$ is determined as follows:

$$x \text{ is simultaneous with } y \text{ if and only if } \kappa^0(x) = \kappa^0(y) \tag{2}$$

moreover,

$$T_\mathcal{S}(t) = \kappa^0(x) \qquad (t \in \mathrm{T}_\mathcal{S},\ x \in t), \tag{3}$$

$$S_U(q) = \boldsymbol{\kappa}(x) \qquad (q \in \mathrm{S}_U,\ x \in q). \tag{4}$$

Proof. If $K = (T_\mathcal{S} \times S_U) \circ \xi_{\mathcal{S},U}$ then (i) is trivial.

κ^0 is constant on the $\mathcal{S}$-instants. In other words, $\mathcal{S}$-instants—more precisely their part in $\mathrm{Dom}\,K$—have the form $\big\{x \in \mathrm{Dom}\,K \,\big|\, \kappa^0(x) = \alpha\big\}$. Then $\big\{\boldsymbol{x} \in \mathbf{M} \,\big|\, \big(\mathrm{D}\kappa^0\big)(x) \cdot \boldsymbol{x} = \mathbf{0}\big\}$ is the tangent space of the corresponding world surface passing through x. Since this tangent space is spacelike, $(\mathrm{D}\kappa^0)(x)$ must be timelike. If $y - x \in T^{\rightarrow}$, then the properties of τ_S and T_S imply that $\kappa^0(y) - \kappa^0(x) > 0$; then $(\mathrm{D}\kappa^0)(x) \cdot (y-x) + \mathrm{ordo}(y-x) > 0$ results in that $(\mathrm{D}\kappa^0)(x) \cdot \boldsymbol{x} > 0$ for all $\boldsymbol{x} \in T^{\rightarrow}$, proving (iii).

As concerns (ii), note that a world line function r satisfies $\dot{r}(s) = \boldsymbol{U}(r(s))$ and takes values in the domain of K if and only if $K(r(s)) = (\kappa^0(r(s)), \boldsymbol{\xi})$ i.e. $r(s) = K^{-1}(\kappa^0(r(s)), \boldsymbol{\xi})$ for a $\boldsymbol{\xi} \in \mathbb{R}^3$ and for all $s \in \mathrm{Dom}\, r$. As a consequence, we have

$$\boldsymbol{U}(r(s)) = \frac{\mathrm{d}}{\mathrm{d}s} K^{-1}(\kappa^0(r(s)), \boldsymbol{\xi}) = \partial_0 K^{-1}(\kappa^0(r(s)), \boldsymbol{\xi}) \cdot (\mathrm{D}\kappa^0)(r(s)) \cdot \dot{r}(s)$$

which, together with condition (iii), implies that $\boldsymbol{U}(x)$ is a positive multiple of $\partial_0 K^{-1}(K(x))$ for all $x \in \mathrm{Dom}\, K$, proving (ii) and equality (1).

Suppose now that $K = (\kappa^0, \boldsymbol{\kappa})$ is a coordinatization that fulfils conditions (ii)–(iii).

Then condition (ii) implies that $\boldsymbol{U}$ *defined* by equality (1) is an observer.

According to (iii), the simultaneity $\mathcal{S}$ is well *defined* by (2) (i.e. the subsets of form $\{x \in \mathrm{Dom}\, K \mid \kappa^0(x) = \alpha\}$ are world surfaces and $\mathcal{S}$ is smooth). Consequently, $T_{\mathcal{S}}$ is well *defined* by formula (3) and it is strictly monotone increasing.

If r is a world line such that $\dot{r}(s) = \boldsymbol{U}(r(s))$ then

$$\frac{\mathrm{d}}{\mathrm{d}s}(\boldsymbol{\kappa}(r(s)) = \mathrm{D}\boldsymbol{\kappa}(r(s)) \cdot \boldsymbol{U}(r(t)) = \mathrm{D}\boldsymbol{\kappa}(r(s)) \cdot \frac{\partial_0 K^{-1}(K(r(s)))}{|\partial_0 K^{-1}(K(r(s)))|} = \mathbf{0}$$

which means that $\boldsymbol{\kappa} \circ r$ is a constant mapping, in other words, $\boldsymbol{\kappa}$ is constant on the $\boldsymbol{U}$-lines; hence S_U is well *defined* by formula (4).

Finally, it is evident that $K = (T_{\mathcal{S}} \times S_U) \circ \xi_{\mathcal{S},U}$. ■

9.2. Lorentzian reference systems

9.2.1. Now we are interested in what kind of affine coordinatization of spacetime can correspond to a reference system.

Let us take an affine coordinatization $K \colon \mathrm{M} \to \mathbb{R}^4$. Then there are

— an $o \in \mathrm{M}$,

— an ordered basis $(\boldsymbol{x}_0, \boldsymbol{x}_1, \boldsymbol{x}_2, \boldsymbol{x}_3)$ of $\mathbf{M}$

such that

$$K(x) = (\boldsymbol{k}^i \cdot (x - o) \mid i = 0,1,2,3) \qquad (x \in \mathrm{M}),$$

$$K^{-1}(\xi) = \sum_{i=0}^{3} \xi^i x_i \qquad (\xi \in \mathbb{R}^4),$$

where $(\boldsymbol{k}^0, \boldsymbol{k}^1, \boldsymbol{k}^2, \boldsymbol{k}^3)$ is the dual of the basis in question.

Proposition. The affine coordinatization K corresponds to a reference system if and only if

(i) $(\boldsymbol{x}_0, \boldsymbol{x}_1, \boldsymbol{x}_2, \boldsymbol{x}_3)$ is a positively oriented basis,

(ii) $\boldsymbol{x}_0$ is a future directed timelike vector,

(iii) $\boldsymbol{x}_1, \boldsymbol{x}_2, \boldsymbol{x}_3$ are spacelike vectors spanning a spacelike linear subspace of $\mathbf{M}$.

Then the corresponding observer is global and inertial, having the constant value

$$\boldsymbol{u} := \frac{\boldsymbol{x}_0}{|\boldsymbol{x}_0|},$$

and the synchronization is given by the hyperplanes directed by the spacelike subspace spanned by $\boldsymbol{x}_1$, $\boldsymbol{x}_2$, $\boldsymbol{x}_3$.

Proof. We show that the present conditions (i)–(iii) correspond to the conditions listed in Proposition 9.1.3.

(i) The coordinatization is orientation preserving if and only if the corresponding basis is positively oriented;

(ii) $\partial_0 K^{-1}(K(x)) = \boldsymbol{x}_0$;

(iii) $-(\mathrm{D}\kappa^0)(x) = -\boldsymbol{k}^0$ for all $x \in \mathrm{M}$. Since $\boldsymbol{k}^0 \cdot \boldsymbol{x}_\alpha = 0$ $(\alpha = 1, 2, 3)$, $-\boldsymbol{k}^0$ is timelike if and only if $\boldsymbol{x}_\alpha$-s span a spacelike linear subspace; then, since $\boldsymbol{k}^0 \cdot \boldsymbol{x}_0 = 1 > 0$ and since $\boldsymbol{x}_0$ is future directed timelike, $-\boldsymbol{k}^0$ must be future directed. ■

According to our result—putting $\boldsymbol{s} := |\boldsymbol{x}_0|$ —, we write an *affine reference system* in the form $(\boldsymbol{u}, o, \boldsymbol{s}, \boldsymbol{x}_1, \boldsymbol{x}_2, \boldsymbol{x}_3)$.

9.2.2. Definition. Let $\boldsymbol{G}$ denote the Lorentz form on $\mathbb{R}^4$ treated in V.4.19 and recall that a linear map $\boldsymbol{L}\colon \mathbf{M} \to \mathbb{R}^4$ is called $\boldsymbol{g} - \boldsymbol{G}$-orthogonal if there is an $\boldsymbol{s} \in \mathbb{T}$ such that $\boldsymbol{G}(\boldsymbol{L} \cdot \boldsymbol{x} \cdot \boldsymbol{L} \cdot \boldsymbol{y}) = \frac{\boldsymbol{g}(\boldsymbol{x},\boldsymbol{y})}{\boldsymbol{s}^2}$ for all $\boldsymbol{x}, \boldsymbol{y} \in \mathbf{M}$.

A coordinatization K is called *Lorentzian* if

— K is affine,

— $\boldsymbol{K}\colon \mathbf{M} \to \mathbb{R}^4$ is $\boldsymbol{g} - \boldsymbol{G}$-orthogonal.

A reference system is Lorentzian if the corresponding coordinatization is Lorentzian. ■

From the previous result we get immediately the following:

Proposition. A coordinatization K is Lorentzian if and only if there are

(i) an $o \in \mathrm{M}$,

(ii) a positively oriented $\boldsymbol{g}$-orthogonal basis $(\boldsymbol{e}_0, \boldsymbol{e}_1, \boldsymbol{e}_2, \boldsymbol{e}_3)$, normed to an $\boldsymbol{s}$, of $\mathbf{M}$ such that $\boldsymbol{e}_0$ is future directed timelike,

and

$$K(x) = \left(\frac{\boldsymbol{e}_i \cdot (x - o)}{\boldsymbol{e}_i^2} \,\middle|\, i = 0, 1, 2, 3 \right) \qquad (x \in \mathrm{M}). \quad ■$$

According to our result, a *Lorentzian reference system* will be given in the form $(\boldsymbol{u}, o, \boldsymbol{s}, \boldsymbol{e}_1, \boldsymbol{e}_2, \boldsymbol{e}_3)$ and we shall use the following names: $\boldsymbol{u}$ is *its velocity value,* o is its *origin,* $\boldsymbol{s}$ is its *spacetime unit,* $(\boldsymbol{e}_1,\ \boldsymbol{e}_2,\ \boldsymbol{e}_3)$ is its *space basis,*. Moreover, putting $\boldsymbol{e}_0 := \boldsymbol{s}\boldsymbol{u}$, we call $(\boldsymbol{e}_0,\ \boldsymbol{e}_1,\ \boldsymbol{e}_2,\ \boldsymbol{e}_3)$ its *spacetime basis.*

9.2.3. Let K be a Lorentzian coordinatization and use the previous notations.

We see from 1.6 that the Lorentzian coordinatization establishes an isomorphism between the spacetime model $(\mathrm{M}, \mathbb{T}, \boldsymbol{g})$ and the arithmetic spacetime model. More precisely, the coordinatization K and the mapping $\mathbb{T} \to \mathbb{R},\ \boldsymbol{t} \mapsto \frac{\boldsymbol{t}}{\boldsymbol{s}}$ constitute an isomorphism.

This isomorphism transforms vectors, covectors and tensors, cotensors etc. into vectors, covectors etc. of the arithmetic spacetime model.

In particular,

$$\boldsymbol{K}\colon\ \mathbf{M} \to \mathbb{R}^4, \qquad \boldsymbol{x} \mapsto \left(\frac{\boldsymbol{e}_i \cdot \boldsymbol{x}}{\boldsymbol{e}_i^2} \,\middle|\, i = 0,1,2,3 \right)$$

is the coordinatization of vectors and

$$\left(\boldsymbol{K}^{-1}\right)^*\colon\ \mathbf{M}^* \to \mathbb{R}^4, \qquad \boldsymbol{k} \mapsto \left(\boldsymbol{k} \cdot \boldsymbol{e}_i \,|\, i = 0,1,2,3\right),$$

is the coordinatization of covectors.

We can generalize the coordinatization for vectors (covectors) of type or cotype $\mathbb{A}$, i.e. for elements in $\mathbf{M} \otimes \mathbb{A}$ or $\frac{\mathbf{M}}{\mathbb{A}}$ $\left(\mathbf{M}^* \otimes \mathbb{A}, \frac{\mathbf{M}^*}{\mathbb{A}}\right)$, too, where $\mathbb{A}$ is a measure line. For instance, elements of $\frac{\mathbf{M}}{\mathbb{T}}$ or $\frac{\mathbf{M}}{\mathbb{T}\otimes\mathbb{T}}$ are coordinatized by the basis $\left(\frac{\boldsymbol{e}_i}{\boldsymbol{s}} \middle| \ i = 0,1,2,3\right)$ and by the basis $\left(\frac{\boldsymbol{e}_i}{\boldsymbol{s}^2} \middle| \ i = 0,1,2,3\right)$, respectively:

$$\frac{\mathbf{M}}{\mathbb{T}} \to \mathbb{R}^4, \qquad \boldsymbol{w} \mapsto \boldsymbol{s}\left(\frac{\boldsymbol{e}_i \cdot \boldsymbol{w}}{\boldsymbol{e}_i^2} \,\middle|\, i = 0,1,2,3 \right),$$

$$\frac{\mathbf{M}}{\mathbb{T}\otimes\mathbb{T}} \to \mathbb{R}^4, \qquad \boldsymbol{p} \mapsto \boldsymbol{s}^2\left(\frac{\boldsymbol{e}_i \cdot \boldsymbol{p}}{\boldsymbol{e}_i^2} \,\middle|\, i = 0,1,2,3 \right).$$

9.2.4. As concerns subscripts and superscripts, we refer to V.4.19.

9.3. Equivalent reference systems

9.3.1. We can repeat, according to the sense, what we said in I.10.5.1.

Recall the notion of automorphisms of the spacetime model (see 1.6.1). An automorphism is a transformation that leaves invariant (preserves) the structure of the spacetime model. Strict automorphisms do not change time periods and distances.

It is quite natural that two objects transformed into each other by a strict automorphism of the spacetime model are considered equivalent (i.e. the same from a physical point of view).

Recalling that $\mathcal{O}(\boldsymbol{g})$ denotes the set of $\boldsymbol{g}$-orthogonal linear maps in $\mathbf{M}$ (see V.2.7) let us introduce the notation

$$\mathcal{P}^{+\to} :=$$
$$\{L\colon \mathrm{M}\to\mathrm{M} \mid L \text{ is affine, } \boldsymbol{L}\in\mathcal{O}(\boldsymbol{g}),\ \boldsymbol{L} \text{ is orientation and arrow preserving}\}$$

and let us call the elements of $\mathcal{P}^{+\to}$ *proper Poincaré transformations.* We shall study these transformations in the next paragraph. For the moment it suffices to know the quite evident fact that $(L, \mathbf{1}_{\mathrm{T}})$ is a strict automorphism if and only if L is a proper Poincaré transformation.

9.3.2. Definition. The coordinatizations K and K' are called *equivalent* if there is a proper Poincaré transformation L such that

$$K'\circ L = K.$$

Two reference systems are *equivalent* if the corresponding coordinatizations are equivalent.

Proposition. The reference systems $(\mathcal{S}, \boldsymbol{U}, T_{\mathcal{S}}, E_{\boldsymbol{U}})$ and $(\mathcal{S}', \boldsymbol{U}', T_{\mathcal{S}'}, E_{\boldsymbol{U}'})$ are equivalent if and only if

(*i*) $\boldsymbol{L}\cdot\boldsymbol{U} = \boldsymbol{U}'\circ L$,

(*ii*) $(T_{\mathcal{S}'}^{-1}\circ T_{\mathcal{S}})\circ\tau_{\mathcal{S}} = \tau_{\mathcal{S}'}\circ L$

(*iii*) $(E_{\boldsymbol{U}'}^{-1}\circ E_{\boldsymbol{U}})\circ C_{\boldsymbol{U}} = C_{\boldsymbol{U}'}\circ L$

Proof. Let K and K' denote the corresponding coordinatizations. It is quite trivial that if the relations above hold, then K and K' are equivalent.

If the relations above hold, for (*i*) we can argue as in I.10.5.3, using $(\boldsymbol{L}\cdot\boldsymbol{x})^2 = \boldsymbol{x}^2$ for all $\boldsymbol{x}\in\mathbf{M}$. As concerns (*ii*) and (*iii*), we can copy the reasoning of (*iii*) in I.10.5.3.

9.3.3. Now we shall see that our definition of equivalence of reference systems is in accordance with the intuitive notion expounded in I.10.5.1.

Proposition. Two Lorentzian reference systems are equivalent if and only if they have the same unit of time (and distance).

Proof. Let the Lorentzian coordinatizations K and K' be defined by the origins o and o' and the spacetime bases $(\boldsymbol{e}_0, \boldsymbol{e}_1, \boldsymbol{e}_2, \boldsymbol{e}_3)$ and $(\boldsymbol{e}_0', \boldsymbol{e}_1', \boldsymbol{e}_2', \boldsymbol{e}_3')$, respectively.

Then $L := K'^{-1} \circ K\colon \mathrm{M} \to \mathrm{M}$ is the affine bijection determined by

$$L(o) = o', \qquad \boldsymbol{L} \cdot \boldsymbol{e}_i = \boldsymbol{e}'_i \qquad (i = 0,1,2,3).$$

Evidently, L is orientation preserving. Moreover, $\boldsymbol{L} \in \mathcal{O}(\boldsymbol{g})$ if and only if $|\boldsymbol{e}_0| = |\boldsymbol{e}'_0|$ and it is arrow-preserving if and only if $\boldsymbol{e}_0$ and $\boldsymbol{e}'_0$ have the same arrow.

9.4. Curve lengths calculated in coordinates

9.4.1. In 6.3.3 we dealt with lengths of curves in the space of an observer $\boldsymbol{U}$ at instants of a synchronization $\mathcal{S}$. It is an interesting question how to calculate these lengths in coordinates corresponding to a coordinatization $K = (T_{\mathcal{S}} \times E_{\boldsymbol{U}}) \circ \xi_{\mathcal{S},\boldsymbol{U}}$.

We shall use the notation $P := K^{-1}$ (P is the parametrization corresponding to the coordinatization K).

Let L and L_t be as in 6.3.3 and let ξ^0 be the coordinate of $t \in \mathrm{T}_{\mathcal{S}}$, i.e. $\xi^0 = \tau_{\mathcal{S}}(t)$.

A parametrization p_t of L_t has the coordinatized form

$$a \mapsto K(p_t(a)) =: \big(\xi^0, (p^\alpha(a) |\, \alpha = 1,2,3)\big) =: \big(\xi^0, \boldsymbol{p}(a)\big)$$

from which we deduce

$$\begin{aligned} p_t &= P(\xi^0, \boldsymbol{p}),\\ \dot{p}_t &= \partial_\alpha P(\xi^0, p)\dot{\boldsymbol{p}}^\alpha \qquad \text{(Einstein summation)}. \end{aligned}$$

Furthermore, we know (see 9.1.3)

$$\boldsymbol{U}(P) = \frac{\partial_0 P}{|\partial_0 P|}.$$

Consequently,

$$\begin{aligned} |\boldsymbol{\pi}_{\boldsymbol{U}(p_t)} \cdot \dot{p}_t|^2 &= |\dot{p}_t|^2 + |\boldsymbol{U}(p_t) \cdot \dot{p}_t|^2 = \\ &= \left(\partial_\alpha P \cdot \partial_\beta P + \frac{(\partial_0 P \cdot \partial_\alpha P)(\partial_0 P \cdot \partial_\beta P)}{|\partial_0 P|^2} \right) (\xi^0, \boldsymbol{p})\dot{p}^\alpha \dot{p}^\beta. \end{aligned}$$

Let us put

$$\boldsymbol{g}_{ik} := \partial_i P \cdot \partial_k P \qquad (i,k = 0,1,2,3).$$

Taking into account that $\boldsymbol{g}_{00}$ is negative, we see that

$$\boldsymbol{h}_{\alpha\beta} := \boldsymbol{g}_{\alpha\beta} - \frac{\boldsymbol{g}_{0\alpha}\boldsymbol{g}_{0\beta}}{\boldsymbol{g}_{00}} \qquad (\alpha, \beta = 1,2,3)$$

is the 'metric tensor' in the $\boldsymbol{U}$-space, i.e. a curve in the $\boldsymbol{U}$-space parametrized by $\boldsymbol{p}$ at an $\mathcal{S}$-instant coordinatized by ξ^0 has length

$$\int \sqrt{\boldsymbol{h}_{\alpha\beta}(\xi^0, \boldsymbol{p}(a))\dot{p}^\alpha(a)\dot{p}^\beta(a)}da.$$

9.4.2. Note that $\boldsymbol{g}_{ik}$ is a function from $\mathbb{R}^4$ into $\mathbb{T}\otimes\mathbb{T}$.

We know that $\big(\partial_i P(\xi) \mid i = 0,1,2,3\big)$ is a basis in $\mathbf{M}$ (the local basis at $P(\xi)$ (see VI.5.6)). Thus, according to V.4.21, $\big(\boldsymbol{g}_{ik}(\xi) \mid i,k = 0,1,2,3\big)$ is the coordinatized form of $\boldsymbol{g}$ corresponding to this basis. More precisely, we get those formulae choosing an $\boldsymbol{s}\in\mathbb{T}^+$ and putting

$$g_{ik}(\xi) := \frac{\boldsymbol{g}_{ik}(\xi)}{\boldsymbol{s}^2}.$$

9.5. Exercises

1. Let K be a coordinatization corresponding to a reference system whose observer is $\boldsymbol{U}$. Demonstrate that the coordinatized form of $\boldsymbol{U}$ according to K is the constant mapping $(1,\mathbf{0})$. (By definition, $(DK\cdot\boldsymbol{U})\circ K^{-1}$ is the coordinatized form of $\boldsymbol{U}$ according to K, see VI.5).

2. Take the uniformly accelerated observer $\boldsymbol{U}$ treated in 6.5. Define a Lorentzian reference system with arbitrary spacetime unit $\boldsymbol{s}$, origin o and with a spacetime basis such that $\boldsymbol{e}_0 := \boldsymbol{s}\boldsymbol{U}(o)$, $\boldsymbol{e}_1 := \boldsymbol{s}\frac{\boldsymbol{a}_o}{|\boldsymbol{a}_o|}$, $\boldsymbol{e}_2$ and $\boldsymbol{e}_3$ are arbitrary. Demonstrate that $\boldsymbol{U}$ will have the coordinatized form

$$(\xi^0,\xi^1,\xi^2,\xi^3) \mapsto \left(\sqrt{1+(\alpha\xi^0)^2}, \alpha\xi^0, 0, 0\right)$$

where α is the number for which $|\boldsymbol{a}_o| = \alpha\frac{1}{\boldsymbol{s}}$ holds.

The $\boldsymbol{U}$-line passing through $o + \sum_{i=0}^{3}\xi^i\boldsymbol{e}_i$ becomes

$$\left\{ \left(\frac{1}{\alpha}\sinh\alpha s,\ \xi^1 + \frac{1}{\alpha}(\cosh\alpha s - 1),\ \xi^2,\ \xi^3 \right) \,\middle|\, s\in\mathbb{R} \right\}.$$

3. Take the uniformly accelerated observer $\boldsymbol{U}$ treated in 6.6. Define a Lorentzian reference system with arbitrary spacetime unit $\boldsymbol{s}$, origin o and with a spacetime basis such that $\boldsymbol{e}_0 := \boldsymbol{s}\boldsymbol{U}(o)$, $\boldsymbol{e}_1 := \boldsymbol{s}\frac{\boldsymbol{a}_o}{|\boldsymbol{a}_o|}$, $\boldsymbol{e}_2$ and $\boldsymbol{e}_3$ are arbitrary. Demonstrate that $\boldsymbol{U}$ will have the coordinatized form

$$\{(\xi^0,\xi^1,\xi^2,\xi^3)\in\mathbb{R}^4 \mid \xi^1 > |\xi^0|\} \to \mathbb{R}^4,$$

$$(\xi^0,\xi^1,\xi^2,\xi^3) \mapsto \frac{1}{\sqrt{-(\xi^0)^2 + (\xi^1)^2}}(\xi^1,\xi^0,0,0).$$

The $\boldsymbol{U}$-line passing through $o+\sum_{i=0}^{3}\xi^i\boldsymbol{e}_i$ becomes

$$\left\{\left(\frac{1}{\xi^1}\sinh\xi^1 s,\ \xi^1+\frac{1}{\xi^1}(\cosh\xi^1 s-1),\ \xi^2,\ \xi^3\right)\ \middle|\ s\in\mathbb{R}\right\}.$$

4. Take the uniformly rotating observer $\boldsymbol{U}$ treated in 6.7. Define a Lorentzian reference system with arbitrary spacetime unit $\boldsymbol{s}$, origin o and with a spacetime basis such that $\boldsymbol{e}_0:=\boldsymbol{s}\boldsymbol{U}(o)$, $\boldsymbol{e}_3$ positively oriented in $\operatorname{Ker}\Omega$, $|\boldsymbol{e}_3|=\boldsymbol{s}$, $\boldsymbol{e}_1$ and $\boldsymbol{e}_2$ arbitrary. Demonstrate that $\boldsymbol{U}$ will have the coordinatized form

$$(\xi^0,\xi^1,\xi^2,\xi^3)\mapsto\left(\sqrt{1+\omega^2\left((\xi^1)^2+(\xi^2)^2\right)},\ -\omega\xi^2,\ \omega\xi^1,\ 0\right)$$

where ω is the number for which $|\Omega|=\omega\frac{1}{\boldsymbol{s}}$ holds.

The $\boldsymbol{U}$-line passing through $o+\sum_{i=0}^{3}\xi^i\boldsymbol{e}_i$ becomes

$$\left\{\left(s\sqrt{1+\omega^2\left((\xi^1)^2+(\xi^2)^2\right)},\ \xi^1\cos\omega s-\xi^2\sin\omega s,\ \xi^1\sin\omega s+\xi^2\cos\omega s,\ \xi^3\right)\ \middle|\ s\in\mathbb{R}\right\}.$$

5. Take the uniformly rotating observer $\boldsymbol{U}$ treated in 6.8. Define a Lorentzian reference system with arbitrary spacetime unit $\boldsymbol{s}$, origin o and with a spacetime basis such that $\boldsymbol{e}_0:=\boldsymbol{s}\boldsymbol{U}(o)$, $\boldsymbol{e}_3$ positively oriented in $\operatorname{Ker}\Omega$, $|\boldsymbol{e}_3|=\boldsymbol{s}$, $\boldsymbol{e}_1$ and $\boldsymbol{e}_2$ arbitrary. Demonstrate that $\boldsymbol{U}$ will have the coordinatized form

$$\left\{(\xi^0,\xi^1,\xi^2,\xi^3)\in\mathbb{R}^4\ \middle|\ \omega^2\left((\xi^1)^2+(\xi^2)^2\right)<1\right\}\to\mathbb{R}^4,$$
$$(\xi^0,\xi^1,\xi^2,\xi^3)\mapsto\frac{1}{\sqrt{1-\omega^2\left((\xi^1)^2+(\xi^2)^2\right)}}(1,-\omega\xi^2,\omega\xi^1,0)$$

where ω is the number for which $|\Omega|=\omega\frac{1}{\boldsymbol{s}}$ holds.

The $\boldsymbol{U}$-line passing through $o+\sum_{i=0}^{3}\xi^i\boldsymbol{e}_i$ becomes

$$\left\{\left(t(s),\ \xi^1\cos\omega t(s)-\xi^2\sin\omega t(s),\ \xi^1\sin\omega t(s)+\xi^2\cos\omega t(s),\ \xi^3\right)\ \middle|\ s\in\mathbb{R}\right\},$$

where

$$t(s):=\frac{s}{\sqrt{1-\omega^2\left((\xi^1)^2+(\xi^2)^2\right)}}.$$

6. Find necessary and sufficient conditions that two affine reference systems be equivalent.
7. Take the uniformly accelerated observer treated in 6.5, consider u_o-synchronization and find a convenient reference system for them.
8. A reference system defined for a uniformly accelerated observer cannot be equivalent to a reference system defined for a uniformly rotating observer.

10. Spacetime groups *

10.1. The Lorentz group

10.1.1. We shall deal with linear maps from $\mathbf{M}$ into $\mathbf{M}$, permanently using the identification $\mathrm{Lin}(\mathbf{M}) \equiv \mathbf{M} \otimes \mathbf{M}^*$.

Recall the notion of g-adjoints, g-orthogonal maps, g-antisymmetric maps (V.1.5, V.2.7).

Definition.

$$\mathcal{L} := \{\boldsymbol{L} \in \mathbf{M} \otimes \mathbf{M}^* \mid \boldsymbol{L}^* \cdot \boldsymbol{L} = \mathbf{1}_{\mathbf{M}}\} = \mathcal{O}(\boldsymbol{g})$$

is called the *Lorentz group*; its elements are the *Lorentz transformations*.

If $\boldsymbol{L}$ is a Lorentz transformation then

$$\mathrm{ar}\,\boldsymbol{L} := \begin{cases} +1 & \text{if } \boldsymbol{L} \text{ is arrow-preserving} \\ -1 & \text{if } \boldsymbol{L} \text{ is arrow-reversing} \end{cases}$$

is the *arrow* of $\boldsymbol{L}$ and

$$\mathrm{sign}\,\boldsymbol{L} := \begin{cases} +1 & \text{if } \boldsymbol{L}|_{\mathbf{S}_u} \text{ is orientation preserving} \\ -1 & \text{if } \boldsymbol{L}|_{\mathbf{S}_u} \text{ is orientation-reversing} \end{cases}$$

is the *sign* of $\boldsymbol{L}$ where $\boldsymbol{u}$ is an arbitrary element of $V(1)$.

Let us put

$$\begin{aligned}
\mathcal{L}^{+\rightarrow} &:= \{\boldsymbol{L} \in \mathcal{L} \mid \mathrm{sign}\,\boldsymbol{L} = \mathrm{ar}\,\boldsymbol{L} = 1\},\\
\mathcal{L}^{+\leftarrow} &:= \{\boldsymbol{L} \in \mathcal{L} \mid \mathrm{sign}\,\boldsymbol{L} = -\mathrm{ar}\,\boldsymbol{L} = 1\},\\
\mathcal{L}^{-\rightarrow} &:= \{\boldsymbol{L} \in \mathcal{L} \mid \mathrm{sign}\,\boldsymbol{L} = -\mathrm{ar}\,\boldsymbol{L} = -1\},\\
\mathcal{L}^{-\leftarrow} &:= \{\boldsymbol{L} \in \mathcal{L} \mid \mathrm{sign}\,\boldsymbol{L} = \mathrm{ar}\,\boldsymbol{L} = -1\}.
\end{aligned}$$

$\mathcal{L}^{+\rightarrow}$ is called the *proper Lorentz group*.

10.1.2. (*i*) From VII.5 we infer that the Lorentz group is a six-dimensional Lie group having the Lie algebra

$$\mathbf{La}(\mathcal{L}) = \mathrm{A}(g) = \{\boldsymbol{H} \in \mathbf{M} \otimes \mathbf{M}^* | \, \boldsymbol{H}^* = -\boldsymbol{H}\}.$$

(*ii*) S, T and L, the set of spacelike vectors, the set of timelike vectors and the set of lightlike vectors are invariant under Lorentz transformations. The arrow of a Lorentz transformation $\boldsymbol{L}$ is $+1$ if and only if $T^{\rightarrow}$, the set of future directed timelike vectors, is invariant for $\boldsymbol{L}$.

(*iii*) The sign of Lorentz transformations is correctly defined. Indeed, if $\boldsymbol{u} \in V(1)$ then $\boldsymbol{L}$ maps $\mathbf{S}_{\boldsymbol{u}}$ onto $\mathbf{S}_{(\mathrm{ar}\boldsymbol{L})\boldsymbol{L}\cdot\boldsymbol{u}}$; these two linear subspaces are oriented according to 1.3.4. It is not hard to see that if the restriction of $\boldsymbol{L}$ onto $\mathbf{S}_{\boldsymbol{u}}$ is orientation preserving for some $\boldsymbol{u}$ then it is orientation preserving for all $\boldsymbol{u}$.

(*iv*) The mappings $\mathcal{L} \to \{-1, 1\}$, $\boldsymbol{L} \mapsto \mathrm{ar}\boldsymbol{L}$ and $\mathcal{L} \to \{-1, 1\}$, $\boldsymbol{L} \mapsto \mathrm{sign}\boldsymbol{L}$ are continuous group homomorphisms. As a consequence, the Lorentz group is disconnected. We shall see in 10.2.4 that the proper Lorentz group $\mathcal{L}^{+\rightarrow}$ is connected. It is quite trivial that if $\boldsymbol{L} \in \mathcal{L}^{+\leftarrow}$ then $\boldsymbol{L} \cdot \mathcal{L}^{+\rightarrow} = \mathcal{L}^{+\leftarrow}$ and similar assertions hold for $\mathcal{L}^{-\rightarrow}$ and $\mathcal{L}^{-\leftarrow}$ as well. Consequently, the Lorentz group has four connected components, the four subsets given in Definition 10.1.1.

From these four components only $\mathcal{L}^{+\rightarrow}$—the proper Lorentz group—is a subgroup; nevertheless, the union of an arbitrary component and of the proper Lorentz group is a subgroup as well.

$\mathcal{L}^{\rightarrow} := \mathcal{L}^{+\rightarrow} \cup \mathcal{L}^{-\rightarrow}$ is called the *orthochronous Lorentz group.*

(*v*) The arrow of $\boldsymbol{L}$ is $+1$ if and only if $T^{\rightarrow}$, the set of future directed timelike vectors is invariant for $\boldsymbol{L}$:

$$\begin{aligned} &\text{if } \mathrm{ar}\boldsymbol{L} = 1 && \text{then } \boldsymbol{L}[T^{\rightarrow}] = T^{\rightarrow}, \quad \boldsymbol{L}[T^{\leftarrow}] = T^{\leftarrow}, \\ &\text{if } \mathrm{ar}\boldsymbol{L} = -1 && \text{then } \boldsymbol{L}[T^{\rightarrow}] = T^{\leftarrow}, \quad \boldsymbol{L}[T^{\leftarrow}] = T^{\rightarrow}. \end{aligned}$$

Moreover, the elements of $\mathcal{L}^{+\rightarrow}$ and $\mathcal{L}^{-\leftarrow}$ preserve the orientation of $\mathbf{M}$, whereas the elements of $\mathcal{L}^{+\leftarrow}$ and $\mathcal{L}^{-\rightarrow}$ reverse the orientation.

10.1.3. $\mathbf{M}$ is of even dimensions, thus $-\mathbf{1}_{\mathbf{M}}$ is orientation-preserving. Evidently, $-\mathbf{1}_{\mathbf{M}}$ is in $\mathcal{L}^{-\leftarrow}$; it is called the *inversion of spacetime vectors.* We have that $\mathcal{L}^{-\leftarrow} = (-\mathbf{1}_{\mathbf{M}}) \cdot \mathcal{L}^{+\rightarrow}$.

We have seen previously that the elements of $\mathcal{L}^{+\leftarrow}$ invert in some sense the timelike vectors and do not invert the spacelike vectors; the elements of $\mathcal{L}^{-\rightarrow}$ invert in some sense the spacelike vectors and do not invert the timelike vectors. However, we cannot select an element of $\mathcal{L}^{+\leftarrow}$ and an element of $\mathcal{L}^{-\rightarrow}$ that we could consider to be the time inversion and the space inversion, respectively.

For each $\boldsymbol{u} \in V(1)$ we can give a $\boldsymbol{u}$-timelike inversion and a $\boldsymbol{u}$-spacelike inversion as follows.

The $\boldsymbol{u}$-*timelike inversion* $\boldsymbol{T_u} \in \mathcal{L}^{+\leftarrow}$ inverts the vectors parallel to $\boldsymbol{u}$ and leaves invariant the spacelike vectors $\boldsymbol{g}$-orthogonal to $\boldsymbol{u}$:

$$\boldsymbol{T_u} \cdot \boldsymbol{u} := -\boldsymbol{u} \quad \text{and} \quad \boldsymbol{T_u} \cdot \boldsymbol{q} := \boldsymbol{q} \quad \text{for} \quad \boldsymbol{q} \in \mathbf{S}_{\boldsymbol{u}}.$$

In general,

$$\boldsymbol{T_u} \cdot \boldsymbol{x} = \boldsymbol{u}(\boldsymbol{u} \cdot \boldsymbol{x}) + \boldsymbol{\pi_u} \cdot \boldsymbol{x} = 2\boldsymbol{u}(\boldsymbol{u} \cdot \boldsymbol{x}) + \boldsymbol{x} \qquad (\boldsymbol{x} \in \mathbf{M}),$$

i.e.

$$\boldsymbol{T_u} = \mathbf{1_M} + 2\boldsymbol{u} \otimes \boldsymbol{u}.$$

The $\boldsymbol{u}$-*spacelike inversion* $\boldsymbol{P_u} \in \mathcal{L}^{-\rightarrow}$ inverts the spacelike vectors $\boldsymbol{g}$-orthogonal to $\boldsymbol{u}$ and leaves invariant the vectors parallel to $\boldsymbol{u}$:

$$\boldsymbol{P_u} \cdot \boldsymbol{u} := \boldsymbol{u} \quad \text{and} \quad \boldsymbol{P_u} \cdot \boldsymbol{q} := -\boldsymbol{q} \quad \text{for} \quad \boldsymbol{q} \in \mathbf{S}_{\boldsymbol{u}}.$$

In general,

$$\boldsymbol{P_u} \cdot \boldsymbol{x} = -\boldsymbol{u}(\boldsymbol{u} \cdot \boldsymbol{x}) - \boldsymbol{\pi_u} \cdot \boldsymbol{x} = -2\boldsymbol{u}(\boldsymbol{u} \cdot \boldsymbol{x}) - \boldsymbol{x} \qquad (\boldsymbol{x} \in \mathbf{M}),$$

i.e.

$$\boldsymbol{P_u} = -\mathbf{1_M} - 2\boldsymbol{u} \otimes \boldsymbol{u}.$$

We easily deduce the following equalities:

$$\boldsymbol{T_u}^{-1} = \boldsymbol{T_u}, \qquad \boldsymbol{P_u}^{-1} = \boldsymbol{P_u},$$
$$-\boldsymbol{T_u} = \boldsymbol{P_u},$$
$$\boldsymbol{T_u} \cdot \boldsymbol{P_u} = \boldsymbol{P_u} \cdot \boldsymbol{T_u} = -\mathbf{1_M}.$$

10.1.4. For $\boldsymbol{u} \in V(1)$ let us consider the Euclidean vector space $(\mathbf{S}_{\boldsymbol{u}}, \mathbb{T}, \boldsymbol{h_u})$ where $\boldsymbol{h_u}$ is the restriction of $\boldsymbol{g}$ onto $\mathbf{S}_{\boldsymbol{u}} \times \mathbf{S}_{\boldsymbol{u}}$. The $\boldsymbol{h_u}$-orthogonal group, $\mathcal{O}(\boldsymbol{h_u})$, called also the group of $\boldsymbol{u}$-*spacelike orthogonal transformations,* can be identified with a subgroup of the Lorentz group:

$$\mathcal{O}(\boldsymbol{h_u}) \equiv \{\boldsymbol{L} \in \mathcal{L}^{\rightarrow} \mid \boldsymbol{L} \cdot \boldsymbol{u} = \boldsymbol{u}\}.$$

The Lorentz group is an analogue of the Galilean group and we have already seen a number of their common properties. However, as concerns their relation to three-dimensional orthogonal groups, they differ significantly.

In the nonrelativistic case there is a *single* three-dimensional orthogonal group in question, $\mathcal{O}(\boldsymbol{h})$, and it can be injected into the Galilean group in different ways according to different velocity values. Moreover, $\boldsymbol{L} \mapsto \boldsymbol{L}|_{\mathbf{S}}$ is a surjective group homomorphism from the Galilean group onto the three-dimensional orthogonal group.

In the relativistic case there are a lot of three-dimensional orthogonal groups, being subgroups of the Lorentz group; one corresponds to each velocity value. Note that, for all $\boldsymbol{u}$, $\boldsymbol{L} \mapsto \boldsymbol{L}|_{\mathbf{S}_{\boldsymbol{u}}}$ is not a surjective group homomorphism from $\mathcal{L}$ onto $\mathcal{O}(\boldsymbol{h}_{\boldsymbol{u}})$; indeed, $\mathbf{S}_{\boldsymbol{u}}$ is invariant for $\boldsymbol{L}$ if and only if $\boldsymbol{L} \cdot \boldsymbol{u} = (\mathrm{ar}\,\boldsymbol{L})\boldsymbol{u}$.

As a consequence, there is not either a 'special Lorentz group' or a '$\boldsymbol{u}$-special Lorentz group' which would be the kernel of the group homomorphism $\boldsymbol{L} \mapsto \boldsymbol{L}|_{\mathbf{S}_{\boldsymbol{u}}}$.

10.1.5. The problem is that, in general, $\mathbf{S}_{\boldsymbol{u}}$ is not invariant for a Lorentz transformation $\boldsymbol{L}$; more closely, $\boldsymbol{L}$ maps $\mathbf{S}_{\boldsymbol{u}}$ onto $\mathbf{S}_{(\mathrm{ar}\,\boldsymbol{L})\boldsymbol{L}\cdot\boldsymbol{u}}$ for all $\boldsymbol{u} \in V(1)$. Let us try to rule out this uneasiness with the aid of the corresponding Lorentz boost $\boldsymbol{L}(\boldsymbol{u}, (\mathrm{ar}\,\boldsymbol{L})\boldsymbol{L}\cdot\boldsymbol{u})$ which maps $\mathbf{S}_{(\mathrm{ar}\,\boldsymbol{L})\boldsymbol{L}\cdot\boldsymbol{u}}$ onto $\mathbf{S}_{\boldsymbol{u}}$ in a 'handsome' way. A simple calculation yields the following result.

Proposition. For all Lorentz transformations $\boldsymbol{L}$ and for all $\boldsymbol{u} \in V(1)$,

$$\boldsymbol{R}(\boldsymbol{L}, \boldsymbol{u}) := (\mathrm{ar}\,\boldsymbol{L})\boldsymbol{L}(\boldsymbol{u}, (\mathrm{ar}\,\boldsymbol{L})\boldsymbol{L}\cdot\boldsymbol{u}) \cdot \boldsymbol{L} =$$

$$= (\mathrm{ar}\,\boldsymbol{L})\boldsymbol{L} + \frac{\big(\boldsymbol{u} + (\mathrm{ar}\,\boldsymbol{L})\boldsymbol{L}\cdot\boldsymbol{u}\big) \otimes \big((\mathrm{ar}\,\boldsymbol{L})\boldsymbol{L})^{-1}\cdot\boldsymbol{u} + \boldsymbol{u}\big)}{1 - \boldsymbol{u}\cdot(\mathrm{ar}\,\boldsymbol{L})\boldsymbol{L}\cdot\boldsymbol{u}} - 2\boldsymbol{u}\otimes\boldsymbol{u}$$

is an element of $\mathcal{O}(\boldsymbol{h}_{\boldsymbol{u}})$. ■

This suggests the idea that an orthochronous Lorentz transformation $\boldsymbol{L}$ should be considered 'special' if $\boldsymbol{R}(\boldsymbol{L}, \boldsymbol{u})|_{\mathbf{S}_{\boldsymbol{u}}} = \mathbf{1}_{\mathbf{S}_{\boldsymbol{u}}}$; then $\boldsymbol{L}(\boldsymbol{u}, (\mathrm{ar}\,\boldsymbol{L})\boldsymbol{L}\cdot\boldsymbol{u}) \cdot \boldsymbol{L} = \mathbf{1}_{\mathrm{M}}$ and consequently $\boldsymbol{L}$ is a Lorentz boost.

Thus Lorentz boosts can be regarded as counterparts of special Galilean transformations. That is why we call them *special Lorentz transformations* as well. However, it is very important that the special Lorentz transformations (Lorentz boosts) *do not form a subgroup* (see 1.3.9).

Note that our result can be formulated as follows: given an arbitrary $\boldsymbol{u} \in V(1)$, every Lorentz transformation $\boldsymbol{L}$ can be decomposed into the product of a special Lorentz transformation and a $\boldsymbol{u}$-spacelike orthogonal transformation, multiplied by the arrow of $\boldsymbol{L}$:

$$\boldsymbol{L} = (\mathrm{ar}\,\boldsymbol{L})\boldsymbol{L}\big((\mathrm{ar}\,\boldsymbol{L})\boldsymbol{L}\cdot\boldsymbol{u}, \boldsymbol{u}\big) \cdot \boldsymbol{R}(\boldsymbol{L}, \boldsymbol{u}).$$

10.1.6. It is worth mentioning that the product of the $\boldsymbol{u}'$-timelike ($\boldsymbol{u}'$-spacelike) inversion and the $\boldsymbol{u}$-timelike ($\boldsymbol{u}$-spacelike) inversion is a special Lorentz transformation. Since

$$\boldsymbol{T}_{\boldsymbol{u}}^{-1} = \boldsymbol{T}_{\boldsymbol{u}} = -\boldsymbol{P}_{\boldsymbol{u}} = \mathbf{1}_{\mathrm{M}} + 2\boldsymbol{u}\otimes\boldsymbol{u},$$

we find—because of $-\boldsymbol{u} - 2(\boldsymbol{u}\cdot\boldsymbol{u}')\boldsymbol{u}' = \boldsymbol{u} - \frac{2\boldsymbol{v}_{\boldsymbol{u}\boldsymbol{u}'}}{\sqrt{1-|\boldsymbol{v}_{\boldsymbol{u}\boldsymbol{u}'}|^2}}$—that $\boldsymbol{T}_{\boldsymbol{u}'} \cdot \boldsymbol{T}_{\boldsymbol{u}}^{-1} = \boldsymbol{P}_{\boldsymbol{u}'} \cdot \boldsymbol{P}_{\boldsymbol{u}}^{-1}$ is the Lorentz boost from $\boldsymbol{u}$ to $\boldsymbol{u} - \frac{2\boldsymbol{v}_{\boldsymbol{u}\boldsymbol{u}'}}{\sqrt{1-|\boldsymbol{v}_{\boldsymbol{u}\boldsymbol{u}'}|^2}}$.

10.1.7. (*i*) Take an $\boldsymbol{u} \in V(1)$ and a $\boldsymbol{0} \neq \boldsymbol{H} \in \mathrm{A}(\boldsymbol{g})$ for which $\boldsymbol{H} \cdot \boldsymbol{u} = \boldsymbol{0}$ holds. Then $\boldsymbol{H}^3 = -|\boldsymbol{H}|^2\boldsymbol{H}$ (V.4.18(*i*)) and we can repeat the proof of I.11.1.8 to have

$$\mathrm{e}^{\boldsymbol{H}} = \left(1_{\mathrm{M}} + \frac{\boldsymbol{H}^2}{|\boldsymbol{H}|^2}\right) + \frac{\boldsymbol{H}^2}{|\boldsymbol{H}|^2}\cos|\boldsymbol{H}| + \frac{\boldsymbol{H}}{|\boldsymbol{H}|}\sin|\boldsymbol{H}|$$

which is an element of $\mathcal{O}(\boldsymbol{h_u})$.

(*ii*) Take an $\boldsymbol{u} \in V(1)$ and a $\boldsymbol{0} \neq \boldsymbol{H} \in \mathrm{A}(\boldsymbol{g})$ whose kernel lies in $\mathbf{S}_{\boldsymbol{u}}$. Then $\boldsymbol{H}^3 = |\boldsymbol{H}|^2\boldsymbol{H}$ (V.4.18(*ii*)) and we can prove as in I.11.1.8 that

$$\mathrm{e}^{\boldsymbol{H}} = \left(1_{\mathrm{M}} - \frac{\boldsymbol{H}^2}{|\boldsymbol{H}|^2}\right) + \frac{\boldsymbol{H}^2}{|\boldsymbol{H}|^2}\cosh|\boldsymbol{H}| + \frac{\boldsymbol{H}}{|\boldsymbol{H}|}\sinh|\boldsymbol{H}|.$$

We can demonstrate this is a Lorentz boost. Recall that there is an $\boldsymbol{n} \in \frac{\mathbf{S}_{\boldsymbol{u}}}{\mathbb{T}}$, $|\boldsymbol{n}| = 1$ such that $\boldsymbol{H} = \alpha\boldsymbol{u}\wedge\boldsymbol{n}$, where $\alpha := |\boldsymbol{H}|$. Then $\boldsymbol{H}^2 = \alpha^2(\boldsymbol{n}\otimes\boldsymbol{n} - \boldsymbol{u}\otimes\boldsymbol{u})$ and executing some calculations we obtain:

Proposition. Let $\boldsymbol{u} \in V(1)$, $\boldsymbol{n} \in \frac{\mathbf{S}_{\boldsymbol{u}}}{\mathbb{T}}$, $|\boldsymbol{n}| = 1$, and $\alpha \in \mathbb{R}$. Then

$$\exp\bigl(\alpha(\boldsymbol{u}\wedge\boldsymbol{n})\bigr) = \boldsymbol{L}(\boldsymbol{u}\cosh\alpha + \boldsymbol{n}\sinh\alpha,\, \boldsymbol{u}).$$

10.1.8. Originally the Lorentz transformations are defined to be linear maps from $\mathbf{M}$ into $\mathbf{M}$. In the usual way, we can consider them to be linear maps from $\frac{\mathbf{M}}{\mathbb{T}}$ into $\frac{\mathbf{M}}{\mathbb{T}}$ as we already did in the preceding paragraphs as well.

$V(1)$ is invariant under orthochronous Lorentz transformations. However, contrary to the nonrelativistic case, here $V(1)$ is not an affine subspace, hence we cannot say anything similar to those in I.11.3.8.

This, too, indicates that the structure of the Lorentz group is more complicated than the structure of the Galilean group.

10.2. The $\boldsymbol{u}$-split Lorentz group

10.2.1. The Lorentz transformations, being elements of $\mathbf{M}\otimes\mathbf{M}^*$, are split by velocity values according to 8.1. These splittings are significantly more complicated than the splittings of Galilean transformations.

Let us start with the splittings of Lorentz boosts. The map $\boldsymbol{\xi}_{\boldsymbol{u}'\boldsymbol{u}}$ defined in 7.1.4 is such a splitting:

$$\boldsymbol{\xi}_{\boldsymbol{u}'\boldsymbol{u}} = \boldsymbol{\xi}_{\boldsymbol{u}} \cdot \boldsymbol{L}(\boldsymbol{u}, \boldsymbol{u}') \cdot \boldsymbol{\xi}_{\boldsymbol{u}}^{-1} \qquad \bigl(\boldsymbol{u}, \boldsymbol{u}' \in V(1)\bigr).$$

For a $\boldsymbol{u} \in V(1)$, it is convenient to introduce the notations

$$B_{\boldsymbol{u}} := \left\{\boldsymbol{v} \in \frac{\mathbf{S}_{\boldsymbol{u}}}{\mathbb{T}}\,\middle|\, |\boldsymbol{v}| < 1\right\}$$

and

$$\kappa(\boldsymbol{v}) := \frac{1}{\sqrt{1-|\boldsymbol{v}|^2}}, \qquad \boldsymbol{D}(\boldsymbol{v}) := \frac{1}{\kappa(\boldsymbol{v})}\left(\mathbf{1}_{\mathbf{S}_u} + \frac{\kappa(\boldsymbol{v})^2}{\kappa(\boldsymbol{v})+1}\boldsymbol{v}\otimes\boldsymbol{v}\right)$$

for $\boldsymbol{v} \in B_{\boldsymbol{u}}$. It is worth mentioning the relation

$$\frac{\kappa(\boldsymbol{v})^2}{\kappa(\boldsymbol{v})+1} = \frac{\kappa(\boldsymbol{v})-1}{|\boldsymbol{v}|^2}.$$

Applying the usual matrix forms we have

$$\xi_{u'u} = \kappa(\boldsymbol{v}_{u'u})\begin{pmatrix} 1 & -\boldsymbol{v}_{u'u} \\ -\boldsymbol{v}_{u'u} & \boldsymbol{D}(\boldsymbol{v}_{u'u}) \end{pmatrix}.$$

A simple calculation yields that

$$\xi_u \cdot \boldsymbol{L}(\boldsymbol{u}',\boldsymbol{u}) \cdot \xi_u^{-1} = \xi_{u'u}^{-1} = \kappa(\boldsymbol{v}_{u'u})\begin{pmatrix} 1 & \boldsymbol{v}_{u'u} \\ \boldsymbol{v}_{u'u} & \boldsymbol{D}(\boldsymbol{v}_{u'u}) \end{pmatrix}.$$

10.2.2. Now taking an arbitrary Lorentz transformation $\boldsymbol{L}$ and a $\boldsymbol{u} \in V(1)$, we make the following manipulation:

$$\xi_u \cdot \boldsymbol{L} \cdot \xi_u^{-1} = \left(\xi_u \cdot \boldsymbol{L}(\boldsymbol{u}, (\operatorname{ar}\boldsymbol{L})\boldsymbol{L}\cdot\boldsymbol{u})^{-1} \cdot \xi_u^{-1}\right) \cdot \left(\xi_u \cdot \boldsymbol{L}(\boldsymbol{u}, (\operatorname{ar}\boldsymbol{L})\boldsymbol{L}\cdot\boldsymbol{u}) \cdot \boldsymbol{L} \cdot \xi_u^{-1}\right).$$

The first factor on the right-hand side equals $\xi^{-1}_{(\operatorname{ar}\boldsymbol{L})\boldsymbol{L}\cdot\boldsymbol{u},\boldsymbol{u}}$. As concerns the second factor, we find that

$$\begin{aligned}\left(\xi_u \cdot \boldsymbol{L}(\boldsymbol{u}, (\operatorname{ar}\boldsymbol{L})\boldsymbol{L}\cdot\boldsymbol{u}) \cdot \boldsymbol{L} \cdot \xi_u^{-1}\right)(\boldsymbol{t},\boldsymbol{q}) &= \left(\xi_u \cdot \boldsymbol{L}(\boldsymbol{u}, (\operatorname{ar}\boldsymbol{L})\boldsymbol{L}\cdot\boldsymbol{u}) \cdot \boldsymbol{L}\right)(\boldsymbol{u}\boldsymbol{t}+\boldsymbol{q}) = \\ = \xi_u \cdot \left((\operatorname{ar}\boldsymbol{L})\boldsymbol{u}\boldsymbol{t} + \boldsymbol{R}(\boldsymbol{L},\boldsymbol{u})\cdot\boldsymbol{q}\right) &= \left((\operatorname{ar}\boldsymbol{L})\boldsymbol{t}, \boldsymbol{R}(\boldsymbol{L},\boldsymbol{u})\cdot\boldsymbol{q}\right)\end{aligned}$$

for all $(\boldsymbol{t},\boldsymbol{q}) \in \mathbb{T}\times\mathbf{S}_u$, i.e. the second factor has the matrix form

$$\begin{pmatrix} \operatorname{ar}\boldsymbol{L} & 0 \\ 0 & \boldsymbol{R}(\boldsymbol{L},\boldsymbol{u}) \end{pmatrix}.$$

As a consequence, we see that the following definition describes the $\boldsymbol{u}$-split form of Lorentz transformations.

Definition. The *$\boldsymbol{u}$-split Lorentz group* is

$$\left\{\kappa(\boldsymbol{v})\begin{pmatrix} 1 & \boldsymbol{v} \\ \boldsymbol{v} & \boldsymbol{D}(\boldsymbol{v}) \end{pmatrix}\begin{pmatrix} \pm 1 & 0 \\ 0 & \pm\boldsymbol{R} \end{pmatrix} \;\middle|\; \boldsymbol{v}\in B_u,\ \boldsymbol{R}\in SO(\boldsymbol{h}_u)\right\}.$$

Its elements are called *$\boldsymbol{u}$-split Lorentz transformations.* ■

The $\boldsymbol{u}$-split Lorentz transformations can be regarded as linear maps $\mathbb{T}\times\mathbf{S}_u \to \mathbb{T}\times\mathbf{S}_u$; the one in the definition makes the correspondence

$$(\boldsymbol{t},\boldsymbol{q}) \mapsto \kappa(\boldsymbol{v})\left(\pm\boldsymbol{t} \pm \boldsymbol{v}\cdot\boldsymbol{R}\cdot\boldsymbol{q},\ \pm\boldsymbol{v}\boldsymbol{t} \pm \boldsymbol{D}(\boldsymbol{v})\cdot\boldsymbol{R}\cdot\boldsymbol{q}\right).$$

The u-split Lorentz group is a six-dimensional Lie group having the Lie algebra

$$\left\{ \begin{pmatrix} 0 & v \\ v & A \end{pmatrix} \middle| \; v \in \frac{\mathbf{S}_u}{\mathbb{T}}, \; A \in \mathrm{A}(h_u) \right\}.$$

10.2.3. The splitting according to u establishes a Lie-group isomorphism between the Lorentz group and the u-split Lorentz group. The isomorphisms corresponding to different u' and u are different.

The difference of splittings can be seen by the usual transformation rule which is rather complicated; since here we need not it, we do not give the details.

10.2.4. The u-splitting sends the proper Lorentz group onto

$$\left\{ \kappa(v) \begin{pmatrix} 1 & v \\ v & D(v) \end{pmatrix} \begin{pmatrix} 1 & 0 \\ 0 & R \end{pmatrix} \middle| \; v \in B_u, \; R \in SO(h_u) \right\}$$

which is evidently a connected set. Since the u-splitting is a Lie group isomorphisms, $\mathcal{L}^{+\rightarrow}$ is connected as well.

10.2.5. We easily verify that

$$\left\{ \kappa(v) \begin{pmatrix} 1 & v \\ v & D(v) \end{pmatrix} \middle| \; v \in B_u \right\}$$

is not a subgroup of the u-split Lorentz group; this reflects the well-known fact that the Lorentz boosts do not form a subgroup.

10.2.6. The Lie algebra of the Lorentz group, too, consists of elements of $\mathbf{M} \otimes \mathbf{M}^*$, thus they are split by velocity values in the same way as the Lorentz transformations; evidently, their split form will be different.

If H is in the Lie algebra of the Lorentz group—i.e. H is a g-antisymmetric tensor—and $u \in V(1)$, then

$$\xi_u \cdot H \cdot \xi_u^{-1} = \begin{pmatrix} 0 & H \cdot u \\ H \cdot u & H - u \wedge H \cdot u \end{pmatrix}.$$

The splitting according to u establishes a Lie algebra isomorphism between the Lie algebra of the Lorentz group and the Lie algebra of the u-split Lorentz group. The isomorphisms corresponding to different u' and u are different.

10.3. Exercises

1. The Lorentz group is not transitive, i.e. for all $x \in \mathbf{M}$, $\{L \cdot x \mid L \in \mathcal{L}\} \neq \mathbf{M}$. What are the orbits of the Lorentz group?

2. The subgroup generated by the Lorentz boosts equals the proper Lorentz group.

3. Prove that the Lie algebra of $\mathcal{O}(\boldsymbol{h_u})$ equals $\{\boldsymbol{H} \in \mathrm{A}(\boldsymbol{g}) \mid \boldsymbol{H} \cdot \boldsymbol{u} = \mathbf{0}\}$ which can be identified with $\mathrm{A}(\boldsymbol{h_u})$.
4. What is the subgroup generated by $\{\boldsymbol{T_u} \mid \boldsymbol{u} \in V(1)\}$?
5. Prove that

$$\boldsymbol{\xi}_u \cdot \boldsymbol{T}_{u'} \cdot \boldsymbol{\xi}_u^{-1} = \boldsymbol{\xi}_{u,u''} \cdot \begin{pmatrix} -1 & 0 \\ 0 & \mathbf{1}_{\mathbf{S}_u} \end{pmatrix},$$

$$\boldsymbol{\xi}_u \cdot \boldsymbol{P}_{u'} \circ \boldsymbol{\xi}_u^{-1} = \boldsymbol{\xi}_{u,u''} \cdot \begin{pmatrix} 1 & 0 \\ 0 & -\mathbf{1}_{\mathbf{S}_u} \end{pmatrix}$$

where $\boldsymbol{u}'' := \boldsymbol{u} - 2\kappa\,(\boldsymbol{v_{uu'}})\,\boldsymbol{v_{uu'}}$.

10.4. The Poincaré group

10.4.1. Now we shall deal with affine maps $L\colon \mathrm{M} \to \mathrm{M}$; as usual, the linear map under L is denoted by $\boldsymbol{L}$.

Definition.

$$\mathcal{P} := \{L\colon \mathrm{M} \to \mathrm{M} \mid L \text{ is affine, } \boldsymbol{L} \in \mathcal{L}\}$$

is called the Poincaré group; its elements are the *Poincaré transformations.*

If L is a Poincaré transformation then

$$\operatorname{ar} L := \operatorname{ar} \boldsymbol{L}, \qquad \operatorname{sign} L := \operatorname{sign} \boldsymbol{L}.$$

$\mathcal{P}^{+\to}$, $\mathcal{P}^{+\leftarrow}$, $\mathcal{P}^{-\to}$ and $\mathcal{P}^{-\leftarrow}$ are the subsets of $\mathcal{P}$ consisting of elements whose underlying linear maps belong to $\mathcal{L}^{+\to}$, $\mathcal{L}^{+\leftarrow}$, $\mathcal{L}^{-\to}$ and $\mathcal{L}^{-\leftarrow}$, respectively.

$\mathcal{P}^{+\to}$ is called the *proper Poincaré group.* ■

According to VII.3.2(*ii*) we can state the following.

Proposition. The Poincaré group is a ten-dimensional Lie group; its Lie algebra consists of the affine maps $H : \mathrm{M} \to \mathbf{M}$ whose underlying linear map is in the Lie algebra of the Lorentz group:

$$\mathbf{La}(\mathcal{P}) = \{H \in \mathrm{Aff}(\mathrm{M}, \mathbf{M}) \mid \boldsymbol{H} \in \mathrm{A}(\boldsymbol{g})\}. \qquad ■$$

The proper Poincaré group is a connected subgroup of the Poincaré group. As regards $\mathcal{P}^{+\leftarrow}$, etc. we can repeat what we said about the components of the Lorentz group.

$\mathcal{P}^{\to} := \mathcal{P}^{+\to} \cup \mathcal{P}^{-\to}$ is called the *orthochronous Poincaré group.*

10.4.2. We can say that the elements of $\mathcal{P}^{-\leftarrow}$ invert spacetime in some sense but there is no element that we could call the spacetime inversion.

For every $o \in \mathbf{M}$ we can give the *o-centered spacetime inversion* in the well-known way (cf. I.11.6.2):

$$I_o(x) := o - (x - o) \qquad (x \in \mathrm{M}).$$

Similarly, we can say that in some sense the elements of $\mathcal{P}^{-\rightarrow}$ contain spacelike inversion and do not contain timelike inversion; the elements of $\mathcal{P}^{+\leftarrow}$ contain timelike inversion and do not contain spacelike inversion. However, *the* space inversion and *the* time inversion do not exist.

For every $o \in \mathrm{M}$ and $\boldsymbol{u} \in V(1)$ we can give the *o-centered* $\boldsymbol{u}$*-timelike inversion* and the *o-centered* $\boldsymbol{u}$*-spacelike inversion* as follows:

$$T_{\boldsymbol{u},o}(x) := o + \boldsymbol{T}_{\boldsymbol{u}} \cdot (x - o),$$
$$P_{\boldsymbol{u},o}(x) := o + \boldsymbol{P}_{\boldsymbol{u}} \cdot (x - o) \qquad (x \in \mathrm{M}).$$

10.4.3. The Poincaré transformations are mappings of spacetime. They play a fundamental role because the proper Poincaré transformations can be considered to be the strict automorphisms of the spacetime model. The following statement is quite trivial.

Proposition. $(F, \mathbf{1}_{\mathbb{T}})$ is a strict automorphism of the special relativistic space time model $(\mathrm{M}, \mathbb{T}, \boldsymbol{g})$ if and only if F is a proper Poincaré transformation.

10.4.4. The Lorentz group is not a subgroup of the Poincaré group. The mapping $\mathcal{P} \to \mathcal{L}$, $L \mapsto \boldsymbol{L}$ is a surjective Lie group homomorphism whose kernel is $\mathcal{T}n(\mathbf{M})$, the translation group of $\mathbf{M}$,

$$\mathcal{T}n(\mathbf{M}) = \{T_{\boldsymbol{a}} \mid \boldsymbol{a} \in \mathbf{M}\} = \{L \in \mathcal{P} \mid \boldsymbol{L} = \mathbf{1}_{\mathbf{M}}\}.$$

As we know, its Lie algebra is $\mathbf{M}$ regarded as the set of constant maps from M into $\mathbf{M}$ (VII.3.3).

For every $o \in \mathrm{M}$,

$$\mathcal{L}_o := \{L \in \mathcal{P} \mid L(o) = o\},$$

called the group of *o-centered Lorentz transformations,* is a subgroup of the Poincaré group; the restriction of the homomorphism $L \mapsto \boldsymbol{L}$ onto $\mathcal{L}_o$ is a bijection between $\mathcal{L}_o$ and $\mathcal{L}$.

In other words, given $o \in \mathrm{M}$, we can assign to every Lorentz transformation $\boldsymbol{L}$ the Poincaré transformation

$$x \mapsto o + \boldsymbol{L} \cdot (x - o),$$

called the *o-centered Lorentz transformation by* $\boldsymbol{L}$.

10.4.5. For every $\boldsymbol{u} \in V(1)$ we can define the subgroup of $\boldsymbol{u}$-timelike translations

$$\mathcal{T}n(\mathbb{T})_{\boldsymbol{u}} := \{T_{\boldsymbol{u}t} \mid \boldsymbol{t} \in \mathbb{T}\} \subset \mathcal{T}n(\mathbf{M})$$

and the subgroup of $\boldsymbol{u}$-spacelike translations

$$\mathcal{T}n(\mathbf{S}_{\boldsymbol{u}}) := \{T_{\boldsymbol{q}} \mid \boldsymbol{q} \in \mathbf{S}_{\boldsymbol{u}}\} \subset \mathcal{T}n(\mathbf{M})$$

10.4.6. For every $\boldsymbol{u} \in V(1)$ and $o \in \mathrm{M}$,

$$\mathcal{O}(\boldsymbol{h}_{\boldsymbol{u}})_o := \{L \in \mathcal{P}^{\rightarrow} \mid L(o) = o,\ \boldsymbol{L} \cdot \boldsymbol{u} = \boldsymbol{u}\},$$

called the group of *o-centered* $\boldsymbol{u}$*-spacelike orthogonal transformations,* is a subgroup of $\mathcal{P}^{\rightarrow}$.

In other words, given $(\boldsymbol{u}, o) \in V(1) \times \mathrm{M}$, we can assign to every $\boldsymbol{R} \in \mathcal{O}(\boldsymbol{h}_{\boldsymbol{u}})$ the Poincaré transformation

$$x \mapsto o - \boldsymbol{u}\big(\boldsymbol{u} \cdot (x - o)\big) + \boldsymbol{R} \cdot \boldsymbol{\pi}_{\boldsymbol{u}} \cdot (x - o),$$

called the *o-centered* $\boldsymbol{u}$*-spacelike orthogonal transformation by* $\boldsymbol{R}$.

10.5. The vectorial Poincaré group

10.5.1. Recall that for an arbitrary world point o, the vectorization of M with origin o, $O_o\colon \mathrm{M} \to \mathbf{M}$, $x \mapsto x - o$ is an affine bijection.

With the aid of such a vectorization we can 'vectorize' the Poincaré group as well: if L is a Poincaré transformation then $O_o \circ L \circ O_o^{-1}$ is an affine transformation of $\mathbf{M}$, represented by the matrix (see VI.2.4(*ii*) and Exercise V I.2.5.2)

$$\begin{pmatrix} 1 & \mathbf{0} \\ L(o) - o & \boldsymbol{L} \end{pmatrix}.$$

The Lie algebra of the Poincaré group consists of affine maps $H\colon \mathrm{M} \to \mathbf{M}$ where $\mathbf{M}$ is considered to be a *vector space* (the *sum* of such maps is a part of the Lie algebra structure). Thus the vectorization $H \circ O_o^{-1}$ is an affine map $\mathbf{M} \to \mathbf{M}$ where the range is regarded as a vector space. Then it is represented by the matrix (see VI.2.4(*iii*))

$$\begin{pmatrix} 0 & \mathbf{0} \\ H(o) & \boldsymbol{H} \end{pmatrix}.$$

10.5.2. Definition. The *vectorial Poincaré group* is

$$\left\{ \begin{pmatrix} 1 & \mathbf{0} \\ a & \boldsymbol{L} \end{pmatrix} \middle|\ a \in \mathbf{M},\ \boldsymbol{L} \in \mathcal{L} \right\}. \qquad \blacksquare$$

The vectorial Poincaré group is a ten-dimensional Lie group, its Lie algebra is the vectorization of the Lie algebra of the Poincaré group:

$$\left\{ \begin{pmatrix} 0 & \mathbf{0} \\ \boldsymbol{a} & \boldsymbol{H} \end{pmatrix} \middle| \; \boldsymbol{a} \in \mathbf{M},\ \boldsymbol{H} \in \mathbf{La}(\mathcal{L}) \right\}.$$

An advantage of this block matrix representation is that the commutator of two Lie algebra elements can be computed as the difference of their two products.

10.5.3. A vectorization of the Poincaré group is a Lie group isomorphism between the Poincaré group and the vectorial Poincaré group. The following transformation rule shows how the vectorizations depend on the world points serving as origins of the vectorization. Let o and o' be two world points; then

$$O_{o'} \circ O_o^{-1} = T_{o-o'} = \begin{pmatrix} 1 & 0 \\ (o-o') & \mathbf{1}_{\mathbf{M}} \end{pmatrix}$$

and

$$T_{o-o'} \cdot \begin{pmatrix} 1 & 0 \\ \boldsymbol{a} & \boldsymbol{L} \end{pmatrix} \cdot T_{o-o'}^{-1} = \begin{pmatrix} 1 & 0 \\ \boldsymbol{a} + (\boldsymbol{L} - \mathbf{1}_{\mathbf{M}}) \cdot (o' - o) & \boldsymbol{L} \end{pmatrix} \qquad (\boldsymbol{a} \in \mathbf{M},\ \boldsymbol{L} \in \mathcal{L}).$$

As concerns the corresponding Lie algebra isomorphisms, we have

$$\begin{pmatrix} 0 & \mathbf{0} \\ \boldsymbol{a} & \boldsymbol{H} \end{pmatrix} \cdot T_{o-o'}^{-1} = \begin{pmatrix} 0 & \mathbf{0} \\ \boldsymbol{a} + \boldsymbol{H} \cdot (o' - o) & \boldsymbol{H} \end{pmatrix} \qquad (\boldsymbol{a} \in \mathbf{M},\ \boldsymbol{H} \in \mathbf{La}(\mathcal{N})).$$

10.6. The $\boldsymbol{u}$-split Poincaré group

10.6.1. With the aid of the splitting corresponding to $\boldsymbol{u} \in V(1)$, we send the transformations of $\mathbf{M}$ into the transformations of $\mathbb{T} \times \mathbf{S}_{\boldsymbol{u}}$. Composing a vectorization and a splitting, we convert Poincaré transformations into affine transformations of $\mathbb{T} \times \mathbf{S}_{\boldsymbol{u}}$.

Embedding the affine transformations of $\mathbb{T} \times \mathbf{S}_{\boldsymbol{u}}$ into the linear transformations of $\mathbb{R} \times (\mathbb{T} \times \mathbf{S}_{\boldsymbol{u}})$ (see VI.2.4(*ii*)) and using the customary matrix representation of such linear maps, we introduce the following notion.

Definition. The $\boldsymbol{u}$*-split Poincaré group* is

$$\left\{ \begin{pmatrix} 1 & 0 & \mathbf{0} \\ \boldsymbol{t} & \kappa(\boldsymbol{v}) & \kappa(\boldsymbol{v})\boldsymbol{v} \\ \boldsymbol{q} & \kappa(\boldsymbol{v})\boldsymbol{v} & \kappa(\boldsymbol{v})\boldsymbol{D}(\boldsymbol{v}) \end{pmatrix} \begin{pmatrix} 1 & 0 & \mathbf{0} \\ \mathbf{0} & \pm 1 & \mathbf{0} \\ \mathbf{0} & \mathbf{0} & \pm \boldsymbol{R} \end{pmatrix} \middle| \; \begin{matrix} \boldsymbol{t} \in \mathbb{T},\ \boldsymbol{q} \in \mathbf{S}_{\boldsymbol{u}}, \\ \boldsymbol{v} \in B_{\boldsymbol{u}},\ \boldsymbol{R} \in \mathcal{SO}(\boldsymbol{h}_{\boldsymbol{u}}) \end{matrix} \right\}. \quad \blacksquare$$

The $\boldsymbol{u}$-split Poincaré group is a ten-dimensional Lie group having the Lie algebra

$$\left\{ \begin{pmatrix} 0 & 0 & \mathbf{0} \\ t & 0 & \boldsymbol{v} \\ q & \boldsymbol{v} & \boldsymbol{A} \end{pmatrix} \;\middle|\; t \in \mathbb{T}, \quad q \in \mathbf{S}_{\boldsymbol{u}}, \quad \boldsymbol{v} \in \frac{\mathbf{S}_{\boldsymbol{u}}}{\mathbb{T}}, \quad \boldsymbol{A} \in \mathrm{A}(\boldsymbol{h}_{\boldsymbol{u}}) \right\}.$$

Keep in mind that the group multiplication of $\boldsymbol{u}$-split Poincaré transformations coincides with the usual matrix multiplication and the commutator of Lie algebra elements is the difference of their two products.

For $\boldsymbol{u} \in V(1)$ and $o \in \mathrm{M}$ put

$$\xi_{\boldsymbol{u},o} := \xi_{\boldsymbol{u}} \circ \mathrm{O}_o : \; \mathrm{M} \to \mathbb{T} \times \mathbf{S}_{\boldsymbol{u}}.$$

Then $L \mapsto \xi_{\boldsymbol{u},o} \circ L \circ \xi_{\boldsymbol{u},o}^{-1}$ is a Lie group isomorphism between the Poincaré group and the $\boldsymbol{u}$-split Poincaré group. Evidently, for different elements of $V(1) \times \mathrm{M}$, the isomorphisms are different. The transformation rule that shows how the isomorphism depends on $(\boldsymbol{u}, o)$ is rather complicated.

Though the Poincaré group and the $\boldsymbol{u}$-split Poincaré group are isomorphic (they have the same Lie group structure), they are not 'identical' : there is no 'canonical' isomorphism between them that we could use to identify them.

The $\boldsymbol{u}$-split Poincaré group is the Poincaré group of the $\boldsymbol{u}$-split special relativistic spacetime model $(\mathbb{T} \times \mathbf{S}_{\boldsymbol{u}}, \mathbb{T}, \boldsymbol{g}_{\boldsymbol{u}})$ (see 1.7). The spacetime model $(\mathrm{M}, \mathbb{T}, \boldsymbol{g})$ and the corresponding $\boldsymbol{u}$-split spacetime model are isomorphic, but they cannot be identified as we pointed out in 1.6.3. Due to the additional structures of the $\boldsymbol{u}$-split spacetime model, the $\boldsymbol{u}$-split Poincaré group has a number of additional structures that the Poincaré group has not.

10.6.2. The $\boldsymbol{u}$-split Poincaré group has the following subgroups:

$$\left\{ \begin{pmatrix} 1 & 0 & \mathbf{0} \\ t & 1 & \mathbf{0} \\ 0 & \mathbf{0} & \mathbf{1}_{\mathbf{S}_{\boldsymbol{u}}} \end{pmatrix} \;\middle|\; t \in \mathbb{T} \right\}, \qquad \left\{ \begin{pmatrix} 1 & 0 & \mathbf{0} \\ 0 & 1 & \mathbf{0} \\ q & \mathbf{0} & \mathbf{1}_{\mathbf{S}_{\boldsymbol{u}}} \end{pmatrix} \;\middle|\; q \in \mathbf{S}_{\boldsymbol{u}} \right\},$$

$$\left\{ \begin{pmatrix} 1 & 0 & \mathbf{0} \\ 0 & 1 & \mathbf{0} \\ 0 & \mathbf{0} & \boldsymbol{R} \end{pmatrix} \;\middle|\; \boldsymbol{R} \in \mathcal{O}(\boldsymbol{h}_{\boldsymbol{u}}) \right\}.$$

In contradistinction to the nonrelativistic case,

$$\left\{ \begin{pmatrix} 1 & 0 & \mathbf{0} \\ \mathbf{0} & \kappa(\boldsymbol{v}) & \kappa(\boldsymbol{v})\boldsymbol{v} \\ \mathbf{0} & \kappa(\boldsymbol{v})\boldsymbol{v} & \kappa(\boldsymbol{v})\boldsymbol{D}(\boldsymbol{v}) \end{pmatrix} \;\middle|\; \boldsymbol{v} \in B_{\boldsymbol{u}} \right\}$$

is not a subgroup of the $\boldsymbol{u}$-split Poincaré group.

The listed $\boldsymbol{u}$-split Poincaré transformations correspond (by the isomorphism established by $(\boldsymbol{u}, o) \in V(1) \times \mathrm{M}$) to the following Poincaré transformations:

$$\begin{aligned}
&x \mapsto x + \boldsymbol{u}\boldsymbol{t} && (\boldsymbol{t} \in \mathbb{T}),\\
&x \mapsto x + \boldsymbol{q} && (\boldsymbol{q} \in \mathbf{S}_{\boldsymbol{u}}),\\
&x \mapsto o - \boldsymbol{u}\big(\boldsymbol{u} \cdot (x - o)\big) + \boldsymbol{R} \cdot \boldsymbol{\pi}_{\boldsymbol{u}} \cdot (x - o) && \big(\boldsymbol{R} \in \mathcal{O}(\boldsymbol{h}_{\boldsymbol{u}})\big),\\
&x \mapsto o + \boldsymbol{L}\big(\kappa(\boldsymbol{v})(\boldsymbol{u} + \boldsymbol{v}),\, \boldsymbol{u}\big) \cdot \boldsymbol{\pi}_{\boldsymbol{u}} \cdot (x - o) && (\boldsymbol{v} \in B_{\boldsymbol{u}}).
\end{aligned}$$

10.6.3. Taking a linear bijection $\mathbb{T} \to \mathbb{R}$ and an orthogonal linear bijection $\mathbf{S}_{\boldsymbol{u}} \to \mathbb{R}^3$, we can transfer the $\boldsymbol{u}$-split Poincaré group into an affine transformation group of $\mathbb{R} \times \mathbb{R}^3$, called the *arithmetic Poincaré group* which is the Poincaré group of the arithmetic spacetime model.

The arithmetic Poincaré transformations can be given in the same form as the $\boldsymbol{u}$-split Poincaré transformations: $\mathbb{R}$, $\mathbb{R}^3$, $\mathcal{O}(3)$ and the open unit ball in $\mathbb{R}^3$ are to be substituted for $\mathbb{T}$, $\mathbf{S}_{\boldsymbol{u}}$, $\mathcal{O}(\boldsymbol{h}_{\boldsymbol{u}})$ and $B_{\boldsymbol{u}}$, respectively.

In conventional treatments one always considers the arithmetic Poincaré group and one speaks about *the* time translation subgroup, *the* space translation subgroup, *the* space rotation subgroup, *the* time inversion etc. which in applications can result in misunderstandings.

Since time and space do not exist and only observer time and observer space make sense, the Poincaré group has no such subgroups; it contains $\boldsymbol{u}$-timelike translations, $\boldsymbol{u}$-spacelike translations, o-centered $\boldsymbol{u}$-spacelike rotations etc.

10.7. Exercises

1. Let L be a Poincaré transformation for which $\boldsymbol{L} = -\mathbf{1}_{\mathrm{M}}$. Then there is a unique $o \in \mathrm{M}$ such that L is the o-centered spacetime inversion.

2. Prove that for all $o \in \mathrm{M}$,

$$O_o \circ \mathcal{L}_o \circ O_o^{-1} = \left\{ \begin{pmatrix} 1 & 0 \\ 0 & \boldsymbol{L} \end{pmatrix} \middle| \; \boldsymbol{L} \in \mathcal{L} \right\}.$$

3. Find $\boldsymbol{\xi}_{\boldsymbol{u},o} \cdot T_{\boldsymbol{u},o} \cdot \boldsymbol{\xi}_{\boldsymbol{u},o}^{-1}$ and $\boldsymbol{\xi}_{\boldsymbol{u},o} \cdot P_{\boldsymbol{u},o} \cdot \boldsymbol{\xi}_{\boldsymbol{u},o}^{-1}$.

4. Prove that the subgroup generated by $\{T_{\boldsymbol{u},o} | \, \boldsymbol{u} \in V(1), o \in \mathrm{M}\}$ equals $\mathcal{P}^{+\to} \cup \mathcal{P}^{+\leftarrow}$.

5. Prove that the derived Lie algebra of the Poincaré group equals the Lie algebra of the Poincaré group, i.e. $[\mathbf{La}(\mathcal{P}), \mathbf{La}(\mathcal{P})] = \mathbf{La}(\mathcal{P})$.

6. Let L be a Poincaré transformation. Consider the real number $\mathrm{ar}\, L$ to be a linear map $\mathbb{T} \to \mathbb{T}$, $\boldsymbol{t} \mapsto (\mathrm{ar}\, L)\boldsymbol{t}$.

If r is a world line function then $L \circ r \circ (\mathrm{ar}\, L)^{-1}$ is a world line function, too.

If C is a world line then $L[\mathrm{C}]$ is a world line, too; moreover, if $\mathrm{C} = \mathrm{Ran}\, r$ then $L[\mathrm{C}] = \mathrm{Ran}\left(L \circ r \circ (\mathrm{ar}\, L)^{-1}\right)$.

11. Relation between the two types of spacetime models

11.1. One often asserts that nonrelativistic physics is the limit of special relativistic physics as the light speed tends to infinity.

Can we give such an exact statement concerning our spacetime models? The answer is no.

We have two different mathematical structures. There is no natural way of introducing a convergence notion even on a class of mathematical structures of the same kind (e.g. on the class of groups and to say that a sequence of groups converges to a given one) and it is quite impossible to introduce a convergence notion on a class consisting of structures of different kinds (e.g. to say that a sequence of groups converges to an algebra).

There is no reasonable limit procedure in which a sequence of special relativistic spacetime models converges to a nonrelativistic spacetime model.

11.2. The following considerations show the real meaning of the usual statements.

Let us fix a special relativistic standard inertial frame with velocity value $\boldsymbol{u}$.

Let us rename $\mathbb{T}$ to $\mathbb{L}$, calling it the measure line of distances. Let us introduce for time periods a new measure line, denoted by $\mathbb{T}$. Let us choose a positive element c of $\frac{\mathbb{L}}{\mathbb{T}}$; it makes the correspondence $\mathbb{T} \to \mathbb{L}$, $\boldsymbol{t} \mapsto c\boldsymbol{t}$.

If $\boldsymbol{u}' \in V(1)$ then $\boldsymbol{v}_{\boldsymbol{u}'\boldsymbol{u}} \in \frac{\mathbf{S}_{\boldsymbol{u}}}{\mathbb{L}}$ and $\boldsymbol{v} := c\boldsymbol{v}_{\boldsymbol{u}'\boldsymbol{u}} \in \frac{\mathbf{S}_{\boldsymbol{u}}}{\mathbb{T}}$ will be considered the relative velocity with respect to the observer. Evidently, $|\boldsymbol{v}| < c$, thus c is the light speed in the new system of measure lines.

Substituting $\frac{\boldsymbol{v}}{c}$ for $\boldsymbol{v}_{\boldsymbol{u}'\boldsymbol{u}}$ and $c\boldsymbol{t}$ for $\boldsymbol{t}$ in the formula 7.1.4 and letting c tend to infinity—which has an exact meaning because elements of finite dimensional vector spaces are involved in that formula—we get the corresponding nonrelativistic transformation rule in I.8.2.4.

Similar statements hold for other formulae that concern relative velocities; e.g. for the addition formula of relative velocities, for the formula of light aberration etc.

However, such a statement, in general, will not be valid for formulae that do not concern relative velocities: e.g. the uniformly accelerated observer treated in 6.6 has no limit as c tends to infinity.

III. FUNDAMENTAL NOTIONS OF GENERAL RELATIVISTIC SPACETIME MODELS

1. As we have already mentioned, the nonrelativistic spacetime model is suitable for describing 'sluggish' mechanical phenomena. To describe 'brisk' mechanical phenomena and electromagnetic phenomena we have to use the special relativistic spacetime model. Of course, the special relativistic spacetime model is good for 'sluggish' mechanical phenomena, too, but their relativistic description is much more complicated than the nonrelativistic one and gives practically the same results.

To avoid misunderstandings, we emphasize that the mechanical effects of electromagnetic phenomena (e.g. the history of charged masspoints in a given electromagnetic field) can be well described nonrelativistically as well, provided that the mechanical phenomena remain 'sluggish' (the relative velocities of masspoints remain much smaller than the light speed). The nonrelativistic spacetime model is not suitable for the description of the electromagnetic phenomena in vacuum: how charges produce electromagnetic field, how an electromagnetic radiation propagates etc.

Gravitational actions are well described in the nonrelativistic spacetime model by absolute scalar potentials. Such potentials do not exist in the special relativistic spacetime model. Other potentials or force fields do not give convenient (experimentally verified) models of gravitational actions.

The problem that faces us is that gravitational actions in 'brisk' mechanical phenomena and electromagnetic phenomena cannot be described and, of course, gravitational phenomena (how do masses produce gravitational fields) cannot be treated in the framework of the special relativistic spacetime model.

There is only one way out: if we want to describe gravitational phenomena as well, then we have to construct a new spacetime model. However, it is not straightforward at all, how we shall do this.

2. Recall what we said about our experience regarding the structure of our space and time: in our space we find straight lines represented by light signals or stretched threads. We know, however, that a thread stretched in the gravitational field of the earth is not straight, it bends; if the thread is short enough and the stretching is strong enough then the curvature of the thread is negligible. However, for longer threads—imagine a thread (wire) across a river—the curvature can be significant.

It seems, that a light signal is better for realizing a straight line. Indeed, in terrestrial distances we do not experience that a light signal is not straight. However, the distances on the earth are small for a light signal. It may happen that light signals turn out to be curved in cosmic distances. Of course, to prove or disprove this possibility we meet great difficulties. A minor problem is that cosmic distances are hardly manageable.

To state that a line is straight or not we have to know what the straight lines are. Straight lines in terrestrial distances are defined in the most convenient way by the trajectories of light signals. Have we a better way to define straight lines in cosmic size? Can we define straight lines in this way? Can we define straight lines at all?

We have to recognize that it makes no sense that a single line in itself is straight or is not straight. We have to relate the trajectories of more light signals and to test whether they satisfy the conditions we expect the set of straight lines have. For instance, if two different light signals meet in more than one point, the trajectories of the signals cannot be straight lines. Unfortunately, it is rather difficult to execute such examinations in cosmic size.

Nevertheless, we have experimental evidence that shows that gravitation influences the propagation of light. The angle between two light beams arriving from two stars have been measured in different circumstances: first the light beams travel 'freely', far from gravitational action; second, they travel near the Sun i.e. under a strong gravitational action. The angles are significantly different.

Light travels along different trajectories in two circumstances. Evidently, the trajectories cannot be straight lines in both cases.

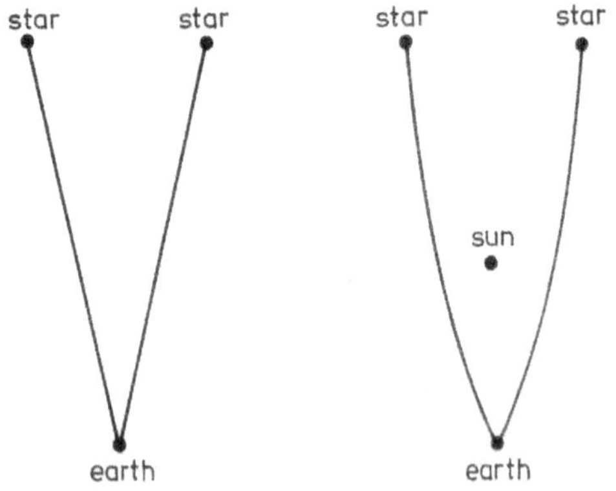

The affine structure of spacetime in the special relativistic model has been based on the straight propagation of light. Thus if we want to construct a spacetime model suitable for the treatment of gravitational phenomena, we have to reject the affine structure.

We have to get accustomed to the strange fact: in general, the notion of a straight line makes no sense. It is worth repeating why. Every notion in our mathematical model must have a physical background. A straight line would be realized by a light beam: we have no better possibility. However, in strong gravitational fields (in cosmic size) the set of light beams maybe does not satisfy the usual conditions imposed on the set of straight lines. One usually says that gravitation 'curves' spacetime. The properties of a curved spacetime can be illustrated as follows: it may happen that two light beams starting simultaneously from the same source in different directions meet again somewhere (this is a 'spacelike curvature') or that two light beams starting from the same source in the same direction in different instants meet again somewhere (this is a 'timelike curvature').

3. According to the idea of Einstein, spacetime models must reflect gravitational actions, a *spacetime model is to be a model of a gravitational action*; the absence of gravitation is modelled by the special relativistic spacetime model.

The theory of gravitation, a deep and large theory, lies out of the scope of this book. That is why only the framework of general relativistic spacetime models will be outlined.

In constructing a general relativistic spacetime model, we do not adhere to the affine structure and we require only that spacetime is a four-dimensional smooth manifold M.

A four-dimensional smooth manifold M is an abstract mathematical structure similar to a four-dimensional smooth submanifold in an affine space; it has the following fundamental properties: every $x \in \mathrm{M}$ has a neighbourhood which can be parametrized by $p\colon \mathbb{R}^4 \rightarrowtail \mathrm{M}$; if p and q are parametrizations then $q^{-1} \circ p$ is smooth. Then to each point x of M a four-dimensional vector space $\mathbf{T}_x(\mathrm{M})$, the tangent space at x, is assigned; every differentiable curve passing through x has its tangent vector in $\mathbf{T}_x(\mathrm{M})$. A neighbourhood of zero of $\mathbf{T}_x(\mathrm{M})$ approximates a neighbourhood of x in M. Smooth submanifolds of an affine space (thus affine spaces themselves) are smooth manifolds.

The gravitational constant makes it possible to use real numbers for measuring spacetime distances in such a way that the gravitational constant is 1.

Our experience that gravitational action in small size does not contradict the notion of a straight line suggests that a general relativistic spacetime model in small size can be 'similar' to a special relativistic spacetime model. That is why we accept that there is a Lorentz form $\boldsymbol{g}_x : \mathbf{T}_x(\mathrm{M}) \times \mathbf{T}_x(\mathrm{M}) \to \mathbb{R}$ is given for all $x \in \mathrm{M}$ in such a way that $x \mapsto \boldsymbol{g}_x$ is smooth in a conveniently defined sense. The assignment $x \mapsto \boldsymbol{g}_x$ is called a *Lorentz field* and is denoted by $\boldsymbol{g}$. Moreover, we assume that every $\boldsymbol{g}_x$ is endowed with an arrow orientation which, too, depends on x in a conveniently defined smooth way.

Definition. *A general relativistic spacetime model* is a pair $(\mathrm{M}, \boldsymbol{g})$ where

— M is a four-dimensional smooth manifold (called *spacetime*or *world*),

— $\boldsymbol{g}$ is an arrow-oriented Lorentz field on M.

Evidently, a special relativistic spacetime model – with $\mathbb{T} = \mathbb{R}$ – is a general relativistic spacetime model: M is an affine space (then every tangent space equals **M**) and $\boldsymbol{g}_x$ is the same for all $x \in \mathrm{M}$.

4. Take a general relativistic spacetime model $(\mathrm{M}, \boldsymbol{g})$. Then S_x, $\mathbf{T}_x$ and L_x, the set of spacelike tangent vectors etc. in $\mathbf{T}_x(\mathrm{M})$ are defined by $\boldsymbol{g}_x$ for all world points x and they have the following meaning:

— a *world line* (the history of a masspoint) is a curve in M whose tangent vectors are timelike (i.e. the tangent vector of a world line C at x is in $\mathbf{T}_x$);

— a *light signal* is a curve in M whose tangent vectors are lightlike.

Let us give an illustration of a general relativistic spacetime model. Let the plane of the page represent the spacetime M, and at the same time, every tangent space is represented by the plane of the page as well. Then we draw the future light cone to every world point.

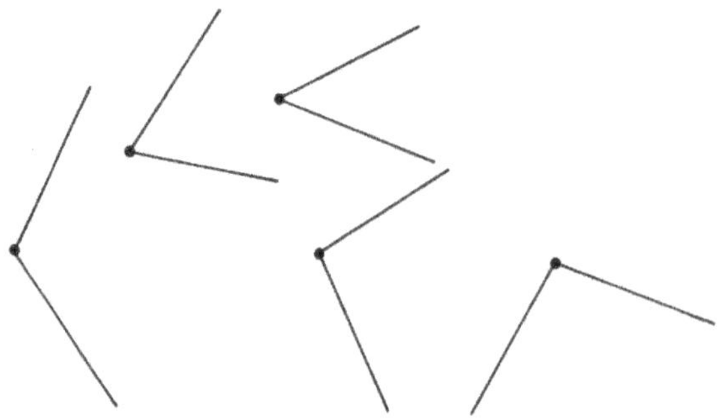

Illustrating the nonrelativistic and the special relativistic spacetime models we have got accustomed to the fact that the Euclidean structure of the plane has to be neglected: the angles and distances in the plane of the page do not reflect, in general, objects of the spacetime model. Now we have to neglect the affine structure of the plane as well: the straight lines of the plane, in general, do not correspond to objects of the spacetime.

We call attention to the fact that in our illustration the spacetime manifold and its tangent spaces which are different sets, are represented by the same plane. The straight lines representing light cones in the previous figure are lines in tangent spaces, they do not lie in the spacetime manifold.

The following figures show a world line and a light signal in the general relativistic spacetime model.

5. As we have said, a general relativistic spacetime model is to be a model of a gravitational action. The theory of gravitation has the task to expound how a gravitational action is modelled by a spacetime model. We know that a special relativistic spacetime model corresponds to the lack of gravitation.

There are different special relativistic spacetime models; however, all of them correspond to the same physical situation: the lack of gravitation. This is reflected in the fact that all special relativistic spacetime models are isomorphic.

It may happen that two general relativistic spacetime models correspond to the same gravitational action; we expect that they must be isomorphic. Now we give the notion of isomorphism.

Definition. The general relativistic spacetime model $(\mathrm{M}, \boldsymbol{g})$ is *isomorphic to* $(\mathrm{M}', \boldsymbol{g}')$ if there is a diffeomorphism $F\colon \mathrm{M} \to \mathrm{M}'$ sucht that

$$\boldsymbol{g}'_{F(x)} \circ \big(\mathrm{D}F(x) \times \mathrm{D}F(x)\big) = \boldsymbol{g}_x \qquad (x \in \mathrm{M}). \quad \blacksquare$$

The phrase F is diffeomorphism means that F is a bijection and both F and F^{-1} are smooth. The derivative of F at x, $\mathrm{D}F(x)$, is a linear map from $\mathbf{T}_x(\mathrm{M})$ into $\mathbf{T}_{F(x)}(\mathrm{M}')$.

6. As examples we give a certain kind of general relativistic spacetime models where the spacetime manifold is a submanifold of an affine space, hence we can use the well-known mathematical tools treated in this book.

Take a special relativistic spacetime model $(\mathrm{M}, \mathbb{R}, \boldsymbol{g})$, select an open subset $\mathrm{M}^{\boldsymbol{A}}$ of M; $\mathrm{M}^{\boldsymbol{A}}$ is an open submanifold of M and $\mathbf{T}_x(\mathrm{M}^{\boldsymbol{A}}) = \mathbf{M}$ for all $x \in \mathrm{M}$. Give a smooth map $\boldsymbol{A}\colon \mathrm{M}^{\boldsymbol{A}} \to \mathcal{GL}(\mathbf{M})$ (i.e. $\boldsymbol{A}(x)$ is a linear bijection $\mathbf{M} \to \mathbf{M}$ for all $x \in \mathrm{M}$). For all $x \in \mathbf{M}^{\boldsymbol{A}}$ we define the Lorentz form $\boldsymbol{g}^{\boldsymbol{A}}_x$ by

$$\boldsymbol{g}^{\boldsymbol{A}}_x(\boldsymbol{x}, \boldsymbol{y}) := \boldsymbol{g}\big(\boldsymbol{A}(x) \cdot \boldsymbol{x},\ \boldsymbol{A}(x) \cdot \boldsymbol{y}\big) \qquad (\boldsymbol{x}, \boldsymbol{y} \in \mathbf{M}).$$

The Lorentz form $\boldsymbol{g}^{\boldsymbol{A}}$ is endowed with an arrow orientation as follows: let $\mathbf{T}^{\to}$ be the future directed timelike cone of $\boldsymbol{g}$; then the future directed timelike cone of $\boldsymbol{g}^{\boldsymbol{A}}_x$ is defined to be $\boldsymbol{A}(x)^{-1}[\mathbf{T}^{\to}]$.

Then $(\mathrm{M}^{\boldsymbol{A}}, \boldsymbol{g}^{\boldsymbol{A}})$ is a general relativistic spacetime model.

PART TWO

MATHEMATICAL TOOLS

IV. TENSORIAL OPERATIONS

In this section $\mathbb{K}$ denotes the field of complex numbers or the field of real numbers, and all vector spaces are given over $\mathbb{K}$.

Tensors and operations with tensors are essential mathematical tools in physics; the simplest physical notions—e.g. meter/secundum—require tensorial operations. Those being familiar with tensors will find no difficulty in reading this book.

0. Identifications

Identifications make easy to handle tensors.

Let $\mathbf{X}$ and $\mathbf{Y}$ be vector spaces over the same field. If there is a linear injection $\boldsymbol{i}: \mathbf{X} \to \mathbf{Y}$ which we find natural ('canonical') from some point of view, we *identify* $\boldsymbol{x}$ and $\boldsymbol{i}(\boldsymbol{x})$ for all $\boldsymbol{x} \in \mathbf{X}$, i.e. we omit $\boldsymbol{i}$ from the notations considering $\mathbf{X}$ to be a linear subspace of $\mathbf{Y}$. Then we write

$$\mathbf{X} \hookrightarrow \mathbf{Y}, \qquad \boldsymbol{x} \equiv \boldsymbol{i}(\boldsymbol{x}),$$

and if $\boldsymbol{i}$ is a bijection,

$$\mathbf{X} \equiv \mathbf{Y}, \qquad \boldsymbol{x} \equiv \boldsymbol{i}(\boldsymbol{x}).$$

In practice, instead of $\boldsymbol{x} \equiv \boldsymbol{i}(\boldsymbol{x})$ an appropriate formula appears that allows us to consider $\boldsymbol{i}$ to be natural.

Of course, 'natural' and 'canonical' are not mathematical notions and it depends on us whether we accept or reject an identification. There are commonly accepted identifications and there are some cases in which some people find a given identification convenient and others do not.

Later, using a lot of identifications, the reader will have the opportunity to see their importance.

1. Duality

1.1. Let $\mathbf{V}$ and $\mathbf{U}$ be vector spaces. Then $\mathrm{Lin}(\mathbf{V}, \mathbf{U})$ denotes the vector space of linear maps $\mathbf{V} \to \mathbf{U}$; $\mathrm{Lin}(\mathbf{V}) := \mathrm{Lin}(\mathbf{V}, \mathbf{V})$.

The value of $\boldsymbol{L} \in \mathrm{Lin}(\mathbf{V},\mathbf{U})$ at $\boldsymbol{v} \in \mathbf{V}$ is denoted by $\boldsymbol{L}\cdot\boldsymbol{v}$.

The composition of linear maps is denoted by a dot as well: for $\boldsymbol{L} \in \mathrm{Lin}(\mathbf{V},\mathbf{U})$, $\boldsymbol{K} \in \mathrm{Lin}(\mathbf{U},\mathbf{W})$ we write $\boldsymbol{K}\cdot\boldsymbol{L}$.

$\mathbf{V}^* := \mathrm{Lin}(\mathbf{V},\mathbb{K})$ is the *dual* of $\mathbf{V}$. The elements of $\mathbf{V}^*$ are often called *linear functionals* or *covectors.*

The dual *separates* the elements of the vector space which means that if $\boldsymbol{v} \in \mathbf{V}$, and $\boldsymbol{p}\cdot\boldsymbol{v} = 0$ for all $\boldsymbol{p} \in \mathbf{V}^*$, then $\boldsymbol{v} = 0$ or, equivalently, if $\boldsymbol{v}_1$ and $\boldsymbol{v}_2$ are different elements of $\mathbf{V}$, then there is a $\boldsymbol{p} \in \mathbf{V}^*$ such that $\boldsymbol{p}\cdot\boldsymbol{v}_1 \neq \boldsymbol{p}\cdot\boldsymbol{v}_2$.

If $\{\boldsymbol{v}_i \mid i \in I\}$ is a basis of $\mathbf{V}$ then there is a set $\{\boldsymbol{p}^i \mid i \in I\}$ in $\mathbf{V}^*$, called the *dual of the basis,* such that

$$\boldsymbol{p}^i\cdot\boldsymbol{v}_j = \begin{cases} 1 & \text{if } i = j \\ 0 & \text{if } i \neq j \end{cases} \qquad (i, j \in I).$$

If $\mathbf{V}$ is finite dimensional, then the dual of a basis is a basis in $\mathbf{V}^*$, hence $\dim(\mathbf{V}^*) = \dim\mathbf{V}$.

Let N denote the (finite) dimension of $\mathbf{V}$. If $\{\boldsymbol{v}_1,\ldots,\boldsymbol{v}_N\}$ is a basis of $\mathbf{V}$ and $\{\boldsymbol{p}^1,\cdots,\boldsymbol{p}^N\}$ is its dual, then for all $\boldsymbol{v} \in \mathbf{V}$ and $\boldsymbol{p} \in \mathbf{V}^*$ we have

$$\boldsymbol{v} = \sum_{i=1}^{N}(\boldsymbol{p}^i\cdot\boldsymbol{v})\boldsymbol{v}_i,$$

$$\boldsymbol{p} = \sum_{i=1}^{N}(\boldsymbol{p}\cdot\boldsymbol{v}_i)\boldsymbol{p}^i.$$

1.2. To every element $\boldsymbol{v}$ of $\mathbf{V}$ we can associate a linear map $\boldsymbol{i}(\boldsymbol{v})\colon \mathbf{V}^* \to \mathbb{K}$, $\boldsymbol{p} \mapsto \boldsymbol{p}\cdot\boldsymbol{v}$, i.e. an element of $\mathbf{V}^{**}$. The correspondence $\mathbf{V} \to \mathbf{V}^{**}$, $\boldsymbol{v} \mapsto \boldsymbol{i}(\boldsymbol{v})$ is a linear injection which seems so natural and simple that we find it convenient to identify $\boldsymbol{v}$ and $\boldsymbol{i}(\boldsymbol{v})$ for all $\boldsymbol{v} \in \mathbf{V}$:

$$\mathbf{V} \subset\!\!\to \mathbf{V}^{**}, \qquad \boldsymbol{v} \equiv \boldsymbol{i}(\boldsymbol{v}),$$

i.e.

$$\boldsymbol{v}\cdot\boldsymbol{p} \equiv \boldsymbol{p}\cdot\boldsymbol{v} \qquad (\boldsymbol{v} \in \mathbf{V},\ \boldsymbol{p} \in \mathbf{V}^*).$$

If $\mathbf{V}$ is finite dimensional then this correspondence is a linear bijection between $\mathbf{V}$ and $\mathbf{V}^{**}$, i.e. the whole dual of $\mathbf{V}^*$ can be identified with $\mathbf{V}$:

$$\mathbf{V} \equiv \mathbf{V}^{**}, \qquad \boldsymbol{v}\cdot\boldsymbol{p} \equiv \boldsymbol{p}\cdot\boldsymbol{v}.$$

1.3. The Cartesian product $\mathbf{V}\times\mathbf{U}$ of the vector spaces $\mathbf{V}$ and $\mathbf{U}$ is a vector space with the pointwise addition and pointwise multiplication by numbers:

$$(\boldsymbol{v}_1,\boldsymbol{u}_1) + (\boldsymbol{v}_2,\boldsymbol{u}_2)\colon = (\boldsymbol{v}_1+\boldsymbol{v}_2, \boldsymbol{u}_1+\boldsymbol{u}_2),$$

$$\alpha(\boldsymbol{v},\boldsymbol{u})\colon = (\alpha\boldsymbol{v}, \alpha\boldsymbol{u})$$

for $\boldsymbol{v}_1,\boldsymbol{v}_2 \in \mathbf{V}$, $\boldsymbol{u}_1,\boldsymbol{u}_2 \in \mathbf{U}$ and $\alpha \in \mathbb{K}$.

We have the identification

$$\mathbf{V}^* \times \mathbf{U}^* \equiv (\mathbf{V} \times \mathbf{U})^*, \qquad (\boldsymbol{p}, \boldsymbol{q}) \cdot (\boldsymbol{v}, \boldsymbol{u}) \equiv \boldsymbol{p} \cdot \boldsymbol{v} + \boldsymbol{q} \cdot \boldsymbol{u}.$$

$$((\boldsymbol{p}, \boldsymbol{q}) \in \mathbf{V}^* \times \mathbf{U}^*, \qquad (\boldsymbol{v}, \boldsymbol{u}) \in \mathbf{V} \times \mathbf{U}).$$

1.4. The *transpose* of $\boldsymbol{L} \in \mathrm{Lin}(\mathbf{V}, \mathbf{U})$ is the linear map

$$\boldsymbol{L}^* : \mathbf{U}^* \to \mathbf{V}^*, \qquad \boldsymbol{f} \mapsto \boldsymbol{f} \circ \boldsymbol{L},$$

i.e.

$$(\boldsymbol{L}^* \cdot \boldsymbol{f}) \cdot \boldsymbol{v} = \boldsymbol{f} \cdot (\boldsymbol{L} \cdot \boldsymbol{v})$$

or, with the identification introduced in 1.2,

$$\boldsymbol{v} \cdot \boldsymbol{L}^* \cdot \boldsymbol{f} = \boldsymbol{f} \cdot \boldsymbol{L} \cdot \boldsymbol{v} \qquad (\boldsymbol{f} \in \mathbf{U}^*, \ \boldsymbol{v} \in \mathbf{V}).$$

If $\boldsymbol{L}, \boldsymbol{K} \in \mathrm{Lin}(\mathbf{V}, \mathbf{U})$, $\alpha \in \mathbb{K}$, then

$$(\boldsymbol{L} + \boldsymbol{K})^* = \boldsymbol{L}^* + \boldsymbol{K}^*,$$
$$(\alpha \boldsymbol{L})^* = \alpha \boldsymbol{L}^*.$$

If $\boldsymbol{L} \in \mathrm{Lin}(\mathbf{V}, \mathbf{U})$, $\boldsymbol{K} \in \mathrm{Lin}(\mathbf{U}, \mathbf{W})$, then

$$(\boldsymbol{K} \cdot \boldsymbol{L})^* = \boldsymbol{L}^* \cdot \boldsymbol{K}^*.$$

If $\mathbf{V}$ and $\mathbf{U}$ are finite dimensional, then
— $\boldsymbol{L}$ is injective if and only if $\boldsymbol{L}^*$ is surjective,
— $\boldsymbol{L}$ is surjective if and only if $\boldsymbol{L}^*$ is injective.
Moreover, in this case—because of the identification $\mathbf{V}^{**} \equiv \mathbf{V}$, $\mathbf{U}^{**} \equiv \mathbf{U}$— we have

$$\boldsymbol{L}^{**} = \boldsymbol{L}.$$

If $\boldsymbol{L}$ is bijective, then

$$\left(\boldsymbol{L}^{-1}\right)^* = (\boldsymbol{L}^*)^{-1}.$$

1.5. Let $\mathbf{V}$ be a finite dimensional vector space and $\boldsymbol{L} \in \mathrm{Lin}(\mathbf{V}, \mathbf{V}^*)$. Then $\boldsymbol{L}^*$ is a linear map from $\mathbf{V}^{**}$ into $\mathbf{V}^*$, i.e. because of the identification $\mathbf{V}^{**} \equiv \mathbf{V}$ we have $\boldsymbol{L}^* \in \mathrm{Lin}(\mathbf{V}, \mathbf{V}^*)$.

The linear map $\boldsymbol{L} : \mathbf{V} \to \mathbf{V}^*$ is called *symmetric or antisymmetric* if $\boldsymbol{L} = \boldsymbol{L}^*$ or $\boldsymbol{L} = -\boldsymbol{L}^*$, respectively.

In general, the *symmetric and antisymmetric parts* of $\boldsymbol{L} \in \mathrm{Lin}(\mathbf{V}, \mathbf{V}^*)$ are

$$\frac{\boldsymbol{L} + \boldsymbol{L}^*}{2} \qquad \text{and} \qquad \frac{\boldsymbol{L} - \boldsymbol{L}^*}{2},$$

respectively.

Similar definitions work well for linear maps $\mathbf{V}^* \to \mathbf{V}$.

On the other hand, the notions of symmetricity, symmetric part etc. make no sense for linear maps $\mathbf{V} \to \mathbf{V}$ *and* $\mathbf{V}^* \to \mathbf{V}^*$.

1.6. $\mathbb{K}^N$, the set of ordered N-tuples of numbers, is a well-known vector space. It is known as well that the linear maps from $\mathbb{K}^N$ into $\mathbb{K}^M$ are identified with the matrices of M rows and N columns, in other words, $\mathrm{Lin}(\mathbb{K}^N, \mathbb{K}^M) \equiv \mathbb{K}^{M\times N}$. As a consequence, we have the identification

$$\left(\mathbb{K}^N\right)^* = \mathrm{Lin}(\mathbb{K}^N, \mathbb{K}) \equiv \mathbb{K}^{1\times N} = \mathbb{K}^N$$

$$\boldsymbol{p}\cdot\boldsymbol{x} \equiv \sum_{i=1}^{N} p_i x^i \qquad (\boldsymbol{p}, \boldsymbol{x} \in \mathbb{K}^N).$$

We adhered to the trick used in physical applications according to which $\left(\mathbb{K}^N\right)^*$ is *identified* with $\mathbb{K}^N$ in such a way that they are *distinguished* in notations as follows.

The components of the elements of $\mathbb{K}^N$ are indexed by superscripts:

$$\boldsymbol{x} = (x^1, \dots, x^N) \in \mathbb{K}^N,$$

and the components of the elements of $\left(\mathbb{K}^N\right)^* \equiv \mathbb{K}^N$ are indexed by subscripts:

$$\boldsymbol{p} = (p_1, \dots, p_N) \in \left(\mathbb{K}^N\right)^*.$$

The identification in question, called the *standard identification,* means that to every $(x^1, \dots, x^N) \in \mathbb{K}^N$ we assign $(x_1, \dots, x_N) \in \left(\mathbb{K}^N\right)^*$ in such a way that $x_i = x^i$ for all $i = 1, \dots, N$.

Moreover, for the sake of simplicity, we often shall not write that the indices run from 1 to N (or to M), denoting the elements in the form (x^i) and (x_i), respectively.

The fundamental rule is that a summation can be carried out only for indices in opposite positions: up and down. Accordingly, the matrix entries are indexed corresponding to the domain and range of the matrix as a linear map:

$$\begin{aligned}
\left(L^i{}_k\right) &: \mathbb{K}^N \to \mathbb{K}^M, \\
\left(L_{ik}\right) &: \mathbb{K}^N \to \left(\mathbb{K}^M\right)^*, \\
\left(L_i{}^k\right) &: \left(\mathbb{K}^N\right)^* \to \left(\mathbb{K}^M\right)^*, \\
\left(L^{ik}\right) &: \left(\mathbb{K}^N\right)^* \to \mathbb{K}^M.
\end{aligned}$$

This trick works well until actual vectors are not involved; this notation does not show for instance whether the ordered pair of numbers $(1,2)$ is an element of $\mathbb{R}^2$ or $\left(\mathbb{R}^2\right)^*$, and whether the matrix

$$\begin{pmatrix} 1 & 0 \\ 2 & 1 \end{pmatrix}$$

maps from $\mathbb{R}^2$ into $\mathbb{R}^2$ or from $\mathbb{R}^2$ into $(\mathbb{R}^2)^*$ etc.

The set of vectors $\chi_1 := (1,0,\dots,0)$, $\chi_2 := (0,1,...,0)$, $\dots$, $\chi_N := (0,0,\dots,1)$ is called the *standard basis* of $\mathbb{K}^N$. In the mentioned identification $(\mathbb{K}^N)^* \equiv \mathbb{K}^N$ the dual of the standard basis is the standard basis itself.

According to this identification the transpose of a matrix as a linear map is the usual matrix transpose.

The above notation shows well that symmetricity, symmetric part etc. make sense only for matrices (L_{ik}) and (L^{ik}).

1.7. The symbol $\mathrm{Bilin}(\mathbf{U}\times\mathbf{V},\mathbb{K})$ stands for the vector space of bilinear maps $\mathbf{U}\times\mathbf{V}\to\mathbb{K}$, often called *bilinear forms.*

We have that

$$\boldsymbol{i}\colon \mathrm{Lin}(\mathbf{V},\mathbf{U}) \to \mathrm{Bilin}(\mathbf{U}^*\times\mathbf{V},\mathbb{K})$$

defined by

$$\big(\boldsymbol{i}(\boldsymbol{L})\big)(\boldsymbol{f},\boldsymbol{v}) := \boldsymbol{f}\cdot\boldsymbol{L}\cdot\boldsymbol{v}$$

$$\big(\boldsymbol{L}\in\mathrm{Lin}(\mathbf{V},\mathbf{U}),\ \boldsymbol{f}\in\mathbf{U}^*,\ \boldsymbol{v}\in\mathbf{V}\big)$$

is a linear injection which we use for the identification

$$\mathrm{Lin}(\mathbf{V},\mathbf{U}) \subset\!\!\to \mathrm{Bilin}(\mathbf{U}^*\times\mathbf{V},\mathbb{K}), \qquad \boldsymbol{f}\cdot\boldsymbol{L}\cdot\boldsymbol{v} \equiv \boldsymbol{L}(\boldsymbol{f},\boldsymbol{v}).$$

If the vector spaces $\mathbf{U}$ and $\mathbf{V}$ have finite dimension then $\boldsymbol{i}$ is a bijection, hence $\equiv$ stands instead of $\subset\!\!\to$.

The reader is asked to examine this identification in the case of matrices i.e. for $\mathrm{Lin}(\mathbb{K}^N,\mathbb{K}^M)$.

1.8. A bilinear form $\boldsymbol{h}\colon \mathbf{V}\times\mathbf{V}\to\mathbb{K}$ is called *symmetric* or *antisymmetric* if $\boldsymbol{h}(\boldsymbol{v},\boldsymbol{u}) = \boldsymbol{h}(\boldsymbol{u},\boldsymbol{v})$ or $\boldsymbol{h}(\boldsymbol{v},\boldsymbol{u}) = -\boldsymbol{h}(\boldsymbol{u},\boldsymbol{v})$, respectively, for all $\boldsymbol{v},\boldsymbol{u}\in\mathbf{V}$.

Similar definitions are accepted for bilinear forms $\mathbf{V}^*\times\mathbf{V}^*\to\mathbb{K}$.

Observe that for finite dimensional $\mathbf{V}$ the notions introduced here and in 1.5 coincide in the identification $\mathrm{Lin}(\mathbf{V},\mathbf{V}^*) \equiv \mathrm{Bilin}(\mathbf{V}^*\times\mathbf{V}^*,\mathbb{K})$.

2. Coordinatization

2.1. Let $\mathbf{V}$ be an N-dimensional vector space over $\mathbb{K}$.

An element $(\boldsymbol{v}_1,\dots,\boldsymbol{v}_N)$ of $\mathbf{V}^N$ is called an *ordered basis* of $\mathbf{V}$ if the set $\{\boldsymbol{v}_1,\dots,\boldsymbol{v}_N\}$ is a basis of $\mathbf{V}$.

An ordered basis of $\mathbf{V}$ induces a linear bijection $\boldsymbol{K}\colon\mathbf{V}\to\mathbb{K}^N$ defined by $\boldsymbol{K}\cdot\boldsymbol{v}_i := \chi_i$ $(i=1,\dots,N)$ where $(\chi_1,\dots\chi_N)$ is the ordered standard basis of $\mathbb{K}^N$. $\boldsymbol{K}$ is called the *coordinatization* of $\mathbf{V}$ corresponding to the given ordered basis. The inverse of the coordinatization, $\boldsymbol{P} := \boldsymbol{K}^{-1}$, is called the *parametrization* of $\mathbf{V}$ corresponding to the given ordered basis. It is quite evident that

$$\boldsymbol{P}\cdot(x^i)=\sum_{i=1}^{N}x^i\boldsymbol{v}_i \qquad \left((x^i)\in\mathbb{K}^N\right).$$

Thus, in view of 1.1 we have

$$\boldsymbol{K}\cdot\boldsymbol{v}=\left(\boldsymbol{p}^i\cdot\boldsymbol{v}|\ i=1,\dots,N\right) \qquad (\boldsymbol{v}\in\mathbf{V})$$

where $(\boldsymbol{p}^1,\dots,\boldsymbol{p}^N)$ is the ordered dual basis of $(\boldsymbol{v}_1,\dots,\boldsymbol{v}_N)$.

Obviously, every linear bijection $\boldsymbol{K}\colon\mathbf{V}\to\mathbb{K}^N$ is a coordinatization in the above sense: the one corresponding to the ordered basis $(\boldsymbol{v}_1,\dots,\boldsymbol{v}_N)$ where $\boldsymbol{v}_i:=\boldsymbol{K}^{-1}\cdot\chi_i$ $(i=1,\dots,N)$.

2.2. A coordinatization of $\mathbf{V}$ determines a coordinatization of $\mathbf{V}^*$, that is induced by the corresponding ordered dual basis. Using the previous notations and denoting the coordinatization in question by $\boldsymbol{C}\colon\mathbf{V}^*\to(\mathbb{K}^N)^*$ we have

$$\boldsymbol{C}\cdot\boldsymbol{p}=(\boldsymbol{p}\cdot\boldsymbol{v}_i|\ i=1,\dots,N) \qquad (\boldsymbol{p}\in\mathbf{V}^*).$$

It is not hard to see that

$$\boldsymbol{C}=\left(\boldsymbol{K}^{-1}\right)^*=\boldsymbol{P}^*.$$

2.3. In the coordinatization $\boldsymbol{K}$, a linear map $\boldsymbol{L}\colon\mathbf{V}\to\mathbf{V}$ is represented by the matrix

$$\boldsymbol{K}\cdot\boldsymbol{L}\cdot\boldsymbol{K}^{-1}=\boldsymbol{K}\cdot\boldsymbol{L}\cdot\boldsymbol{P}=\left(\boldsymbol{p}^i\cdot\boldsymbol{L}\cdot\boldsymbol{v}_k|\ i,k=1,\dots,N\right).$$

To deduce this equality argue as follows:

$$\sum_{k=1}^{N}\left(\boldsymbol{K}\cdot\boldsymbol{L}\cdot\boldsymbol{K}^{-1}\right)^i{}_kx^k=\left(\boldsymbol{K}\cdot\boldsymbol{L}\cdot\boldsymbol{K}^{-1}\cdot x\right)^i=$$

$$=\boldsymbol{p}^i\cdot\boldsymbol{L}\cdot\sum_{k=1}^{N}x^k\boldsymbol{v}_k=\sum_{k=1}^{N}(\boldsymbol{p}^i\cdot\boldsymbol{L}\cdot\boldsymbol{v}_k)x^k.$$

The linear map $\boldsymbol{T}\colon\mathbf{V}\to\mathbf{V}^*$ is represented by the matrix

$$\left(\boldsymbol{K}^{-1}\right)^*\boldsymbol{T}\cdot\boldsymbol{K}^{-1}=\boldsymbol{P}^*\cdot\boldsymbol{T}\cdot\boldsymbol{P}=(\boldsymbol{v}_i\cdot\boldsymbol{T}\cdot\boldsymbol{v}_k|\ i,k=1,\dots,N).$$

It is left to the reader to find the matrix of linear maps $\mathbf{V}^*\to\mathbf{V}$ and $\mathbf{V}^*\to\mathbf{V}^*$.

3. Tensor products

3.1. We start with an abstract definition of tensor products that may seem strange; the properties of tensor products following from this definition will clarify its real meaning.

Definition. Let $\mathbf{V}$ and $\mathbf{U}$ be vector spaces (over the same field $\mathbb{K}$). A *tensor product* of $\mathbf{U}$ and $\mathbf{V}$ is a pair $(\mathbf{Z}, \boldsymbol{h})$, where

(i) $\mathbf{Z}$ is a vector space,

(ii) $\boldsymbol{h}\colon \mathbf{U} \times \mathbf{V} \to \mathbf{Z}$ is a bilinear map having the property that

— if $\mathbf{W}$ is a vector space and $\boldsymbol{c}\colon \mathbf{U} \times \mathbf{V} \to \mathbf{W}$ is a bilinear map,

— then there exists a unique linear map $\boldsymbol{L}\colon \mathbf{Z} \to \mathbf{W}$ such that

$$\boldsymbol{c} = \boldsymbol{L} \circ \boldsymbol{h}.$$

Proposition. The pair $(\mathbf{Z}, \boldsymbol{h})$ satisfying (i) and (ii) is a tensor product of $\mathbf{U}$ and $\mathbf{V}$ if and only if

1) $\mathbf{Z}$ is spanned ($\mathbf{Z}$ is the linear subspace generated) by $\operatorname{Ran} \boldsymbol{h}$,

2) if $\boldsymbol{v}_1, \dots, \boldsymbol{v}_n$ are linearly independent elements of $\mathbf{V}$ and $\boldsymbol{u}_1, \dots, \boldsymbol{u}_n$ are elements of $\mathbf{U}$ then $\sum_{i=1}^{n} \boldsymbol{h}(\boldsymbol{u}_i, \boldsymbol{v}_i) = \mathbf{0}$ implies $\boldsymbol{u}_1 = \dots = \boldsymbol{u}_n = \mathbf{0}$.

Proof. Exclude the trivial cases $\mathbf{V} = \mathbf{0}$ or $\mathbf{U} = \mathbf{0}$.

Suppose 1) is fulfilled. Then every element of $\mathbf{Z}$ is of the form $\sum_{k=1}^{r} \alpha_k \boldsymbol{h}(\boldsymbol{u}_k, \boldsymbol{v}_k)$. Since $\alpha \boldsymbol{h}(\boldsymbol{u}, \boldsymbol{v}) = \boldsymbol{h}(\alpha \boldsymbol{u}, \boldsymbol{v})$, we conclude that the elements of $\mathbf{Z}$ can be written in the form $\sum_{k=1}^{r} \boldsymbol{h}(\boldsymbol{u}_k, \boldsymbol{v}_k)$.

Suppose 2) is fulfilled, too. Take a bilinear map $\boldsymbol{c}\colon \mathbf{U} \times \mathbf{V} \to \mathbf{W}$ and define the map $\boldsymbol{L}\colon \mathbf{Z} \to \mathbf{W}$ by

$$\boldsymbol{L} \cdot \left(\sum_{k=1}^{r} \boldsymbol{h}(\boldsymbol{u}_k, \boldsymbol{v}_k) \right) := \sum_{k=1}^{r} \boldsymbol{c}(\boldsymbol{u}_k, \boldsymbol{v}_k).$$

If $\boldsymbol{L}$ is well-defined, then it is linear, $\boldsymbol{L} \circ \boldsymbol{h} = \boldsymbol{c}$, and it is unique with this property. To demonstrate that $\boldsymbol{L}$ is well-defined, we have to show that

$$\sum_{k=1}^{r} \boldsymbol{h}(\boldsymbol{u}_k, \boldsymbol{v}_k) = \sum_{j=1}^{s} \boldsymbol{h}(\boldsymbol{x}_j, \boldsymbol{y}_j) \quad \text{implies} \quad \sum_{k=1}^{r} \boldsymbol{c}(\boldsymbol{u}_k, \boldsymbol{v}_k) = \sum_{j=1}^{s} \boldsymbol{c}(\boldsymbol{x}_j, \boldsymbol{y}_j),$$

which is eqivalent to

$$\sum_{i=1}^{m} \boldsymbol{h}(\boldsymbol{u}_i, \boldsymbol{v}_i) = \mathbf{0} \qquad \text{implies} \qquad \sum_{i=1}^{m} \boldsymbol{c}(\boldsymbol{u}_i, \boldsymbol{v}_i) = \mathbf{0}.$$

Let us choose a largest set of linearly independent vectors from $\{\boldsymbol{v}_1, \dots, \boldsymbol{v}_m\}$; without loss of generality, we can suppose it is $\{\boldsymbol{v}_1, \dots, \boldsymbol{v}_n\}$ (where, of course, $n \le m$). If $\boldsymbol{v} = \sum_{i=1}^{n} \alpha_i \boldsymbol{v}_i$ then $\boldsymbol{h}(\boldsymbol{u}, \boldsymbol{v}) = \sum_{i=1}^{n} \boldsymbol{h}(\alpha_i \boldsymbol{u}, \boldsymbol{v}_i)$ and a similar formula holds for $\boldsymbol{c}(\boldsymbol{u},\boldsymbol{v})$ as well. Consequently, a rearrangement of the terms in the previous formulae yields that $\boldsymbol{L}$ is well-defined if

$$\sum_{i=1}^{n} \boldsymbol{h}(\boldsymbol{u}_i,\boldsymbol{v}_i) = \boldsymbol{0} \qquad \text{implies} \qquad \sum_{i=1}^{n} \boldsymbol{c}(\boldsymbol{u}_i,\boldsymbol{v}_i) = \boldsymbol{0}$$

whenever $\boldsymbol{v}_1,\ldots,\boldsymbol{v}_n$ are linearly independent which follows from condition 2).

We have proved that conditions 1) and 2) are sufficient for a tensor product.

Since $\boldsymbol{L}\circ\boldsymbol{h} = \boldsymbol{c}$ can define $\boldsymbol{L}$ only on the linear subspace spanned by the range of $\boldsymbol{h}$, condition 1) is necessary for the uniqueness of $\boldsymbol{L}$.

If condition 2) is not satisfied then we can find a bilinear map $\boldsymbol{r}$ such that $\boldsymbol{L}\circ\boldsymbol{h} \neq \boldsymbol{r}$ for all linear maps $\boldsymbol{L}$. Indeed, let the vectors $\boldsymbol{v}_1,\ldots,\boldsymbol{v}_n$ be linearly independent, $\sum\limits_{i=1}^{n} \boldsymbol{h}(\boldsymbol{u}_i,\boldsymbol{v}_i) = \boldsymbol{0}$, and at least one of the $\boldsymbol{u}_i$-s is not zero. Without loss of generality we can assume that $\boldsymbol{u}_1,\ldots,\boldsymbol{u}_m$ (where $m \leq n$) are linearly independent and all the other $\boldsymbol{u}_i$-s are their linear combinations. Complete $\{\boldsymbol{v}_1,\ldots,\boldsymbol{v}_n\}$ to a basis in $\mathbf{V}$ and $\{\boldsymbol{u}_1,\ldots,\boldsymbol{u}_m\}$ to a basis in $\mathbf{U}$. Define the bilinear map $\boldsymbol{r}\colon \mathbf{U}\times\mathbf{V} \to \mathbb{K}$ in such a way that $\boldsymbol{r}(\boldsymbol{u}_1,\boldsymbol{v}_1) := 1$ and $\boldsymbol{r}(\boldsymbol{u},\boldsymbol{v}) := 0$ for all other basis elements $\boldsymbol{u}$ and $\boldsymbol{v}$. Then for all linear maps $\boldsymbol{L}\colon \mathbf{Z} \to \mathbb{K}$ we have $\boldsymbol{L}\cdot\left(\sum\limits_{i=1}^{n} \boldsymbol{h}(\boldsymbol{u}_i,\boldsymbol{v}_i)\right) = 0 \neq 1 = \sum\limits_{i=1}^{n} \boldsymbol{r}(\boldsymbol{u}_i,\boldsymbol{v}_i)$.

3.2. In the next item the existence of tensor products will be proved. Observe that in the case $\mathbf{W} = \mathbf{Z}$, $\boldsymbol{c} = \boldsymbol{h}$, the identity map of $\mathbf{Z}$ fulfils $\boldsymbol{h} = \mathbf{1}_{\mathbf{Z}}\circ\boldsymbol{h}$; according to the definition of the tensor product this is the only possibility, i.e. if $\boldsymbol{L} \in \mathrm{Lin}(\mathbf{Z})$ and $\boldsymbol{h} = \boldsymbol{L}\circ\boldsymbol{h}$ then $\boldsymbol{L} = \mathbf{1}_{\mathbf{Z}}$.

As a consequence, if $(\mathbf{Z}',\boldsymbol{h}')$ is another tensor product of $\mathbf{U}$ and $\mathbf{V}$ then there is a unique linear bijection $\boldsymbol{L}\colon \mathbf{Z} \to \mathbf{Z}'$ such that $\boldsymbol{h}' = \boldsymbol{L}\circ\boldsymbol{h}$. This means that the tensor products of $\mathbf{U}$ and $\mathbf{V}$ are 'canonically isomorphic' or 'essentially the same', hence we speak of *the* tensor product and applying a customary abuse of language we call the corresponding vector space the tensor product ($\mathbf{Z}$ in the definition) denoting it by $\mathbf{U}\otimes\mathbf{V}$, and writing

$$\mathbf{U}\times\mathbf{V} \to \mathbf{U}\otimes\mathbf{V}, \qquad (\boldsymbol{u},\boldsymbol{v}) \mapsto \boldsymbol{u}\otimes\boldsymbol{v}$$

for the corresponding bilinear map ($\boldsymbol{h}$ in the definition); $\boldsymbol{u}\otimes\boldsymbol{v}$ is called the *tensor product* of $\boldsymbol{u}$ and $\boldsymbol{v}$.

An actual given tensor product is called a *realization* of the tensor product and the following symbols are used: $\mathbf{U}\otimes\mathbf{V} \hookrightarrow \mathbf{W}$ or $\mathbf{U}\otimes\mathbf{V} \equiv \mathbf{W}$ denote that the tensor product of $\mathbf{U}$ and $\mathbf{V}$ is realized as a subspace of $\mathbf{W}$ or as the whole vector space $\mathbf{W}$, respectively.

It is worth repeating the results of the previous paragraph in the new notations.

Every element of $\mathbf{U} \otimes \mathbf{V}$ can be written in the form $\sum_{i=1}^{n} \boldsymbol{u}_i \otimes \boldsymbol{v}_i$ where $\boldsymbol{v}_1, \ldots, \boldsymbol{v}_n$ are linearly independent vectors in $\mathbf{V}$. Moreover, if the sum is zero, then $\boldsymbol{u}_1 = \cdots = \boldsymbol{u}_n = \mathbf{0}$. In particular, if $\boldsymbol{u} \neq \mathbf{0}$ and $\boldsymbol{v} \neq \mathbf{0}$ then $\boldsymbol{u} \otimes \boldsymbol{v} \neq \mathbf{0}$.

3.3. For $\boldsymbol{u} \in \mathbf{U}$ and $\boldsymbol{v} \in \mathbf{V}$ we define the linear map

$$\boldsymbol{u} \otimes \boldsymbol{v} \colon \mathbf{V}^* \to \mathbf{U}, \qquad \boldsymbol{p} \mapsto (\boldsymbol{p} \cdot \boldsymbol{v})\boldsymbol{u}.$$

Proposition. $\mathbf{U} \times \mathbf{V} \to \mathrm{Lin}(\mathbf{V}^*, \mathbf{U})$, $(\boldsymbol{u}, \boldsymbol{v}) \mapsto \boldsymbol{u} \otimes \boldsymbol{v}$ is a bilinear map satisfying condition 2) of Proposition 3.1. As a consequence, the linear map $\boldsymbol{u} \otimes \boldsymbol{v}$ is the tensor product of $\boldsymbol{u}$ and $\boldsymbol{v}$ (that is why we used in advance this notation) and $\mathbf{U} \otimes \mathbf{V}$ is realized as a linear subspace of $\mathrm{Lin}(\mathbf{V}^*, \mathbf{U})$ spanned by such elements.

Proof. It is trivial that $(\boldsymbol{u}, \boldsymbol{v}) \mapsto \boldsymbol{u} \otimes \boldsymbol{v}$ is bilinear.

Suppose that $\boldsymbol{v}_1, \ldots, \boldsymbol{v}_n$ are linearly independent vectors in $\mathbf{V}$ and $\sum_{i=1}^{n} \boldsymbol{u}_i \otimes \boldsymbol{v}_i = \mathbf{0}$. Then for arbitrary $\boldsymbol{p} \in \mathbf{V}^*$ and $\boldsymbol{f} \in \mathbf{U}^*$ we have

$$0 = \boldsymbol{f} \cdot \left(\left(\sum_{i=1}^{n} \boldsymbol{u}_i \otimes \boldsymbol{v}_i \right) \cdot \boldsymbol{p} \right) = \sum_{i=1}^{n} (\boldsymbol{f} \cdot \boldsymbol{u}_i)(\boldsymbol{p} \cdot \boldsymbol{v}_i) = \boldsymbol{p} \cdot \left(\sum_{i=1}^{n} (\boldsymbol{f} \cdot \boldsymbol{u}_i) \boldsymbol{v}_i \right).$$

Since $\mathbf{V}^*$ separates the elements of $\mathbf{V}$, this means that $\sum_{i=1}^{n} (\boldsymbol{f} \cdot \boldsymbol{u}_i)\boldsymbol{v}_i = \mathbf{0}$. Because of the linear independence of $\boldsymbol{v}_i$-s this involves $\boldsymbol{f} \cdot \boldsymbol{u}_i = 0$ for all $i = 1, \ldots, n$. Since $\mathbf{U}^*$ separates the elements of $\mathbf{U}$, it follows that $\boldsymbol{u}_1 = \boldsymbol{u}_2 = \cdots = \boldsymbol{u}_n = \mathbf{0}$.

3.4. Proposition. If $\{\boldsymbol{v}_i \mid i \in I\}$ is a basis in $\mathbf{V}$ and $\{\boldsymbol{u}_j \mid j \in J\}$ is a basis in $\mathbf{U}$ then $\{\boldsymbol{u}_j \otimes \boldsymbol{v}_i \mid j \in J,\ i \in I\}$ is a basis in $\mathbf{U} \otimes \mathbf{V}$. ■

According to Propositions 3.3 and 1.7 we have

$$\mathbf{U} \otimes \mathbf{V} \hookrightarrow \mathrm{Lin}(\mathbf{V}^*, \mathbf{U}) \hookrightarrow \mathrm{Bilin}(\mathbf{U}^* \times \mathbf{V}^*, \mathbb{K}).$$

If $\mathbf{U}$ and $\mathbf{V}$ are finite dimensional then

$$\dim(\mathbf{U} \otimes \mathbf{V}) = (\dim \mathbf{U})(\dim \mathbf{V}).$$

Moreover, in this case $\dim(\mathbf{U} \otimes \mathbf{V}) = \dim(\mathrm{Lin}(\mathbf{V}^*, \mathbf{U}))$, hence the present proposition on the bases implies that for finite dimensional vector spaces

$$\mathbf{U} \otimes \mathbf{V} \equiv \mathrm{Lin}(\mathbf{V}^*, \mathbf{U}) \equiv \mathrm{Bilin}(\mathbf{U}^* \times \mathbf{V}^*, \mathbb{K})$$

and because of $\mathbf{V}^{**} \equiv \mathbf{V}$, $\mathbf{U}^{**} \equiv \mathbf{U}$,

$$\mathbf{U} \otimes \mathbf{V}^* \equiv \mathrm{Lin}(\mathbf{V}, \mathbf{U}) \equiv \mathrm{Bilin}(\mathbf{U}^* \times \mathbf{V}, \mathbb{K}),$$
$$\mathbf{U}^* \otimes \mathbf{V} \equiv \mathrm{Lin}(\mathbf{V}^*, \mathbf{U}^*) \equiv \mathrm{Bilin}(\mathbf{U} \times \mathbf{V}^*, \mathbb{K}),$$
$$\mathbf{U}^* \otimes \mathbf{V}^* \equiv \mathrm{Lin}(\mathbf{V}, \mathbf{U}^*) \equiv \mathrm{Bilin}(\mathbf{U} \times \mathbf{V}, \mathbb{K}).$$

3.5. We have the following identifications.

(*i*)
$$\mathbb{K} \otimes \mathbf{V} \equiv \mathbf{V}, \qquad \alpha \otimes \boldsymbol{v} \equiv \alpha \boldsymbol{v},$$

(*ii*)
$$(\mathbf{U} \times \mathbf{V}) \otimes \mathbf{W} \equiv (\mathbf{U} \otimes \mathbf{W}) \times (\mathbf{V} \otimes \mathbf{W},)$$
$$(\boldsymbol{u}, \boldsymbol{v}) \otimes \boldsymbol{w} \equiv (\boldsymbol{u} \otimes \boldsymbol{w}, \boldsymbol{v} \otimes \boldsymbol{w}),$$
$$\mathbf{W} \otimes (\mathbf{U} \times \mathbf{V}) \equiv (\mathbf{W} \otimes \mathbf{U}) \times (\mathbf{W} \otimes \mathbf{V}),$$
$$\boldsymbol{w} \otimes (\boldsymbol{u}, \boldsymbol{v}) \equiv (\boldsymbol{w} \otimes \boldsymbol{u}, \boldsymbol{w} \otimes \boldsymbol{v}),$$

(*iii*) If $\mathbf{U}$ and $\mathbf{V}$ are finite dimensional then

$$\mathbf{U}^* \otimes \mathbf{V}^* \equiv (\mathbf{U} \otimes \mathbf{V})^*, \qquad (\boldsymbol{f} \otimes \boldsymbol{p}) : (\boldsymbol{u} \otimes \boldsymbol{v}) \equiv (\boldsymbol{f} \cdot \boldsymbol{u})(\boldsymbol{p} \cdot \boldsymbol{v}),$$
$$(\boldsymbol{f} \in \mathbf{U}^*, \quad \boldsymbol{p} \in \mathbf{V}^*, \quad \boldsymbol{u} \in \mathbf{U}, \quad \boldsymbol{v} \in \mathbf{V})$$

where we found convenient to write the symbol : for the bilinear map of duality; we shall give an explanation later.

3.6. In mathematical books the tensor product is often said to be commutative which means that we have a unique linear bijection $\mathbf{U} \otimes \mathbf{V} \to \mathbf{V} \otimes \mathbf{U}$, $\boldsymbol{u} \otimes \boldsymbol{v} \mapsto \boldsymbol{v} \otimes \boldsymbol{u}$ admitting an identification. However, we do not find convenient to use this identification because of two reasons:

1) if $\mathbf{U} = \mathbf{V}$, $\boldsymbol{u}, \boldsymbol{v} \in \mathbf{V}$ and $\boldsymbol{u} \neq \boldsymbol{v}$ then, in general, $\boldsymbol{u} \otimes \boldsymbol{v} \neq \boldsymbol{v} \otimes \boldsymbol{u}$;

2) $\boldsymbol{u} \otimes \boldsymbol{v} \in \mathbf{U} \otimes \mathbf{V} \hookrightarrow \mathrm{Lin}(\mathbf{V}^*, \mathbf{U})$, $\boldsymbol{v} \otimes \boldsymbol{u} \in \mathbf{V} \otimes \mathbf{U} \hookrightarrow \mathrm{Lin}(\mathbf{U}^*, \mathbf{V}) \subset \mathrm{Lin}(\mathbf{U}^*, \mathbf{V}^{**})$; it is not hard to see that the transpose of $\boldsymbol{u} \otimes \boldsymbol{v}$ equals $\boldsymbol{v} \otimes \boldsymbol{u}$:

$$(\boldsymbol{u} \otimes \boldsymbol{v})^* = \boldsymbol{v} \otimes \boldsymbol{u}.$$

Hence the unique linear bijection between $\mathbf{U} \otimes \mathbf{V}$ and $\mathbf{V} \otimes \mathbf{U}$ that sends $\boldsymbol{u} \otimes \boldsymbol{v}$ into $\boldsymbol{v} \otimes \boldsymbol{u}$ is the transposing map. We do not want, in general, to identify a linear map with its transpose (e.g. a matrix with its transpose).

However, if one of the vector spaces is one-dimensional, we accept the mentioned identification, i.e.

$$\mathbb{A} \otimes \mathbf{V} \equiv \mathbf{V} \otimes \mathbb{A}, \quad \boldsymbol{a} \otimes \boldsymbol{v} \equiv \boldsymbol{v} \otimes \boldsymbol{a} \qquad \text{if} \quad \dim \mathbb{A} = 1.$$

Moreover, in this case we agree to omit the symbol $\otimes$:

$$\boldsymbol{a}\boldsymbol{v} := \boldsymbol{a} \otimes \boldsymbol{v} \qquad\qquad (\boldsymbol{a} \in \mathbb{A}, \ \boldsymbol{v} \in \mathbf{V}, \ \dim \mathbb{A} = 1).$$

Note that if $\dim \mathbb{A} = 1$ then every element of $\mathbb{A} \otimes \mathbf{V}$ has the form $\boldsymbol{a}\boldsymbol{v}$.

Though, in general, $\mathbb{A} \otimes \mathbf{V} \neq \mathbf{V}$, it makes sense (if $\dim \mathbb{A} = 1$) that an element $\boldsymbol{z}$ of $\mathbb{A} \otimes \mathbf{V}$ is *parallel* to an element $\boldsymbol{v}$ of $\mathbf{V}$: if there is an $\boldsymbol{a} \in \mathbb{A}$ such that $\boldsymbol{z} = \boldsymbol{a}\boldsymbol{v}$.

3.7. It is well known that a linear map $\boldsymbol{L}\colon \mathbf{V}_1 \times \mathbf{V}_2 \to \mathbf{U}_1 \times \mathbf{U}_2$ can be represented in a matrix form:

$$\boldsymbol{L} = \begin{pmatrix} \boldsymbol{L}_{11} & \boldsymbol{L}_{12} \\ \boldsymbol{L}_{21} & \boldsymbol{L}_{22} \end{pmatrix}$$

where $\boldsymbol{L}_{ik} \in \mathrm{Lin}(\mathbf{V}_i, \mathbf{U}_k)$ $(i, k = 1, 2)$ and

$$\boldsymbol{L} \cdot (\boldsymbol{v}_1, \boldsymbol{v}_2) = (\boldsymbol{L}_{11} \cdot \boldsymbol{v}_1 + \boldsymbol{L}_{12} \cdot \boldsymbol{v}_2,\ \boldsymbol{L}_{21} \cdot \boldsymbol{v}_1 + \boldsymbol{L}_{22} \cdot \boldsymbol{v}_2).$$

This corresponds to the finite dimensional identifications (see in particular 3.5(*ii*))

$$\begin{aligned} \mathrm{Lin}(\mathbf{V}_1 \times \mathbf{V}_2, &\mathbf{U}_1 \times \mathbf{U}_2) \equiv \\ &\equiv (\mathbf{U}_1 \times \mathbf{U}_2) \otimes (\mathbf{V}_1 \times \mathbf{V}_2)^* \equiv (\mathbf{U}_1 \times \mathbf{U}_2) \otimes (\mathbf{V}_1^* \times \mathbf{V}_2^*) \equiv \\ &\equiv (\mathbf{U}_1 \otimes \mathbf{V}_1^*) \times (\mathbf{U}_1 \otimes \mathbf{V}_2^*) \times (\mathbf{U}_2 \otimes \mathbf{V}_1^*) \times (\mathbf{U}_2 \otimes \mathbf{V}_2^*) \equiv \\ &\equiv \mathrm{Lin}(\mathbf{V}_1, \mathbf{U}_1) \times \mathrm{Lin}(\mathbf{V}_2, \mathbf{U}_1) \times \mathrm{Lin}(\mathbf{V}_1, \mathbf{U}_2) \times \mathrm{Lin}(\mathbf{V}_2, \mathbf{U}_2). \end{aligned}$$

Accordingly, we find convenient to write

$$(\boldsymbol{u}_1, \boldsymbol{u}_2) \otimes (\boldsymbol{p}_1, \boldsymbol{p}_2) \equiv \begin{pmatrix} \boldsymbol{u}_1 \otimes \boldsymbol{p}_1 & \boldsymbol{u}_1 \otimes \boldsymbol{p}_2 \\ \boldsymbol{u}_2 \otimes \boldsymbol{p}_1 & \boldsymbol{u}_2 \otimes \boldsymbol{p}_2 \end{pmatrix}$$

for $(\boldsymbol{u}_1, \boldsymbol{u}_2) \in (\mathbf{U}_1, \mathbf{U}_2)$ and $(\boldsymbol{p}_1, \boldsymbol{p}_2) \in \mathbf{V}_1^* \times \mathbf{V}_2^*$.

Of course, a similar formula holds for other tensor products, e.g. for the elements of $(\mathbf{U}_1 \times \mathbf{U}_2) \otimes (\mathbf{V}_1 \times \mathbf{V}_2)$:

$$(\boldsymbol{u}_1, \boldsymbol{u}_2) \otimes (\boldsymbol{v}_1, \boldsymbol{v}_2) \equiv \begin{pmatrix} \boldsymbol{u}_1 \otimes \boldsymbol{v}_1 & \boldsymbol{u}_1 \otimes \boldsymbol{v}_2 \\ \boldsymbol{u}_2 \otimes \boldsymbol{v}_1 & \boldsymbol{u}_2 \otimes \boldsymbol{v}_2 \end{pmatrix}.$$

It is not hard to see then (cf. 3.6) that

$$\begin{pmatrix} \boldsymbol{u}_1 \otimes \boldsymbol{v}_1 & \boldsymbol{u}_1 \otimes \boldsymbol{v}_2 \\ \boldsymbol{u}_2 \otimes \boldsymbol{v}_1 & \boldsymbol{u}_2 \otimes \boldsymbol{v}_2 \end{pmatrix}^* = \begin{pmatrix} \boldsymbol{v}_1 \otimes \boldsymbol{u}_1 & \boldsymbol{v}_1 \otimes \boldsymbol{u}_2 \\ \boldsymbol{v}_2 \otimes \boldsymbol{u}_1 & \boldsymbol{v}_2 \otimes \boldsymbol{u}_2 \end{pmatrix}.$$

3.8. If $\mathbb{A}$ is a one-dimensional vector space then $\mathrm{Lin}(\mathbb{A})$ is identified with $\mathbb{K}$: the number α corresponds to the linear map $\boldsymbol{a} \mapsto \alpha\boldsymbol{a}$. As a consequence, we have the following identification, too:

$$\mathbb{A} \otimes \mathbb{A}^* \equiv \mathrm{Lin}(\mathbb{A}) \equiv \mathbb{K}, \qquad \boldsymbol{a}\boldsymbol{\xi} \equiv \boldsymbol{\xi} \cdot \boldsymbol{a} (\equiv \boldsymbol{a} \cdot \boldsymbol{\xi})$$

(remember: $\boldsymbol{a\xi} := \boldsymbol{a} \otimes \boldsymbol{\xi}$). Indeed, by definition, $\boldsymbol{a\xi}: \mathbb{A} \to \mathbb{A},\ \boldsymbol{b} \mapsto (\boldsymbol{\xi} \cdot \boldsymbol{b})\boldsymbol{a}$. If $\boldsymbol{a} = \boldsymbol{0}$ then $\boldsymbol{a\xi} = 0 = \boldsymbol{\xi} \cdot \boldsymbol{a}$. If $\boldsymbol{a} \neq \boldsymbol{0}$ then there is a unique $\frac{b}{a} \in \mathbb{K}$ for all $\boldsymbol{h} \in \mathbb{A}$ such that $\boldsymbol{b} = \frac{b}{a}\boldsymbol{a}$. Thus $(\boldsymbol{\xi} \cdot \boldsymbol{b})\boldsymbol{a} = (\boldsymbol{\xi} \cdot \frac{b}{a}\boldsymbol{a})\boldsymbol{a} = (\boldsymbol{\xi} \cdot \boldsymbol{a})\frac{b}{a}\boldsymbol{a} = (\boldsymbol{\xi} \cdot \boldsymbol{a})\boldsymbol{b}$ and we see that $\boldsymbol{a\xi}$ $(= \boldsymbol{a} \otimes \boldsymbol{\xi})$ equals the multiplication by $\boldsymbol{\xi} \cdot \boldsymbol{a}$.

For one-dimensional vector spaces we prefer the symbol of (tensor) product to the dot for expressing the bilinear map of duality i.e. the symbol $\boldsymbol{a\xi}$ to $\boldsymbol{a} \cdot \boldsymbol{\xi}$.

3.9. Since $\mathbf{V} \times \mathbf{V}^* \to \mathbb{K},\ (\boldsymbol{v}, \boldsymbol{p}) \mapsto \boldsymbol{p} \cdot \boldsymbol{v}$ is a bilinear map, the definition of tensor products ensures the existence of a unique linear map

$$\mathrm{Tr}: \mathbf{V} \otimes \mathbf{V}^* \to \mathbb{K} \qquad \text{such that} \qquad \mathrm{Tr}(\boldsymbol{v} \otimes \boldsymbol{p}) = \boldsymbol{p} \cdot \boldsymbol{v}.$$

If $\mathbf{V}$ is finite dimensional then $\mathbf{V} \otimes \mathbf{V}^* \equiv \mathrm{Lin}(\mathbf{V})$, hence $\mathrm{Tr}\,\mathbf{L}$, called the *trace* of $\boldsymbol{L}$, is defined for all linear maps $\boldsymbol{L}: \mathbf{V} \to \mathbf{V}$.

Since for $\boldsymbol{u}, \boldsymbol{v} \in \mathbf{V}$ and $\boldsymbol{p}, \boldsymbol{q} \in \mathbf{V}^*$ we have $(\boldsymbol{u} \otimes \boldsymbol{p}) \cdot (\boldsymbol{v} \otimes \boldsymbol{q}) = (\boldsymbol{p} \cdot \boldsymbol{v})\boldsymbol{u} \otimes \boldsymbol{q}$, we easily deduce that for all $\boldsymbol{L}, \boldsymbol{K} \in \mathrm{Lin}(\mathbf{V})$ (if $\dim \mathbf{V} < \infty$)

$$\mathrm{Tr}(\boldsymbol{L} \cdot \boldsymbol{K}) = \mathrm{Tr}(\boldsymbol{K} \cdot \boldsymbol{L}).$$

If $\{\boldsymbol{v}_i \mid i = 1, \dots, N\}$ is a basis in $\mathbf{V}$ and $\{\boldsymbol{p}^i \mid i = 1, \dots, N\}$ is its dual then for all $\boldsymbol{v} \in \mathbf{V}$ and $\boldsymbol{p} \in \mathbf{V}^*$

$$\boldsymbol{p} \cdot \boldsymbol{v} = \sum_{i=1}^{N} (\boldsymbol{p}^i \cdot \boldsymbol{v})(\boldsymbol{p} \cdot \boldsymbol{v}_i) = \sum_{i=1}^{N} \boldsymbol{p}^i \cdot (\boldsymbol{v} \otimes \boldsymbol{p}) \cdot \boldsymbol{v}_i,$$

which gives

$$\mathrm{Tr}\,\boldsymbol{L} = \sum_{i=1}^{N} \boldsymbol{p}^i \cdot \boldsymbol{L} \cdot \boldsymbol{v}_i \qquad (\boldsymbol{L} \in \mathrm{Lin}(\mathbf{V})).$$

Note that the trace of linear maps $\mathbf{V} \to \mathbf{V}^*$ *and* $\mathbf{V}^* \to \mathbf{V}$ *makes no sense;* on the other hand, we have (for finite dimensional $\mathbf{V}$)

$$\mathrm{Tr}: \mathrm{Lin}(\mathbf{V}^*) \equiv \mathbf{V}^* \otimes \mathbf{V} \to \mathbb{K}, \qquad \boldsymbol{p} \otimes \boldsymbol{v} \mapsto \boldsymbol{p} \cdot \boldsymbol{v}$$

and we easily see by $(\boldsymbol{v} \otimes \boldsymbol{p})^* = \boldsymbol{p} \otimes \boldsymbol{v}$ that

$$\mathrm{Tr}(\boldsymbol{L}^*) = \mathrm{Tr}\,\boldsymbol{L} \qquad (\boldsymbol{L} \in \mathrm{Lin}(\mathbf{V})).$$

Moreover, if $\mathbf{Z}$ is a finite dimensional vector space, we define

$$\mathrm{Tr}: \mathrm{Lin}(\mathbf{V}, \mathbf{Z} \otimes \mathbf{V}) \equiv \mathbf{Z} \otimes \mathbf{V} \otimes \mathbf{V}^* \to \mathbf{Z}, \qquad \boldsymbol{z} \otimes \boldsymbol{v} \otimes \boldsymbol{p} \mapsto (\boldsymbol{p} \cdot \boldsymbol{v})\boldsymbol{z}.$$

3.10. Let $\mathbf{V}$ be finite dimensional. Then, according to 3.5(*iii*) and $\mathbf{V}^{**} \equiv \mathbf{V}$, we have $\mathbf{V}^* \otimes \mathbf{V} \equiv (\mathbf{V} \otimes \mathbf{V}^*)^*$, $(\boldsymbol{p}' \otimes \boldsymbol{v}'): (\boldsymbol{v} \otimes \boldsymbol{p}) \equiv (\boldsymbol{p}' \cdot \boldsymbol{v})(\boldsymbol{v}' \cdot \boldsymbol{p})$.

It is not hard to see that in other words this reads

$$\mathrm{Lin}(\mathbf{V}^*) \equiv (\mathrm{Lin}(\mathbf{V}))^*, \qquad \boldsymbol{B} : \boldsymbol{L} \equiv \mathrm{Tr}(\boldsymbol{B}^*\boldsymbol{L}),$$

where $\boldsymbol{L} \in \mathrm{Lin}(\mathbf{V})$, $\boldsymbol{B} \in \mathrm{Lin}(\mathbf{V}^*)$ and so $\boldsymbol{B}^* \in \mathrm{Lin}(\mathbf{V})$.

Since a single dot means the composition of linear maps, we denoted the bilinear map of duality by the symbol : to avoid misunderstandings.

3.11. In accordance with our results we have

$$\mathbb{K}^M \otimes \mathbb{K}^N \equiv \mathrm{Lin}\left((\mathbb{K}^N)^*, \mathbb{K}^M\right).$$

By definition, for $\boldsymbol{y} = (y^i) \in \mathbb{K}^M$ and $\boldsymbol{x} = (x^k) \in \mathbb{K}^N$,

$$\boldsymbol{y} \otimes \boldsymbol{x} : (\mathbb{K}^N)^* \to \mathbb{K}^M, \qquad \boldsymbol{p} \mapsto (\boldsymbol{p} \cdot \boldsymbol{x})\boldsymbol{y},$$

from which we deduce that

$$(\boldsymbol{y} \otimes \boldsymbol{x})^{ik} = y^i x^k \qquad (i = 1, \ldots, M,\ k = 1, \ldots, N).$$

Moreover, $\mathbb{K}^N \otimes (\mathbb{K}^N)^* \equiv \mathrm{Lin}(\mathbb{K}^N, \mathbb{K}^N)$, $(\boldsymbol{x} \otimes \boldsymbol{p})^i{}_k = x^i p_k$, and so

$$\mathrm{Tr}\left(L^i{}_k |\ i, k = 1, \ldots, N\right) = \sum_{i=1}^{N} L^i{}_i.$$

Our convention that a summation can be carried out only for a pair of indices in opposite positions shows well that the trace of matrices (L^{ik}) and (L_{ik}) makes no sense.

It can be proved without difficulty that

$$(B^j{}_i) : (L_k{}^l) = \sum_{i,k=1}^{N} B^k{}_i L_k{}^i.$$

3.12. Let $\boldsymbol{L} \in \mathrm{Lin}(\mathbf{U}, \mathbf{X})$ and $\boldsymbol{K} \in \mathrm{Lin}(\mathbf{V}, \mathbf{Y})$. Then $\mathbf{U} \times \mathbf{V} \to \mathbf{X} \otimes \mathbf{Y}$, $(\boldsymbol{u}, \boldsymbol{v}) \mapsto \boldsymbol{L}\boldsymbol{u} \otimes \boldsymbol{K}\boldsymbol{v}$ is a bilinear map, hence there exists a unique linear map $\boldsymbol{L} \otimes \boldsymbol{K} : \mathbf{U} \otimes \mathbf{V} \to \mathbf{X} \otimes \mathbf{Y}$ such that

$$(\boldsymbol{L} \otimes \boldsymbol{K}) \cdot (\boldsymbol{u} \otimes \boldsymbol{v}) = \boldsymbol{L}\boldsymbol{u} \otimes \boldsymbol{K}\boldsymbol{v} \qquad (\boldsymbol{u} \in \mathbf{U},\ \boldsymbol{v} \in \mathbf{V}).$$

It is a simple task to show that $(\boldsymbol{L}, \boldsymbol{K}) \mapsto \boldsymbol{L} \otimes \boldsymbol{K}$ satisfies condition (*ii*) in 3.1, hence $\boldsymbol{L} \otimes \boldsymbol{K}$ is the tensor product of $\boldsymbol{L}$ and $\boldsymbol{K}$, in other words,

$$\mathrm{Lin}(\mathbf{U}, \mathbf{X}) \otimes \mathrm{Lin}(\mathbf{V}, \mathbf{Y}) \hookrightarrow \mathrm{Lin}(\mathbf{U} \otimes \mathbf{V}, \mathbf{X} \otimes \mathbf{Y}).$$

If the vector spaces are finite dimensional then $\equiv$ stands instead of $\subset\!\!\to$. It is not hard to show that

$$(\boldsymbol{L}\otimes\boldsymbol{K})\cdot(\boldsymbol{B}\otimes\boldsymbol{A})=(\boldsymbol{L}\cdot\boldsymbol{B})\otimes(\boldsymbol{K}\cdot\boldsymbol{A})$$

and if both $\boldsymbol{L}$ and $\boldsymbol{K}$ are bijections then $\boldsymbol{L}\otimes\boldsymbol{K}$ is a bijection and

$$(\boldsymbol{L}\otimes\boldsymbol{K})^{-1}=\boldsymbol{L}^{-1}\otimes\boldsymbol{K}^{-1}.$$

3.13. For natural numbers $n\geq 2$ the definition of n-fold tensor products of vector spaces is similar to definition in 3.1, only n-fold linear maps should be taken instead of bilinear ones. We can state the existence and essential uniqueness of n-fold tensor products similarly. We use the notation $\overset{n}{\underset{k=1}{\otimes}}\mathbf{V}_k$ and $\overset{n}{\underset{k=1}{\otimes}}\boldsymbol{v}_k$ for the n-fold tensor product of vector spaces $\mathbf{V}_k$ and vectors $\boldsymbol{v}_k\in\mathbf{V}_k$ $(k=1,\ldots,n)$.

We have the identifications

$$\left(\overset{m}{\underset{k=1}{\otimes}}\mathbf{V}_k\right)\otimes\left(\overset{n}{\underset{k=m+1}{\otimes}}\mathbf{V}_k\right)\equiv\overset{n}{\underset{k=1}{\otimes}}\mathbf{V}_k,$$

$$\left(\overset{m}{\underset{k=1}{\otimes}}\boldsymbol{v}_k\right)\otimes\left(\overset{n}{\underset{k=m+1}{\otimes}}\boldsymbol{v}_k\right)\equiv\overset{n}{\underset{k=1}{\otimes}}\boldsymbol{v}_k.$$

If the vector spaces are finite dimensional then $\overset{n}{\underset{k=1}{\otimes}}\mathbf{V}_k$ is identified with the vector space $\mathrm{Lin}^n(\overset{n}{\underset{k=1}{\times}}\mathbf{V}_k{}^*,\mathbb{K})$ of n-linear maps $\overset{n}{\underset{k=1}{\times}}\mathbf{V}_k{}^*\to\mathbb{K}$, called *$n$-linear forms,* such that

$$\left(\overset{n}{\underset{k=1}{\otimes}}\boldsymbol{v}_k\right)(\boldsymbol{p}^1,\ldots,\boldsymbol{p}^n)\equiv\prod_{k=1}^{n}(\boldsymbol{p}^k\cdot\boldsymbol{v}_k).$$

3.14. For natural numbers $n\geq 2$, the n-fold tensor product of n copies of the vector space $\mathbf{V}$ is denoted by $\overset{n}{\otimes}\mathbf{V}$; for convenience we put $\overset{1}{\otimes}\mathbf{V}:=\mathbf{V}$, $\overset{0}{\otimes}\mathbf{V}:=\mathbb{K}$. Then we have for all natural numbers n and m

$$\left(\overset{n}{\otimes}\mathbf{V}\right)\otimes\left(\overset{m}{\otimes}\mathbf{V}\right)\equiv\overset{n+m}{\otimes}\mathbf{V}.$$

We define the n-fold *symmetric* and *antisymmetric* tensor products of elements of $\mathbf{V}$ as follows:

$$\overset{n}{\underset{k=1}{\vee}}\boldsymbol{v}_k:=\sum_{\pi\in\mathrm{Perm}_n}\overset{n}{\underset{k=1}{\otimes}}\boldsymbol{v}_{\pi(k)},$$

$$\overset{n}{\underset{k=1}{\wedge}}\boldsymbol{v}_k:=\sum_{\pi\in\mathrm{Perm}_n}(\operatorname{sign}\pi)\overset{n}{\underset{k=1}{\otimes}}\boldsymbol{v}_{\pi(k)},$$

where Perm_n denotes the set of permutations of $\{1,\dots,n\}$ and $\mathrm{sign}\,\pi$ is the sign of the permutation π: $\mathrm{sign}\,\pi = 1$ if π is even and $\mathrm{sign}\,\pi = -1$ if π is odd.

For instance,

$$\boldsymbol{v}_1 \vee \boldsymbol{v}_2 = \boldsymbol{v}_1 \otimes \boldsymbol{v}_2 + \boldsymbol{v}_2 \otimes \boldsymbol{v}_1, \qquad \boldsymbol{v}_1 \wedge \boldsymbol{v}_2 = \boldsymbol{v}_1 \otimes \boldsymbol{v}_2 - \boldsymbol{v}_2 \otimes \boldsymbol{v}_1.$$

The linear subspaces of $\overset{n}{\otimes}\,\mathbf{V}$ spanned by the symmetric and antisymmetric tensor products are denoted by $\overset{n}{\vee}\,\mathbf{V}$ and $\overset{n}{\wedge}\,\mathbf{V}$, respectively.

We mention that

$$\frac{1}{n!}\overset{n}{\underset{k=1}{\vee}} \boldsymbol{v}_k \qquad \text{and} \qquad \frac{1}{n!}\overset{n}{\underset{k=1}{\wedge}} \boldsymbol{v}_k$$

are called the *symmetric* and *antisymmetric part* of $\overset{n}{\underset{k=1}{\otimes}} \boldsymbol{v}_k$, respectively. It is worth mentioning that the intersection of $\overset{n}{\wedge}\,\mathbf{V}$ and $\overset{n}{\vee}\,\mathbf{V}$ is the zero subspace; moreover, for $n = 2$ the subspace of antisymmetric tensor products and that of symmetric tensor products span $\mathbf{V} \otimes \mathbf{V}$.

3.15. Let $\mathbf{V}$ be finite dimensional, $\dim \mathbf{V} = N$. Then $\mathbf{V}^{**} \equiv \mathbf{V}$, and we have the following identifications:

$$\overset{n}{\otimes}\,\mathbf{V} \equiv \{n\text{-linear forms on } \mathbf{V}^*\}, \qquad \overset{n}{\otimes}\,\mathbf{V}^* \equiv \{n\text{-linear forms on } \mathbf{V}\},$$
$$\overset{n}{\vee}\,\mathbf{V} \equiv \{\text{ symmetric } n\text{-linear forms on } \mathbf{V}^*\},$$
$$\overset{n}{\vee}\,\mathbf{V}^* \equiv \{\text{ symmetric } n\text{-linear forms on } \mathbf{V}\},$$
$$\overset{n}{\wedge}\,\mathbf{V} \equiv \{\text{ antisymmetric } n\text{-linear forms on } \mathbf{V}^*\},$$
$$\overset{n}{\wedge}\,\mathbf{V}^* \equiv \{\text{ antisymmetric } n\text{-linear forms on } \mathbf{V}\}.$$

It is worth mentioning that

$$\left(\overset{n}{\underset{k=1}{\otimes}} \boldsymbol{v}_k\right)(\boldsymbol{p}^1,\dots,\boldsymbol{p}^n) = \left(\overset{n}{\underset{k=1}{\otimes}} \boldsymbol{p}^k\right)(\boldsymbol{v}_1,\dots,\boldsymbol{v}_n) = \prod_{k=1}^{n} (\boldsymbol{p}^k \cdot \boldsymbol{v}_k),$$

$$\left(\overset{n}{\underset{k=1}{\vee}} \boldsymbol{v}_k\right)(\boldsymbol{p}^1,\dots,\boldsymbol{p}^n) = \left(\overset{n}{\underset{k=1}{\vee}} \boldsymbol{p}^k\right)(\boldsymbol{v}_1,\dots,\boldsymbol{v}_n)$$
$$= \sum_{\pi\in\mathrm{Perm}_n} \prod_{k=1}^{n} (\boldsymbol{p}^{\pi(k)} \cdot \boldsymbol{v}_k),$$

$$\left(\overset{n}{\underset{k=1}{\wedge}} \boldsymbol{v}_k\right)(\boldsymbol{p}^1,\dots,\boldsymbol{p}^n) = \left(\overset{n}{\underset{k=1}{\wedge}} \boldsymbol{p}^k\right)(\boldsymbol{v}_1,\dots,\boldsymbol{v}_n) =$$
$$= \sum_{\pi\in\mathrm{Perm}_n} \mathrm{sign}\,\pi \prod_{k=1}^{n} (\boldsymbol{p}^{\pi(k)} \cdot \boldsymbol{v}_k),$$

for all $\boldsymbol{v}_1, \dots, \boldsymbol{v}_n \in \mathbf{V}$ and $\boldsymbol{p}^1, \dots, \boldsymbol{p}^n \in \mathbf{V}^*$.

If $\{\boldsymbol{v}_i \mid i = 1, \dots, N\}$ is a basis in $\mathbf{V}$ then

$$\left\{ \overset{n}{\underset{k=1}{\otimes}} \boldsymbol{v}_{i_k} \,\middle|\, 1 \le i_k \le N,\ k = 1, \dots, n \right\},$$

$$\left\{ \overset{n}{\underset{k=1}{\vee}} \boldsymbol{v}_{i_k} \,\middle|\, 1 \le i_1 \le i_2 \le \dots \le i_n \le N \right\},$$

$$\left\{ \overset{n}{\underset{k=1}{\wedge}} \boldsymbol{v}_{i_k} \,\middle|\, 1 \le i_1 < i_2 < \dots < i_n \le N \right\}$$

are bases in $\overset{n}{\otimes} \mathbf{V}$, $\overset{n}{\vee} \mathbf{V}$ and $\overset{n}{\wedge} \mathbf{V}$, respectively. Accordingly,

$$\dim\left(\overset{n}{\otimes} \mathbf{V} \right) = N^n, \qquad \dim\left(\overset{n}{\vee} \mathbf{V} \right) = \binom{N+n-1}{n},$$

$$\dim\left(\overset{n}{\wedge} \mathbf{V} \right) = \begin{cases} \binom{N}{n} & \text{if } n \le N \\ 0 & \text{if } n > N \end{cases}$$

Similar statements are true for $\mathbf{V}^*$ instead of $\mathbf{V}$.

3.16. The reader is asked to demonstrate that for $n = 2$ the notions of symmetricity, symmetric part, etc. coincide with those introduced earlier. Moreover, using the formulae in 3.7 we have

$$(\boldsymbol{u}_1, \boldsymbol{u}_2) \wedge (\boldsymbol{v}_1, \boldsymbol{v}_2) = \begin{pmatrix} \boldsymbol{u}_1 \wedge \boldsymbol{v}_1 & \boldsymbol{u}_1 \otimes \boldsymbol{v}_2 - \boldsymbol{v}_1 \otimes \boldsymbol{u}_2 \\ \boldsymbol{u}_2 \otimes \boldsymbol{v}_1 - \boldsymbol{v}_2 \otimes \boldsymbol{u}_1 & \boldsymbol{u}_2 \wedge \boldsymbol{v}_2 \end{pmatrix}$$

for $\boldsymbol{v}_1, \boldsymbol{u}_1 \in \mathbf{V}_1$, $\boldsymbol{v}_2, \boldsymbol{u}_2 \in \mathbf{V}_2$, and a similar equality holds for symmetric tensor products, too.

3.17. We have the following identifications:

$$\overset{n}{\otimes} \mathbf{V}^* \equiv \left(\overset{n}{\otimes} \mathbf{V} \right)^*, \quad \left(\overset{n}{\underset{k=1}{\otimes}} \boldsymbol{p}^k \right) \cdot \left(\overset{n}{\underset{k=1}{\otimes}} \boldsymbol{v}_k \right) \equiv \left(\overset{n}{\underset{k=1}{\otimes}} \boldsymbol{p}^k \right) (\boldsymbol{v}_1, \dots, \boldsymbol{v}_n),$$

$$\overset{n}{\vee} \mathbf{V}^* \equiv \left(\overset{n}{\vee} \mathbf{V} \right)^*, \quad \left(\overset{n}{\underset{k=1}{\vee}} \boldsymbol{p}^k \right) \cdot \left(\overset{n}{\underset{k=1}{\vee}} \boldsymbol{v}_k \right) \equiv \left(\overset{n}{\underset{k=1}{\vee}} \boldsymbol{p}^k \right) (\boldsymbol{v}_1, \dots, \boldsymbol{v}_n),$$

$$\overset{n}{\wedge} \mathbf{V}^* \equiv \left(\overset{n}{\wedge} \mathbf{V} \right)^*, \quad \left(\overset{n}{\underset{k=1}{\wedge}} \boldsymbol{p}^k \right) \cdot \left(\overset{n}{\underset{k=1}{\wedge}} \boldsymbol{v}_k \right) \equiv \left(\overset{n}{\underset{k=1}{\wedge}} \boldsymbol{p}^k \right) (\boldsymbol{v}_1, \dots, \boldsymbol{v}_n).$$

3.18. Let $\mathbf{V}$ be an N-dimensional vector space. If $\mathbf{d}$ is an n-linear (symmetric, antisymmetric) form on $\mathbf{V}$ (i.e. $\mathbf{d}$ is an element of $\overset{n}{\otimes} \mathbf{V}^*$) and $\boldsymbol{L} \in \mathrm{Lin}(\mathbf{V})$, then

$$\mathbf{d} \circ \left(\overset{n}{\times} \boldsymbol{L} \right) : \mathbf{V}^n \to \mathbb{K}, \qquad (\boldsymbol{v}_1, \dots, \boldsymbol{v}_n) \mapsto \mathbf{d}(\boldsymbol{L} \cdot \boldsymbol{v}_1, \dots, \boldsymbol{L} \cdot \boldsymbol{v}_n)$$

is also an n-linear (symmetric, antisymmetric) form.

Since $\overset{N}{\wedge}\mathbf{V}^*$, the vector space of antisymmetric N-linear forms is one-dimensional, for $\boldsymbol{L} \in \operatorname{Lin}(\mathbf{V})$ there is a number (an element of $\mathbb{K}$) $\det \boldsymbol{L}$, called the *determinant* of $\boldsymbol{L}$, such that

$$\boldsymbol{c} \circ \left(\overset{N}{\times} \boldsymbol{L} \right) = (\det \boldsymbol{L})\boldsymbol{c}$$

for all $\boldsymbol{c} \in \overset{N}{\wedge}\mathbf{V}^*$.

Proposition. For all $\boldsymbol{v}_1, \boldsymbol{v}_2, \dots, \boldsymbol{v}_N$ in $\mathbf{V}$ we have

$$\overset{N}{\underset{k=1}{\wedge}} \boldsymbol{L} \cdot \boldsymbol{v}_k = (\det \boldsymbol{L}) \overset{N}{\underset{k=1}{\wedge}} \boldsymbol{v}_k.$$

Proof. $\overset{N}{\underset{k=1}{\wedge}} \boldsymbol{L} \cdot \boldsymbol{v}_k$ is an antisymmetric N-linear form on $\mathbf{V}^*$; 3.15 yields that for all $\boldsymbol{p}^1, \dots, \boldsymbol{p}^N \in \mathbf{V}^*$

$$\left(\overset{N}{\underset{k=1}{\wedge}} \boldsymbol{L} \cdot \boldsymbol{v}_k \right)(\boldsymbol{p}^1, \dots, \boldsymbol{p}^N) = \left(\overset{N}{\underset{k=1}{\wedge}} \boldsymbol{p}^k \right)(\boldsymbol{L} \cdot \boldsymbol{v}_1, \dots, \boldsymbol{L} \cdot \boldsymbol{v}_N) =$$

$$= (\det \boldsymbol{L}) \left(\overset{N}{\underset{k=1}{\wedge}} \boldsymbol{p}^k \right)(\boldsymbol{v}_1, \dots, \boldsymbol{v}_N) = (\det \boldsymbol{L}) \left(\overset{N}{\underset{k=1}{\wedge}} \boldsymbol{v}_k \right)(\boldsymbol{p}^1, \dots, \boldsymbol{p}^N). \blacksquare$$

As a consequence, we have for $\boldsymbol{L}, \boldsymbol{K} \in \operatorname{Lin}(\mathbf{V})$ that

$$\det(\boldsymbol{L} \cdot \boldsymbol{K}) = (\det \boldsymbol{L})(\det \boldsymbol{K}) = \det(\boldsymbol{K} \cdot \boldsymbol{L}).$$

3.19. Let $(\boldsymbol{v}_1, \dots, \boldsymbol{v}_N)$ be an ordered basis of $\mathbf{V}$ and let $(\boldsymbol{p}^1, \dots, \boldsymbol{p}^N)$ be the corresponding dual basis in $\mathbf{V}^*$.
We know that $\left(\overset{N}{\underset{i=1}{\wedge}} \boldsymbol{p}^i \right)(\boldsymbol{v}_1, \dots, \boldsymbol{v}_N) = 1$, thus if $\boldsymbol{L} \in \operatorname{Lin}(\mathbf{V})$ then

$$\det \boldsymbol{L} = (\det \boldsymbol{L}) \left(\overset{N}{\underset{i=1}{\wedge}} \boldsymbol{p}^i \right)(\boldsymbol{v}_1, \dots, \boldsymbol{v}_N) = \left(\overset{N}{\underset{i=1}{\wedge}} \boldsymbol{p}^i \right)(\boldsymbol{L} \cdot \boldsymbol{v}_1, \dots, \boldsymbol{L} \cdot \boldsymbol{v}_N) =$$

$$= \sum_{\pi \in Perm_N} \operatorname{sign} \pi \prod_{i=1}^{N} (\boldsymbol{p}^{\pi(i)} \cdot \boldsymbol{L} \cdot \boldsymbol{v}_i).$$

The last formula is the determinant of the matrix representing $\boldsymbol{L}$ in the coordinatization corresponding to the given ordered basis. Thus for all coordinatizations $\boldsymbol{K}$ of $\mathbf{V}$ we have

$$\det \boldsymbol{L} = \det(\boldsymbol{K} \cdot \boldsymbol{L} \cdot \boldsymbol{K}^{-1}).$$

3.20. Proposition. Let $\mathbf{V}$ and $\mathbf{U}$ be finite dimensional vector spaces. Suppose $\boldsymbol{A}, \boldsymbol{B} \in \operatorname{Lin}(\mathbf{V}, \mathbf{U})$ and $\boldsymbol{B}$ is a bijection. Then

$$\det(\boldsymbol{A} \cdot \boldsymbol{B}^{-1}) = \det(\boldsymbol{B}^{-1} \cdot \boldsymbol{A}).$$

Proof. Observe that if $\mathbf{U} = \mathbf{V}$ then this equality follows from that given at the end of 3.18. However, if $\mathbf{U} \neq \mathbf{V}$, the determinant of $\mathbb{A}$ and $\boldsymbol{B}^{-1}$ make no sense.

$\mathbf{V}$ and $\mathbf{U}$ have the same dimension N since $\boldsymbol{B}$ is a bijection between them. Let $\boldsymbol{K}$ and $\boldsymbol{L}$ be coordinatizations of $\mathbf{V}$ and $\mathbf{U}$, respectively. Then

$$\det(\mathbb{A} \cdot \mathbb{B}^{-1}) = \det(\boldsymbol{L} \cdot \boldsymbol{A} \cdot \boldsymbol{B}^{-1} \cdot \boldsymbol{L}^{-1}).$$

Since $\boldsymbol{L} \cdot \boldsymbol{A} \cdot \boldsymbol{B}^{-1} \cdot \boldsymbol{L}^{-1} = \boldsymbol{L} \cdot \boldsymbol{A} \cdot \boldsymbol{K}^{-1} \cdot \boldsymbol{K} \cdot \boldsymbol{B}^{-1} \cdot \boldsymbol{L}^{-1}$ and both $\boldsymbol{L} \cdot \boldsymbol{A} \cdot \boldsymbol{K}^{-1}$ and $\boldsymbol{K} \cdot \boldsymbol{B}^{-1} \cdot \boldsymbol{L}^{-1}$ are linear maps $\mathbb{K}^N \to \mathbb{K}^N$, hence their determinant is meaningful, we can apply the formula given at the end of 3.18 to get

$$\begin{aligned}\det(\boldsymbol{L} \cdot \boldsymbol{A} \cdot \boldsymbol{B}^{-1} \cdot \boldsymbol{L}^{-1}) &= \det\big((\boldsymbol{L} \cdot \boldsymbol{A} \cdot \boldsymbol{K}^{-1}) \cdot (\boldsymbol{K} \cdot \boldsymbol{B}^{-1} \cdot \boldsymbol{L}^{-1})\big) = \\ &= \det\big((\boldsymbol{K} \cdot \boldsymbol{B}^{-1} \cdot \boldsymbol{L}^{-1}) \cdot (\boldsymbol{L} \cdot \boldsymbol{A} \cdot \boldsymbol{K}^{-1})\big) = \det(\boldsymbol{K} \cdot \boldsymbol{B}^{-1} \cdot \boldsymbol{A} \cdot \boldsymbol{K}^{-1}) \\ &= \det(\boldsymbol{B}^{-1} \cdot \boldsymbol{A}). \qquad \blacksquare\end{aligned}$$

Our result has the following corollary: if $\boldsymbol{L} \in \operatorname{Lin}(\mathbf{V})$ and $\boldsymbol{B} \colon \mathbf{V} \to U$ is a linear bijection then

$$\det(\boldsymbol{B} \cdot \boldsymbol{L} \cdot \boldsymbol{B}^{-1}) = \det \boldsymbol{L}.$$

3.21. For $\boldsymbol{L} \in \operatorname{Lin}(\mathbf{V})$ we define

$$\overset{0}{\otimes} \boldsymbol{L} := \mathbf{1}_{\mathbb{K}},$$

$$\overset{n}{\otimes} \boldsymbol{L} \colon \overset{n}{\otimes} \mathbf{V} \to \overset{n}{\otimes} \mathbf{V}, \qquad \overset{n}{\underset{k=1}{\otimes}} \boldsymbol{v}_k \mapsto \overset{n}{\underset{k=1}{\otimes}} \boldsymbol{L} \cdot \boldsymbol{v}_k.$$

It is trivial that $\overset{n}{\wedge} \mathbf{V}$ and $\overset{n}{\vee} \mathbf{V}$ are invariant for $\overset{n}{\otimes} \boldsymbol{L}$; the restrictions of $\overset{n}{\otimes} \boldsymbol{L}$ onto these linear subspaces will be denoted by $\overset{n}{\wedge} \boldsymbol{L}$ and $\overset{n}{\vee} \boldsymbol{L}$, respectively.

3.22. Exercises

1. Let $\{\boldsymbol{v}_1, \dots, \boldsymbol{v}_N\}$ be a basis of $\mathbf{V}$ and $\{\boldsymbol{p}^1, \dots, \boldsymbol{p}^N\}$ its dual. Then

$$\sum_{i=1}^{N} \boldsymbol{v}_i \otimes \boldsymbol{p}^i \equiv \mathbf{1}_{\mathbf{V}}, \qquad \sum_{i=1}^{N} \boldsymbol{p}^i \otimes \boldsymbol{v}_i \equiv \mathbf{1}_{\mathbf{V}^*},$$

where the symbols on the right-hand sides stand for the identity of $\mathbf{V}$ and of $\mathbf{V}^*$, respectively.

2. The linear subspaces $\mathbf{S}$ and $\mathbf{T}$ of $\mathbf{V}$ are *complementary* if $\mathbf{S} \cap \mathbf{T} = \{0\}$ and the linear subspace spanned by $\mathbf{S} \cup \mathbf{T}$ equals $\mathbf{V}$; then for every $\boldsymbol{v}$ there are uniquely determined elements $\boldsymbol{v}_{\mathbf{S}} \in \mathbf{S}$ and $\boldsymbol{v}_{\mathbf{T}} \in \mathbf{T}$ such that $\boldsymbol{v} = \boldsymbol{v}_{\mathbf{S}} + \boldsymbol{v}_{\mathbf{T}}$. The linear map $\mathbf{V} \to \mathbf{V}$, $\boldsymbol{v} \mapsto \boldsymbol{v}_{\mathbf{S}}$ is called the *projection onto* $\mathbf{S}$ *along* $\mathbf{T}$.

Let $\boldsymbol{v} \in \mathbf{V}$, $\boldsymbol{p} \in \mathbf{V}^*$.

(*i*) If $\boldsymbol{p} \cdot \boldsymbol{v} \neq 0$ then $\frac{\boldsymbol{v} \otimes \boldsymbol{p}}{\boldsymbol{p} \cdot \boldsymbol{v}}$ is the projection onto $\mathbb{K}\boldsymbol{v}$ (the linear subspace spanned by $\boldsymbol{v}$) along $\operatorname{Ker} \boldsymbol{p}$.

(*ii*) If $\alpha \in \mathbb{K}$ such that $\alpha \boldsymbol{p} \cdot \boldsymbol{v} \neq 1$ then $\mathbf{1}_{\mathbf{V}} - \alpha \boldsymbol{v} \otimes \boldsymbol{p}$ is a linear bijection and

$$(\mathbf{1}_{\mathbf{V}} - \alpha \boldsymbol{v} \otimes \boldsymbol{p})^{-1} = \mathbf{1}_{\mathbf{V}} + \frac{\alpha}{1 - \alpha \boldsymbol{p} \cdot \boldsymbol{v}} \boldsymbol{v} \otimes \boldsymbol{p}.$$

3. Demonstrate that

$$\boldsymbol{L} \cdot (\boldsymbol{v} \otimes \boldsymbol{p}) = (\boldsymbol{L} \cdot \boldsymbol{v}) \otimes \boldsymbol{p}, \qquad (\boldsymbol{v} \otimes \boldsymbol{p}) \cdot \boldsymbol{L} = \boldsymbol{v} \otimes \boldsymbol{L}^* \cdot \boldsymbol{p}$$

for $\boldsymbol{v} \in \mathbf{V}$, $\boldsymbol{p} \in \mathbf{V}^*$ and $\boldsymbol{L} \in \operatorname{Lin}(\mathbf{V})$.

4. Prove that

$$\left(\bigwedge_{k=1}^{n} \boldsymbol{p}^k \right) (\boldsymbol{v}_1, \ldots, \boldsymbol{v}_n) = \det \left(\boldsymbol{p}^k \cdot \boldsymbol{v}_i \middle| \, k, i = 1, \ldots, n \right)$$

for $\boldsymbol{p}^1, \ldots, \boldsymbol{p}^n \in V^*$ and $\boldsymbol{v}_1, \ldots, \boldsymbol{v}_n \in \mathbf{V}$.

5. Prove that if $\mathbf{V}$ is a vector space over $\mathbb{K}$ then $\mathbb{K}^N \otimes \mathbf{V} \equiv \mathbf{V}^N$, $\boldsymbol{\xi} \otimes \boldsymbol{v} \equiv (\xi^1 \boldsymbol{v}, \ldots, \xi^N \boldsymbol{v})$.

4. Tensor quotients

4.1. Let $\mathbf{U}, \mathbf{V}$ and $\mathbf{Z}$ be vector spaces (over the same field). A map $\mathsf{q} \colon \mathbf{V} \times (\mathbf{U} \setminus \{0\}) \to \mathbf{Z}$ is called *linear quotient* if

(*i*) $\boldsymbol{v} \mapsto \mathsf{q}(\boldsymbol{v}, \boldsymbol{u})$ is linear for all $\boldsymbol{u} \in \mathbf{U} \setminus \{0\}$,

(*ii*) $\mathsf{q}(\boldsymbol{v}, \alpha \boldsymbol{u}) = \frac{1}{\alpha} \mathsf{q}(\boldsymbol{v}, \boldsymbol{u})$ for all $\boldsymbol{v} \in \mathbf{V}$ and $\boldsymbol{u} \in \mathbf{U} \setminus \{0\}$, $\alpha \in \mathbb{K} \setminus \{0\}$.

Definition. Let $\mathbf{V}$ and $\mathbb{A}$ be vector spaces, $\dim \mathbb{A} = 1$. A *tensor quotient* of $\mathbf{V}$ by $\mathbb{A}$ is a pair $(\mathbf{Z}, \mathsf{q})$ where

(*i*) $\mathbf{Z}$ is a vector space,

(*ii*) $\mathsf{q} \colon \mathbf{V} \times (\mathbb{A} \setminus \{0\}) \to \mathbf{Z}$ is a linear quotient map having the property that

– if $\mathbf{W}$ is a vector space and $\boldsymbol{r} \colon \mathbf{V} \times (\mathbb{A} \setminus \{0\}) \to \mathbf{W}$ is a linear quotient map

—then there exists a unique linear map $\boldsymbol{L} \colon \mathbf{Z} \to \mathbf{W}$ such that

$$\boldsymbol{r} = \boldsymbol{L} \circ \mathsf{q}. \quad \blacksquare$$

Proposition. The pair $(\mathbf{Z}, \mathbf{q})$ is a tensor quotient of $\mathbf{V}$ by $\mathbb{A}$ if and only if
1) $\mathbf{Z} = \operatorname{Ran}\mathbf{q}$,
2) if $\boldsymbol{v} \in \mathbf{V}$, $\boldsymbol{a} \in \mathbb{A} \setminus \{0\}$ and $\mathbf{q}(\boldsymbol{v}, \boldsymbol{a}) = \mathbf{0}$ then $\boldsymbol{v} = \mathbf{0}$.

Proof. Since $\mathbb{A}$ is one-dimensional, for $\boldsymbol{a}, \boldsymbol{b} \in \mathbb{A}$, $\boldsymbol{a} \neq \mathbf{0}$ let $\frac{\boldsymbol{b}}{\boldsymbol{a}}$ denote the number for which $\frac{\boldsymbol{b}}{\boldsymbol{a}}\boldsymbol{a} = \boldsymbol{b}$. Observe that if $\boldsymbol{b} \neq \mathbf{0}$ then $\frac{\boldsymbol{a}}{\boldsymbol{b}}$ is the inverse of $\frac{\boldsymbol{b}}{\boldsymbol{a}}$.

Condition 2) in the proposition is equivalent to the following one: if $\boldsymbol{v}, \boldsymbol{u} \in \mathbf{V}$ and $\boldsymbol{a}, \boldsymbol{b} \in \mathbb{A} \setminus \{0\}$ then $\mathbf{q}(\boldsymbol{v}, \boldsymbol{a}) = \mathbf{q}(\boldsymbol{u}, \boldsymbol{b})$ implies $\boldsymbol{v} = \frac{\boldsymbol{a}}{\boldsymbol{b}}\boldsymbol{u}$. Conversely, it is trivial, that if $\boldsymbol{r}$ is a linear quotient map and $\boldsymbol{v} = \frac{\boldsymbol{a}}{\boldsymbol{b}}\boldsymbol{u}$ then $\boldsymbol{r}(\boldsymbol{v}, \boldsymbol{a}) = \boldsymbol{r}(\boldsymbol{u}, \boldsymbol{b})$. Moreover, $\boldsymbol{r}(\boldsymbol{v}, \boldsymbol{a}) + \boldsymbol{r}(\boldsymbol{u}, \boldsymbol{b}) = \boldsymbol{r}\left(\frac{\boldsymbol{b}}{\boldsymbol{a}}\boldsymbol{v} + \boldsymbol{u}, \boldsymbol{b}\right)$.

Suppose 1) is fulfilled. Then every element of $\mathbf{Z}$ has the form $q(\boldsymbol{v}, \boldsymbol{a})$. If 2) is valid as well and $\boldsymbol{r}$ is a linear quotient map then the formula

$$\boldsymbol{L} \cdot (q(\boldsymbol{v}, \boldsymbol{a})) := \boldsymbol{r}(\boldsymbol{v}, \boldsymbol{a})$$

defines a unique linear map $\boldsymbol{L}$.

If 1) is not fulfilled, the uniqueness of linear maps $\boldsymbol{L}$ for which $\boldsymbol{r} = \boldsymbol{L} \circ \mathbf{q}$ holds fails. If 2) is not valid one can easily construct a linear quotient map for which no linear map exists with the desired composition property.

4.2. We shall see in the next item that tensor quotients exist. In the same way as in the case of tensor products, we can see that the tensor quotients of $\mathbf{V}$ by $\mathbb{A}$ are canonically isomorphic, that is why we speak of *the* tensor product and applying a customary abuse of language we call the corresponding vector space the tensor quotient ($\mathbf{Z}$ in the definition) denoting it by $\frac{\mathbf{V}}{\mathbb{A}}$ and writing

$$\mathbf{V} \times (\mathbb{A} \setminus \{0\}) \to \frac{\mathbf{V}}{\mathbb{A}}, \qquad (\boldsymbol{v}, \boldsymbol{a}) \mapsto \frac{\boldsymbol{v}}{\boldsymbol{a}}$$

for the corresponding linear quotient map ($\mathbf{q}$ in the definition); $\frac{\boldsymbol{v}}{\boldsymbol{a}}$ is called the *tensor quotient* of $\boldsymbol{v}$ by $\boldsymbol{a}$.

We use the term *realization* and the symbol $\equiv$ in the same sense as in the case of tensor products.

It is worth repeating the preceding results in the new notation: every element of $\frac{\mathbf{V}}{\mathbb{A}}$ is of the form $\frac{\boldsymbol{v}}{\boldsymbol{a}}$ and $\frac{\boldsymbol{v}}{\boldsymbol{a}} = \mathbf{0}$ if and only if $\boldsymbol{v} = \mathbf{0}$.

4.3. For $\boldsymbol{v} \in \mathbf{V}$ and $\boldsymbol{a} \in \mathbb{A} \setminus \{0\}$ we define the linear map

$$\frac{\boldsymbol{v}}{\boldsymbol{a}} : \mathbb{A} \to \mathbf{V}, \qquad \boldsymbol{b} \mapsto \frac{\boldsymbol{b}}{\boldsymbol{a}}\boldsymbol{v}$$

where $\frac{\boldsymbol{b}}{\boldsymbol{a}}$ is the number for which $\frac{\boldsymbol{b}}{\boldsymbol{a}}\boldsymbol{a} = \boldsymbol{b}$ holds.

Proposition. $\mathbf{V} \times \mathbb{A} \setminus \{0\} \to \operatorname{Lin}(\mathbb{A}, \mathbf{V})$ is a linear quotient map which satisfies conditions (*i*) and (*ii*) of proposition 4.1. As a consequence, $\frac{\boldsymbol{v}}{\boldsymbol{a}}$ is the tensor quotient of $\boldsymbol{v}$ by $\boldsymbol{a}$ (that is why we used in advance this notation) and $\frac{\mathbf{V}}{\mathbb{A}} \equiv \mathrm{Lin}(\mathbb{A}, \mathbf{V})$. ■

We have $\mathrm{Lin}(\mathbb{A}) \equiv \mathbb{K}$ where $\alpha \in \mathbb{K}$ is identified with the linear map $\boldsymbol{a} \mapsto \alpha\boldsymbol{a}$. Thus, according to the previous result, $\frac{\mathbb{A}}{\mathbb{A}} \equiv \mathbb{K}$ and $\frac{\boldsymbol{b}}{\boldsymbol{a}}$ is the number for which $\frac{\boldsymbol{b}}{\boldsymbol{a}}\boldsymbol{a} = \boldsymbol{b}$ holds, hence our notation in 4.1 used in the present proposition as well, is in accordance with the generally accepted notation for tensor quotients.

4.4. Since for all $\boldsymbol{a} \in \mathbb{A} \setminus \{0\}$ the map $\mathbf{V} \to \frac{\mathbf{V}}{\mathbb{A}}$, $\boldsymbol{v} \mapsto \frac{\boldsymbol{v}}{\boldsymbol{a}}$ is a linear bijection, if $\{\boldsymbol{v}_i \mid i \in I\}$ is a basis in $\mathbf{V}$ then $\left\{\frac{\boldsymbol{v}_i}{\boldsymbol{a}} \,\middle|\, i \in I\right\}$ is a basis in $\frac{\mathbf{V}}{\mathbb{A}}$, and $\dim \frac{\mathbf{V}}{\mathbb{A}} = \dim \mathbf{V}$.

4.5. Let $\mathbf{V}, \mathbf{U}, \mathbb{A}$ and $\mathbb{B}$ be vector spaces, $\dim \mathbb{A} = \dim \mathbb{B} = 1$. We have the following identifications (recall 3.4, 3.5 and 3.8):

(i) $$\frac{\mathbb{K}}{\mathbb{A}} \equiv \mathrm{Lin}(\mathbb{A}, \mathbb{K}) = \mathbb{A}^*, \qquad \frac{\alpha}{\boldsymbol{a}} \cdot \boldsymbol{b} \equiv \alpha \frac{\boldsymbol{b}}{\boldsymbol{a}};$$

(ii) $$\frac{\mathbf{V}}{\mathbb{K}} \equiv \mathrm{Lin}(\mathbb{K}, \mathbf{V})) \equiv \mathbf{V}, \qquad \frac{\boldsymbol{v}}{\alpha} \equiv \frac{1}{\alpha}\boldsymbol{v};$$

(iii) $$\frac{\mathbf{V}}{\mathbb{A}} \equiv \mathrm{Lin}(\mathbb{A}, \mathbf{V}) \equiv \mathbf{V} \otimes \mathbb{A}^*, \qquad \frac{\boldsymbol{v}}{\boldsymbol{a}} \equiv \boldsymbol{v} \otimes \frac{1}{\boldsymbol{a}};$$

(iv) $$\frac{\mathbf{V}^*}{\mathbb{A}^*} \equiv \left(\frac{\mathbf{V}}{\mathbb{A}}\right)^*, \qquad \frac{\boldsymbol{p}}{\boldsymbol{\xi}} \cdot \frac{\boldsymbol{v}}{\boldsymbol{a}} \equiv \frac{\boldsymbol{p} \cdot \boldsymbol{v}}{\boldsymbol{\xi}\boldsymbol{a}};$$

(v) $$\frac{\left(\frac{\mathbf{V}}{\mathbb{A}}\right)}{\mathbb{B}} \equiv \frac{\mathbf{V}}{\mathbb{A} \otimes \mathbb{B}}, \qquad \frac{\left(\frac{\boldsymbol{v}}{\boldsymbol{a}}\right)}{\boldsymbol{h}} \equiv \frac{\boldsymbol{v}}{\boldsymbol{a}\boldsymbol{b}};$$

(vi) $$\frac{\mathbf{V}}{\mathbb{A}} \otimes \frac{\mathbf{U}}{\mathbb{B}} \equiv \frac{\mathbf{V} \otimes \mathbf{U}}{\mathbb{A} \otimes \mathbb{B}} \equiv \frac{\mathbf{V}}{\mathbb{A} \otimes \mathbb{B}} \otimes \mathbf{U} \equiv etc.$$

$$\frac{\boldsymbol{v}}{\boldsymbol{a}} \otimes \frac{\boldsymbol{u}}{\boldsymbol{b}} \equiv \frac{\boldsymbol{v} \otimes \boldsymbol{u}}{\boldsymbol{a}\boldsymbol{b}} \equiv \frac{\boldsymbol{v}}{\boldsymbol{a}\boldsymbol{b}} \otimes \boldsymbol{u} \equiv etc.$$

In particular,

$$\mathbb{A} \otimes \frac{\mathbf{V}}{\mathbb{A}} \equiv \frac{\mathbb{A} \otimes \mathbf{V}}{\mathbb{A}} \equiv \mathbf{V}, \qquad \frac{\mathbb{B}}{\mathbb{A} \otimes \mathbb{B}} \equiv \frac{\mathbb{K}}{\mathbb{A}}.$$

(vii) $$\frac{\mathbf{V} \times \mathbf{U}}{\mathbb{A}} \equiv \frac{\mathbf{V}}{\mathbb{A}} \times \frac{\mathbf{U}}{\mathbb{A}}, \qquad \frac{(\boldsymbol{v}, \boldsymbol{u})}{\boldsymbol{a}} \equiv \left(\frac{\boldsymbol{v}}{\boldsymbol{a}}, \frac{\boldsymbol{u}}{\boldsymbol{a}}\right).$$

Note that according to (v) and (vi) the rules of tensorial multiplication and division coincide with those well known for numbers.

4.6. Let $\mathbf{V}, \mathbf{U}, \mathbb{A}$ and $\mathbb{B}$ be vector spaces, $\dim \mathbb{A} = \dim \mathbb{B} = 1$. If $\boldsymbol{L} \in \mathrm{Lin}(\mathbf{V}, \mathbf{U})$ and $\mathbf{0} \neq \boldsymbol{F} \in \mathrm{Lin}(\mathbb{A}, \mathbb{B})$ then $\mathbf{V} \times (\mathbb{A} \setminus \{0\}) \to \frac{\mathbf{U}}{\mathbb{B}}$, $(\boldsymbol{v}, \boldsymbol{a}) \mapsto \frac{\boldsymbol{L} \cdot \boldsymbol{v}}{\boldsymbol{F} \cdot \boldsymbol{a}}$ is linear quotient, hence there exists a unique linear map $\frac{\boldsymbol{L}}{\boldsymbol{F}} : \frac{\mathbf{V}}{\mathbb{A}} \to \frac{\mathbf{U}}{\mathbb{B}}$ such that

$$\frac{\boldsymbol{L}}{\boldsymbol{F}} \cdot \frac{\boldsymbol{v}}{\boldsymbol{a}} = \frac{\boldsymbol{L} \cdot \boldsymbol{v}}{\boldsymbol{F} \cdot \boldsymbol{a}} \qquad (\boldsymbol{v} \in \boldsymbol{V},\ \boldsymbol{a} \in \mathbb{A} \setminus \{0\}).$$

It is not hard to see that $\frac{\boldsymbol{L}}{\boldsymbol{F}}$ is really the quotient of $\boldsymbol{L}$ by $\boldsymbol{F}$, in other words,

$$\frac{\mathrm{Lin}(\mathbf{V}, \mathbf{U})}{\mathrm{Lin}(\mathbb{A}, \mathbb{B})} \equiv \mathrm{Lin}\left(\frac{\mathbf{V}}{\mathbb{A}}, \frac{\mathbf{U}}{\mathbb{B}}\right).$$

5. Tensorial operations and orientation

In this section $\mathbf{V}$ denotes an N-dimensional real vector space and $\mathbb{A}$ denotes a one-dimensional real vector space.

5.1. Recall that an element $(\boldsymbol{v}_1, \ldots, \boldsymbol{v}_N)$ of $\mathbf{V}^N$ is called an ordered basis of $\mathbf{V}$ if the set $\{\boldsymbol{v}_1, \ldots, \boldsymbol{v}_N\}$ is a basis in $\mathbf{V}$.

Definition. Two ordered bases $(\boldsymbol{v}_1, \ldots, \boldsymbol{v}_N)$ and $(\boldsymbol{v}'_1, \ldots, \boldsymbol{v}'_N)$ of $\mathbf{V}$ are called *equally oriented* if the linear map defined by $\boldsymbol{v}_i \mapsto \boldsymbol{v}'_i$ $(i = 1, \ldots, N)$ has positive determinant. An equivalence class of equally oriented bases is called an *orientation* of $\mathbf{V}$. $\mathbf{V}$ is *oriented* if an orientation of $\mathbf{V}$ is given; the bases in the chosen equivalence class are called *positively oriented,* the other ones are called *negatively oriented.* (More precisely, an oriented vector space is a pair $(\mathbf{V}, \mathrm{o})$ where $\mathbf{V}$ is a vector space and o is one of the equivalence classes of bases.)
A linear bijection between oriented vector spaces is *orientation preserving* or *orientation-reversing* if it sends positively oriented bases into positively oriented ones or into negatively oriented ones, respectively. ■

It is worth mentioning that there are two equivalence classes of equally oriented bases.
Observe that the two bases in the definition are equally oriented if and only if $\bigwedge_{i=1}^{N} \boldsymbol{v}'_i$ is a positive multiple of $\bigwedge_{i=1}^{N} \boldsymbol{v}_i$ (see Proposition 3.18).
If $\mathbf{V}$ is oriented, we orient $\mathbf{V}^*$ by the dual of positively oriented bases of $\mathbf{V}$.
If $\mathbf{U}$ and $\mathbf{V}$ are oriented vector spaces, $\mathbf{U} \times \mathbf{V}$ is oriented by joining positively oriented bases; more closely, if $(\boldsymbol{u}_1, \ldots, \boldsymbol{u}_M)$ and $(\boldsymbol{v}_1, \ldots, \boldsymbol{v}_N)$ are positively oriented bases in $\mathbf{U}$ and in $\mathbf{V}$, respectively, then $\big((\boldsymbol{u}_1, \mathbf{0}), \ldots, (\boldsymbol{u}_M, \mathbf{0})\,,\ (\mathbf{0}, \boldsymbol{v}_1), \ldots, (\mathbf{0}, \boldsymbol{v}_N)\big)$ is defined to be a positively oriented basis in $\mathbf{U} \times \mathbf{V}$.

The reader is asked to verify that the orientation of the dual and the Cartesian products is well-defined.

5.2. Two bases $\boldsymbol{a}$ and $\boldsymbol{a}'$ of the one-dimensional vector space $\mathbb{A}$ are equally oriented if and only if $\boldsymbol{a}'$ is a positive multiple of $\boldsymbol{a}$, in other words, $\frac{\boldsymbol{a}'}{\boldsymbol{a}}$ is a positive number.

If $(\boldsymbol{v}_1,\ldots,\boldsymbol{v}_N)$ and $(\boldsymbol{v}'_1,\ldots,\boldsymbol{v}'_N)$ are equally oriented ordered bases of $\mathbf{V}$, $\boldsymbol{a}$ and $\boldsymbol{a}'$ are equally oriented bases of $\mathbb{A}$, then $(\boldsymbol{a}\boldsymbol{v}_1,\ldots,\boldsymbol{a}\boldsymbol{v}_N)$ and $(\boldsymbol{a}'\boldsymbol{v}'_1,\ldots,\boldsymbol{a}'\boldsymbol{v}'_N)$ are equally oriented bases of $\mathbb{A}\otimes\mathbf{V}$. Indeed, according to our convention $\mathbb{A}\otimes\mathbf{V}\equiv\mathbf{V}\otimes\mathbb{A}$, we have $\bigwedge_{i=1}^{N}(\boldsymbol{a}'\boldsymbol{v}'_i)\equiv(\boldsymbol{a}')^N\bigwedge_{i=1}^{N}\boldsymbol{v}'_i$ which is evidently a positive multiple of $\boldsymbol{a}^N\bigwedge_{i=1}^{N}\boldsymbol{v}_i$.

As a consequence, an orientation of $\mathbf{V}$ and an orientation of $\mathbb{A}$ determine a unique orientation of $\mathbb{A}\otimes\mathbf{V}$; we consider $\mathbb{A}\otimes\mathbf{V}$ to be oriented by this orientation.

We can argue similarly to show that an orientation of $\mathbf{V}$ and an orientation of $\mathbb{A}$ determine a unique orientation of $\frac{\mathbf{V}}{\mathbb{A}}$; we take this orientation of the tensor quotient.

5.3. A nonzero element $\boldsymbol{a}$ of the oriented one-dimensional vector space $\mathbb{A}$ is called *positive*, in symbols $\mathbf{0}<\boldsymbol{a}$, if the corresponding basis is positively oriented.

Moreover, we write $\boldsymbol{a}\le\boldsymbol{b}$ if $\mathbf{0}\le\boldsymbol{b}-\boldsymbol{a}$. It is easily shown that in this way we defined a total ordering on $\mathbb{A}$ for which

(*i*) if $\boldsymbol{a}\le\boldsymbol{b}$ and $\boldsymbol{c}\le\boldsymbol{d}$ then $\boldsymbol{a}+\boldsymbol{c}\le\boldsymbol{b}+\boldsymbol{d}$,

(*ii*) if $\boldsymbol{a}\le\boldsymbol{b}$ and $\alpha\in\mathbb{R}^+$ then $\alpha\boldsymbol{a}\le\alpha\boldsymbol{b}$.

We introduce the notations

$$\mathbb{A}^+ := \{\boldsymbol{a}\in\mathbb{A}\mid \mathbf{0}<\boldsymbol{a}\},\qquad \mathbb{A}_0^+ := \mathbb{A}^+\cup\{\mathbf{0}\}.$$

Furthermore, the absolute value of $\boldsymbol{a}\in\mathbb{A}$ is

$$|\,\boldsymbol{a}| := \begin{cases} \boldsymbol{a} & \text{if } \boldsymbol{a}\in\mathbb{A}^+\\ \mathbf{0} & \text{if } \boldsymbol{a}=\mathbf{0}\\ -\boldsymbol{a} & \text{if } \boldsymbol{a}\notin\mathbb{A}^+. \end{cases}$$

5.4. Even if $\mathbb{A}$ is not oriented, $\mathbb{A}\otimes\mathbb{A}$ has a ‘canonical’ orientation in which the elements of the form $\boldsymbol{a}\otimes\boldsymbol{a}$ are positive. If $\mathbb{A}$ is oriented, the orientation of $\mathbb{A}\otimes\mathbb{A}$ induced by the orientation of $\mathbb{A}$ coincide with the canonical one. Then

$$\mathbb{A}_0^+\to(\mathbb{A}\otimes\mathbb{A})_0^+,\qquad \boldsymbol{a}\mapsto\boldsymbol{a}\otimes\boldsymbol{a} \tag{$*$}$$

is a bijection. Indeed, $\boldsymbol{a}\otimes\boldsymbol{a}=\mathbf{0}$ if and only if $\boldsymbol{a}=\mathbf{0}$. The elements of $(\mathbb{A}\otimes\mathbb{A})^+$ has the form $\boldsymbol{a}\otimes\boldsymbol{b}$ where $\boldsymbol{a},\boldsymbol{b}\in\mathbb{A}^+$; since $\boldsymbol{h}=\lambda\boldsymbol{a}$ for some positive number

λ, we have $\boldsymbol{a}\otimes\boldsymbol{b}=\left(\sqrt{\lambda}\boldsymbol{a}\right)\otimes\left(\sqrt{\lambda}\boldsymbol{a}\right)$, i.e. the above mapping is surjective. If $0\neq\boldsymbol{a}\otimes\boldsymbol{a}=\boldsymbol{b}\otimes\boldsymbol{b}$ then $\boldsymbol{a}\otimes\boldsymbol{a}=\lambda^2\boldsymbol{a}\otimes\boldsymbol{a}$ which implies that $\lambda^2=1$, thus $\lambda=1$, $\boldsymbol{a}=\boldsymbol{b}$: the mapping in question is injective.

In spite of our earlier agreement, in deducing the present result, we preferred not to omit the symbol of tensorial multiplication. However, in applications of the present result we keep our agreement; in particular, we write

$$\boldsymbol{a}^2 := \boldsymbol{a}\boldsymbol{a} \ \ (:= \boldsymbol{a}\otimes\boldsymbol{a}).$$

The inverse of the mapping $*$ is denoted by the symbol $\sqrt{}$ and is called the square root mapping.

Note that

$$\sqrt{\boldsymbol{a}^2} = |\,\boldsymbol{a}| \qquad (\boldsymbol{a}\in\mathbb{A}).$$

V. PSEUDO-EUCLIDEAN VECTOR SPACES

1. Pseudo-Euclidean vector spaces

1.1. Definition. A *pseudo-Euclidean* vector space is a triplet $(\mathbf{V}, \mathbb{B}, \boldsymbol{b})$ where

(*i*) $\mathbf{V}$ is a nonzero finite dimensional real vector space,
(*ii*) $\mathbb{B}$ is a one-dimensional real vector space,
(*iii*) $\boldsymbol{b}\colon \mathbf{V} \times \mathbf{V} \to \mathbb{B} \otimes \mathbb{B}$ is a nondegenerate symmetric bilinear map.

Remarks. (*i*) Nondegenerate means that if $\boldsymbol{b}(\boldsymbol{x}, \boldsymbol{y}) = \mathbf{0}$ for all $\boldsymbol{x} \in \mathbf{V}$ then $\boldsymbol{y} = \mathbf{0}$.

(*ii*) $\boldsymbol{b}(\boldsymbol{x}, \boldsymbol{y})$ is often called the $\boldsymbol{b}$*-product* of $\boldsymbol{x}, \boldsymbol{y} \in \mathbf{V}$. The elements $\boldsymbol{x}$ and $\boldsymbol{y}$ of $\mathbf{V}$ are called $\boldsymbol{b}$*-orthogonal* if their $\boldsymbol{b}$-product is zero.

(*iii*) In mathematical literature one usually considers the case $\mathbb{B} = \mathbb{R}$, i.e. when—because of $\mathbb{R} \otimes \mathbb{R} \equiv \mathbb{R}$—the pseudo-Euclidean form $\boldsymbol{b}$ takes real values. Physical applications require the possibility $\mathbb{B} \neq \mathbb{R}$.

1.2. Definition. A basis $\{\boldsymbol{e}_i \mid i = 1, \ldots, N\}$ of $\mathbf{V}$ is called $\boldsymbol{b}$*-orthogonal* if $\boldsymbol{b}(\boldsymbol{e}_i, \boldsymbol{e}_k) = \mathbf{0}$ for $i \neq k$.

An $\boldsymbol{b}$-orthogonal basis $\{\boldsymbol{e}_i \mid i = 1, \ldots, N\}$ is *normed* to $\boldsymbol{a} \in \mathbb{B}$ if either $\boldsymbol{b}(\boldsymbol{e}_i, \boldsymbol{e}_i) = \boldsymbol{a}^2$ or $\boldsymbol{b}(\boldsymbol{e}_i, \boldsymbol{e}_i) = -\boldsymbol{a}^2$ for all i. If $\mathbb{B} = \mathbb{R}$, an $\boldsymbol{b}$-orthogonal basis normed to 1 is called $\boldsymbol{b}$*-orthonormal.* ■

Since $\mathbb{B} \otimes \mathbb{B}$ has a canonical orientation, it makes sense that $\boldsymbol{b}(\boldsymbol{x}, \boldsymbol{y})$ is negative or positive for $\boldsymbol{x}, \boldsymbol{y} \in \mathbf{V}$.

We can argue like in the case of real-valued bilinear forms to have the following.

Proposition. $\boldsymbol{b}$-orthogonal bases in $\mathbf{V}$ exist and there is a non-negative integer $\operatorname{neg}(\boldsymbol{b})$ such that for every $\boldsymbol{b}$-orthogonal basis $\{\boldsymbol{e}_i \mid i = 1, \ldots, N\}$

$$\boldsymbol{b}(\boldsymbol{e}_i, \boldsymbol{e}_i) < \mathbf{0} \quad \text{for} \quad \operatorname{neg}(\boldsymbol{b}) \quad \text{indices} \quad i,$$
$$\boldsymbol{b}(\boldsymbol{e}_i, \boldsymbol{e}_i) > \mathbf{0} \quad \text{for} \quad N - \operatorname{neg}(\boldsymbol{b}) \quad \text{indices} \quad i. \qquad ■$$

An $\boldsymbol{b}$-orthogonal basis can always be normed to an arbitrary $\mathbf{0} \neq \boldsymbol{a} \in \mathbb{B}$. Further on we deal with $\boldsymbol{b}$-orthogonal bases normed to an element of $\mathbb{B}$ and such a basis will be numbered so that $\boldsymbol{b}$ takes negative values on the first $\mathrm{neg}(\boldsymbol{b})$ elements, i.e.

$$\boldsymbol{b}(\boldsymbol{e}_i, \boldsymbol{e}_i) = \alpha(i)\boldsymbol{a}^2,$$

$$\alpha(i) = \begin{cases} -1 & \text{if } i = 1, \dots, \mathrm{neg}(\boldsymbol{b}) \\ 1 & \text{if } i = \mathrm{neg}(\boldsymbol{b}) + 1, \dots, N. \end{cases}$$

We say that $\boldsymbol{b}$ is *positive definite* if $\boldsymbol{b}(\boldsymbol{x}, \boldsymbol{x}) > \mathbf{0}$ for all nonzero $\boldsymbol{x}$. $\boldsymbol{b}$ is positive definite if and only if $\mathrm{neg}(\boldsymbol{b}) = \mathbf{0}$.

1.3. An important property of pseudo-Euclidean vector spaces is that a natural correspondence exists between $\mathbf{V}^*$ and $\frac{\mathbf{V}}{\mathbb{B}\otimes\mathbb{B}}$. Note that every element of $\frac{\mathbf{V}}{\mathbb{B}\otimes\mathbb{B}}$ is of the form $\frac{\boldsymbol{y}}{\boldsymbol{ab}}$ where $\boldsymbol{y} \in \mathbf{V}$ and $\boldsymbol{a}, \boldsymbol{b} \in \mathbb{B} \setminus \{0\}$. Take such an element of $\frac{\mathbf{V}}{\mathbb{B}\otimes\mathbb{B}}$. Then

$$\mathbf{V} \to \mathbb{R}, \qquad \boldsymbol{x} \mapsto \frac{\boldsymbol{b}(\boldsymbol{y}, \boldsymbol{x})}{\boldsymbol{ab}}$$

is a linear map, i.e. an element of $\mathbf{V}^*$, which we write in the form $\frac{\boldsymbol{b}(\boldsymbol{y},\cdot)}{\boldsymbol{ab}}$.

Proposition. $\frac{\mathbf{V}}{\mathbb{B}\otimes\mathbb{B}} \to \mathbf{V}^*$, $\frac{\boldsymbol{y}}{\boldsymbol{ab}} \mapsto \frac{\boldsymbol{b}(\boldsymbol{y},\cdot)}{\boldsymbol{ab}}$ is a linear bijection.

Proof. It is linear and injective because $\boldsymbol{b}$ is bilinear and nondegenerate, and surjective because the two vector spaces in question have the same dimension. ■

We find this linear bijection so natural that we use it for identifying the vector spaces:

$$\frac{\mathbf{V}}{\mathbb{B}\otimes\mathbb{B}} \equiv \mathbf{V}^*, \qquad \frac{\boldsymbol{y}}{\boldsymbol{ab}} \equiv \frac{\boldsymbol{b}(\boldsymbol{y},\cdot)}{\boldsymbol{ab}}.$$

1.4. (*i*) In the above identification the dual of an $\boldsymbol{b}$-orthogonal basis $\{\boldsymbol{e}_i \mid i = 1, \dots, N\}$ becomes

$$\left\{ \frac{\boldsymbol{e}_i}{\boldsymbol{b}(\boldsymbol{e}_i, \boldsymbol{e}_i)} \,\middle|\, i = 1, \dots, N \right\}$$

which equals

$$\left\{ \frac{\alpha(i)\boldsymbol{e}_i}{\boldsymbol{a}^2} \,\middle|\, i = 1, \dots, N \right\}$$

if the $\boldsymbol{b}$-orthogonal basis is normed to $\boldsymbol{a}$.

As a consequence, for all $\boldsymbol{x} \in \mathbf{V}$ (see IV.1.1),

$$\boldsymbol{x} = \sum_{i=1}^{N} \frac{\boldsymbol{b}(\boldsymbol{e}_i, \boldsymbol{x})}{\boldsymbol{b}(\boldsymbol{e}_i, \boldsymbol{e}_i)} \boldsymbol{e}_i,$$

and $\boldsymbol{x} = \mathbf{0}$ if and only if $\boldsymbol{b}(\boldsymbol{e}_i, \boldsymbol{x}) = \mathbf{0}$ for all $i = 1, \dots, N$.

(*ii*) If $\mathbf{V}$ is oriented then both $\mathbf{V}^*$ and $\frac{\mathbf{V}}{\mathbb{B}\otimes\mathbb{B}}$ are oriented. The above identification is orientation preserving if $\operatorname{neg}(\boldsymbol{b})$ is even and is orientation-reversing if $\operatorname{neg}(\boldsymbol{b})$ is odd.

1.5. Let us take a linear map $\boldsymbol{F}: \mathbf{V} \to \mathbf{V}$. As we know, its transpose is a linear map $\boldsymbol{F}^*: \mathbf{V}^* \to \mathbf{V}^*$; according to the previous identification we can consider it to be a linear map $\boldsymbol{F}^*: \frac{\mathbf{V}}{\mathbb{B}\otimes\mathbb{B}} \to \frac{\mathbf{V}}{\mathbb{B}\otimes\mathbb{B}}$. Consequently, we can define the $\boldsymbol{b}$-adjoint of $\boldsymbol{F}$,

$$\boldsymbol{F}^{\circledast}: \mathbf{V} \to \mathbf{V}, \qquad \boldsymbol{y} \mapsto (\boldsymbol{ab})\left(\boldsymbol{F}^* \cdot \frac{\boldsymbol{y}}{\boldsymbol{ab}}\right).$$

Observe that this is equivalent to

$$\frac{\boldsymbol{F}^{\circledast} \cdot \boldsymbol{y}}{\boldsymbol{ab}} = \boldsymbol{F}^* \cdot \frac{\boldsymbol{y}}{\boldsymbol{ab}} \qquad (\boldsymbol{y} \in \mathbf{V},\ \boldsymbol{a}, \boldsymbol{b} \in \mathbb{B} \setminus \{0\}).$$

According to the definition of the transpose we have

$$\frac{\boldsymbol{y}}{\boldsymbol{ab}} \cdot \boldsymbol{F} \cdot \boldsymbol{x} = \left(\boldsymbol{F}^* \cdot \frac{\boldsymbol{y}}{\boldsymbol{ab}}\right) \cdot \boldsymbol{x} = \frac{\boldsymbol{F}^{\circledast} \cdot \boldsymbol{y}}{\boldsymbol{ab}} \cdot \boldsymbol{x},$$

which means

$$\frac{\boldsymbol{b}(\boldsymbol{y}, \boldsymbol{F} \cdot \boldsymbol{x})}{\boldsymbol{ab}} = \frac{\boldsymbol{b}(\boldsymbol{F}^{\circledast} \cdot \boldsymbol{y}, \boldsymbol{x})}{\boldsymbol{ab}},$$

i.e.

$$\boldsymbol{b}(\boldsymbol{y}, \boldsymbol{F} \cdot \boldsymbol{x}) = \boldsymbol{b}(\boldsymbol{F}^{\circledast} \cdot \boldsymbol{y}, \boldsymbol{x}) = \boldsymbol{b}(\boldsymbol{x}, \boldsymbol{F}^{\circledast} \cdot \boldsymbol{y}) \qquad (\boldsymbol{x}, \boldsymbol{y} \in \mathbf{V}).$$

The definition of $\boldsymbol{b}$-adjoints involves that the formulae in IV. 1.4 remain valid for $\boldsymbol{b}$-adjoints as well: if $\boldsymbol{F}, \boldsymbol{G} \in \operatorname{Lin}(\mathbf{V})$, $\alpha \in \mathbb{R}$, then

$$(\boldsymbol{F} + \boldsymbol{G})^{\circledast} = \boldsymbol{F}^{\circledast} + \boldsymbol{G}^{\circledast},$$
$$(\alpha\boldsymbol{F})^{\circledast} = \alpha\boldsymbol{F}^{\circledast},$$
$$(\boldsymbol{F} \cdot \boldsymbol{G})^{\circledast} = \boldsymbol{G}^{\circledast} \cdot \boldsymbol{F}^{\circledast}.$$

Moreover,

$$\det \boldsymbol{F}^{\circledast} = \det \boldsymbol{F}.$$

1.6. Let $(\mathbf{V}, \mathbb{B}, \boldsymbol{b})$ and $(\mathbf{V}', \mathbb{B}', \boldsymbol{b}')$ be pseudo-Euclidean vector spaces. A linear map $\boldsymbol{L}: \mathbf{V} \to \mathbf{V}'$ is called $\boldsymbol{b}$-$\boldsymbol{b}'$-orthogonal if there is a linear bijection $\boldsymbol{Z}: \mathbb{B} \to \mathbb{B}'$ such that $\boldsymbol{b}' \circ (\boldsymbol{L} \times \boldsymbol{L}) = (\boldsymbol{Z} \otimes \boldsymbol{Z}) \circ \boldsymbol{b}$ i.e.

$$\boldsymbol{b}'(\boldsymbol{L} \cdot \boldsymbol{x}, \boldsymbol{L} \cdot \boldsymbol{y}) = (\boldsymbol{Z} \otimes \boldsymbol{Z})\boldsymbol{b}(\boldsymbol{x}, \boldsymbol{y}) \qquad (\boldsymbol{x}, \boldsymbol{y} \in \mathbf{V}).$$

Note that according to our identification, $\boldsymbol{Z}$ is an element of $\frac{\mathbb{B}'}{\mathbb{B}}$.

It is quite trivial that there is a $\boldsymbol{b}$-$\boldsymbol{b}'$-orthogonal linear map between the pseudo-Euclidean vector spaces if and only if $\dim \mathbf{V} = \dim \mathbf{V}'$ and $\operatorname{neg}(\boldsymbol{b}) = \operatorname{neg}(\boldsymbol{b}')$.

In particular, if $\{\boldsymbol{e}_i \mid i = 1, \ldots, N\}$ is an $\boldsymbol{b}$-orthogonal basis, normed to $\boldsymbol{a}$, of $\mathbf{V}$, and $\{\boldsymbol{e}'_i \mid i = 1, \ldots, N\}$ is an $\boldsymbol{b}'$-orthogonal basis, normed to $\boldsymbol{a}'$, of $\mathbf{V}'$ then

$$\boldsymbol{L} \cdot \boldsymbol{e}_i := \boldsymbol{e}'_i \qquad (i = 1, \ldots, N)$$

determine an $\boldsymbol{b}$-$\boldsymbol{b}'$-orthogonal map for which $\boldsymbol{Z} = \frac{\boldsymbol{a}'}{\boldsymbol{a}}$.

1.7. Let n and N be natural numbers, $N \geq 1$, $n \leq N$. The map

$$\boldsymbol{H}_n : \mathbb{R}^N \times \mathbb{R}^N \to \mathbb{R}, \qquad (\boldsymbol{x}, \boldsymbol{y}) \mapsto -\sum_{i=1}^{n} x^i y^i + \sum_{i=n+1}^{N} x^i y^i = \sum_{i=1}^{N} \alpha(i) x^i y^i$$

(where $\alpha(i) := -1$ for $i = 1, \ldots, n$ and $\alpha(i) := 1$ for $i = n+1, \ldots, N$) is a nondegenerate symmetric bilinear map, i.e. $(\mathbb{R}^N, \mathbb{R}, \boldsymbol{H}_n)$ is a pseudo-Euclidean vector space and $\operatorname{neg}(\boldsymbol{H}_n) = n$.

The standard basis of $\mathbb{R}^N$ is $\boldsymbol{H}_n$-orthonormal.

According to 1.3, we have the identification $(\mathbb{R}^N)^* \equiv \mathbb{R}^N$, but we must pay attention to the fact that if $n \neq 0$ it differs from the standard one mentioned in IV.1.6.

The standard identification is a linear bijection $\boldsymbol{S} : \mathbb{R}^N \to (\mathbb{R}^N)^*$, and the present identification is another one: $\boldsymbol{J}_n : \mathbb{R}^N \to (\mathbb{R}^N)^*$, $\boldsymbol{x} \mapsto \boldsymbol{H}_n(x, \cdot)$. We easily see that

$$(x_i \mid i = 1, \ldots, N) := \boldsymbol{J}_n \cdot \left(x^i \mid i = 1, \ldots, N\right) = \left(\alpha(i) x^i \mid i = 1, \ldots, N\right).$$

The standard identification coincides with $\boldsymbol{J}_0$, the one corresponding to $\boldsymbol{H}_0$.

According to the identification induced by $\boldsymbol{H}_n$, the dual of the standard basis $\{\chi_i \mid i = 1, \ldots, N\}$ is $\{\alpha(i)\chi_i \mid i = 1, \ldots, N\}$.

It is useful to regard $\boldsymbol{H}_n$ as the diagonal matrix in which the first n elements in the diagonal are -1 and the others equal 1.

For the $\boldsymbol{H}_n$-adjoint of the linear map (matrix) $\boldsymbol{F}$ we have $\boldsymbol{x} \cdot \boldsymbol{H}_n \cdot \boldsymbol{F}^{\circledast} \cdot \boldsymbol{y} = (\boldsymbol{F} \cdot \boldsymbol{x}) \cdot \boldsymbol{H}_n \cdot \boldsymbol{y} = \boldsymbol{x} \cdot \boldsymbol{F}^* \cdot \boldsymbol{H}_n \cdot \boldsymbol{y}$ for all $\boldsymbol{x}, \boldsymbol{y} \in \mathbb{R}^N$, where $\boldsymbol{F}^*$ denotes the usual transpose of the matrix $\boldsymbol{F}$; thus $\boldsymbol{F}^* \cdot \boldsymbol{H}_n = \boldsymbol{H}_n \cdot \boldsymbol{F}^{\circledast}$ or

$$\boldsymbol{F}^{\circledast} = \boldsymbol{H}_n \cdot \boldsymbol{F}^* \cdot \boldsymbol{H}_n.$$

1.8. Exercises

1. Let $\boldsymbol{e}_1, \ldots, \boldsymbol{e}_n$ be pairwise $\boldsymbol{b}$-orthogonal vectors in the pseudo-Euclidean vector space $(\mathbf{V}, \mathbb{B}, \boldsymbol{b})$ such that $\boldsymbol{b}(\boldsymbol{e}_i, \boldsymbol{e}_i) \neq \boldsymbol{0}$ for all $i = 1, \ldots, n$. Prove that the following statements are equivalent:

(*i*) $n = \dim \mathbf{V}$ (i.e. the vectors form a basis),

(*ii*) if $\boldsymbol{b}(\boldsymbol{e}_i, \boldsymbol{x}) = \boldsymbol{0}$ for all $i = 1, \ldots, n$ then $\boldsymbol{x} = \boldsymbol{0}$,

(*iii*) $\displaystyle \boldsymbol{x} = \sum_{i=1}^{n} \frac{\boldsymbol{b}(\boldsymbol{e}_i, \boldsymbol{x})}{\boldsymbol{b}(\boldsymbol{e}_i, \boldsymbol{e}_i)} \boldsymbol{e}_i$ for all $\boldsymbol{x} \in \mathbf{V}$,

$$(iv) \qquad \boldsymbol{b}(\boldsymbol{x},\boldsymbol{y}) = \sum_{i=1}^{n} \frac{\boldsymbol{b}(\boldsymbol{e}_i,\boldsymbol{x}) \otimes \boldsymbol{b}(\boldsymbol{e}_i,\boldsymbol{y})}{\boldsymbol{b}(\boldsymbol{e}_i,\boldsymbol{e}_i)} \quad \text{for all} \quad \boldsymbol{x},\boldsymbol{y} \in \boldsymbol{V},$$

$$(v) \qquad \boldsymbol{b}(\boldsymbol{x},\boldsymbol{x}) = \sum_{i=1}^{n} \frac{\boldsymbol{b}(\boldsymbol{e}_i,\boldsymbol{x}) \otimes \boldsymbol{b}(\boldsymbol{e}_i,\boldsymbol{x})}{\boldsymbol{b}(\boldsymbol{e}_i,\boldsymbol{e}_i)} \quad \text{for all} \quad \boldsymbol{x} \in \mathbf{V}.$$

2. Demonstrate that the set $\{\boldsymbol{e}_1,\dots,\boldsymbol{e}_n\}$ of pairwise $\boldsymbol{b}$-orthogonal vectors can be completed to an $\boldsymbol{b}$-orthogonal basis if and only if $\boldsymbol{b}(\boldsymbol{e}_i,\boldsymbol{e}_i) \neq \mathbf{0}$ for all $i = 1,\dots,n$.

2. Tensors of pseudo-Euclidean vector spaces

2.1. Let $\mathbf{V}$ and $\mathbb{A}$ be finite dimensional vector spaces, $\dim \mathbb{A} = 1$. Suppose $\boldsymbol{F}\colon \mathbf{V} \to \mathbf{V}$ is a linear map. Then we can define the linear maps

$$\boldsymbol{F}^{\mathbb{A}}\colon \mathbb{A}\otimes\mathbf{V} \to \mathbb{A}\otimes\mathbf{V}, \qquad \boldsymbol{a}\boldsymbol{v} \mapsto \boldsymbol{a}\otimes(\boldsymbol{F}\cdot\boldsymbol{v}),$$

$$\boldsymbol{F}_{\mathbb{A}}\colon \frac{\mathbf{V}}{\mathbb{A}} \to \frac{\mathbf{V}}{\mathbb{A}}, \qquad \frac{\boldsymbol{v}}{\boldsymbol{a}} \mapsto \frac{\boldsymbol{F}\cdot\boldsymbol{v}}{\boldsymbol{a}}$$

($\boldsymbol{F}^{\mathbb{A}} = \mathbf{1}_{\mathbb{A}}\otimes\boldsymbol{F}$, $\boldsymbol{F}_{\mathbb{A}} = \frac{\boldsymbol{F}}{\mathbf{1}_{\mathbb{A}}}$, see IV.3.12 and IV.4.6).

According to the usual identifications

$$\begin{aligned}\mathrm{Lin}(\mathbb{A}\otimes\mathbf{V}) \equiv &(\mathbb{A}\otimes\mathbf{V})\otimes(\mathbb{A}\otimes\mathbf{V})^* \equiv \mathbb{A}\otimes\mathbf{V}\otimes\mathbb{A}^*\otimes\mathbf{V}^* \equiv \\ \equiv &\mathbb{A}\otimes\mathbb{A}^*\otimes\mathbf{V}\otimes\mathbf{V}^* \equiv \mathbb{R}\otimes\mathbf{V}\otimes\mathbf{V}^* \equiv \mathbf{V}\otimes\mathbf{V}^* \equiv \\ \equiv &\mathrm{Lin}(\mathbf{V}),\end{aligned}$$

we have $\boldsymbol{F}^{\mathbb{A}} \equiv \boldsymbol{F}$ and similarly $\boldsymbol{F}_{\mathbb{A}} \equiv \boldsymbol{F}$. Therefore we shall write $\boldsymbol{F}$ instead of $\boldsymbol{F}^{\mathbb{A}}$ and $\boldsymbol{F}_{\mathbb{A}}$:

$$\text{for} \quad \boldsymbol{s} \in \mathbb{A}\otimes\mathbf{V} \qquad \text{we have} \qquad \boldsymbol{F}\cdot\boldsymbol{s} \in \mathbb{A}\otimes\mathbf{V},$$

$$\text{for} \quad \boldsymbol{n} \in \frac{\mathbf{V}}{\mathbb{A}} \qquad \text{we have} \qquad \boldsymbol{F}\cdot\boldsymbol{n} \in \frac{\mathbf{V}}{\mathbb{A}}.$$

2.2. Let us formulate the previous convention in another way. $\mathbf{V}\otimes\mathbb{A} \equiv \mathrm{Lin}(\mathbb{A}^*,\mathbf{V})$, hence we have the composition $\boldsymbol{F}\cdot\boldsymbol{s}$ of $\boldsymbol{F} \in \mathrm{Lin}(\mathbf{V}) \equiv \mathbf{V}\otimes\mathbf{V}^*$ and $\boldsymbol{s} \in \mathrm{Lin}(\mathbb{A}^*,\mathbf{V}) \equiv \mathbf{V}\otimes\mathbb{A}$.

More generally, if $\mathbf{U}$ and $\mathbf{W}$ are finite dimensional vector spaces, the *dot product* of an element from $\mathbf{U}\otimes\mathbf{V}^*$ and an element from $\mathbf{V}\otimes\mathbf{W}$ is defined to be an element in $\mathbf{U}\otimes\mathbf{W}$; this dot product can be regarded as the composition of the corresponding linear maps and is characterized by

$$(\boldsymbol{u}\otimes\boldsymbol{p})\cdot(\boldsymbol{v}\otimes\boldsymbol{w}) = (\boldsymbol{p}\cdot\boldsymbol{v})\boldsymbol{u}\otimes\boldsymbol{w}.$$

The scheme is worth repeating:

$$\mathbf{U}\otimes\mathbf{V}^* \qquad \text{dot} \qquad \mathbf{V}\otimes\mathbf{W} \qquad \text{results in} \qquad \mathbf{U}\otimes\mathbf{W}.$$

Evidently, we can have $\mathbf{U} = \frac{\mathbb{K}}{\mathbb{A}}$ or $\mathbf{W} = \frac{\mathbb{K}}{\mathbb{A}}$, thus similar formulae are valid for tensor quotients as well.

2.3. What we have said in the previous paragraph concerns any vector spaces. In the following $(\mathbf{V}, \mathbb{B}, \boldsymbol{b})$ denotes a pseudo-Euclidean vector space.

The identification described in 1.3 and the corresponding formula suggests us a new notation: 'removing' the denominator from both sides we arrive at the definition

$$\boldsymbol{x} \cdot \boldsymbol{y} := \boldsymbol{b}(\boldsymbol{x}, \boldsymbol{y}),$$

i.e. in the sequel we omit $\boldsymbol{b}$, denoting the $\boldsymbol{b}$-product of vectors by a simple dot.

The dot product of two elements of $\mathbf{V}$ is an element of $\mathbb{B} \otimes \mathbb{B}$. Then we can extend the previous dot product formalism as follows:

$$\mathbf{U} \otimes \mathbf{V} \quad \text{dot} \quad \mathbf{V} \otimes \mathbf{W} \quad \text{results in} \quad (\mathbb{B} \otimes \mathbb{B}) \otimes \mathbf{U} \otimes \mathbf{W},$$

$$(\boldsymbol{u} \otimes \boldsymbol{v}') \cdot (\boldsymbol{v} \otimes \boldsymbol{w}) := (\boldsymbol{v}' \cdot \boldsymbol{v})\boldsymbol{u} \otimes \boldsymbol{w}.$$

2.4. The $\boldsymbol{b}$-adjoint of $\boldsymbol{F} \in \mathrm{Lin}(\mathbf{V})$ is characterized in the new notation of dot products as follows:

$$\boldsymbol{y} \cdot \boldsymbol{F} \cdot \boldsymbol{x} = (\boldsymbol{F}^* \cdot \boldsymbol{y}) \cdot \boldsymbol{x} = \boldsymbol{x} \cdot \boldsymbol{F}^* \boldsymbol{y} \qquad (\boldsymbol{y}, \boldsymbol{x} \in \mathbf{V}).$$

2.5. According to our convention we have

$$\boldsymbol{n} \cdot \boldsymbol{x} \quad \text{is in} \quad \mathbb{B} \quad \text{for} \quad \boldsymbol{n} \in \frac{\mathbf{V}}{\mathbb{B}} \quad \text{and} \quad \boldsymbol{x} \in \mathbf{V},$$

$$\boldsymbol{n} \cdot \boldsymbol{m} \quad \text{is in} \quad \mathbb{R} \quad \text{for} \quad \boldsymbol{n}, \boldsymbol{m} \in \frac{\mathbf{V}}{\mathbb{B}}.$$

If $\{\boldsymbol{e}_i \mid i = 1, ..., N\}$ is an $\boldsymbol{b}$-orthogonal basis, normed to $\boldsymbol{a} \in \mathbb{B}$, in $\mathbf{V}$, then $\left\{\boldsymbol{n}_i := \frac{\boldsymbol{e}_i}{a} \mid i = 1, ..., N\right\}$ is an $\boldsymbol{b}$-orthonormal basis of $\frac{\mathbf{V}}{\mathbb{B}}$:

$$\boldsymbol{n}_i \cdot \boldsymbol{n}_k = \alpha(i)\delta_{ik} \qquad (i, k = 1, ..., N).$$

It is more convenient to use this basis instead of the original one; for all $\boldsymbol{x} \in \mathbf{V}$ we have

$$\boldsymbol{x} = \sum_{i=1}^{N} \alpha(i)(\boldsymbol{n}_i \cdot \boldsymbol{x})\boldsymbol{n}_i.$$

2.6. (*i*) We have the identifications $\left(\frac{\mathbf{V}}{\mathbb{B}}\right)^* \equiv \frac{\mathbf{V}^*}{\mathbb{B}^*} \equiv \frac{\mathbf{V}}{\mathbb{B}\otimes\mathbb{B}\otimes\mathbb{B}^*} \equiv \frac{\mathbf{V}}{\mathbb{B}}$; the element $\boldsymbol{n}$ of $\frac{\mathbf{V}}{\mathbb{B}}$ is identified with the linear functional $\frac{\mathbf{V}}{\mathbb{B}} \to \mathbb{R}$, $\boldsymbol{m} \mapsto \boldsymbol{n}\cdot\boldsymbol{m}$.

(*ii*) In view of the identifications $\mathrm{Lin}(\mathbf{V}) \equiv \mathbf{V}\otimes\mathbf{V}^* \equiv \mathbf{V}\otimes\frac{\mathbf{V}}{\mathbb{B}\otimes\mathbb{B}} \equiv \frac{\mathbf{V}}{\mathbb{B}}\otimes\frac{\mathbf{V}}{\mathbb{B}}$, or in view of our dot product convention, for $\boldsymbol{n},\boldsymbol{m}\in\frac{\mathbf{V}}{\mathbb{B}}$,

$$\boldsymbol{m}\otimes\boldsymbol{n}\colon \mathbf{V}\to\mathbf{V}, \qquad \boldsymbol{x}\mapsto(\boldsymbol{n}\cdot\boldsymbol{x})\boldsymbol{m}$$

is a linear map, and every element of $\mathrm{Lin}(\mathbf{V})$ is the sum of such linear maps.

Evidently,

$$(\boldsymbol{m}\otimes\boldsymbol{n})\cdot(\boldsymbol{m}'\otimes\boldsymbol{n}') = (\boldsymbol{n}\cdot\boldsymbol{m}')\boldsymbol{m}\otimes\boldsymbol{n}'$$

and

$$(\boldsymbol{m}\otimes\boldsymbol{n})^* = \boldsymbol{n}\otimes\boldsymbol{m}.$$

2.7. Definition. For the pseudo-Euclidean vector space $(\mathbf{V},\mathbb{B},\boldsymbol{b})$ we put

$$\begin{aligned}\mathcal{O}(\boldsymbol{b})\colon &= \{\boldsymbol{L}\in\mathrm{Lin}(\mathbf{V}) \mid \boldsymbol{L}^* = \boldsymbol{L}^{-1}\},\\ \mathrm{A}(\boldsymbol{b})\colon &= \{\boldsymbol{A}\in\mathrm{Lin}(\mathbf{V}) \mid \boldsymbol{A}^* = -\boldsymbol{A}\},\end{aligned}$$

and the elements of $\mathcal{O}(\boldsymbol{b})$ and $\mathrm{A}(\boldsymbol{b})$ are called $\boldsymbol{b}$-*orthogonal* and $\boldsymbol{b}$-*antisymmetric*, respectively.

Proposition. (*i*) For $\boldsymbol{L}\in\mathrm{Lin}(\mathbf{V})$ the following three statements are equivalent:

— $\boldsymbol{L}$ is in $\mathcal{O}(\boldsymbol{b})$,
— $(\boldsymbol{L}\cdot\boldsymbol{y})\cdot(\boldsymbol{L}\cdot\boldsymbol{x}) = \boldsymbol{y}\cdot\boldsymbol{x}$ for all $\boldsymbol{y},\boldsymbol{x}\in\mathbf{V}$,
— $(\boldsymbol{L}\cdot\boldsymbol{x})\cdot(\boldsymbol{L}\cdot\boldsymbol{x}) = \boldsymbol{x}\cdot\boldsymbol{x}$ for all $\boldsymbol{x}\in\mathbf{V}$.

(*ii*) For $\boldsymbol{A}\in\mathrm{Lin}(\mathbf{V})$ the following three statements are equivalent:

— $\boldsymbol{A}$ is in $\mathrm{A}(\boldsymbol{b})$,
— $\boldsymbol{y}\cdot\boldsymbol{A}\cdot\boldsymbol{x} = -(\boldsymbol{A}\cdot\boldsymbol{y})\cdot\boldsymbol{x} = -\boldsymbol{x}\cdot\boldsymbol{A}\cdot\boldsymbol{y}$ for all $\boldsymbol{y},\boldsymbol{x}\in\mathbf{V}$,
— $\boldsymbol{x}\cdot\boldsymbol{A}\cdot\boldsymbol{x} = 0$ for all $\boldsymbol{x}\in\mathbf{V}$.

2.8. Proposition.

(*i*) $|\det\boldsymbol{L}| = 1$ for $\boldsymbol{L}\in\mathcal{O}(\boldsymbol{b})$;
(*ii*) $\mathrm{Tr}\,\boldsymbol{A} = 0$ for $\boldsymbol{A}\in\mathrm{A}(\boldsymbol{b})$.

Proof. It is convenient to regard now the linear maps in question as linear maps $\frac{\mathbf{V}}{\mathbb{B}}\to\frac{\mathbf{V}}{\mathbb{B}}$, according to our identifications described in 2.1. It is not hard to see that this does not influence determinants and traces.

(*i*) Let $\{\boldsymbol{n}_1, \dots, \boldsymbol{n}_N\}$ be a $\boldsymbol{b}$-orthonormal basis in $\frac{\mathbf{V}}{\mathbb{B}}$. According to IV.3.15 and to the identification $\left(\frac{\mathbf{V}}{\mathbb{B}}\right)^* \equiv \frac{\mathbf{V}}{\mathbb{B}}$ we have

$$0 \neq \left(\bigwedge_{k=1}^{N} \boldsymbol{n}_k\right)(\boldsymbol{n}_1, ..., \boldsymbol{n}_N) = \left(\bigwedge_{k=1}^{N} \boldsymbol{L}\cdot\boldsymbol{n}_k\right)(\boldsymbol{L}\cdot\boldsymbol{n}_1, \dots, \boldsymbol{L}\cdot\boldsymbol{n}_N) =$$

$$= (\det \boldsymbol{L})\left(\bigwedge_{k=1}^{N} \boldsymbol{L}\cdot\boldsymbol{n}_k\right)(\boldsymbol{n}_1, \dots, \boldsymbol{n}_N) =$$

$$= (\det \boldsymbol{L})^2\left(\bigwedge_{k=1}^{N} \boldsymbol{n}_k\right)(\boldsymbol{n}_1, \dots, \boldsymbol{n}_N).$$

(*ii*) We know that the dual of the preceding basis is $\{\alpha(i)\boldsymbol{n}_i \mid i = 1, \dots, N\}$, thus in view of IV.3.9,

$$\operatorname{Tr} \boldsymbol{A} = \sum_{i=1}^{N} \alpha(i)\boldsymbol{n}_i \cdot \boldsymbol{A} \cdot \boldsymbol{n}_i = 0.$$

2.9. A linear map $\boldsymbol{S}\colon \mathbf{V} \to \mathbf{V}$ is called $\boldsymbol{b}$-*symmetric* if $\boldsymbol{S}^* = \boldsymbol{S}$ or, equivalently, $\boldsymbol{x}\cdot\boldsymbol{S}\cdot\boldsymbol{y} = \boldsymbol{y}\cdot\boldsymbol{S}\cdot\boldsymbol{x}$ for all $\boldsymbol{x}, \boldsymbol{y} \in \mathbf{V}$. The set of $\boldsymbol{b}$-symmetric linear maps is denoted by $S(\boldsymbol{b})$.

$\mathrm{A}(\boldsymbol{b})$ and $S(\boldsymbol{b})$ are complementary subspaces of $\operatorname{Lin}(\mathbf{V}) \equiv \mathbf{V}\otimes\mathbf{V}^*$, i.e. their intersection is the zero subspace and they span the whole space $\mathbf{V}\otimes\mathbf{V}^*$. Indeed, only the zero linear map is both symmetric and antisymmetric, and for any linear map $\boldsymbol{F}\colon \mathbf{V} \to \mathbf{V}$ we have that

$$\boldsymbol{S} := \frac{\boldsymbol{F} + \boldsymbol{F}^*}{2}, \qquad \boldsymbol{A} := \frac{\boldsymbol{F} - \boldsymbol{F}^*}{2}$$

are $\boldsymbol{b}$-symmetric and $\boldsymbol{b}$-antisymmetric, respectively, and $\boldsymbol{F} = \boldsymbol{S} + \boldsymbol{A}$.

Taking the identification $\mathbf{V}\otimes\mathbf{V}^* \equiv \frac{\mathbf{V}}{\mathbb{B}}\otimes\frac{\mathbf{V}}{\mathbb{B}}$ we can easily see that $\frac{\mathbf{V}}{\mathbb{B}}\vee\frac{\mathbf{V}}{\mathbb{B}} \subset S(\boldsymbol{b})$ and $\frac{\mathbf{V}}{\mathbb{B}}\wedge\frac{\mathbf{V}}{\mathbb{B}} \subset \mathrm{A}(\boldsymbol{b})$; since these subspaces are complementary, too, equalities hold necessarily:

$$\mathbf{V}\vee\mathbf{V}^* := \frac{\mathbf{V}}{\mathbb{B}}\vee\frac{\mathbf{V}}{\mathbb{B}} = S(\boldsymbol{b}), \qquad \mathbf{V}\wedge\mathbf{V}^* := \frac{\mathbf{V}}{\mathbb{B}}\wedge\frac{\mathbf{V}}{\mathbb{B}} = \mathrm{A}(\boldsymbol{b}).$$

As a consequence,

$$\dim S(\boldsymbol{b}) = \frac{N(N+1)}{2}, \qquad \dim \mathrm{A}(\boldsymbol{b}) = \frac{N(N-1)}{2}.$$

Recall that for $\boldsymbol{m}, \boldsymbol{n} \in \frac{\mathbf{V}}{\mathbb{B}}$ we have

$$\boldsymbol{m}\vee\boldsymbol{n} = \boldsymbol{m}\otimes\boldsymbol{n} + \boldsymbol{n}\otimes\boldsymbol{m}, \qquad \boldsymbol{m}\wedge\boldsymbol{n} = \boldsymbol{m}\otimes\boldsymbol{n} - \boldsymbol{n}\otimes\boldsymbol{m}.$$

2.10. Proposition.

$$(\mathbf{V}\otimes\mathbf{V}^*)\times(\mathbf{V}\otimes\mathbf{V}^*) \to \mathbb{R}, \qquad (\boldsymbol{F}, \boldsymbol{G}) \mapsto \boldsymbol{F} : \boldsymbol{G} := \operatorname{Tr}(\boldsymbol{F}^*\cdot\boldsymbol{G})$$

is a nondegenerate symmetric bilinear form, which is positive definite if and only if $\boldsymbol{b}$ is positive definite.

Proof. It is trivially bilinear and symmetric because of the properties of Tr and $\boldsymbol{b}$-adjoints.

Suppose that $\mathrm{Tr}(\boldsymbol{F}^* \cdot \boldsymbol{G}) = 0$ for all $\boldsymbol{F} \in \mathbf{V} \otimes \mathbf{V}^*$, i.e.

$$0 = \sum_{i=1}^{N} \alpha(i) \boldsymbol{n}_i \cdot \boldsymbol{F}^* \cdot \boldsymbol{G} \cdot \boldsymbol{n}_i$$

for all $\boldsymbol{b}$-orthonormal bases $\{\boldsymbol{n}_1, \ldots, \boldsymbol{n}_N\}$ of $\frac{\mathbf{V}}{\mathbb{B}}$. Then taking $\boldsymbol{F} := \boldsymbol{n}_j \otimes \boldsymbol{n}_k$ for all $j, k = 1, \ldots, N$, and using $(\boldsymbol{n}_j \otimes \boldsymbol{n}_k)^* = \boldsymbol{n}_k \otimes \boldsymbol{n}_j$ we conclude that $\boldsymbol{n}_j \cdot \boldsymbol{G} \cdot \boldsymbol{n}_k = 0$ for all j, k which results in $\boldsymbol{G} = \mathbf{0}$.

Since

$$\mathrm{Tr}(\boldsymbol{F}^* \cdot \boldsymbol{F}) = \sum_{i=1}^{N} \alpha(i)(\boldsymbol{F} \cdot \boldsymbol{n}_i) \cdot (\boldsymbol{F} \cdot \boldsymbol{n}_i),$$

we see that if $\boldsymbol{b}$ is positive definite then $\mathrm{Tr}(\boldsymbol{F}^* \cdot \boldsymbol{F}) > 0$; if $\boldsymbol{b}$ is not positive definite then we can easily construct an $\boldsymbol{F}$ such that $\boldsymbol{F} : \boldsymbol{F} < 0$.

Remark. (*i*) Compare the present bilinear form with that of the duality treated in IV.3.10; take into account the identification $\mathbf{V} \otimes \mathbf{V}^* \equiv \mathbf{V} \otimes \frac{\mathbf{V}}{\mathbb{B} \otimes \mathbb{B}} \equiv \frac{\mathbf{V}}{\mathbb{B} \otimes \mathbb{B}} \otimes \mathbf{V} \equiv \mathbf{V}^* \otimes \mathbf{V}$.

(*ii*) The bilinear form is not positive definite, in general, either on the linear subspace of $\boldsymbol{b}$-symmetric linear maps or on the linear subspace of $\boldsymbol{b}$-antisymmetric linear maps.

(*iii*) For $\boldsymbol{k}_1, \boldsymbol{k}_2, \boldsymbol{n}_1, \boldsymbol{n}_2 \in \frac{\mathbf{V}}{\mathbb{B}}$ we have

$$\begin{aligned}
(\boldsymbol{k}_1 \otimes \boldsymbol{n}_1) : (\boldsymbol{k}_2 \otimes \boldsymbol{n}_2) &= (\boldsymbol{k}_1 \cdot \boldsymbol{k}_2)(\boldsymbol{n}_1 \cdot \boldsymbol{n}_2), \\
(\boldsymbol{k}_1 \vee \boldsymbol{n}_1) : (\boldsymbol{k}_2 \vee \boldsymbol{n}_2) &= 2\big((\boldsymbol{k}_1 \cdot \boldsymbol{k}_2)(\boldsymbol{n}_1 \cdot \boldsymbol{n}_2) + (\boldsymbol{k}_1 \cdot \boldsymbol{n}_2)(\boldsymbol{k}_2 \cdot \boldsymbol{n}_1)\big), \\
(\boldsymbol{k}_1 \wedge \boldsymbol{n}_1) : (\boldsymbol{k}_2 \wedge \boldsymbol{n}_2) &= 2\big((\boldsymbol{k}_1 \cdot \boldsymbol{k}_2)(\boldsymbol{n}_1 \cdot \boldsymbol{n}_2) - (\boldsymbol{k}_1 \cdot \boldsymbol{n}_2)(\boldsymbol{k}_2 \cdot \boldsymbol{n}_1)\big),
\end{aligned}$$

which shows that sometimes it is convenient to use the half of this bilinear form for $\boldsymbol{b}$-symmetric and $\boldsymbol{b}$-antisymmetric linear maps:

$$\boldsymbol{F} \bullet \boldsymbol{G} := \frac{1}{2}\boldsymbol{F} : \boldsymbol{G} = \frac{1}{2}\mathrm{Tr}(\boldsymbol{F}^* \cdot \boldsymbol{G}) = \frac{1}{2}\mathrm{Tr}(\boldsymbol{F} \cdot \boldsymbol{G}) \qquad (\boldsymbol{F}, \boldsymbol{G} \in \mathrm{S}(\boldsymbol{b})),$$

$$\boldsymbol{F} \bullet \boldsymbol{G} := \frac{1}{2}\boldsymbol{F} : \boldsymbol{G} = \frac{1}{2}\mathrm{Tr}(\boldsymbol{F}^* \cdot \boldsymbol{G}) = -\frac{1}{2}\mathrm{Tr}(\boldsymbol{F} \cdot \boldsymbol{G}) \qquad (\boldsymbol{F}, \boldsymbol{G} \in \mathrm{A}(\boldsymbol{b})).$$

2.11. Proposition. Let $\boldsymbol{L}$ be an $\boldsymbol{b}$-orthogonal map. Then for all $\boldsymbol{F}, \boldsymbol{G} \in \mathrm{Lin}(\mathbf{V})$

(*i*) $(\boldsymbol{L} \cdot \boldsymbol{F} \cdot \boldsymbol{L}^{-1}) : (\boldsymbol{L} \cdot \boldsymbol{G} \cdot \boldsymbol{L}^{-1}) = \boldsymbol{F} : \boldsymbol{G}$;

(*ii*) if $\boldsymbol{F}$ is $\boldsymbol{b}$-symmetric or $\boldsymbol{b}$-antisymmetric then $\boldsymbol{L} \cdot \boldsymbol{F} \cdot \boldsymbol{L}^{-1}$ is $\boldsymbol{b}$-symmetric or $\boldsymbol{b}$-antisymmetric, respectively.

2.12. Proposition. If $(\boldsymbol{n}_1, \dots, \boldsymbol{n}_N)$ and $(\boldsymbol{n}'_1, \dots, \boldsymbol{n}'_N)$ are equally oriented $\boldsymbol{b}$-orthonormal bases in $\frac{\mathbf{V}}{\mathbb{B}}$ then

$$\bigwedge_{i=1}^{N} \boldsymbol{n}_i = \bigwedge_{i=1}^{N} \boldsymbol{n}'_i.$$

Proof. Evidently, $\boldsymbol{L} \cdot \boldsymbol{n}_i := \boldsymbol{n}'_i$ $(i = 1, \dots, N)$ determines an $\boldsymbol{b}$-orthogonal map $\boldsymbol{L}$ whose determinant is positive since the bases are equally oriented. Then proposition in IV.3.18 gives the desired result. ■

Suppose $\mathbf{V}$ and $\mathbb{B}$ are oriented; then $\frac{\mathbf{V}}{\mathbb{B}}$ is oriented as well and the *Levi-Civita tensor* of $(\mathbf{V}, \mathbb{B}, \boldsymbol{b})$,

$$\boldsymbol{\epsilon} := \bigwedge_{i=1}^{N} \boldsymbol{n}_i = \bigwedge_{i=1}^{N} \frac{\boldsymbol{e}_i}{\boldsymbol{a}} \in \bigwedge_{i=1}^{N} \left(\frac{\mathbf{V}}{\mathbb{B}}\right) \equiv \frac{\overset{N}{\wedge} \mathbf{V}}{\overset{N}{\otimes} \mathbb{B}},$$

is well-defined, where $(\boldsymbol{n}_1, \dots, \boldsymbol{n}_N)$ is a positively oriented orthonormal basis in $\frac{\mathbf{V}}{\mathbb{B}}$, and $(\boldsymbol{e}_1, \dots, \boldsymbol{e}_N)$ is a positively oriented orthogonal basis in $\mathbf{V}$, normed to $\boldsymbol{a} \in \mathbb{B}^+$.

2.13. Exercises

1. According to the theory of tensor quotients, the Levi-Civita tensor can be considered to be a linear map

$$\boldsymbol{\epsilon} : \overset{N}{\otimes} \mathbb{B} \to \overset{N}{\wedge} \mathbf{V}, \qquad \bigotimes_{i=1}^{N} \boldsymbol{a}_i \mapsto \left(\bigotimes_{i=1}^{N} \boldsymbol{a}_i\right) \otimes \boldsymbol{\epsilon}.$$

Prove that $\left(\bigotimes_{i=1}^{N} \boldsymbol{a}_i\right) \otimes \boldsymbol{\epsilon} = \bigwedge_{i=1}^{N} \boldsymbol{e}_i$ where $(\boldsymbol{e}_1, \dots, \boldsymbol{e}_N)$ is a positively oriented $\boldsymbol{b}$-orthogonal basis such that $|\boldsymbol{e}_i \cdot \boldsymbol{e}_i| = |\boldsymbol{a}_i^2|$ $(i = 1, .., N)$.

2. The previous linear map is a bijection whose inverse is $\frac{1}{\boldsymbol{\epsilon}} \in \frac{\overset{N}{\otimes} \mathbb{B}}{\overset{N}{\wedge} \mathbf{V}}$, regarded as a linear map $\overset{N}{\wedge} \mathbf{V} \to \overset{N}{\otimes} \mathbb{B}$, $\bigwedge_{i=1}^{N} \boldsymbol{x}_i \mapsto \frac{\bigwedge_{i=1}^{N} \boldsymbol{x}_i}{\boldsymbol{\epsilon}}$.

Prove that $\frac{\bigwedge_{i=1}^{N} \boldsymbol{x}_i}{\boldsymbol{\epsilon}} = \sum_{\pi \in \mathrm{Perm}\, N} \operatorname{sign} \pi \prod_{i=1}^{N} (\boldsymbol{n}_{\pi(i)} \cdot \boldsymbol{x}_i) =: \boldsymbol{\epsilon}(\boldsymbol{x}_1, \dots, \boldsymbol{x}_N)$.

3. Euclidean vector spaces

3.1. A pseudo-Euclidean vector space $(\mathbf{V}, \mathbb{B}, \boldsymbol{b})$ is called *Euclidean* if $\boldsymbol{b}$ is positive definite or, equivalently, $\operatorname{neg}(\boldsymbol{b}) = 0$.

For a clear distinction, in the following $(\mathbf{S}, \mathbb{L}, \boldsymbol{h})$ denotes a Euclidean vector space.

The notations introduced for pseudo-Euclidean vector spaces will be used, e.g.

$$\boldsymbol{x}\cdot\boldsymbol{y} := \boldsymbol{h}(\boldsymbol{x},\boldsymbol{y}) \qquad (\boldsymbol{x},\boldsymbol{y}\in\mathbf{S});$$

note that if $\boldsymbol{x},\boldsymbol{y}\in\mathbf{S}$ and $\boldsymbol{k},\boldsymbol{n}\in\frac{\mathbf{S}}{\mathbb{L}}$ then

$$\boldsymbol{x}\cdot\boldsymbol{y}\in\mathbb{L}\otimes\mathbb{L}, \qquad \boldsymbol{n}\cdot\boldsymbol{x}\in\mathbb{L}, \qquad \boldsymbol{k}\cdot\boldsymbol{n}\in\mathbb{R}.$$

Moreover, we put

$$|\boldsymbol{x}|^2 := \boldsymbol{h}(\boldsymbol{x},\boldsymbol{x}) \qquad (\boldsymbol{x}\in\mathbf{S}).$$

Lastly, we say orthogonal, adjoint etc. instead of $\boldsymbol{h}$-orthogonal, $\boldsymbol{h}$-adjoint etc.

3.2. Recall that a canonical order is given in $\mathbb{L}\otimes\mathbb{L}$ (IV.5.4) and so in $(\mathbb{L}\otimes\mathbb{L})\otimes(\mathbb{L}\otimes\mathbb{L})$ as well. Thus the absolute value of elements in $\mathbb{L}\otimes\mathbb{L}$ and the square root of elements in $(\mathbb{L}\otimes\mathbb{L})\otimes(\mathbb{L}\otimes\mathbb{L})$ make sense.

Proposition (Cauchy–Schwartz inequality). For all $\boldsymbol{x},\boldsymbol{y}\in\mathbf{S}$ we have

$$|\boldsymbol{x}\cdot\boldsymbol{y}| \leq \sqrt{|\boldsymbol{x}|^2|\boldsymbol{y}|^2}$$

and equality holds if and only if $\boldsymbol{x}$ and $\boldsymbol{y}$ are parallel.

Proof. Exclude the trivial cases $\boldsymbol{x}=\mathbf{0}$ or $\boldsymbol{y}=\mathbf{0}$. Then the positive definiteness of $\boldsymbol{h}$ yields

$$0\leq\left|\boldsymbol{x}-\frac{\boldsymbol{y}\cdot\boldsymbol{x}}{\boldsymbol{y}\cdot\boldsymbol{y}}\boldsymbol{y}\right|^2 = |\boldsymbol{x}|^2 - 2\frac{\boldsymbol{y}\cdot\boldsymbol{x}}{\boldsymbol{y}\cdot\boldsymbol{y}}(\boldsymbol{x}\cdot\boldsymbol{y}) + \left(\frac{\boldsymbol{y}\cdot\boldsymbol{x}}{\boldsymbol{y}\cdot\boldsymbol{y}}\right)^2|\boldsymbol{y}|^2 = |\boldsymbol{x}|^2 - \frac{(\boldsymbol{y}\cdot\boldsymbol{x})(\boldsymbol{x}\cdot\boldsymbol{y})}{|\boldsymbol{y}|^2}$$

where equality holds if and only if $\boldsymbol{x}=\frac{\boldsymbol{y}\cdot\boldsymbol{x}}{\boldsymbol{y}\cdot\boldsymbol{y}}\boldsymbol{y}$. ■

In general, the right-hand side of the Cauchy inequality cannot be written in a simpler form because $|\boldsymbol{x}|$ and $|\boldsymbol{y}|$ make no sense, unless $\mathbb{L}$ is oriented.

3.3. Suppose now that $\mathbb{L}$ is oriented as well. Then we can define the *magnitude* or *length* of $\boldsymbol{x}\in\mathbf{S}$ as a non-negative element of $\mathbb{L}$:

$$|\boldsymbol{x}| := \sqrt{|\boldsymbol{x}|^2}.$$

The following fundamental relations hold:

(*i*) $|\boldsymbol{x}|=0$ if and only if $\boldsymbol{x}=\mathbf{0}$,

(*ii*) $|\alpha\boldsymbol{x}|=|\alpha||\boldsymbol{x}|$,

(*iii*) $|\boldsymbol{x}+\boldsymbol{y}|\leq|\boldsymbol{x}|+|\boldsymbol{y}|$

for all $\boldsymbol{x},\boldsymbol{y}\in\mathbf{S}$ and $\alpha\in\mathbb{R}$. The third relation is called the *triangle inequality* and is proved by the Cauchy–Schwartz inequality.

Moreover, the Cauchy–Schwartz inequality allows us to define the *angle* formed by $\boldsymbol{x} \neq 0$ and $\boldsymbol{y} \neq 0$:

$$\arg(\boldsymbol{x}, \boldsymbol{y}) := \arccos \frac{\boldsymbol{x} \cdot \boldsymbol{y}}{|\boldsymbol{x}||\boldsymbol{y}|}.$$

3.4. The identification

$$\frac{\mathbf{S}}{\mathbb{L} \otimes \mathbb{L}} \equiv \mathbf{S}^*$$

(see 1.3) is a fundamental property of the Euclidean vector space $(\mathbf{S}, \mathbb{L}, \boldsymbol{h})$.

The dual of an orthogonal basis $\boldsymbol{e}_1, \dots, \boldsymbol{e}_N$, normed to $m \in \mathbb{L}$, in this identification becomes $\left\{\frac{\boldsymbol{e}_1}{m^2}, \dots, \frac{\boldsymbol{e}_N}{m^2}\right\}$.

Accordingly, $\boldsymbol{n}_i := \frac{\boldsymbol{e}_i}{m}$ $(i = 1, \dots, N)$ form an orthonormal basis in $\frac{\mathbf{S}}{\mathbb{L}}$ which coincides with its dual basis in the identification $\frac{\mathbf{S}}{\mathbb{L}} \equiv \left(\frac{\mathbf{S}}{\mathbb{L}}\right)^*$:

$$\boldsymbol{n}_i \cdot \boldsymbol{n}_k = \delta_{ik} \qquad (i, k = 1, \dots, N).$$

For all $\boldsymbol{x} \in \mathbf{S}$ we have

$$\boldsymbol{x} = \sum_{i=1}^{N} (\boldsymbol{n}_i \cdot \boldsymbol{x}) \boldsymbol{n}_i.$$

In the following $\frac{\mathbf{S}}{\mathbb{L}}$ will be used frequently, so we find it convenient to introduce a shorter notation:

$$\mathbf{N} := \frac{\mathbf{S}}{\mathbb{L}}.$$

3.5. If $\mathbb{S}$ is a linear subspace of $\mathbf{S}$ then

$$\mathbb{S}^\perp := \{\boldsymbol{x} \in \mathbf{S} \mid \boldsymbol{x} \cdot \boldsymbol{y} = 0 \quad \text{for all} \quad \boldsymbol{y} \in \mathbb{S}\}$$

is called the *orthocomplement* of $\mathbb{S}$. It can be shown that $\mathbb{S}^\perp$ is a linear subspace, complementary to $\mathbb{S}$, i.e. their intersection is the zero subspace and they span the whole $\mathbf{S}$.

Every vector $\boldsymbol{x} \in \mathbf{S}$ can be uniquely decomposed into a sum of two vectors, one in $\mathbb{S}$ and the other in $\mathbb{S}^\perp$, called the *orthogonal projections* of $\boldsymbol{x}$ in $\mathbb{S}$ and in $\mathbb{S}^\perp$, respectively.

Let $\boldsymbol{n}$ be a unit vector in $\mathbf{N}$, i.e. $|\boldsymbol{n}|^2 := \boldsymbol{n} \cdot \boldsymbol{n} = 1$. Then

$$\boldsymbol{n} \otimes \mathbb{L} := \{\boldsymbol{n}d \mid d \in \mathbb{L}\}$$

is a one-dimensional subspace of $\mathbf{S}$. Furthermore, its orthocomplement,

$$\{\boldsymbol{x} \in \mathbf{S} \mid \boldsymbol{n} \cdot \boldsymbol{x} = 0\} = (\boldsymbol{n} \otimes \mathbb{L})^\perp$$

is an $(N-1)$-dimensional subspace. The corresponding projections of $\boldsymbol{x} \in \mathbf{S}$ in $\boldsymbol{n} \otimes \mathbb{L}$ and in $(\boldsymbol{n} \otimes \mathbb{L})^\perp$ are

$$(\boldsymbol{n} \cdot \boldsymbol{x})\boldsymbol{n} \qquad \text{and} \qquad \boldsymbol{x} - (\boldsymbol{n} \cdot \boldsymbol{x})\boldsymbol{n},$$

respectively.

3.6. Proposition. For $\boldsymbol{F} \in \mathrm{Lin}(\mathbf{S})$ we have $\mathrm{Ker}\,\boldsymbol{F}^* = (\mathrm{Ran}\,\boldsymbol{F})^{\perp}$.

Proof. $\boldsymbol{x}$ is in $\mathrm{Ker}\,\boldsymbol{F}^*$, i.e. $\boldsymbol{F}^* \cdot \boldsymbol{x} = \mathbf{0}$ if and only if $\boldsymbol{y} \cdot \boldsymbol{F}^* \cdot \boldsymbol{x} = \mathbf{0}$ for all $\boldsymbol{y} \in \mathbf{S}$, which is equivalent to $\boldsymbol{x} \cdot \boldsymbol{F} \cdot \boldsymbol{y} = \mathbf{0}$ for all $\boldsymbol{y} \in \mathbf{S}$, thus $\boldsymbol{x}$ is in $\mathrm{Ker}\,\boldsymbol{F}^*$ if and only if it is orthogonal to the range of $\boldsymbol{F}$.

3.7. We know that a linear map $\boldsymbol{L} \colon \mathbf{S} \to \mathbf{S}$ is orthogonal, i.e. $\boldsymbol{y} \cdot \boldsymbol{x} = (\boldsymbol{L} \cdot \boldsymbol{y}) \cdot (\boldsymbol{L} \cdot \boldsymbol{x})$ for all $\boldsymbol{x}, \boldsymbol{y} \in \mathbf{S}$ if and only if $|\boldsymbol{L} \cdot \boldsymbol{x}|^2 = |\boldsymbol{x}|^2$ for all $\boldsymbol{x} \in \mathbf{S}$ (see 2.7). Because of the Euclidean structure we need not assume the linearity of $\boldsymbol{L}$, according to the following result.

Proposition. Let $\boldsymbol{L} \colon \mathbf{S} \to \mathbf{S}$ be a map such that $\boldsymbol{L}(\boldsymbol{y}) \cdot \boldsymbol{L}(\boldsymbol{x}) = \boldsymbol{y} \cdot \boldsymbol{x}$ for all $\boldsymbol{x}, \boldsymbol{y} \in \mathbf{S}$. Then $\boldsymbol{L}$ is necessarily linear.

Proof. First of all note that if $\{\boldsymbol{e}_1, \ldots, \boldsymbol{e}_N\}$ is an orthogonal basis in $\mathbf{S}$ then $\{\boldsymbol{L} \cdot \boldsymbol{e}_1, \ldots, \boldsymbol{L} \cdot \boldsymbol{e}_N\}$ is an orthogonal basis as well. As a consequence, $\mathrm{Ran}\,\boldsymbol{L}$ spans $\mathbf{S}$.

If $\boldsymbol{L}(\boldsymbol{y}) = \boldsymbol{L}(\boldsymbol{x})$ then $|\boldsymbol{y}|^2 = |\boldsymbol{x}|^2 = \boldsymbol{y} \cdot \boldsymbol{x}$ and so $|\boldsymbol{y} - \boldsymbol{x}|^2 = \mathbf{0}$, hence $\boldsymbol{L}$ is injective.

Writing $\boldsymbol{x}' := \boldsymbol{L}(\boldsymbol{x})$, $\boldsymbol{y}' := \boldsymbol{L}(\boldsymbol{y})$ and then omitting the prime, we find that $\boldsymbol{L}^{-1}(\boldsymbol{y}) \cdot \boldsymbol{L}^{-1}(\boldsymbol{x}) = \boldsymbol{y} \cdot \boldsymbol{x}$ for all $\boldsymbol{x}, \boldsymbol{y} \in \mathrm{Ran}\,\boldsymbol{L}$ and $\boldsymbol{L}^{-1}(\boldsymbol{y}) \cdot \boldsymbol{x} = \boldsymbol{y} \cdot \boldsymbol{L}(\boldsymbol{x})$ for all $\boldsymbol{x} \in \mathbf{S}$, $\boldsymbol{y} \in \mathrm{Ran}\,\boldsymbol{L}$.

Consequently, for all $\boldsymbol{y} \in \mathrm{Ran}\,\boldsymbol{L}$ and $\boldsymbol{x}_1, \boldsymbol{x}_2 \in \mathbf{S}$ we have

$$\begin{aligned} \boldsymbol{y} \cdot \boldsymbol{L}(\boldsymbol{x}_1 + \boldsymbol{x}_2) &= \boldsymbol{L}^{-1}(\boldsymbol{y}) \cdot (\boldsymbol{x}_1 + \boldsymbol{x}_2) = \boldsymbol{L}^{-1}(\boldsymbol{y}) \cdot \boldsymbol{x}_1 + \boldsymbol{L}^{-1}(\boldsymbol{y}) \cdot \boldsymbol{x}_2 = \\ &= \boldsymbol{y} \cdot \boldsymbol{L}(\boldsymbol{x}_1) + \boldsymbol{y} \cdot \boldsymbol{L}(\boldsymbol{x}_2) = \boldsymbol{y} \cdot (\boldsymbol{L}(\boldsymbol{x}_1) + \boldsymbol{L}(\boldsymbol{x}_2)); \end{aligned}$$

since $\boldsymbol{y}$ is arbitrary in $\mathrm{Ran}\,\boldsymbol{L}$ which spans $\mathbf{S}$, this means that

$$\boldsymbol{L}(\boldsymbol{x}_1 + \boldsymbol{x}_2) = \boldsymbol{L}(\boldsymbol{x}_1) + \boldsymbol{L}(\boldsymbol{x}_2) \qquad (\boldsymbol{x}_1, \boldsymbol{x}_2 \in \mathbf{S}).$$

A similar argument shows that

$$\boldsymbol{L}(\alpha \boldsymbol{x}) = \alpha \boldsymbol{L}(\boldsymbol{x}) \qquad (\alpha \in \mathbb{R}, \boldsymbol{x} \in \mathbf{S}). \quad \blacksquare$$

This has the simple but important consequence that if $\boldsymbol{L} \colon \mathbf{S} \to \mathbf{S}$ is a map such that $|\boldsymbol{L}(\boldsymbol{y}) - \boldsymbol{L}(\boldsymbol{x})|^2 = |\boldsymbol{y} - \boldsymbol{x}|^2$ for all $\boldsymbol{x}, \boldsymbol{y} \in \mathbf{S}$ and $\boldsymbol{L}(\mathbf{0}) = \mathbf{0}$ then $\boldsymbol{L}$ is necessarily linear. The proof is left to the reader as an exercise.

3.8. In the following, assuming that

$$\dim \mathbf{S} = 3,$$

we shall examine the structure of the antisymmetric linear maps of $\mathbf{S}$. As we know, (see 2.9) $\mathrm{A}(\boldsymbol{h}) \equiv \frac{\mathbf{S}}{\mathbb{L}} \wedge \frac{\mathbf{S}}{\mathbb{L}} = \mathbf{N} \wedge \mathbf{N}$ is a three-dimensional vector space endowed (see 2.10) with a real-valued positive definite symmetric bilinear form—an inner product—:

$$A \bullet B = \frac{1}{2}\mathrm{Tr}(A^* \cdot B) = -\frac{1}{2}\mathrm{Tr}(A \cdot B).$$

The magnitude of the antisymmetric linear map A is the real number

$$|A| := \sqrt{A \bullet A}.$$

Recall that for $k_1, k_2, n_1, n_2 \in \mathbf{N}$ we have

$$(k_1 \wedge k_2) \bullet (n_1 \wedge n_2) = (k_1 \cdot k_2)(n_1 \cdot n_2) - (k_1 \cdot n_2)(k_2 \cdot n_1),$$

in particular, if $k_1 = k_2 =: k$, $n_1 = n_2 =: n$,

$$|k \wedge n|^2 = |k|^2 |n|^2 - (k \cdot n)^2.$$

3.9. If $A \in \mathbf{N} \wedge \mathbf{N}$ then $A^* = -A$, thus proposition 3.6 yields that $\mathrm{Ker}\, A$ is the orthogonal complement of $\mathrm{Ran}\, A$.

Proposition. If $0 \neq A \in \mathbf{N} \wedge \mathbf{N}$ then $\mathrm{Ran}\, A$ is two-dimensional, consequently, $\mathrm{Ker}\, A$ is one-dimensional.

Proof. Since $A \neq 0$, there is a $0 \neq x \in \mathrm{Ran}\, A$. Then $x \notin \mathrm{Ker}\, A$, thus $0 \neq A \cdot x \in \mathrm{Ran}\, A$. x and $A \cdot x$ are orthogonal to each other because A is antisymmetric. Consequently, the subspace spanned by x and $A \cdot x$ is two-dimensional in the range of A: $\mathrm{Ran}\, A$ is at least two-dimensional, $\mathrm{Ker}\, A$ is at most one-dimensional. Suppose $\mathrm{Ker}\, A = \{0\}$. Take a $0 \neq y$, orthogonal to both x and $A \cdot x$. $\mathbf{S}$ is three-dimensional, $A \cdot y$ is orthogonal to y, so it lies in the subspace generated by x and $A \cdot x$, i.e. $A \cdot y = \alpha x + \beta A \cdot x$. Multiplying by x and using $x \cdot A \cdot x = 0$, $x \cdot A \cdot y = -y \cdot A \cdot x = 0$, we get $\alpha = 0$. As a consequence, $A \cdot (y - \beta x) = 0$, the vector $y - \beta x$ is in $\mathrm{Ker}\, A$, thus $y - \beta x = 0$, $y = \beta x$, a contradiction.

3.10. Let us take a nonzero $A \in \mathbf{N} \wedge \mathbf{N}$. There is an orthonormal basis $\{n_1, n_2, n_3\}$ in $\mathbf{N}$ such that $n_3 \otimes \mathbb{L} = \mathrm{Ker}\, A$. $\{n_1 \wedge n_2, n_3 \wedge n_1, n_2 \wedge n_3\}$ is a basis in $\mathbf{N} \wedge \mathbf{N}$, thus there are real numbers $\alpha_1, \alpha_2, \alpha_3$ such that

$$A = \alpha_3 (n_1 \wedge n_2) + \alpha_2 (n_3 \wedge n_1) + \alpha_1 (n_2 \wedge n_3).$$

Since $A \cdot n_3 = 0$, we easily find that $\alpha_2 = \alpha_1 = 0$ and, consequently, $|\alpha_3| = |A|$. Renaming n_1 and n_2 and taking their antisymmetric tensor product in a convenient order we arrive at the following result.

Proposition. If $0 \neq \boldsymbol{A} \in \mathbf{N} \wedge \mathbf{N}$ and $\boldsymbol{n}$ is an arbitrary unit vector in $\mathbf{N}$, orthogonal to the kernel of $\boldsymbol{A}$, then there is a unit vector $\boldsymbol{k}$, orthogonal to the kernel of $\boldsymbol{A}$ and to $\boldsymbol{n}$ such that

$$\boldsymbol{A} = |\boldsymbol{A}|\boldsymbol{k} \wedge \boldsymbol{n}. \quad \blacksquare$$

As a consequence, we have

$$\boldsymbol{A}^3 = -|\boldsymbol{A}|^2 \boldsymbol{A}.$$

Moreover, for nonzero $\boldsymbol{A}$ and $\boldsymbol{B}$ in $\mathbf{N} \wedge \mathbf{N}$ the following statements are equivalent:
— $\boldsymbol{A}$ is a multiple of $\boldsymbol{B}$,
— $\operatorname{Ker}\boldsymbol{A} = \operatorname{Ker}\boldsymbol{B}$,
— $\operatorname{Ran}\boldsymbol{A} = \operatorname{Ran}\boldsymbol{B}$.

3.11. If $\boldsymbol{A}, \boldsymbol{B} \in \mathbf{N} \wedge \mathbf{N}$, their commutator

$$[\boldsymbol{A}, \boldsymbol{B}] := \boldsymbol{A} \cdot \boldsymbol{B} - \boldsymbol{B} \cdot \boldsymbol{A}$$

is in $\mathbf{N} \wedge \mathbf{N}$, too. Moreover, the properties of the trace imply that for all $\boldsymbol{C} \in \mathbf{N} \wedge \mathbf{N}$

$$[\boldsymbol{A}, \boldsymbol{B}] \bullet \boldsymbol{C} = [\boldsymbol{C}, \boldsymbol{A}] \bullet \boldsymbol{B} = [\boldsymbol{B}, \boldsymbol{C}] \bullet \boldsymbol{A}.$$

Proposition.

$$|[\boldsymbol{A}, \boldsymbol{B}]|^2 = |\boldsymbol{A}|^2 |\boldsymbol{B}|^2 - (\boldsymbol{A} \bullet \boldsymbol{B})^2.$$

Proof. If $\boldsymbol{A}$ and $\boldsymbol{B}$ are parallel (in particular, if one of them is zero) then the equality holds trivially. If $\boldsymbol{A}$ and $\boldsymbol{B}$ are not parallel, dividing the equality by $|\boldsymbol{A}|^2|\boldsymbol{B}|^2$ we reduce the problem to the case $|\boldsymbol{A}| = |\boldsymbol{B}| = 1$. The ranges of $\boldsymbol{A}$ and $\boldsymbol{B}$ are different two-dimensional subspaces, hence their intersection is a one-dimensional subspace (because $\mathbf{S}$ is three-dimensional). Let $\boldsymbol{n}$ be a unit vector in $\mathbf{N}$ such that $\boldsymbol{n} \otimes \mathbb{L} = \operatorname{Ran}\boldsymbol{A} \cap \operatorname{Ran}\boldsymbol{B}$. Then there are unit vectors $\boldsymbol{k}$ and $\boldsymbol{r}$ in $\mathbf{N}$, orthogonal to $\boldsymbol{n}$, such that $\boldsymbol{A} = \boldsymbol{k} \wedge \boldsymbol{n}$, $\boldsymbol{B} = \boldsymbol{r} \wedge \boldsymbol{n}$. Simple calculations yield

$$[\boldsymbol{A}, \boldsymbol{B}] = \boldsymbol{r} \wedge \boldsymbol{k}, \qquad |[\boldsymbol{A}, \boldsymbol{B}]|^2 = 1 - (\boldsymbol{k} \cdot \boldsymbol{r})^2$$

which gives the desired result in view of 3.8.

3.12. According to the formula cited at the beginning of the previous paragraph, $[\boldsymbol{A}, \boldsymbol{B}]$ is orthogonal to both $\boldsymbol{A}$ and $\boldsymbol{B}$. Consequently, if $\boldsymbol{A}$ and $\boldsymbol{B}$ are orthogonal, $|\boldsymbol{A}| = |\boldsymbol{B}| = 1$ then $\boldsymbol{A}, \mathbb{B}$ and $[\boldsymbol{A}, \boldsymbol{B}]$ form an orthonormal basis in $\mathbf{N} \wedge \mathbf{N}$.

Proposition. For all $\boldsymbol{A}, \boldsymbol{B}, \boldsymbol{C} \in \mathbf{N} \wedge \mathbf{N}$ we have

$$[[\boldsymbol{A}, \boldsymbol{B}], \boldsymbol{C}] = (\boldsymbol{A} \bullet \boldsymbol{C})\boldsymbol{B} - (\boldsymbol{B} \bullet \boldsymbol{C})\boldsymbol{A}.$$

Proof. If $\boldsymbol{A}$ and $\boldsymbol{B}$ are parallel, both sides are zero. If $\boldsymbol{A}$ and $\boldsymbol{B}$ are not parallel (in particular, neither of them is zero) then $\boldsymbol{B} = \alpha\boldsymbol{A} + \mathbb{B}'$ where α is a number and $\mathbb{B}' \neq \mathbf{0}$ is orthogonal to $\mathbb{A}$. $\alpha\boldsymbol{A}$ results in zero on both sides, hence it is sufficient to consider an arbitrary $\boldsymbol{A}$, a $\boldsymbol{B}$ orthogonal to $\boldsymbol{A}$, and three linearly independent elements in the role of $\boldsymbol{C}$; they will be $\boldsymbol{A}$, $\boldsymbol{B}$ and $[\boldsymbol{A}, \boldsymbol{B}]$.

For $\boldsymbol{C} = [\boldsymbol{A}, \boldsymbol{B}]$ the equality is trivial, both sides are zero.

For $\boldsymbol{C} = \boldsymbol{A}$, the right-hand side equals $|\boldsymbol{A}|^2\boldsymbol{B}$; $[[\boldsymbol{A}, \boldsymbol{B}], \boldsymbol{A}]$ on the left-hand side is orthogonal to both $[\boldsymbol{A}, \boldsymbol{B}]$ and $\boldsymbol{A}$, hence it is parallel to $\boldsymbol{B}$: there is a number α such that $[[\boldsymbol{A}, \boldsymbol{B}], \boldsymbol{A}] = \alpha\boldsymbol{B}$. Take the inner product of both sides by $\boldsymbol{B}$, apply the formula at the beginning of 3.11 to have $|[\boldsymbol{A}, \boldsymbol{B}]|^2 = \alpha|\boldsymbol{B}|^2$ which implies $\alpha = |\boldsymbol{A}|^2$ according to the previous result.

A similar argument is applied to $\boldsymbol{C} = \boldsymbol{B}$.

3.13. Let us continue to consider the Euclidean vector space $(\mathbf{S}, \mathbb{L}, \boldsymbol{h})$, $\dim \mathbf{S} = 3$, and suppose that $\mathbf{S}$ and $\mathbb{L}$ are oriented. Then $\mathbf{N} = \frac{\mathbf{S}}{\mathbb{L}}$ is oriented as well (see IV.5.2). According to 2.12, there is a well-defined $\boldsymbol{\epsilon}$ in $\overset{3}{\wedge}\mathbf{N}$ such that

$$\boldsymbol{\epsilon} = \overset{3}{\underset{i=1}{\wedge}} \boldsymbol{n}_i$$

for an arbitrary positively oriented orthonormal basis $(\boldsymbol{n}_1, \boldsymbol{n}_2, \boldsymbol{n}_3)$ of $\mathbf{N}$. $\boldsymbol{\epsilon}$ is called the *Levi-Civita tensor* of $(\mathbf{S}, \mathbb{L}, \boldsymbol{h})$.

The Levi-Civita tensor establishes a linear bijection

$$\boldsymbol{j}: \mathbf{N} \wedge \mathbf{N} \to \mathbf{N}, \qquad \boldsymbol{k} \wedge \boldsymbol{n} \mapsto \boldsymbol{\epsilon}(\cdot, \boldsymbol{k}, \boldsymbol{n}).$$

Let us examine more closely what this is. The dual of $\mathbf{N}$ is identified with $\mathbf{N}$, thus $\boldsymbol{\epsilon}$ can be considered to be a trilinear map

$$\mathbf{N}^3 \to \mathbb{R}, \qquad (\boldsymbol{k}_1, \boldsymbol{k}_2, \boldsymbol{k}_3) \mapsto \sum_{\pi \in \mathrm{Perm}\,3} \operatorname{sign}\pi \prod_{i=1}^{3} \boldsymbol{n}_{\pi(i)} \cdot \boldsymbol{k}_i$$

(see IV.3.15). Thus, for given $\boldsymbol{k}$ and $\boldsymbol{n}$, $\boldsymbol{\epsilon}(\cdot, \boldsymbol{k}, \boldsymbol{n})$ is the linear map $\mathbf{N} \to \mathbb{R}$, $\boldsymbol{r} \mapsto \boldsymbol{\epsilon}(\boldsymbol{r}, \boldsymbol{k}, \boldsymbol{n})$, i.e. it is an element of $\mathbf{N}^* \equiv \mathbf{N}$.

In other words, $\boldsymbol{j}(\boldsymbol{k} \wedge \boldsymbol{n})$ is the element of $\mathbf{N}$ determined by

$$\boldsymbol{r} \cdot \boldsymbol{j}(\boldsymbol{k} \wedge \boldsymbol{n}) = \boldsymbol{\epsilon}(\boldsymbol{r}, \boldsymbol{k}, \boldsymbol{n})$$

for all $\boldsymbol{r} \in \mathbf{N}$.

The Levi-Civita tensor is antisymmetric, hence $\boldsymbol{j}(\boldsymbol{k} \wedge \boldsymbol{n})$ is orthogonal to both $\boldsymbol{k}$ and $\boldsymbol{n}$.

If $(\boldsymbol{n}_1, \boldsymbol{n}_2, \boldsymbol{n}_3)$ is a positively oriented orthonormal basis in $\mathbf{N}$, then

$$\boldsymbol{j}(\boldsymbol{n}_1 \wedge \boldsymbol{n}_2) = \boldsymbol{n}_3, \qquad \boldsymbol{j}(\boldsymbol{n}_2 \wedge \boldsymbol{n}_3) = \boldsymbol{n}_1, \qquad \boldsymbol{j}(\boldsymbol{n}_3 \wedge \boldsymbol{n}_1) = \boldsymbol{n}_2.$$

Proposition. For all $\boldsymbol{A} \in \mathbf{N} \wedge \mathbf{N}$ we have
(*i*) $\boldsymbol{A} \cdot \boldsymbol{j}(\boldsymbol{A}) = \mathbf{0}$,
(*ii*) $|\boldsymbol{j}(\boldsymbol{A})| = |\boldsymbol{A}|$,
(*iii*) if $\boldsymbol{A} \neq 0$ then $(\boldsymbol{j}(\boldsymbol{A}), \boldsymbol{n}, \boldsymbol{A} \cdot \boldsymbol{n})$ is a positively oriented orthogonal basis in $\mathbf{N}$ for arbitrary non zero $\boldsymbol{n}$, orthogonal to $\operatorname{Ker} \boldsymbol{A}$.

Proof. There is a positively oriented orthonormal basis $(\boldsymbol{n}_1, \boldsymbol{n}_2, \boldsymbol{n}_3)$ such that $\boldsymbol{A} = |\boldsymbol{A}|\boldsymbol{n}_1 \wedge \boldsymbol{n}_2$ (and so $\boldsymbol{A} \cdot \boldsymbol{n}_3 = \mathbf{0}$); then $\boldsymbol{j}(\boldsymbol{A}) = -|\boldsymbol{A}|\boldsymbol{n}_3$ from which (*i*) and (*ii*) follow immediately. Moreover, we can choose $\boldsymbol{n}_1 := \frac{\boldsymbol{n}}{|\boldsymbol{n}|}$ where $\boldsymbol{n}$ is an arbitrary nonzero vector orthogonal to $\operatorname{Ker} \boldsymbol{A}$. ■

The kernel of a nonzero $\boldsymbol{A}$ is one-dimensional; according to (*i*), $\boldsymbol{j}(\boldsymbol{A})$ spans the kernel of $\boldsymbol{A}$. The one-dimensional vector space $\operatorname{Ker} \boldsymbol{A}$ will be *oriented* by $\boldsymbol{j}(\boldsymbol{A})$.

Since $\boldsymbol{j}$ is linear, (*ii*) is equivalent to $\boldsymbol{j}(\boldsymbol{A}) \cdot \boldsymbol{j}(\boldsymbol{B}) = \boldsymbol{A} \bullet \boldsymbol{B}$ for all $\boldsymbol{A}$ and $\boldsymbol{B}$ which can also be proved directly using that $\boldsymbol{A} = |\boldsymbol{A}|\boldsymbol{k} \wedge \boldsymbol{n}$, $\boldsymbol{B} = |\boldsymbol{B}|\boldsymbol{r} \wedge \boldsymbol{n}$.

3.14. Definition. The map

$$\mathbf{N} \times \mathbf{N} \to \mathbf{N}, \qquad (\boldsymbol{k}, \boldsymbol{n}) \mapsto \boldsymbol{k} \times \boldsymbol{n} := \boldsymbol{j}(\boldsymbol{k} \wedge \boldsymbol{n})$$

is called the *vectorial product.* ■

It is evident from the properties of $\boldsymbol{j}$ that the vectorial product is an antisymmetric bilinear mapping, $\boldsymbol{k} \times \boldsymbol{n} = \mathbf{0}$ if and only if $\boldsymbol{k}$ and $\boldsymbol{n}$ are parallel, $\boldsymbol{k} \times \boldsymbol{n}$ is orthogonal to both $\boldsymbol{k}$ and $\boldsymbol{n}$,

$$|\boldsymbol{k} \times \boldsymbol{n}|^2 = |\boldsymbol{k}|^2 |\boldsymbol{n}|^2 - (\boldsymbol{k} \cdot \boldsymbol{n})^2.$$

If $\boldsymbol{k}$ and $\boldsymbol{n}$ are not parallel then $\boldsymbol{k}$, $\boldsymbol{n}$ and $\boldsymbol{k} \times \boldsymbol{n}$ form a positively oriented basis in $\mathbf{N}$; moreover, if $(\boldsymbol{n}_1, \boldsymbol{n}_2, \boldsymbol{n}_3)$ is a positively oriented orthonormal basis in $\mathbf{N}$ then

$$\boldsymbol{n}_1 \times \boldsymbol{n}_2 = \boldsymbol{n}_3, \qquad \boldsymbol{n}_2 \times \boldsymbol{n}_3 = \boldsymbol{n}_1, \qquad \boldsymbol{n}_3 \times \boldsymbol{n}_1 = \boldsymbol{n}_2.$$

Proposition.
(*i*)
$$-\boldsymbol{j}([\boldsymbol{A}, \boldsymbol{B}]) = \boldsymbol{j}(\boldsymbol{A}) \times \boldsymbol{j}(\boldsymbol{B}) \qquad (\boldsymbol{A}, \boldsymbol{B} \in \mathbf{N} \wedge \mathbf{N})$$
or, equivalently,
$$\boldsymbol{j}(\boldsymbol{A}) \wedge \boldsymbol{j}(\boldsymbol{B}) = [\boldsymbol{A}, \boldsymbol{B}];$$

(*ii*)
$$-\boldsymbol{A}\cdot\boldsymbol{n} = \boldsymbol{j}(\boldsymbol{A})\times\boldsymbol{n} \qquad (\boldsymbol{A}\in\mathbf{N}\wedge\mathbf{N},\ \boldsymbol{n}\in\mathbf{N})$$
which implies
$$\boldsymbol{A}\cdot\boldsymbol{j}(\boldsymbol{B}) = \boldsymbol{j}([\boldsymbol{A},\boldsymbol{B}]) \qquad (\boldsymbol{A},\boldsymbol{B}\in\mathbf{N}\wedge\mathbf{N}).$$

Proof. There is an orthonormal basis $\{\boldsymbol{n}_1,\boldsymbol{n}_2,\boldsymbol{n}_3\}$ in $\mathbf{N}$ such that $\boldsymbol{A} = |\boldsymbol{A}|\boldsymbol{n}_1\wedge\boldsymbol{n}_2$.

(*i*) $\boldsymbol{n}_1\wedge\boldsymbol{n}_2$, $\boldsymbol{n}_2\wedge\boldsymbol{n}_3$ and $\boldsymbol{n}_3\wedge\boldsymbol{n}_1$ form a basis in $\mathbf{N}\wedge\mathbf{N}$, thus it is sufficient to consider them in the role of $\boldsymbol{B}$; then a simple calculation yields the desired result. bek(*ii*) Take $\boldsymbol{n}_1$, $\boldsymbol{n}_2$ and $\boldsymbol{n}_3$ in the role of $\boldsymbol{n}$. ■

As a consequence of our results we have
$$(\boldsymbol{k}\times\boldsymbol{n})\cdot\boldsymbol{r} = (\boldsymbol{n}\times\boldsymbol{r})\cdot\boldsymbol{k} = (\boldsymbol{r}\times\boldsymbol{k})\cdot\boldsymbol{n} = \boldsymbol{\epsilon}(\boldsymbol{r},\boldsymbol{k},\boldsymbol{n})$$
and
$$(\boldsymbol{k}\times\boldsymbol{n})\times\boldsymbol{r} = (\boldsymbol{k}\cdot\boldsymbol{r})\boldsymbol{n} - (\boldsymbol{n}\cdot\boldsymbol{r})\boldsymbol{k}$$
for all $\boldsymbol{k},\boldsymbol{n},\boldsymbol{r}\in\mathbf{N}$.

3.15. The Levi-Civita tensor establishes another linear bijection as well:
$$\mathrm{j}_\mathrm{o}\colon \overset{3}{\wedge}\mathbf{N}\to\mathbb{R}, \qquad \overset{3}{\underset{i=1}{\wedge}}\boldsymbol{k}_i \mapsto \boldsymbol{\epsilon}(\boldsymbol{k}_1,\boldsymbol{k}_2,\boldsymbol{k}_3).$$
It is quite trivial that ${\mathrm{j}_\mathrm{o}}^{-1}(\alpha) = \alpha\boldsymbol{\epsilon}$ for all $\alpha\in\mathbb{R}$.

3.16. An orthogonal map $\boldsymbol{R}\colon\mathbf{S}\to\mathbf{S}$ (also regarded as an orthogonal map $\mathbf{N}\to\mathbf{N}$, see 2.1) sends orthogonal bases into orthogonal ones, preserves and changes orientation according to whether $\det\boldsymbol{R} = 1$ or $\det\boldsymbol{R} = -1$. In view of IV.3.18,
$$\boldsymbol{\epsilon}\circ\left(\overset{3}{\times}\boldsymbol{R}\right) = (\det\boldsymbol{R})\boldsymbol{\epsilon}.$$

Then one proves without difficulty that
$$\boldsymbol{j}(\boldsymbol{R}\cdot\boldsymbol{k}\wedge\boldsymbol{R}\cdot\boldsymbol{n}) = (\det\boldsymbol{R})\boldsymbol{R}\cdot\boldsymbol{j}(\boldsymbol{k}\wedge\boldsymbol{n}) \qquad (\boldsymbol{k},\boldsymbol{n}\in\mathbf{N}).$$
Since $\boldsymbol{R}\cdot\boldsymbol{k}\wedge\boldsymbol{R}\cdot\boldsymbol{n} = \boldsymbol{R}\cdot(\boldsymbol{k}\wedge\boldsymbol{n})\cdot\boldsymbol{R}^{-1}$, the previous result can be written in the form
$$\boldsymbol{j}(\boldsymbol{R}\cdot\boldsymbol{A}\cdot\boldsymbol{R}^{-1}) = (\det\boldsymbol{R})\boldsymbol{R}\cdot\boldsymbol{j}(\boldsymbol{A}) \qquad (\boldsymbol{A}\in\mathbf{N}\wedge\mathbf{N}).$$

Moreover,
$$\mathrm{j}_\mathrm{o}\left(\overset{3}{\underset{i=1}{\wedge}}\boldsymbol{R}\cdot\boldsymbol{k}_i\right) = (\det\boldsymbol{R})\ \mathrm{j}_\mathrm{o}\left(\overset{3}{\underset{i=1}{\wedge}}\boldsymbol{k}_i\right).$$

3.17. In the usual way, the linear bijection $\boldsymbol{j}$ can be lifted to a linear bijection
$$\mathbf{S}\wedge\mathbf{S}\to\mathbf{S}\otimes\mathbb{L}, \qquad \text{defined by} \qquad \boldsymbol{x}\wedge\boldsymbol{y}\mapsto\boldsymbol{j}\left(\frac{\boldsymbol{x}}{m}\wedge\frac{\boldsymbol{y}}{m}\right)m^2$$
where m is an arbitrary nonzero element of $\mathbb{L}$.

Similarly, the linear bijection $\mathrm{j_o}$ can be lifted to a linear bijection

$$\overset{3}{\wedge}\mathbf{S} \to \overset{3}{\otimes}\mathbb{L}, \qquad \overset{3}{\underset{i=1}{\wedge}} \boldsymbol{x}_i \mapsto \mathrm{j_o}\left(\overset{3}{\underset{i=1}{\wedge}} \frac{\boldsymbol{x}_i}{\boldsymbol{m}}\right)\boldsymbol{m}^3.$$

We have utilized here that $\mathbf{S} = \mathbf{N} \otimes \mathbb{L}$. Evidently, similar formulae are valid for $\mathbf{N} \otimes \mathbb{A}$ where $\mathbb{A}$ is an arbitrary one-dimensional vector space.

3.18. Let us consider the Euclidean vector space $(\mathbb{R}^3, \mathbb{R}, \boldsymbol{H})$ where

$$\boldsymbol{H}(\boldsymbol{x}, \boldsymbol{y}) = \sum_{i=1}^{3} x^i y^i =: \boldsymbol{x} \cdot \boldsymbol{y}$$

(i.e. $\boldsymbol{H} = \boldsymbol{H}_0$ in the notation of 1.7).

The identification $\mathbb{R}^3 \equiv \left(\mathbb{R}^3\right)^*$ is the usual one: the functional corresponding to $\boldsymbol{x}$ and represented by the usual matrix multiplication rule coincides with $\boldsymbol{x}$. In customary notations $\boldsymbol{x}$ considered to be a vector has the components (x^1, x^2, x^3) and $\boldsymbol{x}$ considered to be a covector has the components (x_1, x_2, x_3); the previous statement says that $x_i = x^i$, $i = 1, 2, 3$.

That is why in this case one usually writes only subscripts.

The adjoint of a 3×3 matrix (as a linear map $\mathbb{R}^3 \to \mathbb{R}^3$) coincides with the transpose of the matrix.

$\mathbb{R}^3$ and $\mathbb{R}$ are endowed with the usual orientations: the naturally ordered standard bases are taken to be positively oriented.

The Levi-Civita tensor is given by a matrix of three indices:

$$\boldsymbol{\epsilon} = (\epsilon_{ijk} \mid i, j, k = 1, 2, 3),$$
$$\boldsymbol{\epsilon}(\boldsymbol{x}, \boldsymbol{y}, \boldsymbol{z}) = \sum_{i,j,k=1}^{3} \epsilon_{ijk} x_i y_j z_k,$$

$$\epsilon_{ijk} = \begin{cases} 1 & \text{if } ijk \text{ is an even permutation of } 123 \\ -1 & \text{if } ijk \text{ is an odd permutation of } 123 \\ 0 & \text{otherwise} \end{cases}$$

Then it is an easy task to show that

$$\boldsymbol{j}(L_{jk} \mid j, k = 1, 2, 3) = \left(-\frac{1}{2}\sum_{j,k=1}^{3} \epsilon_{ijk} L_{jk} \mid i = 1, 2, 3\right),$$

$$\boldsymbol{j}^{-1}(x_k \mid k = 1, 2, 3) = \left(-\sum_{k=1}^{3} \epsilon_{ijk} x_k \mid i, j = 1, 2, 3\right),$$

in other notation,

$$\boldsymbol{j}^{-1}(x_1, x_2, x_3) = \begin{pmatrix} 0 & -x_3 & x_2 \\ x_3 & 0 & -x_1 \\ -x_2 & x_1 & 0 \end{pmatrix}.$$

Moreover,

$$\boldsymbol{x} \times \boldsymbol{y} = \left(\sum_{j,k=1}^{3} \epsilon_{ijk} x_j y_k \mid i = 1, 2, 3 \right).$$

3.19. Consider the Euclidean vector space $(\mathbf{S}, \mathbb{L}, \boldsymbol{h})$, $\dim \mathbf{S} = 3$.

A linear coordinatization $\boldsymbol{K}$ of $\mathbf{S}$ is called *orthogonal* if it corresponds to an ordered orthogonal basis $(\boldsymbol{e}_1, \boldsymbol{e}_2, \boldsymbol{e}_3)$ normed to an $\boldsymbol{m} \in \mathbb{L}$.

Since the dual of the basis is $\left(\frac{\boldsymbol{e}_i}{\boldsymbol{m}^2} \mid i = 1, 2, 3\right)$ (see 3.4), we have

$$\boldsymbol{K} \cdot \boldsymbol{x} = \left(\frac{\boldsymbol{e}_i \cdot \boldsymbol{x}}{\boldsymbol{m}^2} \mid i = 1, 2, 3 \right) =: (x^1, x^2, x^3).$$

Consider the identification $\mathbf{S} \equiv \mathbb{L} \otimes \mathbb{L} \otimes \mathbf{S}^* \equiv \left(\frac{\mathbf{S}}{\mathbb{L} \otimes \mathbb{L}}\right)^*$; then $\boldsymbol{x}$, as an element of the dual of $\frac{\mathbf{S}}{\mathbb{L} \otimes \mathbb{L}}$, has the coordinates

$$\left(\boldsymbol{x} \cdot \frac{\boldsymbol{e}_i}{\boldsymbol{m}^2} \mid i = 1, 2, 3 \right) =: (x_1, x_2, x_3).$$

We see, in accordance with the previous paragraph, that $x^i = x_i$ $(i = 1, 2, 3)$ and we can use only subscripts.

Then all the operations regarding the Euclidean structure can be represented by the corresponding operations in $(\mathbb{R}^3, \mathbb{R}, \boldsymbol{H})$, e.g.

—the $\boldsymbol{h}$-product of elements $\boldsymbol{x}, \boldsymbol{y}$ of $\mathbf{S}$ is computed by the inner product of their coordinates in $\mathbb{R}^3$:

$$\text{if} \quad \boldsymbol{K} \cdot \boldsymbol{x} = (x_1, x_2, x_3) \quad \text{and} \quad \boldsymbol{K} \cdot \boldsymbol{y} = (y_1, y_2, y_3)$$

i.e.

$$\boldsymbol{x} = \sum_{i=1}^{3} x_i \boldsymbol{e}_i, \qquad \boldsymbol{y} = \sum_{i=1}^{3} y_i \boldsymbol{e}_i$$

$$\text{then} \quad \boldsymbol{x} \cdot \boldsymbol{y} = \left(\sum_{i=1}^{3} x_i y_i \right) \boldsymbol{m}^2;$$

— the matrix in the coordinatization of an adjoint map will be the transpose of the matrix representing the linear map in question:

$$\begin{aligned} \text{if} \qquad & \boldsymbol{K}\cdot\boldsymbol{L}\cdot\boldsymbol{K}^{-1} = (L_{ik} \mid i,k=1,2,3) \\ \text{then} \qquad & \boldsymbol{K}\cdot\boldsymbol{L}^{*}\cdot\boldsymbol{K}^{-1} = (L_{ki} \mid i,k=1,2,3)\,; \end{aligned}$$

—if both $\mathbf{S}$ and $\mathbb{L}$ are oriented and the basis establishing the coordinatization is positively oriented then the vectorial product can be computed by the vectorial product of coordinates:

$$\boldsymbol{K}\cdot(\boldsymbol{x}\times\boldsymbol{y}) = \left(\sum_{j,k=1}^{3} \epsilon_{ijk}x_j y_k \mid i=1,2,3\right).$$

These statements fail, in general, for a nonorthogonal coordinatization.

3.20. Let $(\boldsymbol{v}_1,\boldsymbol{v}_2,\boldsymbol{v}_3)$ be an arbitrary ordered basis in $\mathbf{S}$, chose a positive element $\boldsymbol{m}$ of $\mathbb{L}$ and put

$$h_{ik} := \frac{\boldsymbol{v}_i\cdot\boldsymbol{v}_k}{\boldsymbol{m}^2}\left(:= \frac{\boldsymbol{h}(\boldsymbol{v}_i,\boldsymbol{v}_k)}{\boldsymbol{m}^2}\right) \in \mathbb{R} \qquad (i,k=1,2,3).$$

The dual of the basis can be represented by vectors $\boldsymbol{r}^1,\boldsymbol{r}^2,\boldsymbol{r}^3$ in $\frac{\mathbf{S}}{\mathbb{L}\otimes\mathbb{L}}$ — usually called the *reciprocal system* of the given basis—in such a way that

$$\boldsymbol{r}^i\cdot\boldsymbol{v}_k = \delta_{ik} \qquad (i,k=1,2,3).$$

It is not hard to see that

$$\boldsymbol{r}^1 := \frac{\boldsymbol{\epsilon}(\cdot,\boldsymbol{v}_2,\boldsymbol{v}_3)}{\boldsymbol{\Delta}}, \qquad \text{etc.}$$

where $\boldsymbol{\Delta} := \boldsymbol{\epsilon}(\boldsymbol{v}_1,\boldsymbol{v}_2,\boldsymbol{v}_3)$.

Put

$$h^{ik} := (\boldsymbol{r}^i\cdot\boldsymbol{r}^k)\boldsymbol{m}^2 \in \mathbb{R} \qquad (i,k=1,2,3).$$

Let us take the coordinatization $\boldsymbol{K}$ of $\mathbf{S}$ defined by the basis $(\boldsymbol{v}_1,\boldsymbol{v}_2,\boldsymbol{v}_3)$. Now we must distinguish between subscripts and superscripts. We agree to write the elements of $\mathbb{R}^3$ in the form (x^i) and the elements of $(\mathbb{R}^3)^*$ in the form (x_i). Then

$$\boldsymbol{K}\cdot\boldsymbol{x} = (\boldsymbol{r}^i\cdot\boldsymbol{x}) =: (x^i).$$

Consider the identification $\mathbf{S}\equiv\mathbb{L}\otimes\mathbb{L}\otimes\mathbf{S}^*\equiv\left(\frac{\mathbf{S}}{\mathbb{L}\otimes\mathbb{L}}\right)^*$; then $\left(\frac{\boldsymbol{v}_i}{\boldsymbol{m}^2} \mid i=1,2,3\right)$ is an ordered basis in $\frac{\mathbf{S}}{\mathbb{L}\otimes\mathbb{L}}$ and $\boldsymbol{x}$, as an element of the dual of $\frac{\mathbf{S}}{\mathbb{L}\otimes\mathbb{L}}$, has the coordinates

$$\left(\boldsymbol{x}\cdot\frac{\boldsymbol{v}_i}{\boldsymbol{m}^2}\right) =: (x_i).$$

Writing $\boldsymbol{x} = \sum_{i=1}^{3} x^k \boldsymbol{v}_k = \sum_{k=1}^{3} x_k \boldsymbol{r}^k \boldsymbol{m}^2$ we find that

$$x_i = \sum_{k=1}^{3} h_{ik} x^k, \qquad x^k = \sum_{i=1}^{3} h^{ki} x_i,$$

i.e., in general, $x^i \neq x_i$.

Now if

$$\boldsymbol{K} \cdot \boldsymbol{x} = (x^i) \qquad \text{and} \qquad \boldsymbol{K} \cdot \boldsymbol{y} = (y^i)$$

i.e.

$$\boldsymbol{x} = \sum_{i=1}^{3} x^i \boldsymbol{v}_i, \qquad \boldsymbol{y} = \sum_{i=1}^{3} y^i \boldsymbol{v}_i$$

then

$$\boldsymbol{x} \cdot \boldsymbol{y} = \left(\sum_{i,k=1}^{3} h_{ik} x^i y^k \right) \boldsymbol{m}^2 = \left(\sum_{k=1}^{3} x_k y^k \right) \boldsymbol{m}^2 = \left(\sum_{i=1}^{3} x^i y_i \right) \boldsymbol{m}^2.$$

3.21. Exercises

In the following we keep assuming that $\dim \mathbf{S} = 3$.

1. Let $\boldsymbol{A}$ be a nonzero element of $\mathrm{A}(\boldsymbol{h})$. If $\boldsymbol{x}$ is a nonzero vector in $\mathbf{S}$, orthogonal to $\operatorname{Ker} \boldsymbol{A}$, then

(*i*) $\boldsymbol{A}^2 \cdot \boldsymbol{x} = -|\boldsymbol{A}|^2 \boldsymbol{x}$,

(*ii*) $|\boldsymbol{A} \cdot \boldsymbol{x}| = |\boldsymbol{A}||\boldsymbol{x}|$,

(*iii*) $\boldsymbol{A} = \frac{(\boldsymbol{A}\cdot\boldsymbol{x})\wedge\boldsymbol{x}}{|\boldsymbol{x}|^2}$.

2. Show that $\operatorname{Ker}(\boldsymbol{A}^2) = \operatorname{Ker} \boldsymbol{A}$ for $\boldsymbol{A} \in \mathbf{N} \wedge \mathbf{N} \equiv \mathrm{A}(\boldsymbol{h})$.

3. Prove that

(*i*)
$$\begin{aligned}\epsilon_{ijk}\epsilon_{rst} = \delta_{ir}\delta_{js}\delta_{kt} + \delta_{is}\delta_{jt}\delta_{kr} + \delta_{it}\delta_{jr}\delta_{ks} \\ - \delta_{it}\delta_{js}\delta_{kr} - \delta_{is}\delta_{jr}\delta_{kt} - \delta_{ir}\delta_{jt}\delta_{ks},\end{aligned}$$

(*ii*)
$$\sum_{k=1}^{3} \epsilon_{ijk}\epsilon_{rsk} = \delta_{ir}\delta_{js} - \delta_{is}\delta_{jr},$$

(*iii*)
$$\sum_{j,k=1}^{3} \epsilon_{ijk}\epsilon_{rjk} = 2\delta_{ir}.$$

4. Let $\boldsymbol{A}$ and $\boldsymbol{B}$ be one-dimensional vector spaces. Define the vectorial product

$$(\mathbf{N} \otimes \mathbb{A}) \times (\mathbf{N} \otimes \mathbb{B}) \to \mathbf{N} \otimes \mathbb{A} \otimes \mathbb{B}.$$

4. Minkowskian vector spaces

4.1. A pseudo-Euclidean vector space $(\mathbf{V}, \mathbb{B}, \boldsymbol{b})$ is called *Minkowskian* if $\dim \mathbf{V} > 1$ and $\operatorname{neg}(\boldsymbol{b}) = 1$.

For a clear distinction, in the following $(\mathbf{M}, \mathbb{T}, \boldsymbol{g})$ denotes a Minkowskian vector space, and

$$\dim \mathbf{M} = 1 + N$$

where $N \geq 1$.

We call attention to the fact that $\boldsymbol{g}$ is usually called a Lorentz metric; since $\boldsymbol{g}$ does not define a metric (distances and angles, see later) we prefer to call it a *Lorentz form.*

The notations introduced for pseudo-Euclidean vector spaces will be used, e.g.

$$\boldsymbol{x} \cdot \boldsymbol{y} := \boldsymbol{g}(\boldsymbol{x}, \boldsymbol{y}) \qquad (\boldsymbol{x}, \boldsymbol{y}) \in \mathbf{M});$$

note that if $\boldsymbol{x}, \boldsymbol{y} \in \mathbf{M}$ and $\boldsymbol{u}, \boldsymbol{v} \in \frac{\mathbf{M}}{\mathbb{T}}$ then

$$\boldsymbol{x} \cdot \boldsymbol{y} \in \mathbb{T} \otimes \mathbb{T}, \qquad \boldsymbol{u} \cdot \boldsymbol{x} \in \mathbb{T}, \qquad \boldsymbol{u} \cdot \boldsymbol{v} \in \mathbb{R}.$$

Moreover, we put

$$\boldsymbol{x}^2 := \boldsymbol{x} \cdot \boldsymbol{x} \qquad (\boldsymbol{x} \in \mathbf{M}).$$

In contradistinction to Euclidean spaces, here we keep saying $\boldsymbol{g}$-orthogonal, $\boldsymbol{g}$-adjoint etc.

The elements of a $\boldsymbol{g}$-orthogonal basis will be numbered from 0 to N, in such a way that for $\{\boldsymbol{e}_0, \boldsymbol{e}_1, \ldots, \boldsymbol{e}_N\}$ we have $\boldsymbol{e}_0{}^2 < \mathbf{0}$, $\boldsymbol{e}_i{}^2 > \mathbf{0}$ if $i = 1, \ldots, N$.

4.2. Recall that there is a canonical orientation on $\mathbb{T} \otimes \mathbb{T}$, hence it makes sense that an element of $\mathbb{T} \otimes \mathbb{T}$ is positive or negative. Let us introduce the notations

$$\begin{aligned} S\colon &= \{\boldsymbol{x} \in \mathbf{M} \mid \boldsymbol{x}^2 > \mathbf{0}\}, & S_0 &:= S \cup \{\mathbf{0}\}, \\ T\colon &= \{\boldsymbol{x} \in \mathbf{M} \mid \boldsymbol{x}^2 < \mathbf{0}\}, & T_0 &:= T \cup \{\mathbf{0}\}, \\ L\colon &= \{\boldsymbol{x} \in \mathbf{M} \mid \boldsymbol{x}^2 = \mathbf{0}, \boldsymbol{x} \neq \mathbf{0}\} & L_0 &:= L \cup \{\mathbf{0}\}. \end{aligned}$$

The elements of S_0, T and L are called *spacelike, timelike* and *lightlike* vectors, respectively.

Neither of S_0, T_0 and L_0 is a linear subspace.

The bilinear map $\boldsymbol{g}$ is continuous (see VI.3.1). Thus S and T are open subsets, L_0 is a closed subset.

4.3. Take a nonzero element $\boldsymbol{x}$ of $\mathbf{M}$. The Lorentz form $\boldsymbol{g}$ is nondegenerate, hence the linear map $\mathbf{M} \to \mathbb{T} \otimes \mathbb{T}$, $\boldsymbol{y} \mapsto \boldsymbol{x} \cdot \boldsymbol{y}$ is a surjection, i.e. it has an N-dimensional kernel. In other words,

$$\mathbf{H}_{\boldsymbol{x}} := \{\boldsymbol{y} \in \mathbf{M} \mid \boldsymbol{x} \cdot \boldsymbol{y} = 0\}$$

is an N-dimensional linear subspace of $\mathbf{M}$. Let $\boldsymbol{g}_{\boldsymbol{x}}$ be the restriction of $\boldsymbol{g}$ onto $\mathbf{H}_{\boldsymbol{x}} \times \mathbf{H}_{\boldsymbol{x}}$; it is an $\mathbb{T} \otimes \mathbb{T}$-valued symmetric bilinear map.

(i) Suppose $\boldsymbol{x} \in S$ or $\boldsymbol{x} \in T$. Then $\boldsymbol{x}$ is not in $\mathbf{H}_{\boldsymbol{x}}$. $\mathbb{R}\boldsymbol{x}$ and $\mathbf{H}_{\boldsymbol{x}}$ are complementary subspaces. As a consequence, $\boldsymbol{g}_{\boldsymbol{x}}$ is nondegenerate, i.e. $(\mathbf{H}_{\boldsymbol{x}}, \mathbb{T}, \boldsymbol{g}_{\boldsymbol{x}})$ is an N-dimensional pseudo-Euclidean vector space. Thus there is a $\boldsymbol{g}_{\boldsymbol{x}}$-orthogonal basis in $\mathbf{H}_{\boldsymbol{x}}$; such a basis, supplemented by $\boldsymbol{x}$, will be a $\boldsymbol{g}$-orthogonal basis in $\mathbf{M}$.

—if $\boldsymbol{x}$ is in S then $\boldsymbol{x}^2 > \mathbf{0}$, so one and only one element of a $\boldsymbol{g}_{\boldsymbol{x}}$-orthogonal basis belongs to T, the other ones belong to S. Consequently, $(\mathbf{H}_{\boldsymbol{x}}, \mathbb{T}, \boldsymbol{g}_{\boldsymbol{x}})$ is an N-dimensional Minkowskian vector space.

—if $\boldsymbol{x}$ is in T then $\boldsymbol{x}^2 < \mathbf{0}$, so all the elements of a $\boldsymbol{g}_{\boldsymbol{x}}$-orthogonal basis belong to S. Consequently, $\mathbf{H}_{\boldsymbol{x}} \subset S_0$ and $(\mathbf{H}_{\boldsymbol{x}}, \mathbb{T}, \boldsymbol{g}_{\boldsymbol{x}})$ is an N-dimensional Euclidean vector space.

(ii) Suppose $\boldsymbol{x} \in L$. Then $\boldsymbol{x}$ itself is in $\mathbf{H}_{\boldsymbol{x}}$, in other words, $\mathbb{R}\boldsymbol{x}$ is contained in $\mathbf{H}_{\boldsymbol{x}}$. One cannot give naturally a subspace complementary to $\mathbf{H}_{\boldsymbol{x}}$. Moreover, $\boldsymbol{g}_{\boldsymbol{x}}$ is degenerate.

Let $\boldsymbol{e}_0$ be an element of T; let $\boldsymbol{s}$ be an element of $\mathbb{T}$ such that $\boldsymbol{e}_0^2 = -\boldsymbol{s}^2$. As we have seen, $\mathbf{H}_{\boldsymbol{e}_0}$ is contained in S_0, so $\boldsymbol{e}_0 \cdot \boldsymbol{x} \neq \mathbf{0}$, and $\boldsymbol{e}_1 := \frac{\boldsymbol{e}_0 \cdot \boldsymbol{e}_0}{\boldsymbol{e}_0 \cdot \boldsymbol{x}} \boldsymbol{x} - \boldsymbol{e}_0$ belongs to S, $\boldsymbol{e}_0 \cdot \boldsymbol{e}_1 = \mathbf{0}$ and ${\boldsymbol{e}_1}^2 = \boldsymbol{s}^2$. $\{\boldsymbol{e}_0, \boldsymbol{e}_1\}$ can be completed to a $\boldsymbol{g}$-orthogonal basis $\{\boldsymbol{e}_0, \boldsymbol{e}_1, \dots, \boldsymbol{e}_N\}$, normed to $\boldsymbol{s}$, of $\mathbf{M}$. The vector $\boldsymbol{x}$ is a linear combination of $\boldsymbol{e}_0$ and $\boldsymbol{e}_1$, thus $\{\boldsymbol{e}_2, \dots, \boldsymbol{e}_N\}$ is contained in $\mathbf{H}_{\boldsymbol{x}}$ and $\{\boldsymbol{x}, \boldsymbol{e}_2, \dots, \boldsymbol{e}_N\}$ is a basis of $\mathbf{H}_{\boldsymbol{x}}$.

This has the immediate consequence that every element of $\mathbf{H}_{\boldsymbol{x}}$ which is not parallel to $\boldsymbol{x}$ belongs to S_0.

4.4. It follows from 4.3(i) that if $\boldsymbol{x} \in T$ then $\boldsymbol{x} \cdot \boldsymbol{y} \neq \mathbf{0}$ for all $\boldsymbol{y} \in T$ and for all $\boldsymbol{y} \in L$.

Moreover, the results in the preceding paragraph imply that

—there are N-dimensional linear subspaces in S_0,

—there are at most one-dimensional linear subspaces in T_0 and L_0 ,

—there is a one-to-one correspondence between N-dimensional linear subspaces in S_0 and one-dimensional linear subspaces in T_0 in such a way that the subspaces in correspondence are $\boldsymbol{g}$-orthogonal to each other.

4.5. The identification

$$\frac{\mathbf{M}}{\mathbb{T}\otimes\mathbb{T}} \equiv \mathbf{M}^*$$

(see 1.3) is a fundamental property of the Minkowskian vector space $(\mathbf{M}, \mathbb{T}, \boldsymbol{g})$.

The dual of a $\boldsymbol{g}$-orthogonal basis $\{\boldsymbol{e}_0, \boldsymbol{e}_1, \dots, \boldsymbol{e}_N\}$, normed to $\boldsymbol{s} \in \mathbb{T}$, in this identification becomes $\left\{-\frac{\boldsymbol{e}_0}{\boldsymbol{s}^2}, \frac{\boldsymbol{e}_1}{\boldsymbol{s}^2}, \dots, \frac{\boldsymbol{e}_N}{\boldsymbol{s}^2}\right\}$.

Accordingly, $\boldsymbol{n}_i := \frac{\boldsymbol{e}_i}{\boldsymbol{s}}$ $(i = 0, 1, \dots, N)$ form a $\boldsymbol{g}$-orthonormal basis in $\frac{\mathbf{M}}{\mathbb{T}}$:

$$\boldsymbol{n}_0 \cdot \boldsymbol{n}_0 = -1, \qquad \boldsymbol{n}_i \cdot \boldsymbol{n}_k = \delta_{ik} \qquad (i, k = 1, \dots, N).$$

The corresponding dual basis in the identification $\frac{\mathbf{M}}{\mathbb{T}} \equiv \left(\frac{\mathbf{M}}{\mathbb{T}}\right)^*$ is $\{-\boldsymbol{n}_0, \boldsymbol{n}_1, \dots, \boldsymbol{n}_N\}$.

For all $\boldsymbol{x} \in \mathbf{M}$ we have

$$\boldsymbol{x} = -(\boldsymbol{n}_0 \cdot \boldsymbol{x})\boldsymbol{n}_0 + \sum_{i=1}^{N} (\boldsymbol{n}_i \cdot \boldsymbol{x})\boldsymbol{n}_i.$$

4.6. The following relation will be a starting point of important results. If $\boldsymbol{x}, \boldsymbol{y} \in T \cup L$, $\boldsymbol{x}$ is not parallel to $\boldsymbol{y}$, $\boldsymbol{z} \in T$, then

$$2(\boldsymbol{x}\cdot\boldsymbol{y})(\boldsymbol{y}\cdot\boldsymbol{z})(\boldsymbol{z}\cdot\boldsymbol{x}) < (\boldsymbol{x}\cdot\boldsymbol{z})^2\boldsymbol{y}^2 + (\boldsymbol{y}\cdot\boldsymbol{z})^2\boldsymbol{x}^2 \leq \mathbf{0}.$$

This is implied by the simple fact that $\boldsymbol{a} := \frac{\boldsymbol{y}\cdot\boldsymbol{z}}{\boldsymbol{x}\cdot\boldsymbol{z}}\boldsymbol{x} - \boldsymbol{y}$ is $\boldsymbol{g}$-orthogonal to $\boldsymbol{z}$, thus $\boldsymbol{a}$ is in S: $\boldsymbol{a}^2 > \mathbf{0}$.

4.7. Since $\mathbb{T}\otimes\mathbb{T}$ is canonically oriented (see IV.5.4), the absolute value of its elements makes sense.

Proposition (reversed Cauchy inequality). If $\boldsymbol{x}, \boldsymbol{y} \in T$ then

$$|\boldsymbol{x}\cdot\boldsymbol{y}| \geq \sqrt{|\boldsymbol{x}^2||\boldsymbol{y}^2|} > \mathbf{0}$$

and equality holds if and only if $\boldsymbol{x}$ and $\boldsymbol{y}$ are parallel.

Proof. Put $\boldsymbol{z} := \boldsymbol{x}$ in the previous formula, and recall that we have the square root mapping from $(\mathbb{T}\otimes\mathbb{T})\otimes(\mathbb{T}\otimes\mathbb{T})$ into $\mathbb{T}\otimes\mathbb{T}$. ■

In general, the right-hand side of this equality cannot be written in a simpler form because $|\boldsymbol{x}|$ and $|\boldsymbol{y}|$ make no sense, unless $\mathbb{T}$ is oriented.

4.8. Definition. The elements $\boldsymbol{x}$ and $\boldsymbol{y}$ of T *have the same arrow* if $\boldsymbol{x}\cdot\boldsymbol{y} < \mathbf{0}$.

Proposition. Having the same arrow is an equivalence relation on T and there are two equivalence classes (called *arrow classes).*

Proof. The relation having the same arrow is evidently reflexive and symmetric. Suppose now that $\boldsymbol{x}$ and $\boldsymbol{y}$ as well as $\boldsymbol{y}$ and $\boldsymbol{z}$ have the same arrow. Then 4.6 implies that $\boldsymbol{x}$ and $\boldsymbol{z}$ have the same arrow as well, hence the relation is transitive.

Let $\boldsymbol{x}$ be an element of T. It is obvious that $\boldsymbol{x}$ and $-\boldsymbol{x}$ have not the same arrow: there are at least two arrow classes. On the other hand, since $\boldsymbol{x}\cdot\boldsymbol{y}\neq \mathbf{0}$ for all $\boldsymbol{y}\in T$, $\boldsymbol{y}$ and $\boldsymbol{x}$ or $-\boldsymbol{y}$ and $\boldsymbol{x}$ have the same arrow: there are at most two arrow classes.

4.9. Proposition. The arrow classes of T are convex cones, i.e. if $\boldsymbol{x}$ and $\boldsymbol{y}$ have the same arrow then $\alpha\boldsymbol{x}+\beta\boldsymbol{y}$ is in their arrow class for all $\alpha,\beta\in\mathbb{R}^+$.

Proof. It is quite evident that $(\alpha\boldsymbol{x}+\beta\boldsymbol{y})^2<\mathbf{0}$ and $(\alpha\boldsymbol{x}+\beta\boldsymbol{y})\cdot\boldsymbol{x}<\mathbf{0}$, thus $\alpha\boldsymbol{x}+\beta\boldsymbol{y}$ is in T, moreover, $\alpha\boldsymbol{x}+\beta\boldsymbol{y}$ and $\boldsymbol{x}$ have the same arrow. ■

The arrow classes are open subsets of $\mathbf{M}$ because the arrow class of $\boldsymbol{x}\in T$ is $\{\boldsymbol{y}\in T\mid \boldsymbol{x}\cdot\boldsymbol{y}<\mathbf{0}\}$.

4.10. Suppose now that $\mathbb{T}$ is oriented. Then we can take the square root of non-negative elements of $\mathbb{T}\otimes\mathbb{T}$, so we define the *pseudo-length* of vectors:

$$|\boldsymbol{x}|:=\sqrt{|\boldsymbol{x}^2|} \qquad (\boldsymbol{x}\in\mathbf{M}).$$

The length of vectors in Euclidean vector spaces has the fundamental properties listed in 3.3. Now we find that

(*i*) $|\boldsymbol{x}|=\mathbf{0}$ if $\boldsymbol{x}=\mathbf{0}$ but $|\boldsymbol{x}|=\mathbf{0}$ does not imply $\boldsymbol{x}=\mathbf{0}$;

(*ii*) $|\alpha\boldsymbol{x}|=|\alpha||\boldsymbol{x}|$ for all $\alpha\in\mathbb{R}$;

(*iii*) there is no definite relation between $|\boldsymbol{x}+\boldsymbol{y}|$ and $|\boldsymbol{x}|+|\boldsymbol{y}|$:

—if $\mathbf{S}\subset S_0$ is a linear subspace then $(\mathbf{S},\mathbb{T},g|_{\mathbf{S}\times\mathbf{S}})$ is a Euclidean vector space, consequently, for $\boldsymbol{x},\boldsymbol{y}\in\mathbf{S}$ the triangle inequality $|\boldsymbol{x}+\boldsymbol{y}|\leq|\boldsymbol{x}|+|\boldsymbol{y}|$ holds,

—for vectors in T a reverse relation can hold, as follows.

Proposition (reversed triangle inequality). If the elements $\boldsymbol{x}$ and $\boldsymbol{y}$ of T have the same arrow then

$$|\boldsymbol{x}+\boldsymbol{y}|\geq|\boldsymbol{x}|+|\boldsymbol{y}|$$

and equality holds if and only if $\boldsymbol{x}$ and $\boldsymbol{y}$ are parallel.

Proof. According to the previous statement $\boldsymbol{x}+\boldsymbol{y}$ belongs to T, thus we can apply the reversed Cauchy inequality:

$$|\boldsymbol{x}+\boldsymbol{y}|^2=-(\boldsymbol{x}+\boldsymbol{y})^2=|\boldsymbol{x}|^2-2(\boldsymbol{x}\cdot\boldsymbol{y})+|\boldsymbol{y}|^2\geq$$
$$\geq|\boldsymbol{x}|^2+2\sqrt{|\boldsymbol{x}|^2|\boldsymbol{y}|^2}+|\boldsymbol{y}|^2=(|\boldsymbol{x}|+|\boldsymbol{y}|)^2. \qquad ■$$

The triangle inequality and 'nonzero vector has nonzero length' are indispensable properties of a length; that is why we use the name pseudo-length.

4.11. Definition. The elements $\boldsymbol{x}$ and $\boldsymbol{y}$ of L *have the same arrow* if $\boldsymbol{x} \cdot \boldsymbol{y} \leq 0$.

Proposition. Having the same arrow is an equivalence relation on L and there are two equivalence classes (called *arrow classes).*

Proof. Argue as in 4.8.

4.12. Now we relate the arrow classes of L to those of T. It is evident that the elements $\boldsymbol{x}$ and $\boldsymbol{y}$ of L have the same arrow if and only if $\alpha\boldsymbol{x} + \beta\boldsymbol{y}$ are in T and have the same arrow for all $\alpha, \beta \in \mathbb{R}^+$.

Proposition. (i) Let $\boldsymbol{x}, \boldsymbol{y} \in L$, $\boldsymbol{z} \in T$. Then $\boldsymbol{x}$ and $\boldsymbol{y}$ have the same arrow if and only if $\boldsymbol{x} \cdot \boldsymbol{z}$ and $\boldsymbol{y} \cdot \boldsymbol{z}$ have the same sign (in the ordered one-dimensional vector space $\mathbb{T} \otimes \mathbb{T}$).

(ii) Let $\boldsymbol{x} \in L$, $\boldsymbol{y}, \boldsymbol{z} \in T$. Then $\boldsymbol{y}$ and $\boldsymbol{z}$ have the same arrow if and only if $\boldsymbol{x} \cdot \boldsymbol{y}$ and $\boldsymbol{x} \cdot \boldsymbol{z}$ have the same sign.

Proof. Apply the inequality in 4.6. ∎

As a consequence, the arrow classes of T and those of L determine each other uniquely. We say that the elements $\boldsymbol{x}$ of L and $\boldsymbol{y}$ of T have the same arrow if $\boldsymbol{x} \cdot \boldsymbol{y} < 0$. According to the previous proposition, if we select an arrow class $T^{\rightarrow}$ from T then there is an arrow class $L^{\rightarrow}$ in L such that all the elements of $T^{\rightarrow}$ and $L^{\rightarrow}$ have the same arrow:

$$L^{\rightarrow} = \{\boldsymbol{y} \in L | \ \boldsymbol{x} \cdot \boldsymbol{y} < \boldsymbol{0}, \ \boldsymbol{x} \in T^{\rightarrow}\}.$$

It can be shown that $L^{\rightarrow} \cup \{\boldsymbol{0}\}$ is the boundary of $T^{\rightarrow}$.

4.13. We say that $(\mathbf{M}, \mathbb{T}, \boldsymbol{g})$ is *arrow-oriented* or an *arrow orientation is associated* with $\boldsymbol{g}$ if we select one of the arrow classes of T. More precisely, an arrow-oriented Minkowskian vector space is $(\mathbf{M}, \mathbb{T}, \boldsymbol{g}, T^{\rightarrow})$ where $(\mathbf{M}, \mathbb{T}, \boldsymbol{g})$ is a Minkowskian vector space and $T^{\rightarrow}$ is one of the arrow classes of T.

A linear isomorphism between arrow-oriented Minkowskian vector spaces is called *arrow-preserving* or *arrow-reversing* if it maps the chosen arrow classes into each other or into the opposite ones, respectively.

4.14. In the following we assume that $\mathbf{M}$ and $\mathbb{T}$ are oriented and $\boldsymbol{g}$ is arrow-oriented; moreover

$$\dim \mathbf{M} = 4.$$

We introduce the notation

$$V(1) := \left\{ \boldsymbol{u} \in \frac{\mathbf{M}}{\mathbb{T}} \mid \boldsymbol{u}^2 = -1,\ \boldsymbol{u} \otimes \mathbb{T}^+ \subset T^{\rightarrow} \right\}.$$

If $\boldsymbol{u} \in V(1)$ then

$$\boldsymbol{u} \otimes \mathbb{T} := \{\boldsymbol{u}t \mid t \in \mathbb{T}\} \subset T_0,$$
$$\mathbf{S}_{\boldsymbol{u}} := \{\boldsymbol{x} \in \mathbf{M} \mid \boldsymbol{u} \cdot \boldsymbol{x} = 0\} \subset S_0$$

are complementary subspaces. The corresponding projections of $\boldsymbol{x} \in \mathbf{M}$ in $\boldsymbol{u}\otimes\mathbb{T}$ and in $\mathbf{S}_{\boldsymbol{u}}$ are

$$-(\boldsymbol{u} \cdot \boldsymbol{x})\boldsymbol{u} \qquad \text{and} \qquad \boldsymbol{x} + (\boldsymbol{u} \cdot \boldsymbol{x})\boldsymbol{u}.$$

Let $\boldsymbol{h}_{\boldsymbol{u}}$ denote the restriction of $\boldsymbol{g}$ onto $\mathbf{S}_{\boldsymbol{u}} \times \mathbf{S}_{\boldsymbol{u}}$; then $(\mathbf{S}_{\boldsymbol{u}}, \mathbb{T}, \boldsymbol{h}_{\boldsymbol{u}})$ is a three-dimensional Euclidean vector space.

4.15. We shall examine the structure of $\boldsymbol{g}$-antisymmetric linear maps of $\mathbf{M}$. As we know (see 2.9) $\mathrm{A}(\boldsymbol{g}) \equiv \frac{\mathbf{M}}{\mathbb{T}} \wedge \frac{\mathbf{M}}{\mathbb{T}}$ is a six-dimensional vector space endowed (see 2.10) with a real-valued nondegenerate symmetric bilinear form:

$$\boldsymbol{H} \bullet \boldsymbol{G} := \frac{1}{2}\mathrm{Tr}(\boldsymbol{H}^* \cdot \boldsymbol{G}) = -\frac{1}{2}\mathrm{Tr}(\boldsymbol{H} \cdot \boldsymbol{G}).$$

In particular, for $\boldsymbol{k}_1, \boldsymbol{k}_2, \boldsymbol{n}_1, \boldsymbol{n}_2 \in \frac{\mathbf{M}}{\mathbb{T}}$ we have

$$(\boldsymbol{k}_1 \wedge \boldsymbol{k}_2) \bullet (\boldsymbol{n}_1 \wedge \boldsymbol{n}_2) = (\boldsymbol{k}_1 \cdot \boldsymbol{n}_1)(\boldsymbol{k}_2 \cdot \boldsymbol{n}_2) - (\boldsymbol{k}_1 \cdot \boldsymbol{n}_2)(\boldsymbol{k}_2 \cdot \boldsymbol{n}_1).$$

If $\{\boldsymbol{n}_0, \boldsymbol{n}_1, \boldsymbol{n}_2, \boldsymbol{n}_3\}$ is a $\boldsymbol{g}$-orthonormal basis of $\frac{\mathbf{M}}{\mathbb{T}}$ then

$$\begin{matrix} \boldsymbol{n}_0 \wedge \boldsymbol{n}_1, & \boldsymbol{n}_0 \wedge \boldsymbol{n}_2, & \boldsymbol{n}_0 \wedge \boldsymbol{n}_3, \\ \boldsymbol{n}_1 \wedge \boldsymbol{n}_2, & \boldsymbol{n}_2 \wedge \boldsymbol{n}_3, & \boldsymbol{n}_3 \wedge \boldsymbol{n}_1 \end{matrix}$$

constitute a basis in $\frac{\mathbf{M}}{\mathbb{T}} \wedge \frac{\mathbf{M}}{\mathbb{T}}$ (see IV.3.15). We can take $\boldsymbol{u} := \boldsymbol{n}_0 \in V(1)$; then every $\boldsymbol{g}$-antisymmetric map can be written in the form $\sum_{i=1}^{3} \alpha_i \boldsymbol{u} \wedge \boldsymbol{n}_i + \sum_{i=1}^{3} \sum_{k<i} \alpha_{ki} \boldsymbol{n}_k \wedge \boldsymbol{n}_i$. The vectors $\boldsymbol{n}_1, \boldsymbol{n}_2, \boldsymbol{n}_3$ span the three-dimensional Euclidean vector space $\frac{\mathbf{S}_{\boldsymbol{u}}}{\mathbb{T}}$ (more precisely, $(\frac{\mathbf{S}_{\boldsymbol{u}}}{\mathbb{T}}, \mathbb{R}, \cdot)$), hence according to the results of the previous chapter there are a real number β and unit vectors $\boldsymbol{k}$ and $\boldsymbol{n}$, $\boldsymbol{g}$-orthogonal to each other and to $\boldsymbol{u}$ such that $\sum_{i=1}^{3} \sum_{k<i} \alpha_{ki} \boldsymbol{n}_k \wedge \boldsymbol{n}_i = \beta \boldsymbol{k} \wedge \boldsymbol{n}$. Furthermore, $\sum_{i=1}^{3} \alpha_i \boldsymbol{u} \wedge \boldsymbol{n}_i = \boldsymbol{u} \wedge \sum_{i=1}^{3} \alpha_i \boldsymbol{n}_i$; thus we arrive at the following result.

Proposition. Let $\boldsymbol{H}$ be an element of $\frac{\mathrm{M}}{\mathbb{T}} \wedge \frac{\mathrm{M}}{\mathbb{T}}$. Then for all $\boldsymbol{u} \in V(1)$ there are $\alpha, \beta \in \mathbb{R}$, $\boldsymbol{r}, \boldsymbol{k}, \boldsymbol{n} \in \frac{\mathrm{S}_u}{\mathbb{T}}$, $\boldsymbol{r}^2 = \boldsymbol{k}^2 = \boldsymbol{n}^2 = 1$, $\boldsymbol{k} \cdot \boldsymbol{n} = 0$ such that

$$\boldsymbol{H} = \alpha \boldsymbol{u} \wedge \boldsymbol{r} + \beta \boldsymbol{k} \wedge \boldsymbol{n}. \quad \blacksquare$$

Observe that then

$$\boldsymbol{H} \bullet \boldsymbol{H} = -\alpha^2 + \beta^2.$$

4.16. Proposition. Take an $\boldsymbol{H} \in \frac{\mathrm{M}}{\mathbb{T}} \wedge \frac{\mathrm{M}}{\mathbb{T}}$ in the form given by the previous proposition. Then $\mathrm{Ker}\,\boldsymbol{H} = \{\mathbf{0}\}$ if and only if $\alpha \neq 0$, $\beta \neq 0$ and $\boldsymbol{r}$ is linearly independent from $\boldsymbol{k}$ and $\boldsymbol{n}$.

Proof. If $\boldsymbol{r}$ is linearly independent from $\boldsymbol{k}$ and $\boldsymbol{n}$ then $\boldsymbol{u}$, $\boldsymbol{r}$, $\boldsymbol{k}$, $\boldsymbol{n}$ are linearly independent vectors. Furthermore, if neither of α and β is zero then for all $\boldsymbol{x} \in \mathrm{M}$

$$\boldsymbol{H} \cdot \boldsymbol{x} = \alpha(\boldsymbol{r} \cdot \boldsymbol{x})\boldsymbol{u} - \alpha(\boldsymbol{u} \cdot \boldsymbol{x})\boldsymbol{r} + \beta(\boldsymbol{n} \cdot \boldsymbol{x})\boldsymbol{k} - \beta(\boldsymbol{k} \cdot \boldsymbol{x})\boldsymbol{n} = \mathbf{0}$$

implies $\boldsymbol{r} \cdot \boldsymbol{x} = \boldsymbol{u} \cdot \boldsymbol{x} = \boldsymbol{n} \cdot \boldsymbol{x} = \boldsymbol{k} \cdot \boldsymbol{x} = \mathbf{0}$; as a consequence, $\boldsymbol{x} = \mathbf{0}$. This means that $\mathrm{Ker}\,\boldsymbol{H} = \{\mathbf{0}\}$.

If $\alpha = 0$ then $\boldsymbol{H} \cdot \boldsymbol{u} = \mathbf{0}$; if $\beta = 0$ then $\boldsymbol{H} \cdot \boldsymbol{m} = \mathbf{0}$ for $\boldsymbol{m} \in \frac{\mathrm{M}}{\mathbb{T}}$, $\boldsymbol{u} \cdot \boldsymbol{m} = 0$, $\boldsymbol{r} \cdot \boldsymbol{m} = 0$. If $\alpha \neq 0$ and $\beta \neq 0$ but $\boldsymbol{r}$ is a linear combination of $\boldsymbol{k}$ and $\boldsymbol{n}$ then $\boldsymbol{H} \cdot \boldsymbol{m} = \mathbf{0}$ for $\boldsymbol{m} \in \frac{\mathrm{M}}{\mathbb{T}}$, $\boldsymbol{u} \cdot \boldsymbol{m} = \boldsymbol{k} \cdot \boldsymbol{m} = \boldsymbol{n} \cdot \boldsymbol{m} = \mathbf{0}$. This means that $\mathrm{Ker}\,\boldsymbol{H} \neq \{\mathbf{0}\}$. $\blacksquare$

Since if $\boldsymbol{k}'$ and $\boldsymbol{n}'$ are $\boldsymbol{g}$-orthogonal unit vectors in the plane spanned by $\boldsymbol{k}$ and $\boldsymbol{n}$ (do not forget that $\frac{\mathrm{S}_u}{\mathbb{T}}$ is a Euclidean vector space) then $\boldsymbol{k}' \wedge \boldsymbol{n}' = \pm \boldsymbol{k} \wedge \boldsymbol{n}$, we can choose $\boldsymbol{n} = \boldsymbol{r}$ if $\mathrm{Ker}\,\boldsymbol{H} \neq \{\mathbf{0}\}$; then for all $\boldsymbol{u} \in V(1)$ there are $\alpha, \beta \in \mathbb{R}$ and $\boldsymbol{k}, \boldsymbol{n} \in \frac{\mathrm{S}_u}{\mathbb{T}}$, $\boldsymbol{k}^2 = \boldsymbol{n}^2 = 1$, $\boldsymbol{k} \cdot \boldsymbol{n} = 0$ such that

$$\boldsymbol{H} = (\alpha \boldsymbol{u} + \beta \boldsymbol{k}) \wedge \boldsymbol{n}.$$

4.17. Proposition. Suppose $\boldsymbol{H} \in \frac{\mathrm{M}}{\mathbb{T}} \wedge \frac{\mathrm{M}}{\mathbb{T}}$, $\mathrm{Ker}\,\boldsymbol{H} \neq \{\mathbf{0}\}$ and put $|\boldsymbol{H}| := \sqrt{|\boldsymbol{H} \bullet \boldsymbol{H}|}$. Then

(*i*) $\boldsymbol{H} \bullet \boldsymbol{H} > 0$ if and only if there are a $\boldsymbol{u} \in V(1)$, $\boldsymbol{k}, \boldsymbol{n} \in \frac{\mathrm{S}_u}{\mathbb{T}}$, $\boldsymbol{k}^2 = \boldsymbol{n}^2 = 1$, $\boldsymbol{k} \cdot \boldsymbol{n} = 0$ such that

$$\boldsymbol{H} = |\boldsymbol{H}| \boldsymbol{k} \wedge \boldsymbol{n}.$$

($\boldsymbol{H}$ is the antisymmetric tensor product of two $\boldsymbol{g}$-orthogonal spacelike vectors.)

(*ii*) $\boldsymbol{H} \bullet \boldsymbol{H} < 0$ if and only if there are a $\boldsymbol{u} \in V(1)$, an $\boldsymbol{n} \in \frac{\mathrm{S}_u}{\mathbb{T}}$, $\boldsymbol{n}^2 = 1$ such that

$$\boldsymbol{H} = |\boldsymbol{H}| \boldsymbol{u} \wedge \boldsymbol{n}.$$

($\boldsymbol{H}$ is the antisymmetric tensor product of a timelike vector and a spacelike vector, $\boldsymbol{g}$-orthogonal to each other.)

(*iii*) $\boldsymbol{H} \bullet \boldsymbol{H} = 0$, $\boldsymbol{H} \neq \boldsymbol{0}$ if and only if there are $\boldsymbol{w}, \boldsymbol{n} \in \frac{\mathrm{M}}{\mathrm{T}}$, $\boldsymbol{w} \neq \boldsymbol{0}$, $\boldsymbol{w}^2 = 0$, $\boldsymbol{n}^2 = 1$, $\boldsymbol{w} \cdot \boldsymbol{n} = 0$ such that

$$\boldsymbol{H} = \boldsymbol{w} \wedge \boldsymbol{n}.$$

($\boldsymbol{H}$ is the antisymmetric tensor product of a lightlike vector and a spacelike vector, $\boldsymbol{g}$-orthogonal to each other.)

Proof. Let us write the formula of the preceding paragraph in the form $\boldsymbol{H} = (\alpha \boldsymbol{u}' + \beta \boldsymbol{k}') \wedge \boldsymbol{n}'$; then $\boldsymbol{H} \bullet \boldsymbol{H} = -\alpha^2 + \beta^2$.

(*i*) Put $\boldsymbol{u} := \dfrac{\beta \boldsymbol{u}' + \alpha \boldsymbol{k}'}{-\alpha^2 + \beta^2}$, $\boldsymbol{k} := \dfrac{\alpha \boldsymbol{u}' + \beta \boldsymbol{k}'}{-\alpha^2 + \beta^2}$, $\boldsymbol{n} := \boldsymbol{n}'$.

(*ii*) Put $\boldsymbol{u} := \dfrac{\alpha \boldsymbol{u}' + \beta \boldsymbol{k}'}{\alpha^2 - \beta^2}$, $\boldsymbol{n} := \boldsymbol{n}'$.

(*iii*) Put $\boldsymbol{w} := \alpha \boldsymbol{u}' + \beta \boldsymbol{k}'$, $\boldsymbol{n} := \boldsymbol{n}'$.

4.18. (*i*) We see that $\operatorname{Ker} \boldsymbol{H} \neq \{0\}$ and $\boldsymbol{H} \bullet \boldsymbol{H} > 0$ is equivalent to the statement that there is a $\boldsymbol{u} \in V(1)$ such that $\boldsymbol{H} \cdot \boldsymbol{u} = \boldsymbol{0}$.

Note that then $\boldsymbol{H}^3 = -|\boldsymbol{H}|^2 \boldsymbol{H}$.

(*ii*) On the contrary, if $\operatorname{Ker} \boldsymbol{H} \neq \{0\}$ and $\boldsymbol{H} \bullet \boldsymbol{H} < 0$ then $\boldsymbol{H}^3 = |\boldsymbol{H}|^2 \boldsymbol{H}$.

4.19. According to our convention introduced in 4.1, let us number the coordinates of elements of $\mathbb{R}^{1+3}$ from 0 to 3 and let us consider the Minkowskian vector space $(\mathbb{R}^{1+3}, \mathbb{R}, \boldsymbol{G})$ where

$$\boldsymbol{G}(\boldsymbol{x}, \boldsymbol{y}) = -x^0 y^0 + \sum_{i=1}^{3} x^i y^i =: \boldsymbol{x} \cdot \boldsymbol{y}$$

(i.e. $\boldsymbol{G} = \boldsymbol{H}_1$ in the notation of 1.7).

Now the identification $\mathbb{R}^{1+3} \equiv (\mathbb{R}^{1+3})^*$ induced by $\boldsymbol{G}$ is described in usual notations as follows. $\boldsymbol{x} \in \mathbb{R}^{1+3}$ regarded as a vector has the components (x^0, x^1, x^2, x^3); $\boldsymbol{x}$ regarded as a covector has the components (x_0, x_1, x_2, x_3), and the values of the linear functional $\boldsymbol{x}$ are computed by the usual matrix multiplication:

$$\boldsymbol{x} \cdot \boldsymbol{y} = \sum_{i=0}^{3} x_i y^i.$$

Then we have

$$x_0 = -x^0, \qquad x_i = x^i \qquad (i = 1, 2, 3).$$

As usual, we apply the symbols (x^i) and (x_i) for the vectors and covectors and we accept the *Einstein summation rule:* a summation from 0 to 3 is to be carried out for equal subscripts and superscripts.

Introducing

$$G_{ik} := G^{ik} := \begin{cases} -1 & \text{if } i = k = 0 \\ 1 & \text{if } i = k \in \{1, 2, 3\} \\ 0 & \text{if } i \neq k \end{cases}$$

we can write that

$$x^i = G^{ik} x_k, \qquad x_i = G_{ik} x^k \qquad \text{(summation!)}.$$

Observe that $G_{ik} = \boldsymbol{G}(\chi_i, \chi_k)$ where $\{\chi_0, \chi_1, \chi_2, \chi_3\}$ is the standard basis of $\mathbb{R}^{1+3}$.

According to the identification induced by $\boldsymbol{G}$, the dual of the standard basis $\{\chi_0, \chi_1, \chi_2, \chi_3\}$ is $\{-\chi_0, \chi_1, \chi_2, \chi_3\}$.

It is useful to regard $\boldsymbol{G}$ as the diagonal matrix in which the first ('zeroth') element in the diagonal is -1 and the other ones equal 1.

For the $\boldsymbol{G}$-adjoint $\boldsymbol{L}^{\circledast}$ of the linear map (matrix) $\boldsymbol{L}$ we have

$$\boldsymbol{L}^{\circledast} = \boldsymbol{G} \cdot \boldsymbol{L}^* \cdot \boldsymbol{G}$$

where $\boldsymbol{L}^*$ is the transpose of $\boldsymbol{L}$ (see 1.7).

A linear map $\boldsymbol{L}\colon \mathbb{R}^{1+3} \to \mathbb{R}^{1+3}$ is given by its matrix $\left(L^i{}_k\right)$,

a linear map $\boldsymbol{P}\colon (\mathbb{R}^{1+3})^* \to (\mathbb{R}^{1+3})^*$ is given by its matrix $\left(P_i{}^k\right)$ etc. see IV.1.6.

For the transpose of $L\colon \mathbb{R}^{1+3} \to \mathbb{R}^{1+3}$

$$(L^*)_i{}^k = L^k{}_i,$$

holds, thus for the $\boldsymbol{G}$-adjoint we have

$$(L^{\circledast})^i{}_k = G^{im} L^n{}_m G_{nk} \qquad \text{(summation!)}.$$

Consequently, a $\boldsymbol{G}$-antisymmetric linear map has the form

$$\begin{pmatrix} 0 & \alpha_1 & \alpha_2 & \alpha_3 \\ \alpha_1 & 0 & -\beta_3 & \beta_2 \\ \alpha_2 & \beta_3 & 0 & -\beta_1 \\ \alpha_3 & -\beta_2 & \beta_1 & 0 \end{pmatrix}$$

If (x^i) is in T, i.e. $x^i \cdot x_i < 0$ then $x^0 \neq 0$. It is not hard to see that (x^i) and (y^i) have the same arrow if and only if x^0 and y^0 have the same sign. As a consequence, an arrow class is characterized by the sign of the zeroth component of its element. One usually takes the arrow orientation in such a way that

$$T^{\rightarrow} := \{(x^i) \in \mathbb{R}^{1+3} \mid x^i x_i < 0\,,\ x^0 > 0\}.$$

4.20. Consider the four-dimensional Minkowskian vector space $(\mathbf{M}, \mathbb{T}, \boldsymbol{g})$. A linear coordinatization $\boldsymbol{K}$ of $\mathbf{M}$ is called $\boldsymbol{g}$-orthogonal if it corresponds to an ordered $\boldsymbol{g}$-orthogonal basis $(\boldsymbol{e}_0, \boldsymbol{e}_1, \boldsymbol{e}_2, \boldsymbol{e}_3)$ normed to an $\boldsymbol{s} \in \mathbb{T}$. According to the identification $\mathbf{M}^* \equiv \frac{\mathbf{M}}{\mathbb{T}\otimes\mathbb{T}}$, the basis in question has the dual $\left(-\frac{\boldsymbol{e}_0}{\boldsymbol{s}^2}, \frac{\boldsymbol{e}_1}{\boldsymbol{s}^2}, \frac{\boldsymbol{e}_2}{\boldsymbol{s}^2}, \frac{\boldsymbol{e}_3}{\boldsymbol{s}^2}\right)$; thus we have

$$\boldsymbol{K}\cdot\boldsymbol{x} = \left(-\frac{\boldsymbol{e}_0\cdot\boldsymbol{x}}{\boldsymbol{s}^2}, \frac{\boldsymbol{e}_1\cdot\boldsymbol{x}}{\boldsymbol{s}^2}, \frac{\boldsymbol{e}_2\cdot\boldsymbol{x}}{\boldsymbol{s}^2}, \frac{\boldsymbol{e}_3\cdot\boldsymbol{x}}{\boldsymbol{s}^2}\right) =: (x^i) \qquad (\boldsymbol{x} \in \mathbf{M}).$$

Consider the identification $\mathbf{M} \equiv \mathbb{T}\otimes\mathbb{T}\otimes\mathbf{M}^* \equiv \left(\frac{\mathbf{M}}{\mathbb{T}\otimes\mathbb{T}}\right)^*$; then $\boldsymbol{x}$, as an element of the dual of $\frac{\mathbf{M}}{\mathbb{T}\otimes\mathbb{T}}$, has the coordinates

$$\left(\boldsymbol{x}\cdot\frac{\boldsymbol{e}_i}{\boldsymbol{s}^2} \mid i = 0,1,2,3\right) =: (x_i).$$

We see, in accordance with the previous paragraph, that $x^0 = -x_0$ and $x^i = x_i$ $(i = 1,2,3)$.

Then all the operations regarding the Minkowskian structure can be represented by the corresponding operations in $(\mathbb{R}^{1+3}, \mathbb{R}, \boldsymbol{G})$, e.g.

—the $\boldsymbol{g}$-product of elements $\boldsymbol{x}, \boldsymbol{y}$ of $\mathbf{M}$ is computed by the $\boldsymbol{G}$-product of their coordinates in $\mathbb{R}^{1+3}$:

$$\text{if} \qquad \boldsymbol{K}\cdot\boldsymbol{x} = (x^i) \qquad \text{and} \qquad \boldsymbol{K}\cdot\boldsymbol{y} = (y^i)$$

i.e.

$$\boldsymbol{x} = \sum_{0=1}^{3} x^i \boldsymbol{e}_i, \qquad \boldsymbol{y} = \sum_{0=1}^{3} y^i \boldsymbol{e}_i$$

$$\text{then} \qquad \boldsymbol{x}\cdot\boldsymbol{y} = \left(-x^0 y^0 + \sum_{i=1}^{3} x^i y^i\right)\boldsymbol{s}^2;$$

—the matrix in the coordinatization of a $\boldsymbol{g}$-adjoint map will be the $\boldsymbol{G}$-adjoint of the matrix representing the linear map in question:

$$\text{if} \quad \boldsymbol{K}\cdot\boldsymbol{L}\cdot\boldsymbol{K}^{-1} = \left(L^i{}_k \mid i,k = 1,2,3\right)$$
$$\text{then} \quad \boldsymbol{K}\cdot\boldsymbol{L}^*\cdot\boldsymbol{K}^{-1} = \left(L^k{}_i \mid i,k = 1,2,3\right).$$

4.21. Let $(\boldsymbol{v}_0, \boldsymbol{v}_1, \boldsymbol{v}_2, \boldsymbol{v}_3)$ be an arbitrary ordered basis in $\mathbf{M}$, choose a positive element $\boldsymbol{s}$ of $\mathbb{T}$ and put

$$G_{ik} := \frac{\boldsymbol{v}_i \cdot \boldsymbol{v}_k}{\boldsymbol{s}^2} \left(:= \frac{\boldsymbol{g}(\boldsymbol{v}_i, \boldsymbol{v}_k)}{\boldsymbol{s}^2} \right) \in \mathbb{R} \qquad (i, k = 0, 1, 2, 3).$$

The dual of the basis can be represented by vectors $\boldsymbol{r}^0, \boldsymbol{r}^1, \boldsymbol{r}^2, \boldsymbol{r}^3$ in $\frac{\mathbf{M}}{\mathbb{T}\otimes\mathbb{T}}$ —usually called the *reciprocal system* of the given basis—in such a way that

$$\boldsymbol{r}^i \cdot \boldsymbol{v}_k = \delta_{ik} \qquad (i, k = 0, 1, 2, 3).$$

It is not hard to see that

$$\boldsymbol{r}^0 := \frac{\boldsymbol{\epsilon}(\cdot, \boldsymbol{v}_1, \boldsymbol{v}_2, \boldsymbol{v}_3)}{\boldsymbol{\Delta}} \qquad \text{etc.}$$

where $\boldsymbol{\epsilon}$ is the Levi-Civita tensor (see V.2.12) and $:= \boldsymbol{\epsilon}(\boldsymbol{v}_0, \boldsymbol{v}_1, \boldsymbol{v}_2, \boldsymbol{v}_3)$.

Put

$$G^{ik} := (\boldsymbol{r}^i \cdot \boldsymbol{r}^k)\boldsymbol{s}^2 \in \mathbb{R} \qquad (i, k = 0, 1, 2, 3).$$

Let us take the coordinatization $\boldsymbol{K}$ of $\mathbf{M}$ defined by the basis $(\boldsymbol{v}_0, \boldsymbol{v}_1, \boldsymbol{v}_2, \boldsymbol{v}_3)$. Then

$$\boldsymbol{K} \cdot \boldsymbol{x} = \left(\boldsymbol{r}^i \cdot \boldsymbol{x}\right) =: \left(x^i\right).$$

Consider the identification $\mathbf{M} \equiv \mathbb{T} \otimes \mathbb{T} \otimes \mathbf{M}^* \equiv \left(\frac{\mathbf{M}}{\mathbb{T}\otimes\mathbb{T}}\right)^*$; then $\left(\frac{\boldsymbol{v}_i}{\boldsymbol{s}^2} \middle| i = 0, 1, 2, 3\right)$ is an ordered basis in $\frac{\mathbf{M}}{\mathbb{T}\otimes\mathbb{T}}$, and $\boldsymbol{x}$, as an element of the dual of $\frac{\mathbf{M}}{\mathbb{T}\otimes\mathbb{T}}$, has the coordinates

$$\left(\boldsymbol{x} \cdot \frac{\boldsymbol{v}_i}{\boldsymbol{s}^2}\right) =: (x_i).$$

Writing $\boldsymbol{x} = \sum_{i=0}^{3} x^k \boldsymbol{v}_k = \sum_{k=0}^{3} x_k \boldsymbol{r}^k \boldsymbol{s}^2$ we find that

$$x_i = G_{ik} x^k, \qquad x^k = G^{ki} x_i \qquad \text{(summation!)}.$$

Now if

$$\boldsymbol{K} \cdot \boldsymbol{x} = \left(x^i\right) \qquad \text{and} \qquad \boldsymbol{K} \cdot \boldsymbol{y} = \left(y^i\right)$$

i.e.

$$\boldsymbol{x} = \sum_{i=0}^{3} x^i \boldsymbol{v}_i, \qquad \boldsymbol{y} = \sum_{i=0}^{3} y^i \boldsymbol{v}_i$$

then

$$\boldsymbol{x} \cdot \boldsymbol{y} = \left(G_{ik} x^i y^k\right) \boldsymbol{s}^2 = \left(x_k y^k\right) \boldsymbol{s}^2 = \left(x^i y_i\right) \boldsymbol{s}^2.$$

4.22. Exercises

1. Let $T^{\rightarrow}$ and $L^{\rightarrow}$ be the arrow classes corresponding to each other according to 4.12. Prove that

$$T^{\rightarrow} + L^{\rightarrow} = T^{\rightarrow},$$
$$L^{\rightarrow} + L^{\rightarrow} = T^{\rightarrow} \cup L^{\rightarrow}.$$

2. If $\boldsymbol{x}$ is in $T^{\rightarrow}$ and $\boldsymbol{y}$ is in S then $\boldsymbol{x}+\boldsymbol{y}$ is not in $T^{\leftarrow}$. (Hint: suppose $-(\boldsymbol{x}+\boldsymbol{y}) \in T^{\rightarrow}$ and use $T^{\rightarrow} + T^{\rightarrow} = T^{\rightarrow}$.)

3. If $\mathbf{H}$ is a linear subspace in $S_{\mathbf{0}}$ then $(\mathbf{H}, \mathbb{T}, \boldsymbol{g}|_{\mathbf{H}\times\mathbf{H}})$ is a Euclidean vector space. Consequently, the length of vectors and the angle between vectors in $\mathbf{H}$ makes sense. Since every $\boldsymbol{x} \in S_{\mathbf{0}}$ belongs to some linear subspace in $S_{\mathbf{0}}$ (e.g. to the linear subspace generated by $\boldsymbol{x}$), the length of every element in S_0 makes sense. On the other hand, if $\boldsymbol{x}, \boldsymbol{y} \in S_{\mathbf{0}}$, the linear subspace generated by $\boldsymbol{x}$ and $\boldsymbol{y}$ need not be contained in $S_{\mathbf{0}}$; as a consequence, the angle between two elements of $S_{\mathbf{0}}$ may not be meaningful.

Take a $\boldsymbol{g}$-orthogonal basis $\{\boldsymbol{e}_0, \boldsymbol{e}_1, \dots, \boldsymbol{e}_N\}$, normed to $\boldsymbol{s} \in \mathbb{T}$. Then $\boldsymbol{e}_1$ and $\boldsymbol{x} := 2\boldsymbol{e}_1 + \boldsymbol{e}_0$ are vectors in S that do not satisfy the Cauchy inequality and the triangle inequality.

4. Suppose $\dim \mathbf{M} = 4$, $\mathbf{M}$ and $\mathbb{T}$ are oriented. Then the Levi-Civita tensor of $(\mathbf{M}, \mathbb{T}, \boldsymbol{g})$ can be defined by $\boldsymbol{\epsilon} := \bigwedge_{i=0}^{3} \frac{\boldsymbol{e}_i}{\boldsymbol{s}}$ where $(\boldsymbol{e}_0, \boldsymbol{e}_1, \boldsymbol{e}_2, \boldsymbol{e}_3)$ is a positively oriented ordered basis, normed to $\boldsymbol{s} \in \mathbb{T}$.

Prove that

$$\frac{\mathbf{M}}{\mathbb{T}} \to \overset{3}{\wedge} \frac{\mathbf{M}}{\mathbb{T}}, \qquad \boldsymbol{n} \mapsto \boldsymbol{\epsilon}(\cdot,\cdot,\cdot,\boldsymbol{n})$$

and

$$\boldsymbol{J}\colon \frac{\mathbf{M}}{\mathbb{T}} \wedge \frac{\mathbf{M}}{\mathbb{T}} \to \frac{\mathbf{M}}{\mathbb{T}} \wedge \frac{\mathbf{M}}{\mathbb{T}}, \qquad \boldsymbol{k} \wedge \boldsymbol{n} \mapsto \boldsymbol{\epsilon}(\cdot,\cdot,\boldsymbol{k},\boldsymbol{n})$$

are linear bijections. Moreover,

$$\boldsymbol{J}(\boldsymbol{J}(\boldsymbol{H})) = -\boldsymbol{H}, \qquad \boldsymbol{H} \bullet \boldsymbol{J}(\boldsymbol{G}) = \boldsymbol{J}(\boldsymbol{G}) \bullet \boldsymbol{G},$$

thus $\boldsymbol{J}(\boldsymbol{H}) \bullet \boldsymbol{J}(\boldsymbol{G}) = -\boldsymbol{H} \bullet \boldsymbol{G}$ for all $\boldsymbol{H}, \boldsymbol{G} \in \frac{\mathbf{M}}{\mathbb{T}} \wedge \frac{\mathbf{M}}{\mathbb{T}}$.

5. Give the actual form of the previous bijections in the case $(\mathbb{R}^{1+3}, \mathbb{R}, \boldsymbol{G})$.

6. Let $\boldsymbol{\epsilon}$ be the Levi-Civita tensor of the Minkowskian vector space $(\mathbf{M}, \mathbb{T}, \boldsymbol{g})$, $\dim \mathbf{M} = 4$. If $\boldsymbol{u} \in V(1)$ then $\boldsymbol{\epsilon}(\boldsymbol{u},\cdot,\cdot,\cdot)$ is the Levi-Civita tensor of the three-dimensional Euclidean vector space $(\mathbf{S}_{\boldsymbol{u}}, \mathbb{T}, \boldsymbol{h}_{\boldsymbol{u}})$ where $\boldsymbol{h}_{\boldsymbol{u}}$ is the restriction of $\boldsymbol{g}$ onto $\mathbf{S}_{\boldsymbol{u}} \times \mathbf{S}_{\boldsymbol{u}}$.

VI. AFFINE SPACES

1. Fundamentals

1.1. Definition. An *affine space* is a triplet $(\mathrm{V}, \mathbf{V}, -)$ where

(*i*) V is a nonvoid set,

(*ii*) $\mathbf{V}$ is a vector space,

(*iii*) $-$ is a map from $\mathrm{V} \times \mathrm{V}$ into $\mathbf{V}$, denoted by

$$(x, y) \mapsto x - y,$$

having the properties

1) for every $o \in \mathrm{V}$ the map $O_o \colon \mathrm{V} \to \mathbf{V}$, $x \mapsto x - o$ is bijective,

2) $(x - y) + (y - z) + (z - x) = \mathbf{0}$ for all $x, y, z \in \mathrm{V}$. ■

O_o is often called the *vectorization* of V *with origin* o.

As usual, we shall denote an affine space by a single letter; we say that V is an affine space over the vector space $\mathbf{V}$ and we call the map $-$ *subtraction.*

The *dimension* of an affine space V is, by definition, the dimension of the underlying vector space $\mathbf{V}$. V is *oriented* if $\mathbf{V}$ is oriented (in this case $\mathbf{V}$ is necessarily a finite-dimensional real vector space).

Proposition. Let V be an affine space. Then

(*i*) $x - y = \mathbf{0}$ if and only if $x = y$ $(x, y \in \mathrm{V})$,

(*ii*) $x - y = -(y - x)$ $(x, y \in \mathrm{V})$,

(*iii*) for a natural number $n \geq 3$ and $x_1, x_2, \ldots, x_n \in \mathrm{V}$,

$$(x_1 - x_2) + (x_2 - x_3) + \ldots. + (x_n - x_1) = \mathbf{0}.$$

Proof. (*i*) Put $z := x$, $y := x$ in 2) of the above definition to have $x - x = \mathbf{0}$. Property 1) says then that $x - y \neq \mathbf{0}$ for $x \neq y$.

(*ii*) Put $z := x$ in 2) and use the previous result.

(*iii*) Starting with 2) we can prove by induction. ■

As a consequence, we can rearrange the parentheses as follows:

$$(x - y) + (u - v) = (x - v) + (u - y) \qquad (x, y, u, v \in \mathrm{V}). \tag{1}$$

1.2. Observe that the sign $-$ in (*ii*) of the previous proposition denotes two different objects. Inside the parantheses it means the subtraction in the affine space, outside it means the subtraction in the underlying vector space. This ambiguity does not cause confusion if we are careful. We even find it convenient to increase a bit the ambiguity.

For given $y \in \mathrm{V}$, the inverse of the map O_y is denoted by

$$\mathbf{V} \to \mathrm{V}, \qquad \boldsymbol{x} \mapsto y + \boldsymbol{x}. \tag{2}$$

Hence, by definition,

$$y + (x - y) = x \qquad\qquad (x, y \in \mathrm{V}), \tag{3}$$

and a simple reasoning shows that

$$(x + \boldsymbol{x}) + \boldsymbol{y} = x + (\boldsymbol{x} + \boldsymbol{y}) \qquad\qquad (x \in \mathrm{V}, \boldsymbol{x}, \boldsymbol{y} \in \mathbf{V}). \tag{4}$$

Here the symbol $+$ on the left-hand side stands twice for the operation introduced by (2), on the right-hand side first it denotes this operation and then the addition of vectors.

Keep in mind the followings:

(i) the sum and the difference of two vectors, the multiple of a vector are meaningful, they are vectors;

(ii) the difference of two elements of the affine space is meaningful, it is a vector (sums and multiples make no sense);

(iii) the sum of an affine space element and of a vector is meaningful, it is an element of the affine space.

According to (1)–(4), we can apply the usual rules of addition and subtraction paying always attention to that the operations be meaningful; for instance, the rearrangement $(y + x) - y$ in (3) makes no sense.

1.3. Linear combinations of affine space elements cannot be defined in general, for multiples and sums make no sense. However, a good trick allows us to define convex combinations.

Proposition. Let $x_1, \dots, x_n$ be elements of the affine space V and let $\alpha_1, \dots, \alpha_n$ be non-negative real numbers such that $\sum_{k=1}^{n} \alpha_k = 1$. Then there is a unique $x_o \in \mathrm{V}$ for which

$$\sum_{k=1}^{n} \alpha_k (x_k - x_o) = \mathbf{0}. \tag{$*$}$$

Proof. Let x be an arbitrary element of V; then a simple calculation based on $x_k - x_o = (x_k - x) + (x - x_o)$ shows that $x_o := x + \sum_{k=1}^{n} \alpha_k(x_k - x)$ satisfies equality $(*)$. Suppose y_o is another element with this property. Then

$$\mathbf{0} = \sum_{k=1}^{n} \alpha_k(x_k - x_o) - \sum_{k=1}^{n} \alpha_k(x_k - y_o) = \sum_{k=1}^{n} \alpha_k\left((x_k - x_o) - (x_k - y_o)\right) =$$
$$= \sum_{k=1}^{n} \alpha_k(y_o - x_o) = y_o - x_o. \quad \blacksquare$$

Remove formally the parentheses in $(*)$ to arrive at the following definition.

Definition. The element x_o in the previous proposition is called the *convex combination* of the elements $x_1, \ldots, x_n$ with coefficients $\alpha_1, \ldots, \alpha_n$ and is denoted by

$$\sum_{k=1}^{n} \alpha_k x_k. \quad \blacksquare$$

Correspondingly we can define convex subsets and the convex hull of subsets in affine spaces as they are defined in vector spaces.

If $\beta_1, \ldots, \beta_n$ are non-negative real numbers, $\beta := \sum_{k=1}^{n} \beta_k > 0$, then we can take the convex combination of $x_1, \ldots, x_n$ with the coefficients $\alpha_k := \frac{\beta_k}{\beta}$ $(k = 1, \ldots, n)$ which will be denoted by

$$\frac{\sum_{k=1}^{n} \beta_k x_k}{\sum_{k=1}^{n} \beta_k}.$$

1.4. (i) A vector space $\mathbf{V}$, endowed with the vectorial subtraction, is an affine space over itself.

(ii) If $\mathbf{M}$ is a nontrivial linear subspace of the vector space $\mathbf{V}$ and $\boldsymbol{x} \in \mathbf{V}$, $\boldsymbol{x} \notin \mathbf{M}$ then $\boldsymbol{x} + \mathbf{M} := \{\boldsymbol{x} + \boldsymbol{y} \mid \boldsymbol{y} \in \mathrm{M}\}$ endowed with the vectorial subtraction is an affine space but is not a vector space regarding the vectorial operations in $\mathbf{V}$.

(iii) If I is an arbitrary nonvoid set and V_i is an affine space over $\mathbf{V}_i$ $(i \in I)$ then $\underset{i\in I}{\times} \mathrm{V}_i$, endowed with the subtraction

$$(x_i)_{i\in I} - (y_i)_{i\in I} := (x_i - y_i)_{i\in I}$$

is an affine space over $\underset{i\in I}{\times} \mathbf{V}_i$.

1.5. Definition. A nonvoid subset S of an affine space V is called an *affine subspace* if there is a linear subspace $\mathbf{S}$ of $\mathbf{V}$ such that $\{x - y \mid x, y \in S\} = \mathbf{S}$. S is called *directed* by $\mathbf{S}$ and the dimension of S is that of $\mathbf{S}$.

One-dimensional and two-dimensional affine subspaces of a real affine space are called *straight lines* and *planes,* respectively. *Hyperplanes* are affine subspaces having the dimension of V but one, in a finite dimensional affine space V.

Two affine subspaces are said to be *parallel* if they are directed by the same linear subspace. ■

An affine subspace S directed by $\mathbf{S}$, endowed with the subtraction inherited from V, is an affine space over $\mathbf{S}$.

If $\mathbf{S}$ is a linear subspace of $\mathbf{V}$ and $x \in \mathrm{V}$ then $x + \mathbf{S} := \{x + \boldsymbol{s} \mid \boldsymbol{s} \in \mathbf{S}\}$ is the unique affine subspace containing x and directed by $\mathbf{S}$.

Points of V are zero-dimensional affine subspaces.

1.6. A pseudo-Euclidean (Euclidean, Minkowskian) affine space is a triplet $(\mathrm{V}, \mathbb{B}, \boldsymbol{b})$ where V is an affine space over the vector space $\mathbf{V}$ and $(\mathbf{V}, \mathbb{B}, \boldsymbol{b})$ is a pseudo-Euclidean (Euclidean, Minkowskian) vector space.

1.7. Exercises

1. Prove that the following definition of affine spaces is equivalent to that given in 1.1.

A triplet $(\mathrm{V}, \mathbf{V}, +)$ is an *affine space* if

(*i*) V is a nonvoid set,

(*ii*) $\mathbf{V}$ is a vector space,

(*iii*) $+$ is a map from $\mathrm{V} \times \mathbf{V}$ into V, denoted by

$$(x, \boldsymbol{x}) \mapsto x + \boldsymbol{x}$$

having the properties

1) $(x + \boldsymbol{x}) + \boldsymbol{y} = x + (\boldsymbol{x} + \boldsymbol{y}) \qquad (x \in \mathrm{V}, \boldsymbol{x}, \boldsymbol{y} \in \mathbf{V})$,

2) for every $x \in \mathrm{V}$ the map $\mathbf{V} \to \mathrm{V}$, $\boldsymbol{x} \mapsto x + \boldsymbol{x}$ is bijective.

2. Let V be an affine space over $\mathbf{V}$. Let $\mathrm{V}/\mathbf{N}$ denote the set of affine subspaces in V, directed by a given linear subspace $\mathbf{N}$ of $\mathbf{V}$. If $\mathbf{M}$ is a linear subspace complementary to $\mathbf{N}$ then $\mathrm{V}/\mathbf{N}$ becomes an affine space over $\mathbf{M}$ if we define the subtraction by

$$S - T := x - y \qquad (x \in S,\ y \in T,\ x - y \in \mathbf{M}).$$

In other words, if $\boldsymbol{P}$ denotes the projection onto $\mathbf{M}$ along $\mathbf{N}$ then

$$(x + \mathbf{N}) - (y + \mathbf{N}) := \boldsymbol{P} \cdot (x - y).$$

Illustrate this fact by $\mathrm{V} := \mathbb{R}^2$, $\mathbf{N} := \{(\alpha, 0) \mid \alpha \in \mathbb{R}\}$, $\mathbf{M} := \{(\alpha, m\alpha) \mid \alpha \in \mathbb{R}\}$ where m is a given nonzero number.

3. Prove that the intersection of affine subspaces is an affine subspace, thus the affine subspace generated by a subset of an affine space is meaningful.

4. Let $\mathbf{V}$ be a vector space over the field $\mathbb{K}$. Then $\{1\} \times \mathbf{V}$ is an affine subspace of the vector space $\mathbb{K} \times \mathbf{V}$.

5. Let T be a one-dimensional oriented affine space over the vector space $\mathbb{T}$. Then an order can be defined on T by $a < b$ if and only if $a - b < \mathbf{0}$. Define the intervals of T.

2. Affine maps

2.1. Definition. Let V and U be affine spaces over $\mathbf{V}$ and $\mathbf{U}$, respectively. A map $L\colon \mathrm{V} \to \mathrm{U}$ is called *affine* if there is a linear map $\boldsymbol{L}\colon \mathbf{V} \to \mathbf{U}$ such that

$$L(y) - L(x) = \boldsymbol{L} \cdot (y - x) \qquad (x, y, \in \mathrm{V}).$$

We say that L is an affine map over $\mathbf{L}$. If L is a bijection, V and U are oriented, L is called *orientation preserving* or *orientation-reversing* if $\boldsymbol{L}$ has that property. ■

The formula above is equivalent to

$$L(x + \boldsymbol{x}) = L(x) + \boldsymbol{L} \cdot \boldsymbol{x} \qquad (x \in V, \boldsymbol{x} \in \mathbf{V}).$$

It is easy to show that the linear map $\boldsymbol{L}$ in the definition is unique.

2.2. Proposition. Let $L\colon \mathrm{V} \to \mathrm{U}$ be an affine map. Then

(*i*) L is injective or surjective if and only if $\boldsymbol{L}$ is injective or surjective, respectively; if L is bijective then L^{-1} is an affine bijection over $\boldsymbol{L}^{-1}$;

(*ii*) $\boldsymbol{L} = \mathbf{0}$ if and only if L is a constant map;

(*iii*) $\mathrm{Ran}\, L$ is an affine subspace of U, directed by $\mathrm{Ran}\, \boldsymbol{L}$;

(*iv*) if Z is an affine subspace of U, directed by $\mathbf{Z}$, and $(\mathrm{Ran}\, L) \cap \mathrm{Z} \neq \emptyset$ then $\overset{-1}{L}(\mathrm{Z})$ is an affine subspace of V, directed by $\overset{-1}{\boldsymbol{L}}(\mathbf{Z})$;

(*v*) L preserves convex combinations. ■

Observe that according to (*iv*), for all $u \in \mathrm{Ran}\, L$, $\overset{-1}{L}(\{u\})$ is an affine subspace of V, directed by $\mathrm{Ker}\, \boldsymbol{L}$.

2.3. Proposition. (*i*) If L and K are affine maps such that $K \circ L$ exists then $K \circ L$ is affine map over $\boldsymbol{K} \cdot \boldsymbol{L}$;

(*ii*) Let I be a nonvoid set. If $L_i\colon \mathrm{V}_i \to \mathrm{U}_i$ $(i \in I)$ are affine maps, then $\underset{i \in I}{\times} L_i$ is an affine map over $\underset{i \in I}{\times} \boldsymbol{L}_i$;

(*iii*) If $L_i\colon \mathrm{V} \to \mathrm{U}_i$ $(i \in I)$ are affine maps then $(L_i)_{i \in I}$ is an affine map over $(\boldsymbol{L}_i)_{i \in I}$;

(*iv*) If L and K are affine maps from V into U then

$$K - L\colon V \to \boldsymbol{U}, \qquad x \mapsto K(x) - L(x)$$

is an affine map over $\boldsymbol{K} - \boldsymbol{L}$ (recall that the vector space $\mathbf{U}$ is an affine space over itself).

2.4. (*i*) Let $\mathbf{V}$ and $\mathbf{U}$ be vector spaces and consider them to be affine spaces. Take a linear map $\boldsymbol{L}\colon \mathbf{V} \to \mathbf{U}$ and an $\boldsymbol{a} \in \mathrm{U}$; then $\mathbf{V} \to \mathbf{U},\ \boldsymbol{x} \mapsto \boldsymbol{a} + \boldsymbol{L}\cdot\boldsymbol{x}$ is an affine map over $\boldsymbol{L}$.

Conversely, suppose $L\colon \mathbf{V} \to \mathbf{U}$ is an affine map over the linear map $\boldsymbol{L}$. Put $\boldsymbol{a} := L(\mathbf{0})$. Then $L(\boldsymbol{x}) = \boldsymbol{a} + \boldsymbol{L}\cdot\boldsymbol{x}$ for all $\boldsymbol{x} \in \mathbf{V}$.

Thus we have proved:

Proposition. We can identify the set of affine maps from $\mathbf{V}$ into $\mathbf{U}$ with $\{(\boldsymbol{a}, \boldsymbol{L}) \mid \boldsymbol{a} \in \mathbf{U}, \boldsymbol{L} \in \mathrm{Lin}(\mathbf{V}, \mathbf{U})\} = \mathbf{U} \times \mathrm{Lin}(\mathbf{V}, \mathbf{U})$ in such a way that

$$(\boldsymbol{a}, \boldsymbol{L})(\boldsymbol{x}) := \boldsymbol{a} + \boldsymbol{L}\cdot\boldsymbol{x} \qquad\qquad (\boldsymbol{x} \in \boldsymbol{L}). \quad ■$$

Such an affine map $(\boldsymbol{a}, \boldsymbol{L})\colon \mathbf{V} \to \mathbf{U}$ can be represented in more suitable ways, as follows.

(*ii*) $\mathbf{V} \to \mathbb{K}\times\mathbf{V},\ \boldsymbol{x} \mapsto (1, \boldsymbol{x})$ (see Exercise 1.7.4) is an affine injection. We often find convenient to identify $\mathbf{V}$, considered to be an affine space, with $\{1\} \times \mathbf{V}$.

Take an affine map $(\boldsymbol{a}, \boldsymbol{L})\colon \mathbf{V} \to \mathbf{U}$ and consider it to be an affine map from $\{1\} \times \mathbf{V}$ into $\{1\} \times \mathbf{U}$. It can be uniquely extended to a linear map $\mathbb{K} \times \mathbf{V} \to \mathbb{K} \times \mathbf{U},\ (\alpha, \boldsymbol{x}) \mapsto (\alpha, \alpha\boldsymbol{a} + \boldsymbol{L}\cdot\boldsymbol{x})$.

Representing the linear maps from $\mathbb{K}\times\mathbf{V}$ into $\mathbb{K}\times\mathbf{U}$ by a matrix (see IV.3.7), we can write the extension of the affine map $(\boldsymbol{a}, \boldsymbol{L})$ in the form

$$\begin{pmatrix} 1 & 0 \\ a & L \end{pmatrix}.$$

(*iii*) It often occurs that the vector space $\mathbf{V}$ is regarded as an affine space (i.e. we use only its affine structure, the subtraction of vectors) but the vector space $\mathbf{U}$ is continued to be regarded as a vector space (i.e. we use its vectorial structure, the sum of vectors and the multiple of vectors).

In this case we identify $\mathbf{V}$ with $\{1\} \times \mathbf{V}$ and $\mathbf{U}$ with $\{0\} \times \mathbf{U}$, and so we can conceive that $(\boldsymbol{a}, \boldsymbol{L})$ maps from $\{1\} \times \mathbf{V}$ into $\{0\} \times \mathbf{U}$. This map can be uniquely extended to a linear map $\mathbb{K} \times \mathbf{V} \to \mathbb{K} \times \mathbf{U},\ (\alpha, \boldsymbol{x}) \mapsto (0, \alpha\boldsymbol{a} + \boldsymbol{L}\cdot\boldsymbol{x})$. Then the affine map $(\boldsymbol{a}, \boldsymbol{L})$ in a matrix representation has the form

$$\begin{pmatrix} 0 & 0 \\ a & L \end{pmatrix}.$$

2.5. Exercises

1. Let o be an element of the affine space V. Then $O_o\colon \mathrm{V} \to \mathbf{V},\ x \mapsto x - o$ is an affine map over $\mathbf{1_V}$.

Consequently, if V is N-dimensional then there are affine bijections $\mathrm{V} \to \mathbb{K}^N$.

2. Let $L\colon \mathrm{V} \to \mathrm{V}$ be an affine map and o an element of V. Then $O_o \circ L \circ O_o{}^{-1}$ is an affine map $\mathbf{V} \to \mathbf{V}$. Using the matrix form given in the preceding paragraph show that

$$O_o \circ L \circ O_o{}^{-1} = \begin{pmatrix} 1 & 0 \\ L(o) - o & \boldsymbol{L} \end{pmatrix}.$$

3. Let $H\colon \mathrm{V} \to \mathbf{V}$ be an affine map and o an element of V. Then $H \circ O_o{}^{-1}$ is an affine map $\mathbf{V} \to \mathbf{V}$. Then the vector space $\mathbf{V}$ as the domain of this affine map is considered to be an affine space (representing the affine space V); and $\mathbf{V}$ as the range is considered to be a vector space. Using the matrix form given in the preceding paragraph show that

$$H \circ O_o{}^{-1} = \begin{pmatrix} 0 & \mathbf{0} \\ H(o) & \boldsymbol{H} \end{pmatrix}.$$

4. The matrix forms of affine maps $\mathbf{V} \to \mathbf{V}$ are extremely useful for obtaining the composition of such maps because we can apply the usual matrix multiplication rule. Find the composition of

(i) $(\boldsymbol{a}, \boldsymbol{L})\colon \{1\} \times \mathbf{V} \to \{1\} \times \mathbf{V}$ and $(\boldsymbol{b}, \boldsymbol{K})\colon \{1\} \times \mathbf{V} \to \{1\} \times \mathbf{V}$,

(ii) $(\boldsymbol{a}, \boldsymbol{L})\colon \{1\} \times \mathbf{V} \to \{1\} \times \mathbf{V}$ and $(\boldsymbol{b}, \boldsymbol{K})\colon \{1\} \times \mathbf{V} \to \{0\} \times \mathbf{V}$.

5. Let V be an affine space over $\mathbf{V}$.

(i) If $\boldsymbol{a} \in \mathbf{V}$ then $T_{\boldsymbol{a}}\colon \mathrm{V} \to \mathrm{V},\ x \mapsto x + \boldsymbol{a}$ is an affine map over $\mathbf{1_V}$.

(ii) If $L\colon \mathrm{V} \to \mathrm{V}$ is an affine map over $\mathbf{1_V}$ then there is an $\boldsymbol{a} \in \mathbf{V}$ such that $L = T_{\boldsymbol{a}}$.

(iii) For all $x, y \in \mathrm{V}$ we have $O_y \circ O_x{}^{-1} = T_{x-y}$.

6. If $L\colon \mathrm{V} \to \mathrm{V}$ is an affine map over $-\mathbf{1_V}$ then there is an $o \in \mathrm{V}$ such that $L(x) = o - (x - o)\ (x \in \mathrm{V})$.

7. Let K and L be affine maps between the same affine spaces. Show that $\boldsymbol{K} = \boldsymbol{L}$ if and only if $K - L$ is a constant map.

8. Let $K, L, A\colon \mathrm{V} \to \mathrm{V}$ be affine maps. Show that $A \circ K - A \circ L = \boldsymbol{A} \circ (K - L)$.

3. Differentiation

3.1. Let $\mathbf{V}$ be a vector space. A *norm* on $\mathbf{V}$ is a map

$$\|\cdot\|\colon \mathbf{V} \to \mathbb{R}_0^+ \quad , \quad \boldsymbol{x} \mapsto \|\boldsymbol{x}\|$$

for which (*i*) $\|\boldsymbol{x}\| = 0$ if and only if $\boldsymbol{x} = \mathbf{0}$,

(*ii*) $\|\alpha \boldsymbol{x}\| = |\alpha| \|\boldsymbol{x}\|$ for all $\alpha \in \mathbb{K}$, $\boldsymbol{x} \in \mathbf{V}$,

(*iii*) $\|\boldsymbol{x} + \boldsymbol{y}\| \leq \|\boldsymbol{x}\| + \|\boldsymbol{y}\|$ for all $\boldsymbol{x}, \boldsymbol{y} \in \mathbf{V}$.

The distance of $\boldsymbol{x}, \boldsymbol{y} \in \mathbf{V}$ is defined to be $\|\boldsymbol{x} - \boldsymbol{y}\|$; the map

$$\mathbf{V} \times \mathbf{V} \to \mathbb{R}, \qquad (\boldsymbol{x}, \boldsymbol{y}) \mapsto \|\boldsymbol{x} - \boldsymbol{y}\|$$

is called the *metrics* associated with the norm.

The reader is supposed to be familiar with the fundamental notions of analysis connected with metrics: open subsets, closed subsets, convergence, continuity, etc.

It is important that if $\mathbf{V}$ is finite-dimensional then all the norms on $\mathbf{V}$ are equivalent, i.e. they determine the same open subsets, closed subsets, convergent series, continuous functions etc.

As a consequence, in finite-dimensional vector spaces—e.g. in pseudo-Euclidean vector spaces—we can speak about open subsets, closed subsets, continuity etc. without giving an actual norm. Linear, bilinear, multilinear maps between finite-dimensional vector spaces are continuous.

3.2. If V is an affine space over $\mathbf{V}$ and there is a norm on $\mathbf{V}$ then

$$\mathrm{V} \times \mathrm{V} \to \mathbb{R}, \qquad (x, y) \mapsto \|x - y\|$$

is a metrics on V. Then the open subsets, closed subsets, convergence etc. are defined in V.

In the following we deal with finite dimensional real affine spaces; hence we speak about the fundamental notions of analysis without specifying norms on the underlying vector spaces.

As usual, if $\mathbf{V}$ and $\mathbf{U}$ are finite-dimensional vector spaces, $\mathrm{ordo} : \mathbf{V} \rightarrowtail \mathbf{U}$ denotes a function such that

(*i*) it is defined in a neighbourhood of $\mathbf{0} \in \mathbf{V}$,

(*ii*) $\lim_{\boldsymbol{x} \to \mathbf{0}} \frac{\mathrm{ordo}(\boldsymbol{x})}{\|\boldsymbol{x}\|} = \mathbf{0}$ for some (hence for every) norm $\|\cdot\|$ on $\mathbf{V}$.

3.3. Definition. Let V and U be affine spaces. A map $F : \mathrm{V} \rightarrowtail \mathrm{U}$ is called *differentiable* at an interior point x of $\mathrm{Dom} F$ if there is a linear map $\mathrm{D}F(x) : \mathbf{V} \to \mathbf{U}$ and a neighbourhood $\mathcal{N}(x) \subset \mathrm{Dom} F$ of x such that

$$F(y) - F(x) = \mathrm{D}F(x) \cdot (y - x) + \mathrm{ordo}(y - x) \qquad (y \in \mathcal{N}(x)).$$

$\mathrm{D}F(x)$ is the derivative of F at x.

F is *differentiable on a subset* S of $\mathrm{Dom} F$ if it is differentiable at every point of S. F is *differentiable* if it is differentiable on its domain (which is necessarily open in this case). F is *continuously differentiable* if it is differentiable and $\mathrm{Dom} F \to \mathrm{Lin}(\mathbf{V}, \mathbf{U})$, $x \mapsto \mathrm{D}F(x)$ is continuous.

If the real affine spaces V and U are oriented, a differentiable mapping $F: \mathrm{V} \rightarrowtail \mathrm{U}$ is called *orientation preserving* if $\mathrm{D}F(x): \mathbf{V} \to \mathbf{U}$ is an orientation preserving linear bijection for all $x \in \mathrm{Dom}\, F$. ■

The differentiability of F at x is equivalent to the following: there is a neighbourhood $\mathcal{N}$ of $\mathbf{0} \in \mathbf{V}$ such that $x + \mathcal{N} \subset \mathrm{Dom} F$ and

$$F(x + \boldsymbol{x}) - F(x) = \mathrm{D}F(x) \cdot \boldsymbol{x} + \mathrm{ordo}(\boldsymbol{x}) \qquad (\boldsymbol{x} \in \mathcal{N}).$$

This form shows immediately that $\mathrm{D}F(x)$ is uniquely determined.

3.4. If the affine spaces in question are actually vector spaces, i.e. F is a map between vector spaces then the above definition coincides with the one in standard analysis. Hence in the case of vector spaces we can apply the well-known results regarding differentiability. Moreover, for affine spaces one proves without difficulty that

(*i*) a differentiable map is continuous;

(*ii*) if $F: \mathrm{V} \rightarrowtail \mathrm{U}$ and $G: \mathrm{U} \rightarrowtail \mathrm{W}$ are differentiable then $G \circ F$ is differentiable, too, and

$$\mathrm{D}(G \circ F)(x) = \mathrm{D}G(F(x)) \cdot \mathrm{D}F(x) \qquad (x \in \mathrm{Dom}\,(G \circ F));$$

(*iii*) if $F, G: \mathrm{V} \rightarrowtail \mathrm{U}$ are differentiable then $F - G: \mathrm{V} \rightarrowtail \mathbf{U}$, $x \mapsto F(x) - G(x)$ is differentiable and

$$\mathrm{D}(F - G)(x) = \mathrm{D}F(x) - \mathrm{D}G(x) \qquad (x \in \mathrm{Dom} F \cap \mathrm{Dom} G).$$

(*iv*) An affine map $L: \mathrm{V} \to \mathrm{U}$ is differentiable, its derivative at every x equals the underlying linear map:

$$\mathrm{D}L(x) = \boldsymbol{L} \qquad (x \in \mathrm{V}).$$

3.5. Let V and U be affine spaces. If $F: \mathrm{V} \rightarrowtail \mathrm{U}$ is differentiable and its derivative map $\mathrm{V} \rightarrowtail \mathrm{Lin}\,(\mathbf{V}, \mathbf{U})$, $x \mapsto \mathrm{D}F(x)$ is differentiable then F is called *twice differentiable.*

Differentiability of higher order is defined similarly. An infinitely many times differentiable map is called *smooth.*

The second derivative of F at x is denoted by $\mathrm{D}^2F(x)$; by definition, it is an element of $\mathrm{Lin}\,(\mathbf{V}, \mathrm{Lin}\,(\mathbf{V}, \mathbf{U}))$.

The n-th derivative of F at x, $\mathrm{D}^nF(x)$ is an element of
$\mathrm{Lin}\,(\mathbf{V}, \mathrm{Lin}\,(\mathbf{V}, \ldots, \mathrm{Lin}\,(\mathbf{V}, \mathbf{U}) \ldots))$.

This rather complicated object is significantly simplified with the aid of tensor products.

We know that $\mathrm{Lin}\,(\mathbf{V}, \mathbf{U}) \equiv \mathbf{U} \otimes \mathbf{V}^*$. Thus $\mathrm{D}F(x) \in \mathbf{U} \otimes \mathbf{V}^*$.

Further,

$$\mathrm{Lin}\,(\mathbf{V}, \mathrm{Lin}\,(\mathbf{V}, \mathbf{U})) \equiv \mathrm{Lin}\,(\mathbf{V}, \mathbf{U} \otimes \mathbf{V}^*) \equiv (\mathbf{U} \otimes \mathbf{V}^*) \otimes \mathbf{V}^* \equiv \mathbf{U} \otimes \mathbf{V}^* \otimes \mathbf{V}^*,$$

thus $\mathrm{D}^2F(x) \in \mathbf{U} \otimes \mathbf{V}^* \otimes \mathbf{V}^*$.

Similarly we have that $\mathrm{D}^n F(x) \in \mathbf{U} \otimes \left(\overset{n}{\otimes} \mathbf{V}^* \right)$.

Moreover, a well-known theorem states that the n-th derivative is symmetric, i.e. $\mathrm{D}^n F(x) \in \mathbf{U} \otimes \left(\overset{n}{\vee} \mathbf{V}^* \right)$.

3.6. We often need the following particular result.

Proposition. Let V, U and Z be affine spaces, $A\colon \mathrm{V} \to \mathrm{Z}$ an affine surjection. A mapping $f\colon \mathrm{Z} \rightarrowtail \mathrm{U}$ is k times (continuously) differentiable if and only if $f \circ A$ is k times (continuously) differentiable ($k \in \mathbb{N}$).

Proof. The first part of the statement is trivial.

Suppose that $F := f \circ A$ is k times (continuously) differentiable. We know that there is a linear injection $\boldsymbol{L}\colon \mathbf{Z} \to \mathbf{V}$ such that $\boldsymbol{A} \cdot \boldsymbol{L} = \mathbf{1}_{\mathbf{Z}}$. Then for $z \in \mathrm{Dom} f \subset \mathrm{Z}$, $\boldsymbol{\xi}$ in a neighbourhood of $\mathbf{0} \in \mathbf{Z}$ we have

$$f(z+\boldsymbol{\xi}) - f(z) = F(x + \boldsymbol{L}\cdot\boldsymbol{\xi}) - F(x) = \mathrm{D}F(x)\cdot\boldsymbol{L}\cdot\boldsymbol{\xi} + \mathrm{ordo}(\boldsymbol{L}\cdot\boldsymbol{\xi})$$

if $A(x) = z$. Since $\mathrm{ordo}(\boldsymbol{L}\cdot\boldsymbol{\xi}) = \mathrm{ordo}(\boldsymbol{\xi})$, we see that f is (continuously) differentiable and

$$\mathrm{D}f(z) = \mathrm{D}F(x)\cdot\boldsymbol{L} \qquad (z \in \mathrm{Dom} f,\ x \in \overset{-1}{A}\{z\}).$$

Moreover, $\mathrm{D}f\colon \mathrm{Z} \rightarrowtail \mathbf{U} \otimes \mathbf{Z}^*$ is a mapping such that $\mathrm{D}f \circ A = \mathrm{D}F \cdot \boldsymbol{L}$ and we can repeat the previous arguments to obtain that if F is twice (continuously) differentiable (i.e. $\mathrm{D}F$ is (continuously) differentiable) then f is twice (continuously) differentiable (i.e. $\mathrm{D}f$ is (continuously) differentiable).

Proceeding in this way we can demonstrate k times (continuously) differentiability.

3.7. (*i*) Let $\boldsymbol{C}\colon \mathrm{V} \rightarrowtail \mathbf{V}$ be a differentiable mapping (a *vector field* in V). Then $\mathrm{D}\boldsymbol{C}(x) \in \mathbf{V} \otimes \mathbf{V}^*$ for all $x \in \mathrm{Dom}\,\boldsymbol{C}$, thus we can take its trace:

$$\mathrm{D}\cdot\boldsymbol{C}(x) := \mathrm{Tr}\,(\mathrm{D}\boldsymbol{C}(x)).$$

The mapping $\mathrm{V} \rightarrowtail \mathbb{R}$, $x \mapsto \mathrm{D}\cdot\boldsymbol{C}(x)$ is called the *divergence* of $\boldsymbol{C}$.

If $\mathbf{Z}$ is a vector space, the divergence of differentiable mappings $\mathrm{V} \rightarrowtail \mathbf{Z} \otimes \mathbf{V}$ is defined similarly according to IV.3.9.

(*ii*) Let $\boldsymbol{S}\colon \mathrm{V} \rightarrowtail \mathbf{V}^*$ be a differentiable mapping (a *covector field* in V). Then $\mathrm{D}\boldsymbol{S}(x) \in \mathbf{V}^* \otimes \mathbf{V}^*$ for all $x \in \mathrm{Dom}\,\boldsymbol{S}$, and we can take

$$\mathrm{D}\wedge\boldsymbol{S}(x) := (\mathrm{D}\boldsymbol{S}(x))^* - \mathrm{D}\boldsymbol{S}(x).$$

The mapping $\mathrm{V} \rightarrowtail \mathbf{V}^* \wedge \mathbf{V}^*$, $x \mapsto \mathrm{D}\wedge\boldsymbol{S}(x)$ is called the *curl* of $\boldsymbol{S}$.

(*iii*) *Keep in mind that a vector field has no curl and a covector field has no divergence.*

3.8. (*i*) Let V_1, V_2 and U be affine spaces and consider a differentiable mapping $F\colon V_1 \times V_2 \rightarrowtail U$. Take an $(x_1, x_2) \in \operatorname{Dom} F$ and fix x_2. Then $V_1 \rightarrowtail U$, $y_1 \mapsto F(y_1, x_2)$ is a differentiable mapping; its derivative at x_1 is called the *first partial derivative* of F at (x_1, x_2) and is denoted by $D_1F(x_1, x_2)$. By definition, $D_1F(x_1, x_2)$ is a linear map $\mathbf{V}_1 \to \mathbf{U}$.

The second partial derivative $D_2F(x_1, x_2)$ of F is defined similarly, and an evident generalization can be made for the k-th partial derivative $(k = 1, \dots, n)$ of a mapping $\underset{k=1}{\overset{n}{\times}} V_k \rightarrowtail U$.

For a vector field $\boldsymbol{C}\colon V_1 \times V_2 \rightarrowtail \mathbf{V}_1 \times \mathbf{V}_2$ we define the components $\boldsymbol{C}^i\colon V_1 \times V_2 \rightarrowtail \mathbf{V}_i$ $(i = 1, 2)$ such that $\boldsymbol{C} = (\boldsymbol{C}^1, \boldsymbol{C}^2)$. Then $D\boldsymbol{C}(x_1, x_2)$ is an element of $(\mathbf{V}_1 \times \mathbf{V}_2) \otimes (\mathbf{V}_1 \times \mathbf{V}_2)^* \equiv (\mathbf{V}_1 \times \mathbf{V}_2) \otimes (\mathbf{V}_1^* \times \mathbf{V}_2^*)$. It is not hard to see that using a matrix form corresponding to the convention introduced in IV.3.7 we have

$$D\boldsymbol{C}(x_1, x_2) = \begin{pmatrix} D_1\boldsymbol{C}^1 & D_2\boldsymbol{C}^1 \\ D_1\boldsymbol{C}^2 & D_2\boldsymbol{C}^2 \end{pmatrix}(x_1, x_2),$$

where the symbol (x_1, x_2) after the matrix means that every entry is to be taken at (x_1, x_2); shortly,

$$D\boldsymbol{C} = \begin{pmatrix} D_1\boldsymbol{C}^1 & D_2\boldsymbol{C}^1 \\ D_1\boldsymbol{C}^2 & D_2\boldsymbol{C}^2 \end{pmatrix}.$$

Furthermore, we easily find that

$$D \cdot \boldsymbol{C} = D_1 \cdot \boldsymbol{C}^1 + D_2 \cdot \boldsymbol{C}^2.$$

(*ii*) Similar notations for a covector field $\boldsymbol{S} = (\boldsymbol{S}_1, \boldsymbol{S}_2)\colon V_1 \times V_2 \rightarrowtail (\mathbf{V}_1 \times \mathbf{V}_2)^* \equiv \mathbf{V}_1^* \times \mathbf{V}_2^*$ yield

$$D\boldsymbol{S} = \begin{pmatrix} D_1\boldsymbol{S}_1 & D_2\boldsymbol{S}_1 \\ D_1\boldsymbol{S}_2 & D_2\boldsymbol{S}_2 \end{pmatrix}$$

and

$$D \wedge \boldsymbol{S} = \begin{pmatrix} D_1 \wedge \boldsymbol{S}_1 & (D_1\boldsymbol{S}_2)^* - D_2\boldsymbol{S}_1 \\ (D_2\boldsymbol{S}_1)^* - D_1\boldsymbol{S}_2 & D_2 \wedge \boldsymbol{S}_2 \end{pmatrix}.$$

3.9. A vector field $\boldsymbol{C}\colon \mathbb{R}^N \rightarrowtail \mathbb{R}^N$ is given by its components $C^i\colon \mathbb{R}^N \rightarrowtail \mathbb{R}$ $(i = 1, \dots, N)$, $\boldsymbol{C} = (C^1, \dots, C^N)$. Its derivative at ξ is a linear map $\mathbb{R}^N \to \mathbb{R}^N$; one easily finds for its matrix entries

$$(D\boldsymbol{C}(\xi))^i{}_k = \partial_k C^i(\xi) \qquad (i, k = 1, \dots, N)$$

where ∂_k denotes the k-th partial differentiation.

Then

$$D \cdot \boldsymbol{C} = \sum_{i=1}^{N} \partial_i C^i.$$

A covector field $\boldsymbol{S}\colon \mathbb{R}^N \rightarrowtail (\mathbb{R}^N)^*$ is given by its components $S_i\colon \mathbb{R}^N \rightarrowtail \mathbb{R}$ $(i = 1,\dots,N)$, $\boldsymbol{S} = (S_1,\dots,S_N)$. We have

$$(\mathrm{D}\boldsymbol{S}(\xi))_{ik} = \partial_k S_i(\xi)$$

and

$$(\mathrm{D}\wedge \boldsymbol{S})_{ik} = \partial_i S_k - \partial_k S_i$$

for $i,k = 1,\dots,N$.

3.10. If T is a one-dimensional affine space, V is an affine space and $r\colon \mathrm{T} \rightarrowtail \mathrm{V}$ is differentiable, then, for $t \in \mathrm{Dom}\,r$, $\mathrm{D}r(t)$ is an element of $\mathbf{V}\otimes\mathbb{T}^* \equiv \frac{\mathbf{V}}{\mathbb{T}}$.

It is not hard to see that in this case

$$\frac{\mathrm{d}r(t)}{\mathrm{d}t} := \dot{r}(t) := \mathrm{D}r(t) = \lim_{\substack{\boldsymbol{t}\to 0\\ \boldsymbol{t}\in\mathbb{T}}} \frac{r(t+\boldsymbol{t}) - r(t)}{\boldsymbol{t}}.$$

Similarly we arrive at $\frac{\mathrm{d}^2 r(t)}{\mathrm{d}t^2} := \ddot{r}(t) := \mathrm{D}^2 r(t) \in \frac{\mathbf{V}}{\mathbb{T}\otimes\mathbb{T}}$.

3.11. Let V and T as before and suppose T is real and oriented. Recall that then $\mathbb{T}^+$ and $\mathbb{T}^-$ denote the sets of positive and negative elements of $\mathbb{T}$, respectively.

Then $r\colon \mathrm{T} \rightarrowtail \mathrm{V}$ is called *differentiable on the right* at an interior point t of $\mathrm{Dom}\,r$ if there exists

$$\dot{r}^+(t) := \lim_{\substack{\boldsymbol{t}\to 0\\ \boldsymbol{t}\in\mathbb{T}^+}} \frac{r(t+\boldsymbol{t}) - r(t)}{\boldsymbol{t}},$$

called the *right derivative* of r at t.

The differentiability on the left and the left derivative $\dot{r}^-(t)$ are defined similarly.

Definition. Let V and T be as before, T is oriented. A function $r\colon \mathrm{T} \rightarrowtail \mathrm{V}$ is called *piecewise differentiable* if it is

(i) continuous,

(ii) differentiable with the possible exception of finite points where r is differentiable both on the right and on the left.

r is called *piecewise twice differentiable* if

(i) it is piecewise differentiable,

(ii) it is twice differentiable where it is differentiable,

(iii) if a is a point where r is not differentiable then there exist

$$\lim_{\substack{\boldsymbol{t}\to 0\\ \boldsymbol{t}\in\mathbb{T}^+}} \frac{\dot{r}(a+\boldsymbol{t}) - \dot{r}^+(a)}{\boldsymbol{t}} \qquad \text{and} \qquad \lim_{\substack{\boldsymbol{t}\to 0\\ \boldsymbol{t}\in\mathbb{T}^-}} \frac{\dot{r}(a+\boldsymbol{t}) - \dot{r}^-(a)}{\boldsymbol{t}}.$$

3.12. Recall that for a finite dimensional vector space $\mathbf{V}$, $\mathrm{Lin}(\mathbf{V}) \equiv \mathbf{V}\otimes\mathbf{V}^*$ is a finite-dimensional vector space as well. Hence the differentiability of a function $\boldsymbol{R}\colon \mathrm{T} \rightarrowtail \mathrm{Lin}(\mathbf{V})$ makes sense. It can be shown without difficulty that $\boldsymbol{R}$ is differentiable (and then its derivative at t is $\dot{\boldsymbol{R}}(t) \in \frac{\mathbf{V}\otimes\mathbf{V}^*}{\mathbb{T}} \equiv \mathrm{Lin}\left(\frac{\mathbf{V}}{\mathbb{T}}, \mathbf{V}\right)$) if and only if $\mathrm{T} \rightarrowtail \mathbf{V}$, $t \mapsto \boldsymbol{R}(t)\cdot\boldsymbol{v}$ is differentiable for all $\boldsymbol{v} \in \mathbf{V}$ and then

$$\frac{\mathrm{d}}{\mathrm{d}t}(\boldsymbol{R}(t)\cdot\boldsymbol{v}) = \left(\frac{\mathrm{d}}{\mathrm{d}t}\boldsymbol{R}(t)\right)\cdot\boldsymbol{v}.$$

Moreover, if $\boldsymbol{r}\colon \mathrm{T} \rightarrowtail \mathbf{V}$ is a differentiable function then $\boldsymbol{R}\cdot\boldsymbol{r}$ is differentiable and

$$(\boldsymbol{R}\cdot\boldsymbol{r})^{\boldsymbol{\cdot}} = \dot{\boldsymbol{R}}\cdot\boldsymbol{r} + \boldsymbol{R}\cdot\dot{\boldsymbol{r}}.$$

4. Submanifolds in affine spaces

In this section the affine spaces are real and finite dimensional.

4.1. The inverse mapping theorem and the implicit mapping theorem are important and well-known results of analysis. Now we formulate them for affine spaces in a form convenient for our application.

The inverse mapping theorem. Let V and U be affine spaces, $\dim\mathrm{U} = \dim\mathrm{V}$. If $F\colon \mathrm{V} \rightarrowtail \mathrm{U}$ is $n \geq 1$ times continuously differentiable, $e \in \mathrm{Dom}\,F$ and $\mathrm{D}F(e)\colon \mathbf{V} \to \mathbf{U}$ is a linear bijection, then there is a neighbourhood $\mathcal{N}$ of e, $\mathcal{N} \subset \mathrm{Dom}\,F$, such that

(*i*) $F|_{\mathcal{N}}$ is injective,

(*ii*) $F[\mathcal{N}]$ is open in U,

(*iii*) $(F|_{\mathcal{N}})^{-1}$ is n times continuously differentiable.

The implicit mapping theorem. Let V and U be affine spaces, $\dim\mathrm{U} < \dim\mathrm{V}$. Suppose $S\colon \mathrm{V} \rightarrowtail \mathrm{U}$ is $n \geq 1$ times continuously differentiable, $e \in \mathrm{Dom}\,S$ and $\mathrm{D}S(e)$ is surjective.

Let $\mathbf{V}_1$ be a linear subspace of $\mathbf{V}$ such that the restriction of $\mathrm{D}S(e)$ onto $\mathbf{V}_1$ is a bijection between $\mathbf{V}_1$ and $\mathbf{U}$ and suppose $\mathbf{V}_0$ is a subspace complementary to $\mathbf{V}_1$.

Then there are

— neighbourhoods $\mathcal{N}_0$ and $\mathcal{N}_1$ of the zero in $\mathbf{V}_0$ and in $\mathbf{V}_1$, respectively, $e + \mathcal{N}_0 + \mathcal{N}_1 \subset \mathrm{Dom}\,S$,

— a uniquely determined, n times continuously differentiable mapping $G\colon \mathcal{N}_0 \to \mathcal{N}_1$ such that

$$S(e + \boldsymbol{x}_0 + G(\boldsymbol{x}_0)) = S(e) \qquad (\boldsymbol{x}_0 \in \mathcal{N}_0). \quad \blacksquare$$

Observe that $\mathbf{V}_0 := \mathrm{Ker}\,\mathrm{D}S(e)$ and a subspace $\mathbf{V}_1$ complementary to $\mathbf{V}_0$ satisfy the above requirements.

4.2. Definition. Let V be an affine space, $\dim \mathrm{V} := N \geq 2$. Let M and n be natural numbers, $1 \leq M \leq N$, $n \geq 1$. A subset $\mathcal{H}$ of V is called an *M-dimensional n times differentiable simple submanifold* in V if there are

— an M-dimensional affine space D,

— a mapping $p\colon \mathrm{D} \rightarrowtail \mathrm{V}$, called a *parametrization* of $\mathcal{H}$, such that

(*i*) $\mathrm{Dom}\,p$ is open and connected, $\mathrm{Ran}\,p = \mathcal{H}$,

(*ii*) p is n times continuously differentiable and $\mathrm{D}p(\xi)$ is injective for all $\xi \in \mathrm{Dom}\,p$,

(*iii*) p is injective and p^{-1} is continuous. ■

Recall that $\mathrm{D}p(\xi) \in \mathrm{Lin}(\mathbb{L}, \mathbf{V})$.

Since p is differentiable, it is continuous.

The parametrization of $\mathcal{H}$ is not unique. For instance, if S is an affine space and $L\colon \mathrm{S} \to \mathrm{D}$ is an affine bijection then $p \circ L$ is a parametrization, too. In particular, we can take $\mathrm{S} := \mathbb{R}^M$ (see Exercise 2.5.2); as a consequence, D can be replaced by $\mathbb{R}^M$ in the definition.

The inverse mapping theorem implies that the N-dimensional, n times differentiable simple submanifolds are the connected open subsets of V.

Evidently, an M-dimensional affine subspace of V is an M-dimensional n times differentiable simple manifold for all n.

4.3. Definition. Let $N \geq 2$. A subset $\mathcal{H}$ of the N-dimensional affine space V is called an *M-dimensional n times differentiable submanifold* if every $x \in \mathcal{H}$ has a neighbourhood $\mathcal{N}(x)$ in V such that $\mathcal{N}(x) \cap \mathcal{H}$ is an M-dimensional n times differentiable simple submanifold.

A subset which is an n times differentiable submanifold for all $n \in \mathcal{N}$ is a *smooth submanifold.*

A *submanifold* means an$n$times differentiable submanifold for some$n$.

A submanifold which is a closed subset of V is called a *closed submanifold.*

One-dimensional submanifolds, two-dimensional submanifolds and $(N-1)$-dimensional submanifolds are called *curves* or *lines, surfaces* and *hypersurfaces,* respectively. ■

By definition, every point of a submanifold has a neighbourhood in the submanifold that can be parametrized. A parametrization of such a neighbourhood is called a *local parametrization* of the manifold.

4.4. Proposition. Let $\mathcal{H}$ be an M-dimensional n times differentiable submanifold in V, $M < N$, and let $p\colon \mathbb{R}^M \rightarrowtail \mathrm{V}$ be a local parametrization of $\mathcal{H}$. If $e \in \mathrm{Ran}\,p$ then there are

— a neighbourhood $\mathcal{N}$ of e in V,

— continuously n times differentiable mappings

$$F:\mathcal{N}\to\mathbb{R}^M, \qquad S:\mathcal{N}\to\mathbb{R}^{N-M}$$

such that

(*i*) $\mathcal{N}\cap\mathcal{H}\subset \operatorname{Ran}p$;

(*ii*) $F(p(\xi))=\xi,\ S(p(\xi))=0$ for all $\xi\in\operatorname{Dom}p,\ p(\xi)\in\mathcal{N}$;

(*iii*) $\mathrm{D}S(x)$ is surjective for all $x\in\mathcal{N}$.

Proof. There is a unique $\alpha\in\operatorname{Dom}p$ for which $p(\alpha)=e$. $\mathrm{D}p(\alpha)\colon\mathbb{R}^M\to \mathrm{V}$ is a linear injection, hence $\mathbf{V}_1:=\operatorname{Ran}\mathrm{D}p(\alpha)$ is an M-dimensional linear subspace. Let $\mathbf{V}_0$ be a linear subspace, complementary to $\mathbf{V}_1$. Evidently, $\dim\mathbf{V}_0=N-M$.

Let $\boldsymbol{P}\colon\mathbf{V}\to\mathbf{V}$ be the projection onto $\mathbf{V}_1$ along $\mathbf{V}_0$ (i.e. $\boldsymbol{P}$ is linear and $\boldsymbol{P}\cdot\boldsymbol{x}_1=\boldsymbol{x}_1$ for $\boldsymbol{x}\in\mathbf{V}_1$ and $\boldsymbol{P}\cdot\boldsymbol{x}_0=\mathbf{0}$ for $\boldsymbol{x}\in\mathbf{V}_0$). Then

$$\boldsymbol{P}\cdot(p-e)\colon\mathbb{R}^M\rightarrowtail\mathbf{V}_1, \qquad \xi\mapsto\boldsymbol{P}\cdot(p(\xi)-e)$$

is n times continuously differentiable, its derivative at α equals $\boldsymbol{P}\cdot\mathrm{D}p(\alpha)$; it is a linear bijection from $\mathbb{R}^M$ onto $\mathbf{V}_1$. Thus, according to the inverse mapping theorem, there is a neighbourhood $\boldsymbol{\Omega}$ of α such that $\boldsymbol{P}\cdot(p-e)|_{\boldsymbol{\Omega}}$ is injective, its inverse is continuously differentiable, $(\boldsymbol{P}\cdot(p-e))[\boldsymbol{\Omega}]=\boldsymbol{P}[p[\boldsymbol{\Omega}]-e]$ is open in $\mathbf{V}_1$.

For the sake of simplicity and without loss of generality we can suppose $\boldsymbol{\Omega}=\operatorname{Dom}p$ (considering $p|_{\boldsymbol{\Omega}}$ instead of p).

Then the continuity of $\boldsymbol{P}$ involves that $\overset{-1}{\boldsymbol{P}}(\boldsymbol{P}[p[\boldsymbol{\Omega}]-e])$ is an open subset of $\mathbf{V}$ and so $e+\overset{-1}{\boldsymbol{P}}(\boldsymbol{P}[p[\boldsymbol{\Omega}]-e])$ is an open subset of V. Since p^{-1} is continuous, $p[\boldsymbol{\Omega}]$ is open in $\operatorname{Ran}p$ and $p[\boldsymbol{\Omega}]\subset e+\overset{-1}{\boldsymbol{P}}(\boldsymbol{P}[p[\boldsymbol{\Omega}]-e])$; thus there is an open subset $\mathcal{N}$ in $e+\overset{-1}{\boldsymbol{P}}(\boldsymbol{P}[p[\boldsymbol{\Omega}]-e])\subset\mathrm{V}$ such that $p[\boldsymbol{\Omega}]=\mathcal{H}\cap\mathcal{N}$.

Let $\boldsymbol{L}\colon\mathbf{V}_0\to\mathbb{R}^{N-M}$ be a linear bijection and

$$F:=(\boldsymbol{P}\cdot(p-e))^{-1}\circ\boldsymbol{P}\cdot(\mathbf{1}_{\mathrm{V}}-e)|_{\mathcal{N}}, \qquad S:=\boldsymbol{L}\circ(\mathbf{1}_{\mathrm{V}}-p\circ F).$$

$\mathcal{N}\subset e+\overset{-1}{\boldsymbol{P}}(\boldsymbol{P}[p[\boldsymbol{\Omega}]-e])$ implies $\boldsymbol{P}[e+\mathcal{N}]\subset\boldsymbol{P}[p[\boldsymbol{\Omega}]-e]=\operatorname{Dom}(\boldsymbol{P}\cdot(p-e))^{-1}$, hence both F and S are defined on $\mathcal{N}$. It is left to the reader to prove that properties (*ii*) and (*iii*) in the proposition hold.

4.5. Proposition. Let $p\colon\mathbb{R}^M\rightarrowtail\mathrm{V}$ and $q\colon\mathbb{R}^M\rightarrowtail\mathrm{V}$ be local parameterizations of the M-dimensional n times differentiable submanifold $\mathcal{H}$ such that $\operatorname{Ran}p\cap\operatorname{Ran}q\neq\emptyset$. Then $p^{-1}\circ q\colon\mathbb{R}^M\rightarrowtail\mathbb{R}^M$ is n times continuously differentiable and

$$\mathrm{D}(p^{-1}\circ q)(q^{-1}(x))=[\mathrm{D}p(p^{-1}(x))]^{-1}\cdot\mathrm{D}q(q^{-1}(x)) \qquad (x\in\operatorname{Ran}p\cap\operatorname{Ran}q).$$

Proof. If $M = N$ then the inverse mapping theorem implies that p^{-1} is n times continuously differentiable; as a consequence, $p^{-1} \circ q$ is n times continuously differentiable as well and the above formula is valid in view of the well-known rule of differentiation of composite mappings.

If $M < N$, the differentiability of p^{-1} makes no sense because $\mathcal{H}$ contains no open subsets in V. Nevertheless, $p^{-1} \circ q$ is continuously differentiable as we shall see below.

Let e be an arbitrary point of $\operatorname{Ran} p \cap \operatorname{Ran} q$. According to the previous proposition, there are a neighbourhood $\mathcal{N}$ of e and an n times continuously differentiable mapping F for which $\mathcal{N} \cap \mathcal{H} \subset \operatorname{Ran} p$ and $F \circ p \subset \mathbf{1}_{\mathbb{R}^M}$ holds.

Then $\Omega := \overset{-1}{q}(\mathcal{N}) \subset \operatorname{Dom}(p^{-1} \circ q)$ is open and

$$(p^{-1} \circ q)\big|_\Omega = (F \circ p) \circ (p^{-1} \circ q)\big|_\Omega = F \circ q|_\Omega ;$$

the mapping on the right-hand side is n times continuously differentiable being a composition of two such mappings. Thus we have shown that each point of $\operatorname{Dom}(p^{-1} \circ q)$ has a neighbourhood Ω in which $p^{-1} \circ q$ is n times continuously differentiable.

Let x be an element of $\operatorname{Ran} p \cap \operatorname{Ran} q$, $\xi := q^{-1}(x)$ and $\Phi := p^{-1} \circ q$. Then $p \circ \Phi \subset q$ and $\xi \in \operatorname{Dom}(p \circ \Phi)$. Thus

$$\mathrm{D}q(q^{-1}(x)) = \mathrm{D}q(\xi) = \mathrm{D}p(\Phi(\xi)) \cdot \mathrm{D}\Phi(\xi) = \mathrm{D}p(p^{-1}(x)) \cdot \mathrm{D}\Phi(\xi), \qquad (*)$$

which gives immediately the desired equality. ∎

Evidently, then $q^{-1} \circ p$ is n times continuously differentiable as well. Since $q^{-1} \circ p = (p^{-1} \circ q)^{-1}$, this means that the derivative of $p^{-1} \circ q$ at every point is a linear bijection $\mathbb{R}^M \to \mathbb{R}^M$.

As a consequence, the dimension of a submanifold is uniquely determined. Supposing that a submanifold is both M-dimensional and M'-dimensional we get $M = M'$.

We have proved the statement for parametrizations from $\mathbb{R}^M$. Obviously, the same is true for parametrizations with domains in affine spaces.

4.6. Proposition. Let p and q be local parametrizations of a submanifold such that $\operatorname{Ran} p \cap \operatorname{Ran} q \neq \emptyset$. If $x \in \operatorname{Ran} p \cap \operatorname{Ran} q$ then

$$\operatorname{Ran}\left(\mathrm{D}p(p^{-1}(x))\right) = \operatorname{Ran}\left(\mathrm{D}q(q^{-1}(x))\right).$$

Proof. Equality $(*)$ in the preceding paragraph involves that the range of $\mathrm{D}q(q)^{-1}(x))$ is contained in the range of $\mathrm{D}p(p^{-1}(x))$. A similar argument yields that the range of $\mathrm{D}p(p^{-1}(x))$ is contained in the range of $\mathrm{D}q(q)^{-1}(x))$.

Definition. Let $\mathcal{H}$ be an M-dimensional submanifold, $x \in \mathcal{H}$. Then

$$\mathbf{T}_x(\mathcal{H}) := \operatorname{Ran}\left(\mathrm{D}p(p^{-1}(x))\right)$$

is called the *tangent space* of $\mathcal{H}$ at x where p is a parametrization of $\mathcal{H}$ such that $x \in \operatorname{Ran} p$. The elements of $\mathbf{T}_x(\mathcal{H})$ are called *tangent vectors* of $\mathcal{H}$ at x. ■

The preceding proposition says that the tangent space, though it is defined by a parametrization, is independent of the parametrization.

The tangent space is an M-dimensional linear subspace of $\mathbf{V}$. $x + \mathbf{T}_x(\mathcal{H})$ is an affine subspace of V which we call the *geometric tangent space* of $\mathcal{H}$ at x.

4.7. Let $M < N$. We have seen in Proposition 4.4. that every point e of an M-dimensional n times differentiable submanifold $\mathcal{H}$ has a neighbourhood $\mathcal{N}$ in V and an n times continuously differentiable mapping $S: \mathcal{N} \to \mathbb{R}^{N-M}$ such that $\mathcal{N} \cap \mathcal{H} = \overset{-1}{S}(\{0\})$ and $\mathrm{D}S(x)$ is surjective. Evidently, $\mathbb{R}^{N-M}$ and $0 \in \mathbb{R}^{N-M}$ can be replaced by an arbitrary affine space U, $\dim \mathrm{U} = N - M$, and a point $o \in \mathrm{U}$, respectively.

Now we prove a converse statement.

Proposition. Let V and U be affine spaces, $\dim \mathrm{V} =: N$, $\dim \mathrm{U} =: N - M$, and $S: \mathrm{V} \rightarrowtail \mathrm{U}$ an n times continuously differentiable mapping. Suppose $o \in \operatorname{Ran} S$. Then

$$\mathcal{H} := \{x \in \overset{-1}{S}(\{o\}) \mid \operatorname{Ran} \mathrm{D}S(x) \quad \text{is} \quad (N-M)\text{-dimensional}\}$$

is either void or an M-dimensional n times differentiable submanifold of V.

Proof. Suppose $\mathcal{H}$ is not void and e belongs to it. Then $\mathbf{V}_0 := \operatorname{Ker} \mathrm{D}S(e)$ is an M-dimensional linear subspace of $\mathbf{V}$. Let $\mathbf{V}_1$ be a linear subspace, complementary to $\mathbf{V}_0$. Then we can apply the implicit mapping theorem: there are neighbourhoods $\mathcal{N}_0$ and $\mathcal{N}_1$ of the zero in $\mathbf{V}_0$ and in $\mathbf{V}_1$, respectively, an n times continuously differentiable mapping $G: \mathcal{N}_0 \to \mathcal{N}_1$ such that

$$S(e + \boldsymbol{x}_0 + G(\boldsymbol{x}_0)) = S(e) \qquad (\boldsymbol{x}_0 \in \mathcal{N}_0).$$

Let us define

$$p: \mathbf{V}_0 \rightarrowtail \mathrm{V}, \quad \boldsymbol{x}_0 \mapsto e + \boldsymbol{x}_0 + G(\boldsymbol{x}_0) \qquad (\boldsymbol{x}_0 \in \mathcal{N}_0).$$

Evidently, p is n times continuously differentiable and $\operatorname{Ran} p \subset \overset{-1}{S}(\{o\})$.

We can easily see that p is injective, its inverse is $x \mapsto \boldsymbol{P} \cdot (x - e)$ where $\boldsymbol{P}$ is the projection onto $\mathbf{V}_0$ along $\mathbf{V}_1$. Consequently, p^{-1} is continuous.

These mean that p is a parametrization of $\mathcal{H}$ in a neighbourhood of e.

4.8. Proposition. Let $\mathcal{H} \neq \emptyset$ be the submanifold described in the previous proposition. Then

$$\mathbf{T}_x(\mathcal{H}) = \operatorname{Ker} \mathrm{D}S(x) \qquad (x \in \mathcal{H}).$$

Proof. Let p be a local parametrization of $\mathcal{H}$. Then $S \circ p = \text{const.}$, thus for $x \in \operatorname{Ran} p$ we have $\mathrm{D}S(x) \cdot \mathrm{D}p(p^{-1}(x)) = 0$ from which we deduce immediately that $\mathbf{T}_x(\mathcal{H}) := \operatorname{Ran} \mathrm{D}p(p^{-1}(x)) \subset \operatorname{Ker} \mathrm{D}S(x)$. Since both linear subspaces on the two sides of $\subset$ are M-dimensional, equality occurs necessarily.

4.9. Definition. Let p and q be two local parametrizations of a submanifold, $\operatorname{Dom} p \subset \mathbb{R}^M$, $\operatorname{Dom} q \subset \mathbb{R}^M$ and $\operatorname{Ran} p \cap \operatorname{Ran} q \neq \emptyset$. Then p and q are said to be *equally oriented* if the determinant of $\mathrm{D}(p^{-1} \circ q)(\xi)$ is positive for all $\xi \in \operatorname{Dom}(p^{-1} \circ q)$.

A family $(p_i)_{i \in I}$ of local parametrizations of a submanifold $\mathcal{H}$ is *orienting* if $\mathcal{H} = \cup_{i \in I} \operatorname{Ran} p_i$ and, in the case $\operatorname{Ran} p_i \cap \operatorname{Ran} p_j \neq \emptyset$, p_i and p_j are equally oriented $(i, j \in I)$.

Two orienting parametrization families are called *equally orienting* if their union is orienting as well.

The submanifold is *orientable* if it has an orienting parametrization family. ■

To be equally orienting is an equivalence relation. If the submanifold is connected, there are exactly two equivalence classes.

An orientable submanifold together with one of the equivalence classes of the orienting local parametrization families is an *oriented submanifold.* A local parametrization of an oriented submanifold is called *positively oriented* if it belongs to a family of the chosen equivalence class.

A simple submanifold is obviously orientable.

Connected N-dimensional submanifolds—i.e. connected open subsets—are orientable.

4.10. Let p be a local parametrization of the submanifold $\mathcal{H}$, $\operatorname{Dom} p \subset \mathbb{R}^M$. If $(\chi_1, \dots, \chi_M)$ is the standard ordered basis of $\mathbb{R}^M$ then $\mathrm{D}p(\xi) \cdot \chi_i = \partial_i p(\xi)$ $(i = 1, \dots, M)$ for $\xi \in \operatorname{Dom} p$. This means that $((\partial_1 p(\xi), \dots, \partial_M p(\xi))$ is an ordered basis in $\mathbf{T}_{p(\xi)}(\mathcal{H})$.

In other words, $\big(\partial_1 p(p^{-1}(x)), \dots, \partial_M p(p^{-1}(x))\big)$ is an ordered basis in $\mathbf{T}_x(\mathcal{H})$ $(x \in \operatorname{Ran} p)$.

If q is another local parametrization, with domain in $\mathbb{R}^M$, and $x \in \operatorname{Ran} p \cap \operatorname{Ran} q \neq \emptyset$ then $\big(\partial_1 q(q)^{-1}(x)), \dots, \partial_M q(q^1(x))\big)$ is another ordered basis in $\mathbf{T}_x(\mathcal{H})$.

Evidently,

$$\partial_i q(q^{-1}(x)) = \mathrm{D}q(q^{-1}(x)) \cdot \left[\mathrm{D}p(p^{-1}(x))\right]^{-1} \cdot \partial_i p(p^{-1}(x))$$

for all $i = 1, .., M$.

We know from 4.5 and IV.3.20 that

$$\det\left(\mathrm{D}(p^{-1}\circ q)(q^{-1}(x))\right) = \det\left([\mathrm{D}p(p^{-1}(x))]^{-1}\cdot \mathrm{D}q(q^{-1}(x))\right) = \\ = \det\left(\mathrm{D}q(q^{-1}(x))\cdot [\mathrm{D}p(p^{-1}(x))]^{-1}\right).$$

We have proved the following statement.

Proposition. Let p and q be local parametrizations of the submanifold $\mathcal{H}$, $\mathrm{Dom}\, p\subset \mathbb{R}^M$, $\mathrm{Dom}\, q\subset \mathbb{R}^M$ and $\mathrm{Ran}\, p\cap \mathrm{Ran}\, q\neq \emptyset$. p and q are equally oriented if and only if the ordered bases

$$\left(\partial_1 p(p^{-1}(x)),\dots,\partial_M p(p^{-1}(x))\right)$$

and

$$\left(\partial_1 q(q^{-1}(x)),\dots,\partial_M q(q^{-1}(x))\right)$$

in $\mathbf{T}_x(\mathcal{H})$ are equally oriented for all $x\in \mathrm{Ran}\, p\cap \mathrm{Ran}\, q$.

4.11 Observe that in the case $M=1$, i.e. when the submanifold is a curve, instead of partial derivatives we have a single derivative of p, denoted usually by $\dot{p}$. Then $\dot{p}(p^{-1}(x))$ spans the (one-dimensional) tangent space at x.

Two local parametrizations p and q are equally oriented if and only if one of the following three conditions is fulfilled:

(*i*) $(p^{-1}\circ q)^{\cdot}(\alpha)>0$ for all $\alpha\in \mathrm{Dom}\,(p^{-1}\circ q)$,
(*ii*) $p^{-1}\circ q\colon \mathbb{R}\rightarrowtail\mathbb{R}$ is strictly monotone increasing,
(*iii*) $\dot{p}(p^{-1}(x))$ is a positive multiple of $\dot{q}(q^{-1}(x))$ for all $x\in \mathrm{Ran}\, p\cap \mathrm{Ran}\, q$.

4.12. The following notion concerning curves appears frequently in application.

Let x and y be different elements of V. We say that the curve $\mathcal{C}$ *connects* x and y if these points form the boundary of $\mathcal{C}$, i.e. $\{x,y\}=\overline{\mathcal{C}}\setminus\mathcal{C}$ where $\overline{\mathcal{C}}$ is the closure of $\mathcal{C}$. We can conceive that x and y are the extremities of a curve connecting them.

4.13. Definition. Let $\mathcal{H}$ and $\mathcal{F}$ be M-dimensional and K-dimen- sional submanifolds of V and U, respectively. A mapping $F\colon\mathcal{H}\rightarrowtail\mathcal{F}$ is called *differentiable* at x if there are local parametrizations q of $\mathcal{H}$ and p of $\mathcal{F}$ for which $x\in\mathrm{Ran}\, q$, $F(x)\in\mathrm{Ran}\, p$, and the function $p^{-1}\circ F\circ q\colon\mathbb{R}^M\rightarrowtail\mathbb{R}^K$ is differentiable at $q^{-1}(x)$.

The derivative of F at x is defined to be the linear map $\mathrm{D}F(x)\colon \mathbf{T}_x(\mathcal{H})\to \mathbf{T}_{F(x)}(\mathcal{F})$ that satisfies

$$\mathrm{D}p(F(x))^{-1}\cdot \mathrm{D}F(x)\cdot \mathrm{D}q(q^{-1}(x)) = \mathrm{D}(p^{-1}\circ F\circ q)(q^{-1}(x)).$$

F is *differentiable* if it is differentiable at each point of its domain.

If $\mathcal{H}$ and $\mathcal{F}$ are n times differentiable submanifolds, we define F to be k times (continuously) differentiable, for $0 \le k \le n$, if $p^{-1} \circ F \circ q$ is k times (continuously) differentiable.

4.14. Exercises

1. Let V and U be affine spaces. The graph of an n times continuously differentiable mapping $F\colon \mathrm{V} \rightarrowtail \mathrm{U}$ —i.e. the set $\{(x, F(x)) \mid x \in \operatorname{Dom} F\}$ —is a $(\dim V)$-dimensional n times differentiable submanifold in $\mathrm{V} \times \mathrm{U}$. Give its tangent space at an arbitrary point.

2. Prove that the mapping $(F, S)\colon \mathrm{V} \rightarrowtail \mathbb{R}^M \times \mathbb{R}^{N-M} \equiv \mathbb{R}^N$ described in 4.4 is injective and its inverse is $(\xi, \eta) \mapsto p(\xi) + \boldsymbol{L}^{-1}\eta$.

3. Let $(\mathbf{V}, \mathbb{B}, \boldsymbol{b})$ be a pseudo-Euclidean vector space, $\mathbf{0} \neq \boldsymbol{a} \in \mathbb{B}$. Prove that

$$\left\{\boldsymbol{x} \in \mathbf{V} \,\middle|\, \boldsymbol{x} \cdot \boldsymbol{x} = \boldsymbol{a}^2\right\} \quad \text{and} \quad \left\{\boldsymbol{x} \in \mathbf{V} \,\middle|\, \boldsymbol{x} \cdot \boldsymbol{x} = -\boldsymbol{a}^2\right\}$$

are either void or hypersurfaces in $\mathbf{V}$ whose tangent space at $\boldsymbol{x}$ equals

$$\{\boldsymbol{y} \in \mathbf{V} \mid \boldsymbol{x} \cdot \boldsymbol{y} = 0\}.$$

(The derivative of the map $\mathbf{V} \to \mathbb{B} \otimes \mathbb{B}$, $\boldsymbol{x} \mapsto \boldsymbol{x} \cdot \boldsymbol{x}$ at $\boldsymbol{x}$ is $2\boldsymbol{x}$ regarded as the linear map $\mathbf{V} \to \mathbb{B} \otimes \mathbb{B}$, $\boldsymbol{y} \mapsto 2\boldsymbol{x} \cdot \boldsymbol{y}$.) Why is the statement not true for $\boldsymbol{a} = \mathbf{0}$?

4. A linear bijection $\mathbb{R}^M \to \mathbb{R}^M$ has a positive determinant if and only if it is orientation preserving. On the basis of this remark define that two local parametrizations $p\colon \mathrm{D} \rightarrowtail \mathrm{V}$ and $q\colon \mathrm{E} \rightarrowtail \mathrm{V}$ of a submanifold are equally oriented where D and E are oriented affine spaces.

5. Use the notations of 4.13. Prove that

(*i*) $p^{-1} \circ F \circ q$ is differentiable for some p and q if and only if it is differentiable for all p and q;

(*ii*) the derivative of F is uniquely defined.

(*iii*) if F is the restriction of a k times (continuously) differentiable mapping $G\colon \mathrm{V} \rightarrowtail \mathrm{U}$ then F is k times (continuously) differentiable and $\mathrm{D}F(x)$ is the restriction of $\mathrm{D}G(x)$ onto $\mathbf{T}_x(\mathcal{H})$.

5. Coordinatization

5.1. Let V be an N-dimensional real affine space. Take an $o \in \mathrm{V}$ and an ordered basis $(\boldsymbol{x}_1, \ldots, \boldsymbol{x}_N)$ of $\mathbf{V}$. The affine map $K\colon \mathrm{V} \to \mathbb{R}^N$ determined by $K(o + \boldsymbol{x}_i) := \chi_i$ $(i = 1, \ldots, N)$ where $(\chi_1, \ldots, \chi_N)$ is the ordered standard basis of $\mathbb{R}^N$ is called the *coordinatization* of V corresponding to o and $(\boldsymbol{x}_1, \ldots, \boldsymbol{x}_N)$.

The inverse of the coordinatization, $P := K^{-1}$, is called the corresponding *parametrization* of V. It is quite evident that

$$P(\xi) = o + \sum_{i=1}^{N} \xi^i \boldsymbol{x}_i \qquad (\xi \in \mathbb{R}^N).$$

Moreover, if $(\boldsymbol{p}^1, \dots, \boldsymbol{p}^N)$ is the dual of the basis in question, then

$$K(x) = \big(\boldsymbol{p}^i \cdot (x - o) \,|\, i = 1, \dots, N\big) \qquad (x \in \mathrm{V}).$$

Obviously, every affine bijection $K\colon \mathrm{V} \to \mathbb{R}^N$ is a coordinatization in the above sense: the one corresponding to $o := K^{-1}(0)$ and $(\boldsymbol{x}_1, \dots, \boldsymbol{x}_N)$ where $\boldsymbol{x}_i := K^{-1}(\chi_i) - o \ \ (i = 1, \dots, N)$.

Such a parametrization maps straight lines into straight lines. More closely, if $\alpha \in \mathbb{R}^N$ then P maps the straight line passing through α and parallel to χ_i into the straight line passing through $P(\alpha)$ and parallel to $\boldsymbol{x}_i$:

$$P[\alpha + \mathbb{R}\chi_i] = P(\alpha) + \mathbb{R}\boldsymbol{x}_i \qquad (i = 1, \dots, N).$$

This is why affine coordinatizations are generally called *rectilinear.*

5.2. In application we often need nonaffine coordinatizations as well. Coordinatization means in general that we represent the elements of the affine space by ordered N-tuples of real numbers (i.e. by elements of $\mathbb{R}^N$) in a smooth way.

Definition. Let V be an N-dimensional affine space. A mapping $K\colon \mathrm{V} \rightarrowtail \mathbb{R}^N$ is called a *local coordinatization* of V if

(*i*) K is injective,
(*ii*) K is smooth,
(*iii*) $\mathrm{D}K(x)$ is injective for all $x \in \mathrm{Dom}\,K$. ■

Evidently, $\mathrm{D}K(x)$ is bijective since the dimensions of its domain and range are equal; thus the inverse mapping theorem implies that also the inverse of K has the properties (*i*)–(*ii*)–(*iii*;) $P := K^{-1}$ is called a *local parametrization* of V. We often omit the adjective 'local'.

5.3. If $\alpha \in \mathrm{Ran}\,K = \mathrm{Dom}\,P$ then $P[\alpha + \mathbb{R}\chi_i]$ is a smooth curve in V; a parametrization of this curve is $p_i\colon \mathbb{R} \rightarrowtail \mathrm{V}, \ a \mapsto P(\alpha + a\chi_i) \ \ (i = 1, ..., N)$. The parametrization maps straight lines into curves, that is why such coordinatizations are often called *curvilinear.*

The curves corresponding to parallel straight lines do not intersect each other. The curves corresponding to meeting straight lines intersect each other *transversally,* i.e. their tangent spaces at the point of intersection do not coincide. For instance, using the previous notations we have that $\dot{p}_i(0) = \mathrm{D}P(\alpha)\cdot\chi_i = \partial_i P(\alpha)$ is the tangent vector of the curve $P[\alpha + \mathbb{R}\chi_i]$ at $P(\alpha)$; if $i \neq k$ then $\dot{p}_i(0) \neq \dot{p}_k(0)$.

If $x \in \operatorname{Dom} K$ then $P[K(x)+\mathbb{R}\chi_i]$ is called the i-th *coordinate line* passing through x.

5.4. Recall Proposition 4.4: if $\mathcal{H}$ is an M-dimensional smooth submanifold of V then for every $e \in \mathcal{H}$ there is a coordinatization $K := (F, S)$ of V in a neighbourhood of e such that the first M coordinate lines run in $\mathcal{H}$. In other words, if P is the corresponding parametrization of V then $\mathbb{R}^M \rightarrowtail \mathrm{V}$, $\zeta \mapsto P(\zeta, 0)$ is a parametrization of $\mathcal{H}$.

5.5. The most frequently used curvilinear coordinatizations are the polar coordinatization, the cylindrical coordinatization and the spherical coordinatization. We give them as coordinatizations in $\mathbb{R}^2$ and $\mathbb{R}^3$; composed with affine coordinatizations they result in curvilinear coordinatizations of two- and three-dimensional affine spaces.

(*i*) *Polar coordinatization*

$$K: \mathbb{R}^2 \backslash \{(x_1, 0) \mid x_1 \le 0\} \to \mathbb{R}^+ \times]-\pi, \pi[\,,$$
$$x = (x_1, x_2) \mapsto \left(|x|,\ \operatorname{sign}(x_2)\ \arccos \frac{x_1}{|x|} \right);$$

its inverse is

$$P: \mathbb{R}^+ \times]-\pi, \pi[\to \mathbb{R}^2 \setminus \{(x_1, 0) \mid x_1 \le 0\},$$
$$(r, \varphi) \mapsto (r \cos\varphi,\ r \sin\varphi),$$

for which

$$\mathrm{D}P(r, \varphi) = \begin{pmatrix} \cos\varphi & -r\sin\varphi \\ \sin\varphi & r\cos\varphi \end{pmatrix},$$
$$\det\left(\mathrm{D}P(r, \varphi)\right) = r.$$

(*ii*) *Cylindrical coordinatization*

$$K: \mathbb{R}^3 \setminus \{(x_1, 0, x_3) \mid x_1 \le 0,\ x_3 \in \mathbb{R}\} \to \mathbb{R}^+ \times]-\pi, \pi[\times \mathbb{R},$$
$$x = (x_1, x_2, x_3) \mapsto \left(\sqrt{x_1^2 + x_2^2},\ \operatorname{sign}(x_2) \arccos \frac{x_1}{\sqrt{x_1^2 + x_2^2}},\ x_3 \right);$$

its inverse is

$$P: \mathbb{R}^+ \times]-\pi, \pi[\times \mathbb{R} \to \mathbb{R}^3 \setminus \{(x_1, 0, x_3,) \mid x_1 \le 0,\ x_3 \in \mathbb{R}\},$$
$$(\rho, \varphi, z) \mapsto (\rho\cos\varphi,\ \rho\sin\varphi,\ z),$$

for which

$$\mathrm{D}P(\rho,\varphi,z)=\begin{pmatrix}\cos\varphi & -\rho\sin\varphi & o\\ \sin\varphi & \rho\cos\varphi & 0\\ 0 & 0 & 1\end{pmatrix},$$

$$\det\left(\mathrm{D}P(\rho,\varphi,z)\right)=\rho.$$

(*iii*) *Spherical coordinatization*

$$K:\mathbb{R}^3\setminus\{(x_1,0,x_3)\mid x_1\le 0,\ x_3\in\mathbb{R}\}\to\mathbb{R}^+\times]0,\pi[\times]-\pi,\pi[,$$

$$x=(x_1,x_2,x_3)\mapsto\left(|x|,\ \arccos\frac{x_3}{|x|},\ \operatorname{sign}(x_2)\arccos\frac{x_1}{\sqrt{x_1^2+x_2^2}}\right);$$

its inverse is

$$P:\mathbb{R}^+\times]0,\pi[\times[-\pi,\pi[\to\mathbb{R}^3\setminus\{(x_1,0,x_3)\mid x_1\le 0,\ x_3\in\mathbb{R}\},$$

$$(r,\vartheta,\varphi)\mapsto(r\sin\vartheta\cos\varphi,\ r\sin\vartheta\sin\varphi,\ r\cos\vartheta),$$

for which

$$\mathrm{D}P(r,\vartheta,\varphi)=\begin{pmatrix}\sin\vartheta\cos\varphi & r\cos\vartheta\cos\varphi & -r\sin\vartheta\sin\varphi\\ \sin\vartheta\sin\varphi & r\cos\vartheta\sin\varphi & r\sin\vartheta\cos\varphi\\ \cos\vartheta & -r\sin\vartheta & 0\end{pmatrix},$$

$$\det\left(\mathrm{D}P(r,\vartheta,\varphi)\right)=r^2\sin\vartheta.$$

5.6. Let $K\colon \mathrm{V}\rightarrowtail\mathbb{R}^N$ be a coordinatization. Then for all $x\in\operatorname{Dom}K$ the tangent vectors of the coordinate lines passing through x form a basis in $\mathbf{V}$. More closely, if P is the corresponding parametrization then $\partial_iP(P^{-1}(x))=\mathrm{D}P(P^{-1}(x))\cdot\chi_i$ $(i=1,\dots,N)$ form a basis in $\mathbf{V}$ which is called the *local basis* at x corresponding to K.

Note that $\mathrm{D}K(x)\colon\mathbf{V}\to\mathbb{R}^N$ is the linear bijection that sends the local basis into the standard basis of $\mathbb{R}^N$, i.e. $\mathrm{D}K(x)$ is the coordinatization of $\mathbf{V}$ corresponding to the local basis at x.

We shall often use the relation

$$[\mathrm{D}K(P(\xi))]^{-1}=\mathrm{D}P(\xi)\qquad(\xi\in\operatorname{Dom}P)$$

which will be written in the form

$$\mathrm{D}K(P)^{-1}=\mathrm{D}P.$$

(*i*) A vector field $\boldsymbol{C}\colon\mathrm{V}\rightarrowtail\mathbf{V}$ is coordinatized in such a way that for $x\in\operatorname{Dom}\boldsymbol{C}\cap\operatorname{Dom}K$ the vector $\boldsymbol{C}(x)$ is given by its coordinates with respect to the local basis at x and x is represented by the coordinatization in question; the *coordinatized form* of $\boldsymbol{C}$ is the function

$$\mathrm{D}K(P)\cdot\boldsymbol{C}(P)\colon \mathbb{R}^N \rightarrowtail \mathbb{R}^N, \qquad \xi \mapsto \mathrm{D}K(P(\boldsymbol{\xi}))\cdot\boldsymbol{C}(P(\boldsymbol{\xi})).$$

(*ii*) A covector field $\boldsymbol{S}\colon \mathrm{V} \rightarrowtail \mathbf{V}^*$ is coordinatized similarly, with the aid of the dual of the local bases (see IV.2.2); the *coordinatized form* of $\boldsymbol{S}$ is the function

$$\mathrm{D}P^*\cdot\boldsymbol{S}(P)\colon \mathbb{R}^N \to (\mathbb{R}^N)^*, \qquad \xi \mapsto \mathrm{D}P(\xi)^*\cdot\boldsymbol{S}(P(\xi)) = \boldsymbol{S}(P(\xi))\cdot\mathrm{D}P(\xi).$$

(*iii*) Accordingly (see IV.2.3), the *coordinatizated forms* of the tensor fields $\boldsymbol{L}\colon \mathrm{V} \rightarrowtail \mathbf{V}\otimes\mathbf{V}^* \equiv \mathrm{Lin}(\mathbf{V})$ and $\boldsymbol{F}\colon \mathrm{V} \rightarrowtail \mathbf{V}^*\otimes\mathbf{V}^* \equiv \mathrm{Lin}(\mathbf{V},\mathbf{V}^*)$ are

$$\mathrm{D}K(P)\cdot\boldsymbol{L}(P)\cdot\mathrm{D}P\colon\ \mathbb{R}^N \rightarrowtail \mathbb{R}^N\otimes\left(\mathbb{R}^N\right)^*, \quad \xi \mapsto \mathrm{D}K(P(\xi))\cdot\boldsymbol{L}(P(\xi))\cdot\mathrm{D}P(\xi),$$
$$\mathrm{D}P^*\cdot\boldsymbol{F}(P)\cdot\mathrm{D}P\colon\ \mathbb{R}^N \rightarrowtail \left(\mathbb{R}^N\right)^*\otimes\left(\mathbb{R}^N\right)^*, \quad \xi \mapsto \mathrm{D}P(\xi)^*\cdot\boldsymbol{F}(P(\xi))\cdot\mathrm{D}P(\xi).$$

5.7. If $K\colon \mathrm{V} \to \mathbb{R}^N$ is an affine coordinatization then $\mathrm{D}K(x) = \boldsymbol{K}$ for all $x \in \mathrm{V}$ where $\boldsymbol{K}$ is the linear map under K. Similarly, $\mathrm{D}P(\xi) = \boldsymbol{P}$ for all $\xi \in \mathbb{R}^N$.

In this case the vector field $\boldsymbol{C}$ and the covector field $\mathbf{S}$ have the coordinatized form

$$\xi \mapsto \boldsymbol{K}\cdot\boldsymbol{C}(P(\xi)), \tag{1}$$
$$\xi \mapsto \boldsymbol{P}^*\cdot\boldsymbol{S}(P(\xi)). \tag{2}$$

The derivative of $\boldsymbol{C}$ is the mixed tensor field $\mathrm{D}\boldsymbol{C}\colon \mathrm{V} \rightarrowtail \mathbf{V}\otimes\mathbf{V}^*$, $x \mapsto \mathrm{D}\boldsymbol{C}(x)$, and the derivative of $\boldsymbol{S}$ is the cotensor field $\mathrm{D}\boldsymbol{S}\colon \mathrm{V} \rightarrowtail \mathbf{V}^*\otimes\mathbf{V}^*$, $x \mapsto \mathrm{D}\boldsymbol{S}(x)$. Now they have the coordinatized forms

$$\xi \mapsto \boldsymbol{K}\cdot\mathrm{D}\boldsymbol{C}(P(\xi))\cdot\boldsymbol{P}, \tag{3}$$
$$\xi \mapsto \boldsymbol{P}^*\cdot\mathrm{D}\boldsymbol{S}(P(\xi))\cdot\boldsymbol{P}. \tag{4}$$

A glance at the previous formulae convinces us that (3) and (4) are the derivatives of (1) and (2), respectively.

Thus in the case of a rectilinear coordinatization the order of differentiation and coordinatization can be interchanged: taking coordinates first and then differentiating is the same as differentiating first and then taking coordinates.

5.8. In the case of curvilinear coordinates, in general, the order of differentiation and coordinatization cannot be interchanged.

To get a rule, how to compute the coordinatized form of the derivative of a vector field or a covector field from the coordinatized form of these fields, we introduce a new notation.

Without loss of generality we suppose that $\mathrm{V}=\mathbb{R}^N$, since every curvilinear coordinatization $\mathrm{V}\rightarrowtail\mathbb{R}^N$ can be obtained as the composition of a rectilinear coordinatization $\mathrm{V}\to\mathbb{R}^N$ and a curvilinear one $\mathbb{R}^N\rightarrowtail\mathbb{R}^N$.

For the components of elements in $\mathrm{V}=\mathbb{R}^N$, Latin subscripts and superscripts: $i,j,k,\dots$; for the components of the curvililinear coordinates in $\mathbb{R}^N$, Greek subscripts and superscripts: $\alpha,\beta,\gamma,\dots$ are used. Moreover, we agree that all indices run from 1 to N and we accept the Einstein summation rule: for equal subscripts and superscripts a summation is to be taken from 1 to N.

Thus for K we write K^α, for P we write P^i; moreover, for any function $\phi:\mathbb{R}^N\rightarrowtail\mathbb{R}$ we find it convenient to write $\phi(P)$ instead of $\phi\circ P$. The rule of differentiation of composite functions will be used frequently,

$$\partial_\alpha\left(\phi(P)\right)=(\partial_i\phi)(P)\partial_\alpha P^i,$$

as well as the relations

$$\begin{aligned}\partial_\gamma P^i\partial_j K^\gamma(P)&=\delta^i{}_j,\\ \partial_j K^\alpha(P)\partial_\beta P^j&=\delta^\alpha{}_\beta.\end{aligned}\tag{$*$}$$

The second one implies

$$\partial_i\partial_j K^\alpha(P)\partial_\gamma P^i\partial_\beta P^j+\partial_i K^\alpha(P)\partial_\gamma\partial_\beta P^i=0.$$

We put

$$\Gamma^\alpha{}_{\beta\gamma}:=\partial_\gamma\partial_\beta P^i\partial_i K^\alpha(P)=-\partial_i\partial_j K^\alpha(P)\partial_\gamma P^i\partial_\beta P^j$$

and we call it the *Christoffel symbol* of the coordinatization in question.

The Christoffel symbol is a mapping defined on $\operatorname{Dom}P$; for $\xi\in\operatorname{Dom}P$, $\Gamma(\xi)$ is a bilinear map from $\mathbb{R}^N\times\mathbb{R}^N$ into $\left(\mathbb{R}^N\right)^*$:

$$(\boldsymbol{\zeta},\boldsymbol{\eta})\mapsto\left(\Gamma^\alpha{}_{\beta\gamma}(\xi)\zeta^\beta\eta^\gamma\,\middle|\,\alpha=1,\dots,N\right).$$

It is usually emphasized that the Christoffel symbol is not a tensor of third order though it has three indices. This means that in general there is no mapping $\mathrm{V}\rightarrowtail\operatorname{Bilin}(\mathbf{V}\times\mathbf{V},\mathbf{V}^*)$ (third order tensor field) whose coordinatized form would be the Christoffel symbol.

5.9. The coordinatized form of

$f:V\rightarrowtail\mathbb{R}$	is	$f(P)$,
$\boldsymbol{C}:\mathrm{V}\rightarrowtail\mathbf{V}$	is	$\partial_i K^\alpha(P)C^i(P)=:\mathcal{C}^\alpha$,
$\boldsymbol{S}:\mathrm{V}\rightarrowtail\mathbf{V}^*$	is	$\partial_\alpha P^i S_i(P)=:\mathcal{S}_\alpha$,
$\boldsymbol{L}:\mathrm{V}\rightarrowtail\mathbf{V}\otimes\mathbf{V}^*$	is	$\partial_i K^\alpha(P)L^i{}_k(P)\partial_\beta P^k=:\mathcal{L}^\alpha{}_\beta$,
$\boldsymbol{F}:\mathrm{V}\rightarrowtail\mathbf{V}^*\otimes\mathbf{V}^*$	is	$\partial_\alpha P^i T_{ik}(P)\partial_\beta P^k=:\mathcal{T}_{\alpha\beta}$.

(i) The coordinatized form of $\mathrm{D}f : \mathrm{V} \rightarrowtail \mathbf{V}^*$ is $\partial_\alpha P^i \partial_i f(P) = \partial_\alpha (f(P))$; thus for a real-valued function the order of differentiation and coordinatization can be interchanged even in the case of curvilinear coordinatization.
(ii) The coordinatized form of $\mathrm{D}\boldsymbol{C} : \mathrm{V} \rightarrowtail \mathbf{V} \otimes \mathbf{V}^*$ is

$$(\mathcal{D}\mathcal{C})^\alpha{}_\beta := \partial_i K^\alpha(P)(\partial_k C^i)(P)\partial_\beta P^k,$$

whereas the derivative of the coordinatized form of $\boldsymbol{C}$ reads

$$\partial_\beta \left(\partial_i K^\alpha(P) C^i(P)\right) = (\partial_i \partial_k K^\alpha)(P)\partial_\beta P^k C^i(P) + \partial_i K^\alpha(P)(\partial_k C^i)(P)\partial_\beta P^k.$$

The second term equals the coordinatized form of $\mathrm{D}\boldsymbol{C}$; with the aid of relation $(*)$ in 5.8, the first term is transformed into an expression containing the Christoffel symbol and the coordinatized form of $\boldsymbol{C}$. In this way we get

$$(\mathcal{D}\mathcal{C})^\alpha{}_\beta = \partial_\beta \mathcal{C}^\alpha + \Gamma^\alpha{}_{\beta\gamma}\mathcal{C}^\gamma.$$

(iii) Similarly, if $(\mathcal{D}\mathcal{S})_{\alpha\beta}$ denotes the coordinatized form of $\mathrm{D}\boldsymbol{S}$ then

$$(\mathcal{D}\mathcal{S})_{\alpha\beta} = \partial_\beta \mathcal{S}_\alpha - \Gamma^\gamma{}_{\alpha\beta}\mathcal{S}_\gamma.$$

5.10. Now we shall examine the coordinatizated form of two-times differentiable functions $\mathrm{T} \rightarrowtail \mathrm{V}$ where T is a one-dimensional affine space.

A useful notation will be applied: functions $\mathrm{T} \rightarrowtail \mathrm{V}$ and elements of V will be denoted by the same letter. If necessary, supplementary remarks rule out ambiguity.

For the sake of simplicity and without loss of generality we suppose that $\mathrm{T} = \mathbb{R}$.

Let $K : \mathrm{V} \rightarrowtail \mathbb{R}^N$ be a coordinatization, $P := K^{-1}$.

For $x \in \mathrm{V}$ let $\xi := K(x)$; then $x = P(\xi)$.

For $x : \mathrm{T} \rightarrowtail \mathrm{V}$ we put $\xi := K(x) := K \circ x$; then $x = P(\xi) := P \circ \xi$.

Denoting the differentiation by a dot we deduce

$$\begin{aligned} \dot{\xi} = \mathrm{D}K(x) \cdot \dot{x}, \qquad \dot{x} = \mathrm{D}P(\xi) \cdot \dot{\xi}, \\ \ddot{x} = \mathrm{D}^2 P(\xi)(\dot{\xi}, \dot{\xi}) + \mathrm{D}P(\xi) \cdot \ddot{\xi}, \\ \ddot{\xi} = \mathrm{D}^2 K(x)(\dot{x}, \dot{x}) + \mathrm{D}K(x) \cdot \ddot{x}, \end{aligned} \tag{$**$}$$

from which we obtain

$$\mathrm{D}K(x) \cdot \ddot{x} = \ddot{\xi} - \Gamma(\xi)(\dot{\xi}, \dot{\xi}) \tag{$***$}$$

where

$$\Gamma(\xi) := \mathrm{D}^2 K(P(\xi)) \circ (\mathrm{D}P(\xi) \times \mathrm{D}P(\xi))$$

is exactly the Christoffel symbol of the coordinatization.

In view of physical application, x, $\dot{x}$ and $\ddot{x}$ will be called *position, velocity* and *acceleration,* respectively.

The velocity at $t \in \mathbb{R}$, $\dot{x}(t)$ is in $\mathbf{V}$; it is represented by its coordinates corresponding to the local basis at $x(t)$, i.e. by $\mathrm{D}K(x(t)) \cdot \dot{x}(t)$. Thus $(**)$ tells us that the coordinatized form of velocity coincides with the derivative of the coordinatization of position.

Similarly, $\mathrm{D}K(x(t)) \cdot \ddot{x}(t)$ gives the coordinates of acceleration in the local basis at $x(t)$. Thus $(***)$ shows that the coordinatized form of acceleration does not coincide with the second derivative of the coordinatization of position.

5.11. Now we consider the coordinatizations treated in 5.5. They are *orthogonal* which means that every local basis is orthogonal with respect to the usual inner product in $\mathbb{R}^N$ $(N = 2, 3)$; in other words, if $\{\boldsymbol{\chi}_1, \dots, \boldsymbol{\chi}_N\}$ is the standard basis in $\mathbb{R}^N$ then $\{\mathrm{D}P(\xi) \cdot \boldsymbol{\chi}_i \mid i = 1, \dots, N\}$ is an orthogonal basis (the local basis at $P(\xi)$).

Introducing the notation

$$\alpha_i(\xi) := |\mathrm{D}P(\xi) \cdot \boldsymbol{\chi}_i| \qquad (i = 1, \dots, N)$$

we define the linear map $T(\xi) \colon \mathbb{R}^N \to \mathbb{R}^N$ by

$$T(\xi) \cdot \boldsymbol{\chi}_i := \alpha_i(\xi) \boldsymbol{\chi}_i \qquad (i = 1, \dots, N)$$

and then

$$\mathrm{D}P(\xi) = R(\xi) \cdot T(\xi)$$

where $R(\xi) \colon \mathbb{R}^N \to \mathbb{R}^N$ is an orthogonal linear map.

In usual physical applications one prefers orthonormal local bases, i.e. one takes $\frac{\mathrm{D}P(\xi) \cdot \boldsymbol{\chi}_i}{\alpha_i(\xi)} = \mathrm{D}P(\xi) \cdot T(\xi)^{-1} \cdot \boldsymbol{\chi}_i = R(\xi) \cdot \boldsymbol{\chi}_i$ instead of $\mathrm{D}P(\xi) \cdot \boldsymbol{\chi}_i$ $(i = 1, .., N)$.

The vector $\boldsymbol{y} \in \mathbb{R}^N$ at $P(\xi)$ has the coordinates $R(\xi)^{-1} \cdot \boldsymbol{y}$ in the local basis $\{R(\xi) \cdot \boldsymbol{\chi}_i \mid i = 1, \dots, N\}$.

Take $x \colon \mathbb{R} \rightarrowtail \mathbb{R}^N$, $\xi := K(x)$, $x = P(\xi)$ as in the previous paragraph. Then

$$\dot{x} = R(\xi) \cdot T(\xi) \cdot \dot{\xi}$$

from which we derive

$$\begin{aligned} \ddot{x} &= R(\xi)^{\cdot} \cdot T(\xi) \cdot \dot{\xi} + R(\xi) \cdot T(\xi)^{\cdot} \cdot \dot{\xi} + R(\xi) \cdot T(\xi) \cdot \ddot{\xi} = \\ &= R(\xi) \cdot \left(\left(R(\xi)^{-1} \cdot R(\xi)^{\cdot} \cdot T(\xi) + T(\xi)^{\cdot} \right) \cdot \dot{\xi} + T(\xi) \cdot \ddot{\xi} \right). \end{aligned}$$

According to the foregoings, the coordinates of velocity in the orthonormal local basis at $P(\xi)$ are

$$T(\xi) \cdot \dot{\xi}$$

and the coordinates of acceleration in the orthonormal local basis at $P(\xi)$ are

$$\Big(R(\xi)^{-1}\cdot R(\xi)^{\cdot}\cdot T(\xi)+T(\xi)^{\cdot}\Big)\cdot\dot{\xi}+T(\xi)\cdot\ddot{\xi}.$$

5.12. (*i*) For polar coordinates $\xi=(r,\varphi)$,

$$R(r,\varphi)=\begin{pmatrix}\cos\varphi & -\sin\varphi\\ \sin\varphi & \cos\varphi\end{pmatrix}=:R(\varphi)$$
$$T(r,\varphi)=\begin{pmatrix}1 & 0\\ 0 & r\end{pmatrix}=:T(r).$$

Furthermore

$$R(\varphi)^{\cdot}=\dot{\varphi}R(\varphi)\cdot R(\pi/2),\qquad T(r)^{\cdot}=T(\dot{r}),$$

and so velocity and acceleration in the local orthonormal basis at (r,φ) are

$$(\dot{r},r\dot{\varphi})\qquad\text{and}\qquad(\ddot{r}-r\dot{\varphi}^2,\ r\ddot{\varphi}+2\dot{r}\dot{\varphi}),$$

respectively.

(*ii*) For cylindrical coordinates $\xi=(\rho,\varphi,z)$,

$$R(\rho,\varphi,z)=\begin{pmatrix}\cos\varphi & -\sin\varphi & 0\\ \sin\varphi & \cos\varphi & 0\\ 0 & 0 & 1\end{pmatrix}=:R(\varphi),$$
$$T(\rho,\varphi,z)=\begin{pmatrix}1 & 0 & 0\\ 0 & \rho & 0\\ 0 & 0 & 1\end{pmatrix}=:T(\rho)$$

and we deduce as previously that velocity and acceleration in the local orthonormal basis at (ρ,φ,z) are

$$(\dot{\rho},\ \rho\dot{\varphi},\ \dot{z})\qquad\text{and}\qquad(\ddot{\rho}-\rho\dot{\varphi}^2,\ \rho\ddot{\varphi}+2\dot{\rho}\dot{\varphi},\ \ddot{z}),$$

respectively.

(*iii*) For spherical coordinates $\xi=(r,\vartheta,\varphi)$,

$$R(r,\vartheta,\varphi)=\begin{pmatrix}\sin\vartheta\cos\varphi & \cos\vartheta\cos\varphi & -\sin\varphi\\ \sin\vartheta\sin\varphi & \cos\varphi\sin\varphi & \cos\varphi\\ \cos\vartheta & -\sin\vartheta & 0\end{pmatrix}=:R(\vartheta,\varphi),$$
$$T(r,\vartheta,\varphi)=\begin{pmatrix}1 & 0 & 0\\ 0 & r & 0\\ 0 & 0 & r\sin\vartheta\end{pmatrix}=:T(r,\vartheta).$$

The components of velocity in the local orthonormal basis at (r, ϑ, φ) are

$$(\dot{r},\ r\dot{\vartheta},\ r\sin\vartheta\ \dot{\varphi}).$$

The components of acceleration are given by rather complicated formulae; the ambitious reader is asked to perform the calculations.

5.13. Exercises

1. Give the polar coordinatized form of the linear map (vector field) $\mathbb{R}^2 \to \mathbb{R}^2$ whose matrix is

$$\begin{pmatrix} \cos\alpha & -\sin\alpha \\ \sin\alpha & \cos\alpha \end{pmatrix}.$$

2. Give the cylindrical and the spherical coordinates of the following vector fields:
(*i*) $\boldsymbol{L}\colon \mathbb{R}^3 \to \mathbb{R}^3$ is a linear map;
(*ii*) $\mathbb{R}^3 \to \mathbb{R}^3$, $x \mapsto |x|\boldsymbol{v}$ where $\boldsymbol{v}$ is a given nonzero element of $\mathbb{R}^3$.
3. Find the coordinatized form of
(*i*) the divergence of a vector field,
(*ii*) the curl of a covector field.

6. Differential equations

6.1. Definition. Let V be a finite-dimensional affine space over the vector space $\mathbf{V}$.

Suppose $\boldsymbol{C}\colon \mathrm{V} \rightarrowtail \mathbf{V}$ is a differentiable vector field, $\mathrm{Dom}\,\boldsymbol{C}$ is connected.

Then a *solution* of the *differential equation*

$$(x\colon \mathbb{R} \rightarrowtail \mathrm{V})? \qquad \dot{x} = \boldsymbol{C}(x)$$

is a differentiable function $r\colon \mathbb{R} \rightarrowtail \mathrm{V}$ such that
(*i*) $\mathrm{Dom}\,r$ is an interval,
(*ii*) $\mathrm{Ran}\,r \subset \mathrm{Dom}\,\boldsymbol{C}$,
(*iii*) $\dot{r}(t) = \boldsymbol{C}(r(t))$ for $t \in \mathrm{Dom}\,r$.

The range of a solution is called an *integral curve* of $\boldsymbol{C}$. An integral curve is *maximal* if it is not contained properly in an integral curve. ■

An integral curve, in general, is not a curve in the sense of our definition in 4.3, i.e. it is not necessarily a submanifold.

6.2. Definition. Let $\boldsymbol{C}$ be as before and let x_o be an element of $\mathrm{Dom}\,\boldsymbol{C}$. A *solution* of the *initial value problem*

$$(x\colon \mathbb{R} \rightarrowtail \mathrm{V})? \qquad \dot{x} = \boldsymbol{C}(x), \qquad x(t_o) = x_o \qquad (*)$$

is a solution r of the corresponding differential equation such that

$$t_o \in \operatorname{Dom} r \qquad \text{and} \qquad r(t_o) = x_o.$$

The range of the solution of the initial value problem is called the *integral curve* of $\boldsymbol{C}$ *passing through* x_o. ■

The well-known existence and local uniqueness theorem asserts that solutions of the initial value problem exist and two solutions coincide on the intersection of their domain; consequently there is a single maximal integral curve of $\boldsymbol{C}$ passing through x_o.

6.3. Let U be another affine space over the vector space $\mathbf{U}$, $\dim \mathrm{U} = \dim \mathrm{V}$. Suppose $L\colon \mathrm{V} \rightarrowtail \mathrm{U}$ is a continuously differentiable injection whose inverse is continuously differentiable as well, and $\operatorname{Dom} \boldsymbol{C} \subset \operatorname{Dom} L$.

Put

$$\boldsymbol{G}\colon \mathrm{U} \rightarrowtail \mathbf{U}, \qquad y \mapsto \mathrm{D}L\left(L^{-1}(y)\right) \cdot \boldsymbol{C}(L^{-1}(y)).$$

Then r is a solution of the initial value problem $(*)$ if and only if $L \circ r$ is a solution of the initial value problem

$$(y\colon \mathbb{R} \rightarrowtail \mathrm{U})? \qquad \dot{y} = \boldsymbol{G}(y), \qquad y(t_o) = L(x_o) \qquad (**).$$

That is why we call $(**)$ the *transformation of* $(*)$ by L.

6.4. Proposition. Let $\boldsymbol{C}$ be a differentiable vector field in V and let $\mathcal{H}$ be a submanifold in the domain of $\boldsymbol{C}$. If $\boldsymbol{C}(x) \in \mathbf{T}_x(\mathcal{H})$ for all $x \in \mathcal{H}$ and $x_o \in \mathcal{H}$ then every solution r of the initial value problem $(*)$ runs in $\mathcal{H}$, i.e. $\operatorname{Ran} r \subset \mathcal{H}$.

Proof. The element x_o has a neighbourhood $\mathcal{N}$ in V and there are continuously differentiable functions $F\colon \mathcal{N} \to \mathbb{R}^M$, $S\colon \mathcal{N} \to \mathbb{R}^{N-M}$ such that $S(x) = 0$ for $x \in \mathcal{H} \cap \mathcal{N}$, and $K := (F, S)\colon \mathrm{V} \rightarrowtail \mathbb{R}^M \times \mathbb{R}^{N-M} \equiv \mathbb{R}^N$ is a local coordinatization of V. For $P := K^{-1}$ (the corresponding local parametrization of V) $\zeta \mapsto P(\zeta, 0)$ is a parametrization of $\mathcal{H} \cap \mathcal{N}$. Thus the tangent space of $\mathcal{H}$ at $P(\zeta, 0)$ is $\operatorname{Ker} \mathrm{D}S(P(\zeta, 0))$ (see 4.4 and 4.8).

The coordinatized form of $\boldsymbol{C}$ becomes

$$(\Phi, \Psi)\colon \mathbb{R}^M \times \mathbb{R}^{N-M} \rightarrowtail \mathbb{R}^M \times \mathbb{R}^{N-M}$$

where

$$\Phi(\zeta, \eta)\colon = \mathrm{D}F(P(\zeta, \eta)) \cdot \boldsymbol{C}(P(\zeta, \eta)),$$
$$\Psi(\zeta, \eta)\colon = \mathrm{D}S(P(\zeta, \eta)) \cdot \boldsymbol{C}(P(\zeta, \eta)).$$

Then the coordinatization transforms the initial value problem $(*)$ into the following one:

$$(\dot{\zeta},\dot{\eta}) = (\Phi(\zeta,\eta),\Psi(\zeta,\eta)), \qquad \zeta(t_o) = F(x_o), \qquad \eta(t_o) = 0. \qquad (***)$$

This means that r is a solution of $(*)$ if and only if $(F\circ r, S\circ r)$ is a solution of $(***)$, or (ρ,σ) is a solution of $(***)$ if and only if $P\circ(\rho,\sigma)$ is a solution of $(*)$.

Since $\boldsymbol{C}(x)\in \mathbf{T}_x(\mathcal{H})$ for $x\in\mathcal{H}$, $\boldsymbol{C}\big(P(\zeta,0)\big)$ is in the kernel of $\mathrm{D}S\big(P(\zeta,0)\big)$, i.e. $\Psi(\zeta,0)=0$ for all possible $\zeta\in\mathbb{R}^M$. Then if ρ is a solution of the initial value problem

$$\dot{\zeta} = \Phi(\zeta,0), \qquad \zeta(t_o) = F(x_o)$$

then $(\rho,0)$ is a solution of $(***)$. Then the uniqueness of solutions of initial value problems implies that every solution of $(***)$ has the form $(\rho,0)$. Consequently, $t\mapsto P(\rho(t),0)$, a solution of $(*)$, takes values in $\mathcal{H}$.

6.5. Physical application requires differential equations for functions $\mathrm{T}\rightarrowtail\mathrm{V}$ where T is a one-dimensional real affine space. Since the derivative of such functions takes values in $\frac{\mathbf{V}}{\mathbb{T}}$, we start with a differentiable mapping $\boldsymbol{C}\colon \mathrm{V}\rightarrowtail\frac{\mathbf{V}}{\mathbb{T}}$. A solution of the differential equation

$$(x\colon \mathrm{T}\rightarrowtail\mathrm{V})? \qquad \dot{x} = \boldsymbol{C}(x)$$

is a differentiable function $r\colon \mathrm{T}\rightarrowtail\mathrm{V}$ for which (i)–(ii)–(iii) of definition 6.1 holds.

Integral curves, solutions of initial value problems etc. are formulated as previously.

7. Integration on curves

7.1. Let T be an oriented one-dimensional affine space over the vector space $\mathbb{T}$. Suppose $\mathbb{A}$ is a one-dimensional vector space and $f\colon \mathrm{T}\rightarrowtail\mathbb{A}$ is a continuous function defined on an interval (see Exercise 1.7.5). If $a,b\in\operatorname{Dom} f$, $a<b$, then

$$\int_a^b f(t)\mathrm{d}t \in \mathbb{A}\otimes\mathbb{T}$$

is defined by some limit procedure, in the way well-known in standard analysis of real functions, using the integral approximation sums of the form

$$\sum_{k=1}^{n} f(t_k)(t_{k+1}-t_k).$$

7.2. Let V be an affine space over the vector space $\mathbf{V}$ and let $\mathbb{A}$ be a one-dimensional vector space. Suppose $F: \mathrm{V} \times \mathbf{V} \to \mathbb{A}$ is a continuous function, positively homogeneous in the second variable, i.e.

$$F(x, \lambda \boldsymbol{x}) = \lambda F(x, \boldsymbol{x}) \qquad (x \in \mathrm{V}, \lambda \in \mathbb{R}_0^+, \boldsymbol{x} \in \mathbf{V}).$$

Let C be a connected curve in V.

Proposition. Let $p, q: \mathbb{R} \rightarrowtail \mathrm{V}$ be equally oriented parametrizations of C, $x, y \in \mathrm{Ran}\, p \cap \mathrm{Ran}\, q$. Then

$$\int_{p^{-1}(x)}^{p^{-1}(y)} F(p(t), \dot{p}(t))\, \mathrm{d}t = \int_{q^{-1}(x)}^{q^{-1}(y)} F(q(s), \dot{q}(s))\, \mathrm{d}s.$$

Proof. We know that $\Phi := p^{-1} \circ q: \mathbb{R} \to \mathbb{R}$ is differentiable and $\dot{\Phi} > 0$ (see 4.11). Consequently, $q = p \circ \Phi$, $\dot{q}(s) = \dot{p}(\Phi(s)) \cdot \dot{\Phi}(s)$ and

$$\int_{q^{-1}(x)}^{q^{-1}(y)} F(q(s), \dot{q}(s))\, \mathrm{d}s = \int_{\Phi^{-1}(p^{-1}(x))}^{\Phi^{-1}(p^{-1}(y))} F(p(\Phi(s)),\ \dot{p}(\Phi(s)))\, \dot{\Phi}(s) \mathrm{d}s,$$

which gives the desired result by the well-known formula of integration by substitution. ∎

7.3. Suppose C is oriented. Then, according to the previous result, we introduce the notation

$$\int_x^y F(\cdot, \mathrm{d}C) := \int_{p^{-1}(x)}^{p^{-1}(y)} F(p(t), \dot{p}(t)) \mathrm{d}t$$

where p is an arbitrary positively oriented parametrization of C such that $x, y \in \mathrm{Ran}\, p$.

Note that according to the definition we have

$$\int_y^x F(\cdot, \mathrm{d}C) = -\int_x^y F(\cdot, \mathrm{d}C).$$

If C is not oriented, we shall use the symbol

$$\int_{[x,y]} F(\cdot, \mathrm{d}C) := \left| \int_{p^{-1}(x)}^{p^{-1}(y)} F(p(t), \dot{p}(t)) \mathrm{d}t \right|$$

where p is an arbitrary parametrization.

We frequently meet the particular case when F does not depend on the elements of V, i.e. there is a positively homogeneous $f\colon \mathbf{V} \to \mathbb{A}$ such that $F(x, \boldsymbol{x}) = f(\boldsymbol{x})$ for all $x \in \mathrm{V}$, $\boldsymbol{x} \in \mathbf{V}$. Then we use the symbol

$$\int_x^y f(\mathrm{d}\mathcal{C}) \qquad \text{and} \qquad \int_{[x,y]} f(\mathrm{d}\mathcal{C})$$

for the corresponding integrals.

7.4. We can generalize the previous result for a parametrization $r\colon \mathrm{T} \rightarrowtail \mathrm{V}$ where T is an oriented one-dimensional affine space over the vector space $\mathbb{T}$. Then $\dot{r}(t)$ is in $\frac{\mathbf{V}}{\mathbb{T}}$ and accepting the definition $F\left(x, \frac{\boldsymbol{x}}{\boldsymbol{t}}\right) := \frac{F(x,\boldsymbol{x})}{|\boldsymbol{t}|}$ $(x \in \mathrm{V}, \boldsymbol{x} \in \mathbf{V}, \mathbf{0} \neq \boldsymbol{t} \in \mathbb{T})$ we have

$$\int_x^y F(\cdot, \mathrm{d}\mathcal{C}) = \int_{r^{-1}(x)}^{r^{-1}(y)} F\left(r(t), \dot{r}(t)\right) \mathrm{d}t$$

if r is positively oriented.

7.5. Let $(\mathrm{V}, \mathbb{B}, \boldsymbol{b})$ be a pseudo-Euclidean affine space (i.e. V is an affine space over $\mathbf{V}$ and $(\mathbf{V}, \mathbb{B}, \boldsymbol{b})$ is a pseudo-Euclidean vector space). Supposing $\mathbb{B}$ is oriented, we have the square root mapping $(\mathbb{B} \otimes \mathbb{B})_0^+ \to \mathbb{B}_0^+$ and

$$\mathbf{V} \to \mathbb{B}, \qquad \boldsymbol{x} \mapsto |\boldsymbol{x}| := \sqrt{|\boldsymbol{x} \cdot \boldsymbol{x}|}$$

is a positively homogeneous function. Thus if $\mathcal{C}$ is an oriented curve in V, then

$$\int_x^y |\mathrm{d}\mathcal{C}|$$

is meaningful for all $x, y \in \mathcal{C}$. In the Euclidean case it is regarded as the signed *length* of the curve segment between x and y; in the nonEuclidean case it is interpreted as the *pseudo-length* of the curve segment.

Proposition. Suppose that $|\boldsymbol{x}| \neq 0$ for all nonzero tangent vectors $\boldsymbol{x}$ of $\mathcal{C}$. Then for all $x_o \in \mathcal{C}$,

$$\mathcal{C} \to \mathbb{B}, \qquad x \mapsto \int_{x_o}^x |\mathrm{d}\mathcal{C}|$$

is a continuous injection whose inverse is a positively oriented parametrization of $\mathcal{C}$.

Proof. Let Z denote the above mapping and choose a positively oriented parametrization $p\colon \mathbb{R} \rightarrowtail \mathrm{V}$ and put $t_o := p^{-1}(x_o)$. Then

$$(Z \circ p)(t) = \int_{t_o}^{t} |\dot{p}(s)| \mathrm{d}s \qquad (t \in \mathrm{Dom}\, p);$$

consequently, $Z \circ p\colon \mathbb{R} \rightarrowtail \mathbb{B}$ is continuously differentiable and $(Z \circ p)^{\cdot}(t) = |\dot{p}(t)| > 0$ for all $t \in \mathrm{Dom}\, p$. Thus $Z \circ p$ is strictly monotone increasing: it is injective and its inverse $(Z \circ p)^{-1}$ is continuously differentiable as well, and according to the well-known rule,

$$\left((Z \circ p)^{-1}\right)^{\cdot} = \frac{1}{(Z \circ p)^{\cdot}\left((Z \circ p)^{-1}\right)} > 0.$$

As a consequence, introducing the notation $r := Z^{-1}$, we have that $r = p \circ (Z \circ p)^{-1}$ is continuously differentiable, too, and

$$\dot{r} \circ r^{-1} = \frac{\dot{p}}{|\dot{p}|} \circ p^{-1}.$$

This means that r is a parametrization of $\mathcal{C}$ and $r^{-1} \circ p$ $(= Z \circ p)$ has everywhere positive derivative, i.e. r and p are equally oriented. ■

It is worth noting that $|\dot{r}| = 1$.

VII. LIE GROUPS

We treat only a special type of Lie groups appearing in physics; so we avoid the application of the theory of smooth manifolds.

1. Groups of linear bijections

1.1. Let $\mathbf{V}$ be an N-dimensional real vector space, $N \neq 0$.

Then $\mathrm{Lin}(\mathbf{V})$ is an N^2-dimensional real vector space.

Now the symbol of composition between elements of $\mathrm{Lin}(\mathbf{V})$ will be omitted, i.e. we write $\boldsymbol{AB} := \boldsymbol{A} \circ \boldsymbol{B}$ for $\boldsymbol{A}, \boldsymbol{B} \in \mathrm{Lin}(\mathbf{V})$.

Since $\mathbf{V}$ is finite dimensional, all norms on it are equivalent, i.e. all norms give the same open subsets. Given a norm $\|\ \|$ on $\mathbf{V}$, a norm is defined on $\mathrm{Lin}(\mathbf{V})$ by

$$\|\boldsymbol{A}\| := \sup_{\|\boldsymbol{v}\|=1} \|\boldsymbol{A} \cdot \boldsymbol{v}\|$$

for which $\|\boldsymbol{AB}\| \le \|\boldsymbol{A}\|\ \|\boldsymbol{B}\|$ holds $(\boldsymbol{A}, \boldsymbol{B} \in \mathrm{Lin}(\mathbf{V}))$.

We introduce the notation

$$\mathcal{GL}(\mathbf{V}) := \{\boldsymbol{F} \in \mathrm{Lin}(\mathbf{V}) |\ \boldsymbol{F} \text{ is bijective}\}.$$

Endowed with the multiplication $(\boldsymbol{F}, \boldsymbol{G}) \mapsto \boldsymbol{FG}$ (composition), $\mathcal{GL}(\mathbf{V})$ is a group whose identity (neutral element) is

$$\boldsymbol{I} := \mathbf{1}_{\mathbf{V}}.$$

1.2. One can prove without difficulty that if $\boldsymbol{A} \in \mathrm{Lin}(\mathbf{V})$, $\|\boldsymbol{A}\| < 1$, then $\boldsymbol{I} - \boldsymbol{A} \in \mathcal{GL}(\mathbf{V})$ and

$$(\boldsymbol{I} - \boldsymbol{A})^{-1} = \sum_{n=0}^{\infty} \boldsymbol{A}^n.$$

In other words, if $\boldsymbol{K} \in \operatorname{Lin}(\mathbf{V})$, $\|\boldsymbol{I}-\boldsymbol{K}\|<1$, then $\boldsymbol{K} \in \mathcal{GL}(\mathbf{V})$ and

$$\boldsymbol{K}^{-1}=\sum_{n=0}^{\infty}(\boldsymbol{I}-\boldsymbol{K})^{n}.$$

Proposition. Let $\boldsymbol{F} \in \mathcal{GL}(\mathbf{V})$. If $\boldsymbol{L} \in \operatorname{Lin}(\mathbf{V})$ and $\|\boldsymbol{F}-\boldsymbol{L}\|<\frac{1}{\|\boldsymbol{F}^{-1}\|}$ then $\boldsymbol{L} \in \mathcal{GL}(\mathbf{V})$.

Proof. $\|\boldsymbol{I}-\boldsymbol{F}^{-1}\boldsymbol{L}\|=\|\boldsymbol{F}^{-1}(\boldsymbol{F}-\boldsymbol{L})\| \leq \|\boldsymbol{F}^{-1}\|\ \|\boldsymbol{F}-\boldsymbol{L}\|<1$, thus $\boldsymbol{F}^{-1}\cdot\boldsymbol{L}$ is bijective. $\boldsymbol{F}$ is bijective by assumption, hence $\boldsymbol{F}(\boldsymbol{F}^{-1}\boldsymbol{L})=\boldsymbol{L}$ is bijective as well. ■

As a corollary of this result we have that $\mathcal{GL}(\mathbf{V})$ is an open subset of $\operatorname{Lin}(\mathbf{V})$.

1.3. The proof of the following statement is elementary. The mappings

$$\mathrm{m}\colon \mathcal{GL}(\mathbf{V}) \times \mathcal{GL}(\mathbf{V}) \to \mathcal{GL}(\mathbf{V}), \qquad (\boldsymbol{F},\boldsymbol{G}) \mapsto \boldsymbol{F}\boldsymbol{G},$$

$$\mathrm{j}\colon \mathcal{GL}(\mathbf{V}) \to \mathcal{GL}(\mathbf{V}), \qquad \boldsymbol{F} \mapsto \boldsymbol{F}^{-1}$$

are smooth and

$$\mathrm{Dm}(\boldsymbol{F},\boldsymbol{G})\colon \operatorname{Lin}(\mathbf{V}) \times \operatorname{Lin}(\mathbf{V}) \to \operatorname{Lin}(\mathbf{V}), \qquad (\boldsymbol{A},\boldsymbol{B}) \mapsto \boldsymbol{A}\boldsymbol{G}+\boldsymbol{F}\boldsymbol{B},$$

$$\mathrm{Dj}(\boldsymbol{F})\colon \operatorname{Lin}(\mathbf{V}) \mapsto \operatorname{Lin}(\mathbf{V}), \qquad \boldsymbol{A} \mapsto -\boldsymbol{F}^{-1}\boldsymbol{A}\boldsymbol{F}^{-1}.$$

1.4. It is a well-known fact, too, that for $\boldsymbol{A} \in \operatorname{Lin}(\mathbf{V})$

$$\exp \boldsymbol{A} := e^{\boldsymbol{A}} := \sum_{n=0}^{\infty} \frac{\boldsymbol{A}^{n}}{n!}$$

is meaningful, it is an element of $\mathcal{GL}(\mathbf{V})$ and

$$e^{\mathbf{0}}=\boldsymbol{I}, \qquad \left(e^{\boldsymbol{A}}\right)^{-1}=e^{-\boldsymbol{A}}.$$

Moreover, the *exponential mapping,*

$$\operatorname{Lin}(\mathbf{V}) \to \mathcal{GL}(\mathbf{V}), \qquad \boldsymbol{A} \mapsto e^{\boldsymbol{A}}$$

is smooth, its derivative at $\mathbf{0} \in \operatorname{Lin}(\mathbf{V})$ is the identity map $\operatorname{Lin}(\mathbf{V}) \to \operatorname{Lin}(\mathbf{V})$.

The inverse mapping theorem implies that the exponential mapping is injective in a neighbourhood of $\mathbf{0}$, its inverse regarding this neighbourhood is smooth as well.

If $\boldsymbol{A}, \boldsymbol{B} \in \operatorname{Lin}(\mathbf{V})$ and $\boldsymbol{AB} = \boldsymbol{BA}$ then $e^{\boldsymbol{A}}e^{\boldsymbol{B}} = e^{\boldsymbol{B}}e^{\boldsymbol{A}} = e^{\boldsymbol{A}+\boldsymbol{B}}$. In particular, $e^{t\boldsymbol{A}}e^{s\boldsymbol{A}} = e^{s\boldsymbol{A}}e^{t\boldsymbol{A}} = e^{(t+s)\boldsymbol{A}}$ for $t, s \in \mathbb{R}$.

1.5. For $\boldsymbol{A} \in \operatorname{Lin}(\mathbf{V})$, the function $\mathbb{R} \to \mathcal{GL}(\mathbf{V})$, $t \mapsto e^{t\boldsymbol{A}}$ is smooth and

$$\frac{\mathrm{d}}{\mathrm{d}t}\left(e^{t\boldsymbol{A}}\right) = \boldsymbol{A}e^{t\boldsymbol{A}} = e^{t\boldsymbol{A}}\boldsymbol{A}.$$

As a consequence, the initial value problem

$$(\boldsymbol{X}: \mathbb{R} \rightarrowtail \operatorname{Lin}(\mathbf{V}))? \qquad \dot{\boldsymbol{X}} = \boldsymbol{X}\boldsymbol{A}, \qquad \boldsymbol{X}(0) = \boldsymbol{I}$$

has the unique maximal solution

$$\boldsymbol{R}(t) = e^{t\boldsymbol{A}} \qquad (t \in \mathbb{R}).$$

2. Groups of affine bijections

2.1. Let V be an affine space over the N-dimensional real vector space $\mathbf{V}$. Then

$$\operatorname{Aff}(\mathrm{V}, \mathbf{V}) := \{A : \mathrm{V} \to \mathbf{V} \mid A \quad \text{is affine}\},$$

endowed with the pointwise operations, is a real vector space.

Given $o \in \mathrm{V}$, the correspondence

$$\operatorname{Aff}(\mathrm{V}, \mathbf{V}) \to \mathbf{V} \times \operatorname{Lin}(\mathbf{V}), \qquad A \mapsto (A(o), \boldsymbol{A})$$

(where $\boldsymbol{A}$ is the linear map under A) is a linear bijection; it is evidently linear and injective and it is surjective because the affine map $\mathrm{V} \to \mathbf{V}$, $x \mapsto \boldsymbol{A} \cdot (x - o) + \boldsymbol{a}$ corresponds to $(\boldsymbol{a}, \boldsymbol{A}) \in \mathbf{V} \times \operatorname{Lin}(\mathbf{V})$.

As a consequence, $\operatorname{Aff}(\mathrm{V}, \mathbf{V})$ is an $(N + N^2)$-dimensional vector space.

2.2. We easily find that

$$\operatorname{Aff}(\mathrm{V}) := \{L : \mathrm{V} \to \mathrm{V} \mid L \quad \text{is affine}\},$$

endowed with the pointwise subtraction (see VI.2.3(*iv*)), is an affine space over $\operatorname{Aff}(\mathrm{V}, \mathbf{V})$. Thus, according to the previous paragraph, $\operatorname{Aff}(\mathrm{V})$ is $(N + N^2)$-dimensional.

Two elements K and L of $\mathrm{Aff}(\mathrm{V})$, as well as an element A of $\mathrm{Aff}(\mathrm{V}, \mathbf{V})$ and an element L of $\mathrm{Aff}(\mathrm{V})$ can be composed; the symbol of compositions will be omitted, i.e. $KL := K \circ L$ and $AL := A \circ L$.

We introduce

$$\mathcal{GA}(\mathrm{V}) := \{F \in \mathrm{Aff}(\mathrm{V}) \mid F \quad \text{is bijective}\}.$$

Endowed with the multiplication $(F, G) \mapsto FG$ (composition), $\mathcal{GA}(\mathrm{V})$ is a group whose identity (neutral element) is

$$I := 1_{\mathrm{V}}.$$

2.3. Given $o \in \mathrm{V}$, the mapping

$$\mathrm{Aff}(\mathrm{V}) \to \mathbf{V} \times \mathrm{Lin}(\mathbf{V}), \qquad L \mapsto (L(o) - o, \boldsymbol{L})$$

is an affine bijection over the linear bijection given in 2.1. Evidently, this bijection maps $\mathcal{GA}(\mathrm{V})$ onto $\mathbf{V} \times \mathcal{GL}(\mathbf{V})$. As a consequence, $\mathcal{GA}(\mathrm{V})$ is an open subset of $\mathrm{Aff}(\mathrm{V})$.

2.4. The mappings

$$\mathrm{m}\colon \mathcal{GA}(\mathrm{V}) \times \mathcal{GA}(\mathrm{V}) \to \mathcal{GA}(\mathrm{V}), \qquad (F, G) \mapsto FG,$$

$$\mathrm{j}\colon \mathcal{GA}(\mathrm{V}) \to \mathcal{GA}(\mathrm{V}), \qquad F \mapsto F^{-1}$$

are smooth and

$$\mathrm{Dm}(F, G)\colon \mathrm{Aff}(\mathrm{V}, \mathbf{V}) \times \mathrm{Aff}(\mathrm{V}, \mathbf{V}) \to \mathrm{Aff}(\mathrm{V}, \mathbf{V}), \qquad (A, B) \mapsto \boldsymbol{A}G + \boldsymbol{F}B,$$

$$\mathrm{Dj}(F)\colon \mathrm{Aff}(\mathrm{V}, \mathbf{V}) \to \mathrm{Aff}(\mathrm{V}, \mathbf{V}), \qquad A \mapsto -\boldsymbol{F}^{-1}AF^{-1}.$$

2.5. If $\boldsymbol{P} \in \mathcal{GL}(\mathbf{V})$ then

$$\ell_{\boldsymbol{P}}\colon \mathrm{Aff}(\mathrm{V}, \mathbf{V}) \to \mathrm{Aff}(\mathrm{V}, \mathbf{V}), \qquad A \to \boldsymbol{P}A$$

is a linear bijection, $(\ell_{\boldsymbol{P}})^{-1} = \ell_{\boldsymbol{P}^{-1}}$.

If $P \in \mathcal{GA}(\mathrm{V})$ then

$$\ell_P\colon \mathrm{Aff}(\mathrm{V}) \to \mathrm{Aff}(\mathrm{V}), \qquad L \mapsto PL$$

is an affine bijection over $\ell_{\boldsymbol{P}}$, where $\boldsymbol{P}$ is the linear map under P; moreover, $(\ell_P)^{-1} = \ell_{P^{-1}}$.

2.6. If $A \in \mathrm{Aff}(\mathrm{V}, \mathbf{V})$ and $\boldsymbol{A} \in \mathrm{Lin}(\mathbf{V})$ is the linear map under A then

$$\exp A := e^A := I + \sum_{n=1}^{\infty} \frac{\boldsymbol{A}^{n-1} A}{n!}$$

is meaningful, it is an element of $\mathcal{GA}(\mathrm{V})$,

$$e^0 = I, \qquad \left(e^A\right)^{-1} = e^{-A}$$

and the linear map under e^A is $e^{\boldsymbol{A}}$.

Moreover, the *exponential mapping*

$$\mathrm{Aff}(\mathrm{V}, \mathbf{V}) \to \mathcal{GA}(\mathrm{V}), \qquad A \mapsto e^A$$

is smooth, its derivative at $0 \in \mathrm{Aff}(\mathrm{V}, \mathbf{V})$ is the identity map $\mathrm{Aff}(\mathrm{V}, \mathbf{V}) \to \mathrm{Aff}(\mathrm{V}, \mathbf{V})$.

The inverse mapping theorem implies that the exponential mapping is injective in a neighbourhood of 0, its inverse regarding this neighbourhood is smooth as well.

If $A, B \in \mathrm{Aff}(\mathrm{V}, \mathbf{V})$ and $\boldsymbol{A}\boldsymbol{B} = \boldsymbol{B}\boldsymbol{A}$ then $e^A e^B = e^B e^A = e^{A+B}$. In particular, $e^{tA} e^{sA} = e^{sA} e^{tA} = e^{(t+s)A}$ for $t, s \in \mathbb{R}$.

2.7. For $A \in \mathrm{Aff}(\mathrm{V}, \mathbf{V})$, the function $\mathbb{R} \to \mathcal{GA}(\mathrm{V})$, $t \mapsto e^{tA}$ is smooth and

$$\frac{\mathrm{d}}{\mathrm{d}t}\left(e^{tA}\right) = e^{t\boldsymbol{A}} A = A e^{tA}.$$

As a consequence, the initial value problem

$$(X : \mathbb{R} \rightarrowtail \mathrm{Aff}(\mathrm{V}, \mathbf{V}))? \qquad \dot{X} = AX, \qquad X(0) = I$$

has the unique maximal solution

$$R(t) = e^{tA} \qquad (t \in \mathbb{R}).$$

3. Lie groups

3.1. Definition Let V be an N-dimensional real affine space. A subgroup $\mathcal{G}$ of $\mathcal{GA}(\mathrm{V})$ which is an M-dimensional smooth submanifold of $\mathcal{GA}(\mathrm{V})$ is called an *M-dimensional plain Lie group.* ■

The group multiplication $\mathcal{G}\times\mathcal{G}\to\mathcal{G}$, $(F,G)\mapsto FG$ and the inversion $\mathcal{G}\to\mathcal{G}$, $F\mapsto F^{-1}$ are smooth mappings (see 2.4 and VI.4.13, Exercise VI.4.14.5(*iii*)).

Observe that by definition $0 < M \le N + N^2$. $(N+N^2)$-dimensional plain Lie groups are $\mathcal{GA}(\mathrm{V})$ and its open subgroups.

Remark. In general, a Lie group is defined to be a group endowed with a smooth structure in such a way that the group multiplication and the inversion are smooth mappings.

Since we shall deal only with plain Lie groups, we shall omit the adjective 'plain'. By the way, all the results we shall derive for plain Lie groups are valid for arbitrary Lie groups as well.

3.2. (*i*) For $\boldsymbol{a}\in\mathbf{V}$ we defined the affine bijection $T_{\boldsymbol{a}}\colon \mathrm{V}\to\mathrm{V}$, $x\mapsto x+\boldsymbol{a}$ (VI.2.4.3), the *translation by* $\boldsymbol{a}$. It is quite evident that $T_{\boldsymbol{a}} = T_{\boldsymbol{b}}$ if and only if $\boldsymbol{a}=\boldsymbol{b}$ and so

$$\mathcal{T}n(\mathbf{V}) := \{T_{\boldsymbol{a}} \mid \boldsymbol{a}\in\mathbf{V}\},$$

called the *translation group* of V, is an N-dimensional Lie group. The group multiplication in $\mathcal{T}n(\mathbf{V})$ corresponds exactly to the addition in $\mathbf{V}$ that is why one often says that $\mathbf{V}$—in particular $\mathbb{R}^N$—endowed with the addition as a group multiplication is an N-dimensional Lie group.

(*ii*) If the vector space $\mathbf{V}$ is considered to be an affine space then $\mathcal{GL}(\mathbf{V})$ is a subgroup and an N^2-dimensional submanifold of $\mathcal{GA}(\mathbf{V})$, thus $\mathcal{GL}(\mathbf{V})$ is an N^2-dimensional Lie group.

3.3. It is obvious that

$$\mathcal{GA}(\mathrm{V})\to\mathcal{GL}(\mathbf{V}),\qquad L\mapsto \boldsymbol{L}\qquad (\boldsymbol{L}\ \text{ is the linear map under } L)$$

is a smooth group homomorphism whose kernel is $\mathcal{T}n(\mathbf{V})$ ($\boldsymbol{L}=\boldsymbol{I}$ if and only if $L\in\mathcal{T}n(\mathbf{V})$, see VI.2.5.6).

(*i*) Take a Lie group $\mathcal{G}\subset\mathcal{GA}(\mathrm{V})$. Then

$$\mathrm{under}(\mathcal{G}) := \{\boldsymbol{F}\in\mathcal{GL}(\mathbf{V}) \mid \boldsymbol{F}\ \text{ is under an } F\in\mathcal{G}\},$$

i.e. the image of $\mathcal{G}$ by the above group homomorphism is a Lie group.

(*ii*) Conversely, if $\boldsymbol{\mathcal{G}}\subset\mathcal{GL}(\mathbf{V})$ is an M-dimensional Lie group, then

$$\mathrm{over}(\boldsymbol{\mathcal{G}}) := \{F\in\mathcal{GA}(\mathrm{V}) \mid F\ \text{ is over an } \boldsymbol{F}\in\boldsymbol{\mathcal{G}}\},$$

the pre-image of $\boldsymbol{\mathcal{G}}$ by the above group homomorphism, is an $(M+N)$-dimensional Lie group.

3.4. Recall that the tangent spaces of $\mathcal{G}$ are linear subspaces of $\mathrm{Aff}(\mathrm{V},\mathbf{V})$. Every tangent space of $\mathcal{G}$ is obtained quite simply from the tangent space at I: $\mathbf{T}_F(\mathcal{G})$ is the 'translation' by F of $\mathbf{T}_I(\mathcal{G})$.

Proposition. Let $\mathcal{G} \subset \mathcal{GA}(\mathrm{V})$ be a Lie group. Then

$$\mathbf{T}_F(\mathcal{G}) = \boldsymbol{F}[\mathbf{T}_I(\mathcal{G})] = \{\boldsymbol{F}A \mid A \in \mathbf{T}_I(\mathcal{G})\} \qquad (F \in \mathcal{G}).$$

Proof. Let $\mathcal{G}$ be M-dimensional. There is a neighbourhood $\mathcal{N}$ of I in $\mathcal{GA}(\mathrm{V})$, a smooth mapping $S\colon \mathcal{N} \to \mathbb{R}^{N+N^2-M}$ such that $\mathcal{G} \cap \mathcal{N} = \overset{-1}{S}(\{0\})$ (see VI.4.4), and $\mathbf{T}_I(\mathcal{G}) = \operatorname{Ker} \mathrm{D}S(I)$ (see VI.4.8).

Let F be an arbitrary element of $\mathcal{G}$. Then $\mathcal{G}$ is invariant under the affine bijection $\ell_{F^{-1}} = \ell_F{}^{-1}$, thus $S \circ \ell_F{}^{-1}|_{\mathcal{G}} = 0$. Consequently, if P is in the domain of $S \circ \ell_F{}^{-1}$, i.e. $\ell_F{}^{-1}P = F^{-1}P$ is in $\mathcal{N}$, recalling that $\ell_F{}^{-1}$ is an affine map over $\boldsymbol{\ell_F}{}^{-1}$, hence $\mathrm{D}\ell_F{}^{-1}(P) = \boldsymbol{\ell_F}{}^{-1}$, we have

$$\begin{aligned}\mathbf{T}_P(\mathcal{G}) &= \operatorname{Ker} \mathrm{D}(S \circ \ell_F{}^{-1})(P) = \operatorname{Ker}\left(\mathrm{D}S(\ell_F{}^{-1}P) \cdot \mathrm{D}\ell_F{}^{-1}(P)\right) = \\ &= \operatorname{Ker}\left(\mathrm{D}S(F^{-1}P)\boldsymbol{\ell_F}{}^{-1}\right) = \{A \in \operatorname{Aff}(\mathrm{V}, \mathbf{V}) \mid \mathrm{D}S(F^{-1}P)\boldsymbol{F}^{-1}A = \mathbf{0}\} = \\ &= \left\{\boldsymbol{F}B \mid B \in \operatorname{Ker} \mathrm{D}S(F^{-1}P)\right\} = \boldsymbol{F}\left(\operatorname{Ker} \mathrm{D}S(F^{-1}P)\right).\end{aligned}$$

We can take $P := F$ to have the desired result. ∎

The tangent space of $\mathcal{G}$ at I plays an important role; for convenience we introduce the notation

$$\mathbf{La}(\mathcal{G}) := \mathbf{T}_I(\mathcal{G}).$$

Note that $\mathbf{La}(\mathcal{GA}(\mathrm{V})) = \operatorname{Aff}(\mathrm{V}, \mathbf{V})$, $\mathbf{La}(\mathcal{GL}(\mathbf{V}) = \operatorname{Lin}(\mathbf{V})$.

Moreover, $\mathbf{La}(\mathcal{T}n(\mathbf{V})) = \mathbf{V}$ where $\mathbf{V}$ is identified with the constant maps $\mathrm{V} \to \mathbf{V}$.

3.5. Definition. A smooth function $R\colon \mathbb{R} \to \mathcal{G} \subset \mathcal{GA}(\mathrm{V})$ is called a *one-parameter subgroup* in the Lie group $\mathcal{G}$ if

$$R(t+s) = R(t)R(s) \qquad (t, s \in \mathbb{R}).$$ ∎

In other words, a one-parameter subgroup is a smooth group homomorphism $R\colon \mathcal{T}n(\mathbb{R}) \to \mathcal{G}$. Evidently, $R(0) = I$ and $R(-t) = R(t)^{-1}$.

There are three possibilities.

(*i*) There is a neighbourhood of $0 \in \mathbb{R}$ such that $R(t) = I$ for all t in that neighbourhood; then R is a constant function, $R(t) = I$ for all $t \in \mathbb{R}$.

(*ii*) There is a $T \in \mathbb{R}^+$ such that $R(T) = I$ but $R(t) \neq I$ for $0 < t < T$; then R is periodic, $R(t+T) = R(t)$ for all $t \in \mathbb{R}$.

(*iii*) $R(t) \neq I$ for all $0 \neq t \in \mathbb{R}$.

3.6. If $\boldsymbol{R}(t)$ denotes the linear map under $R(t)$ then $\boldsymbol{R}\colon \mathbb{R} \to \operatorname{under}(\mathcal{G})$ is a one-parameter subgroup; $\boldsymbol{R}(0) = \boldsymbol{I}$.

Differentiating with respect to s in the defining equality of R and then putting $s = 0$ we get

$$\dot{R}(t) = \mathbf{R}(t)\dot{R}(0) = \dot{R}(0)R(t) \qquad (t \in \mathbb{R})$$

which shows that if $\operatorname{Ran} R$ is not a single point (if R is not constant) then it is a one-dimensional submanifold and a subgroup in $\mathcal{GA}(V)$. Thus $\operatorname{Ran} R$ is either the singleton $\{I\}$ or a one-dimensional Lie group. In the case (*ii*) treated in the preceding paragraph, the restriction of R to an interval shorter than T is a local parametrization of $\operatorname{Ran} R$; in the case (*iii*) R is a parametrization of $\operatorname{Ran} R$.

3.7. Proposition. Every one-parameter subgroup R in $\mathcal{G}$ has the form

$$R(t) = e^{tA} \qquad (t \in \mathbb{R})$$

where $A = \dot{R}(0) \in \mathbf{La}(\mathcal{G})$.

Conversely, if $A \in \mathbf{La}(\mathcal{G}) \subset \operatorname{Aff}(\mathrm{V}, \mathbf{V})$ then $t \mapsto e^{tA}$ is a one-parameter subgroup in $\mathcal{G}$.

Proof. According to the previous paragraph, the one-parameter subgroup R is the solution of the initial value problem

$$(X : \mathbb{R} \rightarrowtail \mathcal{G}a(\mathrm{V}))? \qquad \dot{X} = \mathbf{X}A, \qquad X(0) = I$$

where $A := \dot{R}(0)$. Apply 2.7 to obtain the first statement.

Conversely, $t \mapsto e^{tA}$ is a one-parameter subgroup in $\mathcal{GA}(\mathrm{V})$; we have to show only that $e^{tA} \in \mathcal{G}$ for all $t \in \mathbb{R}$ which follows from VI.6.4. ∎

The assertions are true for *local one-parameter subgroups* as well, i.e. for smooth functions $R : \mathbb{R} \to \mathcal{G}$ defined on an interval around $0 \in \mathbb{R}$ such that $R(t + s) = R(t)R(s)$ whenever $t, s, t + s$ are in $\operatorname{Dom} R$.

3.8. The previous result involves that $e^A \in \mathcal{G}$ for $A \in \mathbf{La}(\mathcal{G})$, i.e. the restriction of the exponential mapping onto $\mathbf{La}(\mathcal{G})$ takes values in $\mathcal{G}$. Since the exponential mapping is smooth and injective in a neighbourhood of 0, its inverse regarding this neighbourhood is smooth as well (in particular continuous), we can state:

Proposition. Let $\mathcal{G}$ be a Lie group. Then

$$\mathbf{La}(\mathcal{G}) \to \mathcal{G}, \qquad A \mapsto e^A$$

is a parametrization of $\mathcal{G}$ in a neighbourhood of the identity I.

In particular, every element in a neighbourhood of I belongs to a one-parameter subgroup.

3.9. Proposition. Every element of $\mathcal{G}$ in a neighbourhood of the identity is a product of elements taken from one-parameter subgroups corresponding to a basis of $\mathbf{La}(\mathcal{G})$.

Proof. Let $A_1, \dots A_M$ be a basis of $\mathbf{La}(\mathcal{G})$ and complete it to a basis $A_1, \dots, A_P$ of $\mathrm{Aff}(\mathrm{V}, \mathbf{V})$ where $P := N + N^2$. Then

$$\Phi : \mathbb{R}^P \to \mathcal{GA}(\mathrm{V}), \qquad (t_1, t_2, \dots, t_P) \mapsto \exp(t_1 A_1) \exp(t_2 A_2) \dots \exp(t_P A_P)$$

is a smooth map, $\Phi(0, 0, \dots, 0) = I$, $\partial_k \Phi(0, 0, \dots, 0) = A_k$ $(k = 1, \dots, P)$. We can state on the basis of the inverse mapping theorem that Φ is injective in a neighbourhood of $(0, 0, \dots, 0)$, its inverse regarding this neighbourhood is smooth as well.

Thus the restriction of Φ onto $\mathbb{R}^M$ regarded as the subspace of $\mathbb{R}^P$ consisting of elements whose i-th components are zero for $i = M + 1, \dots, P$ is a parametrization of $\mathcal{G}$ in a neighbourhood of I. ■

Note that in general

$$\exp(t_1 A_1) \exp(t_2 A_2) \dots \exp(t_P A_P) \neq \exp\left(\sum_{k=1}^{P} t_k A_k \right).$$

3.10. If $\mathcal{G}$ is connected, every element of $\mathcal{G}$ is a product of elements in a neighbourhood of I, hence every element is a product of elements taken from one-parameter subgroups corresponding to a basis of $\mathbf{La}(\mathcal{G})$, since the following proposition is true.

Proposition. If $\mathcal{G}$ is connected and $\mathcal{V}$ is a neighbourhood of the identity I in $\mathcal{G}$, then

$$\mathcal{G} = \bigcup_{n \in \mathbb{N}} \mathcal{V}^n$$

where $\mathcal{V}^n := \{F_1 F_2 \dots F_n \mid F_k \in \mathcal{V},\ k = 1, \dots, n\}$.

Proof. Given $F \in \mathcal{G}$, the mapping $\mathcal{G} \to \mathcal{G}$, $G \mapsto FG$ is bijective, continuous, its inverse is continuous as well. Thus for all $F \in \mathcal{G}$, $F\mathcal{V} := \{FG \mid G \in \mathcal{V}\}$ is open, so $\mathcal{V}^2 = \bigcup_{F \in \mathcal{V}} F\mathcal{V}$ is open as well. Consequently, $\mathcal{V}^n$ is open for all n and thus $\mathcal{H} := \bigcup_{n \in \mathcal{N}} \mathcal{V}^n$ is open, too. We shall show that the closure of $\mathcal{H}$ in $\mathcal{G}$ equals $\mathcal{H}$; thus $\mathcal{H}$, being open and closed, equals $\mathcal{G}$.

Let L be an element of the closure of $\mathcal{H}$ in $\mathcal{G}$. Since $L\mathcal{V}^{-1}$ is a neighbourhood of L, there is an $F \in \mathcal{H}$ such that $F \in L\mathcal{V}^{-1}$ which implies $L \in F\mathcal{V}$; since $F\mathcal{V} \subset \mathcal{H}\mathcal{V} = \mathcal{H}$, the proof is complete.

4. The Lie algebra of a Lie group

4.1. Recall that if $\mathcal{G}$ is a Lie group in $\mathcal{GA}(\mathrm{V})$ then $\mathbf{La}(\mathcal{G})$, the tangent space of $\mathcal{G}$ at $I = \mathbf{1}_{\mathrm{V}}$ is a linear subspace of $\mathrm{Aff}(\mathrm{V}, \mathbf{V})$. If $A \in \mathrm{Aff}(\mathrm{V}, \mathbf{V})$ then $\boldsymbol{A}$ denotes the underlying linear map $\mathbf{V} \to \mathbf{V}$.

Proposition. Let $\mathcal{G}$ be a Lie group. If $A, B \in \mathbf{La}(\mathcal{G})$ then

$$\boldsymbol{A}B - \boldsymbol{B}A \in \mathbf{La}(\mathcal{G}).$$

Proof. Take a neighbourhood $\mathcal{N}$ of I in $\mathcal{GA}(\mathrm{V})$ and a smooth map S defined on $\mathcal{N}$ such that $\overset{-1}{S}(\{0\}) = \mathcal{G} \cap \mathcal{N}$ and $\mathbf{La}(\mathcal{G}) = \mathrm{Ker}\,\mathrm{D}S(I)$ (see the proof of 3.3).

Then

$$t \mapsto S\left(e^{tA}e^{tB}\right) = 0 \qquad \text{and} \qquad t \mapsto S\left(e^{tB}e^{tA}\right) = 0$$

for t in a neighbourhood of $0 \in \mathbb{R}$. Differentiating the first function with respect to t we get

$$t \mapsto \mathrm{D}S\left(e^{tA}e^{tB}\right) \cdot \left(e^{t\boldsymbol{A}}Ae^{tB} + e^{t\boldsymbol{A}}e^{t\boldsymbol{B}}B\right) = 0.$$

Again differentiating and then taking $t = 0$ we deduce

$$\mathrm{D}^2S(I)\,(A + B, A + B) + \mathrm{D}S(I) \cdot (\boldsymbol{A}A + 2\boldsymbol{A}B + \boldsymbol{B}B) = 0.$$

Similarly we derive from the second function that

$$\mathrm{D}^2S(I)\,(B + A, B + A) + \mathrm{D}S(I) \cdot (\boldsymbol{B}B + 2\boldsymbol{B}A + \boldsymbol{A}A) = 0.$$

Let us subtract the equalities from each other to have

$$\mathrm{D}S(I) \cdot (\boldsymbol{A}B - \boldsymbol{B}A) = 0$$

which ends the proof.

4.2. According to the previous proposition we are given the *commutator mapping*

$$\mathbf{La}(\mathcal{G}) \times \mathbf{La}(\mathcal{G}) \to \mathbf{La}(\mathcal{G}), \qquad (A, B) \mapsto \boldsymbol{A}B - \boldsymbol{B}A =: [A, B].$$

Proposition. The commutator mapping
(*i*) is bilinear,
(*ii*) is antisymmetric,
(*iii*) satisfies the *Jacobian identity:*

$$[[A, B], C] + [[B, C], A] + [[C, A], B] = 0 \qquad (A, B, C \in \mathbf{La}(\mathcal{G})).$$

Definition. $\mathbf{La}(\mathcal{G})$ endowed with the commutator mapping is called the *Lie algebra of* $\mathcal{G}$. ■

We deduce without difficulty that for $A, B \in \mathbf{La}(\mathcal{G})$

$$[A, B] = \frac{1}{2}\left(\frac{\mathrm{d}^2}{\mathrm{d}t^2}\left(e^{tA}e^{tB}e^{-tA}e^{-tB}\right)\right)_{t=0}.$$

4.3. The Lie algebra of $\mathcal{GA}(\mathrm{V})$ is $\mathrm{Aff}(\mathrm{V}, \mathbf{V})$. We have seen that if a linear subspace $\mathbf{L}$ of $\mathrm{Aff}(\mathrm{V}, \mathbf{V})$ is the tangent space at I of a Lie group then the commutator of elements from $\mathbf{L}$ belongs to $\mathbf{L}$, too; in other words, $\mathbf{L}$ is a Lie subalgebra of $\mathrm{Aff}(\mathrm{V}, \mathbf{V})$.

Conversely, if $\mathbf{L}$ is a Lie subalgebra of $\mathrm{Aff}(\mathrm{V}, \mathbf{V})$ then there is a Lie group $\mathcal{G}$ such that $\mathbf{La}(\mathcal{G}) = \mathbf{L}$: the subgroup generated by $\{e^A \mid A \in \mathbf{L}\}$. It is not so easy to verify that this subgroup is a submanifold.

4.4. Definition. Let $\mathcal{G}$ and $\mathcal{H}$ be Lie groups. A mapping $\Phi\colon \mathcal{G} \rightarrowtail \mathcal{H}$ is called a local Lie group homomorphism if

(*i*) $\mathrm{Dom}\,\Phi$ is a neighbourhood of the identity of $\mathcal{G}$,

(*ii*) Φ is smooth,

(*iii*) $\Phi(FG) = \Phi(F)\Phi(G)$ whenever $F, G, FG \in \mathrm{Dom}\,\Phi$.

If Φ is injective and Φ^{-1} is smooth as well, then Φ is a *local Lie group isomorphism.* ■

4.5. For a local Lie group homomorphism $\Phi\colon \mathcal{G} \rightarrowtail \mathcal{H}$ we put

$$\boldsymbol{\Phi} := \mathrm{D}\Phi(I) \in \mathrm{Lin}\left(\mathbf{La}(\mathcal{G}), \mathbf{La}(\mathcal{H})\right).$$

If $A \in \mathbf{La}(\mathcal{G})$, then $t \mapsto \Phi(e^{tA})$ is a local one-parameter subgroup in $\mathcal{H}$ and

$$\left(\frac{\mathrm{d}}{\mathrm{d}t}\Phi(e^{tA})\right)_{t=0} = \boldsymbol{\Phi}(A),$$

which implies

$$\Phi(e^{tA}) = e^{t\boldsymbol{\Phi}(A)}$$

for t in a neighbourhood of $0 \in \mathbb{R}$.

Proposition. $\boldsymbol{\Phi}\colon \mathbf{La}(\mathcal{G}) \to \mathbf{La}(\mathcal{H})$ is a Lie algebra homomorphism, i.e. it is linear and

$$[\boldsymbol{\Phi}(A), \boldsymbol{\Phi}(B)] = \boldsymbol{\Phi}\left([A, B]\right) \qquad (A, B \in \mathbf{La}(\mathcal{G})).$$

Proof. Start with

$$[\mathbf{\Phi}(A), \mathbf{\Phi}(B)] = \left(\frac{\mathrm{d}^2}{\mathrm{d}t^2}\left(e^{t\mathbf{\Phi}(A)}e^{t\mathbf{\Phi}(B)} - e^{t\mathbf{\Phi}(B)}e^{t\mathbf{\Phi}(A)}\right)\right)_{t=0} =$$
$$= \left(\frac{\mathrm{d}^2}{\mathrm{d}t^2}\left(\Phi(e^{tA}e^{tB}) - \Phi(e^{tB}e^{tA})\right)\right)_{t=0}$$

and then apply the formulae in the proof of 4.1 putting Φ in place of S.

4.6. The previous proposition involves that locally isomorphic Lie groups have isomorphic Lie algebras. One can prove the converse, too, a fundamental theorem of the theory of Lie groups: if the Lie algebras of two Lie groups are isomorphic then the Lie groups are locally isomorphic.

5. Pseudo-orthogonal groups

Let $(\mathbf{V}, \mathbb{B}, \boldsymbol{b})$ be a pseudo-Euclidean vector space. Recall the notations (see V.2.7)

$$\mathcal{O}(\boldsymbol{b}) := \left\{\boldsymbol{L} \in \mathcal{GL}(\mathbf{V}) \;\middle|\; \boldsymbol{L}^* \cdot \boldsymbol{L} = \boldsymbol{I}\right\},$$
$$\mathrm{A}(\boldsymbol{b}) := \left\{\boldsymbol{A} \in \mathrm{Lin}(\mathbf{V}) \;\middle|\; \boldsymbol{A}^* = -\boldsymbol{A}\right\}.$$

Proposition. If $\dim \mathbf{V} = N$ then $\mathcal{O}(\boldsymbol{b})$ is an $\frac{N(N-1)}{2}$-dimensional Lie group having $\mathrm{A}(\boldsymbol{b})$ as its Lie algebra.

Proof. It is evident that $\mathcal{O}(\boldsymbol{b})$ is a subgroup of $\mathcal{GL}(\mathbf{V})$.

We know that $\mathrm{A}(\boldsymbol{b})$ and $S(\boldsymbol{b}) := \{\boldsymbol{S} \in \mathrm{Lin}(\mathbf{V}) | \; \boldsymbol{S}^* = \boldsymbol{S}\}$ are complementary subspaces, $\dim S(\boldsymbol{b}) = \frac{N(N+1)}{2}$, $\dim \mathrm{A}(\boldsymbol{b}) = \frac{N(N-1)}{2}$ (see V.2.9).

Let us consider the mapping

$$\varphi\colon \mathcal{GL}(\mathbf{V}) \to S(\boldsymbol{b}), \qquad \boldsymbol{L} \mapsto \boldsymbol{L}^* \cdot \boldsymbol{L}.$$

Since the $\boldsymbol{b}$-adjunction $\boldsymbol{L} \mapsto \boldsymbol{L}^*$ is linear and the multiplication in $\mathrm{Lin}(\mathbf{V})$ is bilinear, φ is smooth. Moreover, the equality

$$(\boldsymbol{L} + \boldsymbol{H})^* \cdot (\boldsymbol{L} + \boldsymbol{H}) - \boldsymbol{L}^* \cdot \boldsymbol{L} = \boldsymbol{L}^* \cdot \boldsymbol{H} + \boldsymbol{H}^* \cdot \boldsymbol{L} + \boldsymbol{H}^* \cdot \boldsymbol{H}$$

shows that

$$\mathrm{D}\varphi(\boldsymbol{L}) \cdot \boldsymbol{H} = \boldsymbol{L}^* \cdot \boldsymbol{H} + \boldsymbol{H}^* \cdot \boldsymbol{L} \qquad (\boldsymbol{L} \in \mathcal{GL}(\mathbf{V}),\ \boldsymbol{H} \in \mathrm{Lin}(\mathbf{V})).$$

We have $\mathcal{O}(\boldsymbol{b}) = \{\boldsymbol{L} \in \mathcal{GL}(\mathbf{V}) \mid \varphi(\boldsymbol{L}) = \boldsymbol{I}\}$ and $\mathrm{D}\varphi(\boldsymbol{L})$ is surjective if $\boldsymbol{L}$ is in $\mathcal{O}(\boldsymbol{b})$: if $\boldsymbol{S} \in S(\boldsymbol{b})$ then $\mathrm{D}\varphi(\boldsymbol{L}) \cdot \frac{\boldsymbol{L} \cdot \boldsymbol{S}}{2} = \boldsymbol{S}$. Consequently, $\mathcal{O}(\boldsymbol{b})$ is a smooth submanifold of $\mathcal{GL}(\mathbf{V})$ (see VI.4.7).

Finally, $\mathrm{D}\varphi(\boldsymbol{I})\cdot\boldsymbol{H}=\boldsymbol{0}$ if and only if $\boldsymbol{H}\in\mathrm{A}(\boldsymbol{b})$, hence

$$\mathbf{La}(\mathcal{O}(\boldsymbol{b}))=\operatorname{Ker}\mathrm{D}\varphi(\boldsymbol{I})=\mathrm{A}(\boldsymbol{b}).$$

6. Exercises

1. Let $\mathcal{G}$ be a Lie group, $A,B\in\mathbf{La}(\mathcal{G})$. Prove that $[A,B]=0$ if and only if $e^{tA}e^{tB}=e^{tB}e^{tA}$ for all t in an interval around $0\in\mathbb{R}$.

Consequently, $\mathcal{G}$ is commutative (Abelian) if and only if $\mathbf{La}(\mathcal{G})$ is commutative (the commutator mapping on $\mathbf{La}(\mathcal{G})$ is zero).

2. Using the definition of exponentials (see 2.6) demonstrate that

$$[A,B]=\lim_{t\to 0}\frac{1}{t^2}\left(e^{tA}e^{tB}-e^{tB}e^{tA}\right)=\lim_{t\to 0}\frac{1}{t^2}\left(e^{tA}e^{tB}e^{-tA}e^{-tB}-I\right).$$

3. Let $\mathbf{V}$ be a finite dimensional real vector space and make the identification

$$\mathrm{Aff}(\mathbf{V})\equiv\mathbf{V}\times\mathrm{Lin}(\mathbf{V}),\qquad A\equiv(A(0),\boldsymbol{A}),$$

i.e. $(\boldsymbol{a},\boldsymbol{A})\in\mathbf{V}\times\mathrm{Lin}(\mathbf{V})$ is considered to be the affine map

$$\mathbf{V}\to\mathbf{V},\qquad \boldsymbol{x}\mapsto\boldsymbol{A}\boldsymbol{x}+\boldsymbol{a}.$$

Then the composition of such affine maps is

$$(\boldsymbol{a},\boldsymbol{A})(\boldsymbol{b},\boldsymbol{B})=(\boldsymbol{a}+\boldsymbol{A}\boldsymbol{b},\boldsymbol{A}\boldsymbol{B}).$$

In this way we have $\mathcal{GA}(\mathbf{V})\equiv\mathbf{V}\times\mathcal{GL}(\mathbf{V})$.

Prove that

$$e^{(\boldsymbol{a},\boldsymbol{0})}=(\boldsymbol{a},\boldsymbol{I}),\qquad e^{(\boldsymbol{0},\boldsymbol{A})}=(\boldsymbol{0},e^{\boldsymbol{A}}).$$

4. Let n be a positive integer. Prove that

$$\mathcal{O}(n):=\{\boldsymbol{L}\in\mathrm{Lin}(\mathbb{R}^n)\mid\ \boldsymbol{L}^*\boldsymbol{L}=\boldsymbol{I}\}$$
$$\mathcal{SO}(n):=\{\boldsymbol{L}\in\mathcal{O}(n)\mid\ \det\boldsymbol{L}=1\}$$

are $\frac{n(n-1)}{2}$-dimensional Lie groups having the same Lie algebra:

$$\{\boldsymbol{A}\in\mathrm{Lin}(\mathbb{R}^n)\mid\ \boldsymbol{A}^*=-\boldsymbol{A}\}$$

(*cf.* Proposition in Section 5).

Give a local Lie group isomorphism between $\mathcal{O}(n)$ and $\mathcal{SO}(n)$.

5. A complex vector space and its complex linear maps can be considered to be a real vector space and real linear maps.

Demonstrate that

$$\mathcal{SL}(2,\mathbb{C}) := \{\boldsymbol{L} \in \mathrm{Lin}(\mathbb{C}^2) \mid \det \boldsymbol{L} = 1\}$$

is a six-dimensional Lie group having

$$\{\boldsymbol{A} \in \mathrm{Lin}(\mathbb{C}^2) \mid \mathrm{Tr}\,\boldsymbol{A} = \mathbf{0}\}$$

as its Lie algebra.

6. Let n be a positive integer. Prove that

$$\mathcal{U}(n) := \{\boldsymbol{L} \in \mathrm{Lin}(\mathbb{C}^n) \mid \boldsymbol{L}^*\boldsymbol{L} = \boldsymbol{I}\},$$
$$\mathcal{SU}(n) := \{\boldsymbol{L} \in \mathcal{U}(n) \mid \det \boldsymbol{L} = 1\}$$

are an n^2-dimensional and an (n^2-1)-dimensional Lie group, respectively. (The star denotes adjoint with respect to the usual complex inner product; in other words, if $\boldsymbol{L}$ is regarded as a matrix then $\mathbf{L}^*$ is the conjugate of the transpose of $\boldsymbol{L}$.) Verify that they have the Lie algebras

$$\{\boldsymbol{A} \in \mathrm{Lin}(\mathbb{C}^n) \mid \boldsymbol{A}^* = -\boldsymbol{A}\},$$
$$\{\boldsymbol{A} \in \mathrm{Lin}(\mathbb{C}^n) \mid \boldsymbol{A}^* = -\boldsymbol{A},\ \mathrm{Tr}\,\boldsymbol{A} = \mathbf{0}\},$$

respectively.

7. Prove that

$$\mathcal{U}(1) := \{\boldsymbol{L} \in \mathrm{Lin}(\mathbb{C}) \mid \boldsymbol{L}^*\boldsymbol{L} = \boldsymbol{I}\} \equiv \{\alpha \in \mathbb{C} \mid |\alpha| = 1\}$$

is a one-dimensional Lie group, locally isomorphic but not isomorphic to $\mathcal{T}n(\mathbb{R})$.

8. Let $\mathcal{G} \subset \mathcal{GA}(\mathrm{V})$ be a Lie group. An *orbit* of $\mathcal{G}$ is a nonvoid subset P of V such that $\{L(x) \mid L \in \mathcal{G}\} = \mathrm{P}$ for some—hence for all— $x \in \mathrm{P}$.

Prove that distinct orbits are disjoint. V is the union of orbits of $\mathcal{G}$. In other words, the relation $\sim$ on V defined by $x \sim y$ if x and y are in the same orbit of $\mathcal{G}$ is an equivalence relation.

9. Find the orbits of $\mathcal{GA}(\mathrm{V})$, $\mathcal{GL}(\mathbf{V})$, $\mathcal{T}n(\mathbf{V})$, $\mathcal{O}(n)$, $\mathcal{SO}(n)$, $\mathcal{U}(n)$, $\mathcal{SU}(n)$.

SUBJECT INDEX

LIST OF SYMBOLS

1. Basic notation

$\blacksquare$	marks the end of a proposition, a proof or a definition, if necessary
$:=$ $=:$	defining equalities; the symbol on the side of the colon is defined to equal the other one
$\emptyset$	the void set
$\mathbb{N}$	the set of non-negative integers
$\mathbb{R}$	the set of real numbers
$\mathbb{R}^+$	the set of positive real numbers
$\mathbb{R}_0^+$	the set of non-negative real numbers
X^n	the n-fold Cartesian product of the set X with itself ($n \in \mathbb{N}$)
$\mathrm{Dom}\, f$	the *domain* of the map f
$\mathrm{Ran}\, f$	the *range* of the map f
$f\colon \mathrm{X} \to \mathrm{Y}$	f is a map with $\mathrm{Dom}\, f = \mathrm{X}$, $\mathrm{Ran}\, f \subset Y$
$f\colon \mathrm{X} \rightarrowtail \mathrm{Y}$	f is a map with $\mathrm{Dom}\, f \subset \mathrm{X}$, $\mathrm{Ran}\, f \subset \mathrm{Y}$
$\mapsto$	the symbol showing a mapping rule
$f\vert_{\mathrm{A}}$	the *restriction* of the map f onto $\mathrm{A} \cap \mathrm{Dom}\, f$
$f \subset g$	the map g is an *extension* of f, i.e. $\mathrm{Dom}\, f \subset \mathrm{Dom}\, g$, $g\vert_{\mathrm{Dom}\, f} = f$
$\overset{-1}{f}$	the *total inverse* of the map $f\colon \mathrm{X} \rightarrowtail \mathrm{Y}$, i.e. if $\mathrm{H} \subset \mathrm{Y}$ then $\overset{-1}{f}(\mathrm{H}) = \{x \in \mathrm{Dom}\, f \mid f(x) \in \mathrm{H}\}$
$g \circ f$	the *composition* of the maps $g\colon \mathrm{Y} \rightarrowtail \mathrm{Z}$ and $f\colon \mathrm{X} \rightarrowtail \mathrm{Y}$, $\mathrm{Dom}\,(g \circ f) := \overset{-1}{f}(\mathrm{Dom}\, g) \cap \mathrm{Dom}\, f$, $x \mapsto g(f(x))$
$\underset{i \in \mathrm{T}}{\times} f_i$	the *Cartesian product* of the maps $f_i\colon \mathrm{X}_i \mapsto \mathrm{Y}_i$ $(i \in \mathrm{T})$: $\left(\underset{i\in\mathrm{T}}{\times} \mathrm{X}_i\right) \to \left(\underset{i\in\mathrm{T}}{\times} \mathrm{Y}_i\right)$, $(x_i)_{i\in\mathrm{T}} \mapsto \big(f_i(x_i)\big)_{i\in\mathrm{T}}$
$\overset{n}{\times} f$	the n-fold Cartesian product of f with itself ($n \in \mathbb{N}$)

$(f_i)_{i\in \mathrm{T}}$	the *joint* of the maps $f_i\colon \mathrm{X}\to \mathrm{Y}_i$, $X \to \underset{i\in \mathrm{T}}{\times} \mathrm{Y}_i,\ x\mapsto \big(f_i(x)\big)_{i\in \mathrm{T}}$
$\mathrm{Ker}\,\boldsymbol{L}$	$:= \overset{-1}{\boldsymbol{L}}\{\mathbf{0}\}$, the *kernel* of the linear map $\boldsymbol{L}$
pr^k	$\mathbb{R}^N \to \mathbb{R}$, the k-th coordinate projection
$\mathbf{1}_{\mathbf{X}}$	the identity map $\mathbf{X}\to\mathbf{X}$, $\boldsymbol{x}\mapsto\boldsymbol{x}$ for a vector space $\mathbf{X}$
1_{X}	the identity map $\mathrm{X}\to\mathrm{X}$, $x\mapsto x$ for an affine space X

2. Other notations

$*$	marks the dual of vector spaces and the transpose of linear maps, IV.1.1, IV.1.4
$\circledast$	marks adjoints of linear maps, V.1.5
$\otimes$	tensor product, IV.3.2
$\wedge$	antisymmetric tensor product, IV.3.14
$\vee$	symmetric tensor product, IV.3.14
$\star$	$q\star t$ is the unique element in the intersection of q and t, I.2.2, II.3.6.2
ar	the arrow of spacetime transformations, I.11.3.1, II.10.1.1
$\mathrm{A}(\boldsymbol{b})$	the set of $\boldsymbol{b}$-antisymmetric maps, V.2.7 in particular, $\mathrm{A}(\boldsymbol{h})$: V.3.8 $\mathrm{A}(\boldsymbol{g})$: V.4.15
$\boldsymbol{h}$	Euclidean form, V.3.1, I.1.2.1
$\boldsymbol{h}_{\boldsymbol{u}}$	Euclidean form on $\mathbf{S}_{\boldsymbol{u}}$, II.1.3.3
$B_{\boldsymbol{u}}$	the set of relative velocities with respect to $\boldsymbol{u}$, II.4.2.5
C_U	$C_U(x)$ is the $\boldsymbol{U}$-line passing through x ($\boldsymbol{U}$-space point that x is incident with), I.3.1.6, II.3.1.3
$\mathbf{S}$	the set of spacelike vectors, I.1.2.1
$\mathbf{S}_{\boldsymbol{u}}$	the set of vectors $\boldsymbol{g}$-orthogonal to $\boldsymbol{u}$, II.1.3.2
S_U	$\boldsymbol{U}$-space, I.3.2.1, II.3.1.3
$\mathbf{S}_U$	the set of space vectors of a rigid observer $\boldsymbol{U}$, I.4.3.4
det	determinant, IV.3.18
$\mathrm{D}F$	derivative of F, VI.3.3
$\boldsymbol{g}$	Lorentz form, V.4.1, II.1.2.1
$\mathcal{G}$	Galilean group, I.11.3.1
ξ_U	splitting according to $\boldsymbol{U}$, I.3.3.3,
$\xi_{\boldsymbol{u},o}$	splitting according to $(\boldsymbol{u},o)$, I.4.1.4, II.3.6.3
$\boldsymbol{\xi}_{\boldsymbol{u}}$	vector splitting, I.8.2.1, II.7.1.1
$\xi_{S,U}$	splitting according to a reference frame, II.3.2.3

$\xi_{u'u}$	the vector transformation rule, II.8.2.4, II.7.1.4
i	embedding of $\mathbf{S}$ into $\mathbf{M}$, I.1.2.1
i_u	embedding of $\mathbf{S}_u$ into $\mathbf{M}$, II.1.3.2
$\mathrm{T}_{\mathcal{S}}$	time according to the simultaneity $\mathcal{S}$, II.3.2.1
$\mathcal{L}$	Lorentz group, II.10.1.1
$\mathbf{La}(\ )$	Lie algebra of a Lie group, VII.3.3, VII.4.2
$\boldsymbol{L}(\boldsymbol{u}', \boldsymbol{u})$	Lorentz boost, II.1.3.8
	special Galilean transformation, I.11.3.7
$\mathcal{N}$	Noether group, I.11.6.1
$\mathbf{N}$	$:= \frac{\mathbb{S}}{\mathbb{L}}$, I.1.3.5
O_o	vectorization with origin o, VI.1.1
$\mathcal{O}(\boldsymbol{h})$	group of orthogonal transformations, I.11.1.1
$\mathcal{O}(\boldsymbol{h}_{\boldsymbol{u}})$	group of orthogonal transformations, II.10.1.4
$\boldsymbol{\pi}_{\boldsymbol{u}}$	projection along $\boldsymbol{u}$, I.1.3.8, II.1.3.2
$\mathcal{P}$	Poincaré group, II.10.3.1
$\boldsymbol{P}_{\boldsymbol{u}}$	$\boldsymbol{u}$-spacelike inversion I.11.3.4, II.10.1.3
$\boldsymbol{R}_U(t, t_o)$	rotation of a rigid observer, I.4.2.2
$\boldsymbol{\eta}_{\boldsymbol{u}}$	covector splitting, I.8.3.1, II.7.2.1
$\boldsymbol{\eta}_{\boldsymbol{u}'\boldsymbol{u}}$	the covector transformation rule, I.8.3.2, II.7.2.2
sign	sign of permutations, IV.3.14
	sign of spacetime transformations, I.11.3.1,II.10.1.1
$\mathcal{SO}(\boldsymbol{h})$	group of rotations, I.11.1.2
τ	time evaluation, I.1.2.2
$\tau_{\mathcal{S}}$	time evaluation of a synchronization, II.3.2.1
$\tau_{\boldsymbol{u}}$	standard $\boldsymbol{u}$-time evaluation, II.3.6.1
$\mathbf{T}_x(\ )$	tangent space at x, VI.4.6
$\mathcal{T}n(\ )$	translation group, VII.3.1
$T_{\boldsymbol{a}}$	translation by $\boldsymbol{a}$, VII.3.1
$\boldsymbol{T}_{\boldsymbol{u}}$	$\boldsymbol{u}$-timelike inversion, I.11.3.4, II.10.1.3
$\boldsymbol{v}_{\boldsymbol{u}'\boldsymbol{u}}$	relative velocity, I.6.2.2, II.4.2.2
$V(1)$	the set of absolute velocities, I.1.3.7, II.1.3.1
$V(0)$	the set of lightlike velocities, II.4.7.1

COMMENTS AND BIBLIOGRAPHY

The fundamental notions of space and time appear in all branches of physics, giving a general background of phenomena. Nowadays the mathematical way of thinking and speaking becomes general in physics; that is why it is indispensable to construct mathematically exact models of spacetime.

Since 1976, an educational and research programme has been in progress at the Department of Applied Analysis, Eötvös Loránd University, Budapest, to build up a mathematical theory of physics in which only mathematically defined notions appear. In this way we can rule out tacit assumptions and the danger of confusions, and physics can be put on a firm basis.

The first results of this work were published in two books:

[1] Matolcsi, T.: *A Concept of Mathematical Physics, Models for Spacetime,* Akadémiai Kiadó, 1984,

[2] Matolcsi, T.: *A Concept of Mathematical Physics, Models in Mechanics,* Akadémiai Kiadó, 1986.

Since that time our teaching experience revealed that a mathematical treatment of spacetime could claim more interest than we had thought it earlier. The notions of the spacetime models throw new light on the whole physics, a number of relations become clearer, simpler and more understandable; e.g. the old problem of material objectivity in continuum physics has been completely solved, as discussed in

[3] Matolcsi, T.: *On material frame-indifference,* Archive for Rational Mechanics and Analysis, **91** (1986), 99–118.

That is why it seems necessary that spacetime models be formulated in a way more familiar to physicists; so they can acquire and apply the notions and results more easily. The present work is an enlarged and more detailed version of [1]. The notations (due to the dot product) became simpler. The amount of applied mathematical tools decreased (by omitting some marginal facts, the theory of smooth manifolds could be eliminated), the material, the explanations and the number of the illustrative examples increased.

There is only one point where the new version contradicts the former one because of the following reason. In the literature one usually distinguishes between the Lorentz group (a group of linear transformations of $\mathbb{R}^4$) and the Poincaré group, called also the 'inhomogeneous Lorentz group' (the Lorentz group together with the translations of $\mathbb{R}^4$). In our terminology, one considers the arithmetic Lorentz group which is a subgroup of the arithmetic Poincaré group. However, we know that in the absolute treatment the Poincaré group consists of transformations of the affine space M, whereas the Lorentz group consists of transformations of the vector space $\mathbf{M}$; the Lorentz group is not a subgroup of the Poincaré group. Special Lorentz transformations play a fundamental role in usual treatments in connection with transformation rules.

The counterpart of the Poincaré group in the nonrelativistic case is usually called the Galilean group and one does not determine its vectorial subgroup that corresponds to the Lorentz group. The special Galilean transformations play a fundamental role in connection with transformation rules. In the absolute treatment we must distinguish between the transformation group of the affine space M and the transformation group of the vector space $\mathbf{M}$ which is not a subgroup of the former group. The special Galilean transformations turn to be transformations of $\mathbf{M}$; that is why I found it convenient to call the corresponding linear transformation group the *Galilean group* and to introduce the name *Noether group* for the group of affine transformations.

In the former version I used these names interchanged because then group representations (applied in mechanical models) were in my mind and it escaped my attention that from the point of view of transformation rules—which have a fundamental importance—the present names are more natural.

The present treatment of spacetime is somewhat different from the usual ones; of course, there are works in which elements of the present models appear. First of all, in

[4] Weyl, H.: *Space–Time–Matter,* Dover publ. 1922

spacetime is stated to be a four-dimensional affine space, the bundle structure of nonrelativistic spacetime (i.e. spacetime, time and time evaluation) and the Euclidean structure on a hyperplane of simultaneous world points appear as well. However, all these are not collected to form a clear mathematical structure; moreover, the advantages of affine spaces are not used, immediately coordinates and indices are taken; thus the possibility of an absolute description is not utilized.

A similar structure ('neoclassical spacetime': spacetime and time elapse) is expounded in

[5] Noll, W.: *Lectures on the foundation of continuum mechanics and thermodynamics,* Arch. Rat. Mech. **52** (1973) 62–92.

In these works time periods and distances are considered to be real numbers. The notion of observer remains undefined; even this undefined notion is used to introduce e.g. differentiability in the "neoclassical spacetime".

When comparing our notions, results and formulae with those of other treatments, using the phrases 'in most of the textbooks', 'in conventional treatments' we refer e.g. to the following books:

[6] French, A.P. *Special Relativity,* Norton, New York, 1968,

[7] Essen, L.: *The Special Theory of Relativity,* Clarendon, Oxford, 1971,

[8] Møller, C.: *The Theory of Relativity,* Oxford University Press, 1972,

[9] Taylor, J.G.: *Special Relativity,* Clarendon, Oxford, 1975,

[10] Bergmann, P.G.: *Introduction to the Theory of Relativity,* Dover publ., New York, 1976.

General relativity, i.e. the theory of gravitation is one of the most beautiful and mathematically well elaborated area of physics which is treated in a number of excellent books, e.g.

[11] Misner, C.W.–Thorne, K.S.–Wheeler, J.A.: *Gravitation,* W.H.Freeman & Co., 1973,

[12] Adler, R.–Bazin, M.–Schiffer, M.: *Introduction to General Relativity,* McGraw-Hill, 1975,

[13] Ohanian, H.C.: *Gravitation and Spacetime,* W.W.Norton & Co., 1976,

[14] Rindler, W.: *Essential Relativity. Special, General and Cosmological,* McGraw-Hill, 1977,

[15] Wald, R.: *Space, Time and Gravity,* Chicago Press, 1977.

To understand the nonrelativistic and special relativistic spacetime models, it is sufficient to have some elementary knowledge in linear algebra and analysis. Tensors and tensorial operations are the main mathematical tools used throughout the present book. Those familiar with tensors will have no difficulty in reading the book. The necessary mathematical tools are summarized in its second part where the reader can find a long and detailed chapter on tensors.

The book uses the basic notions and theorems of linear algebra (linear combination, linear independence, linear subspace etc.) without explanation. There are many excellent books on linear algebra from which the reader can acquire the necessary knowledge, e.g.,

[16] Halmos, P.R.: *Finite Dimensional Vector Spaces,* Springer, 1974,

[17] Smith, L.: *Linear Algebra, Springer,* 1978,

[18] Grittel, D.H.: *Linear Algebra and its Applications,* Harwood, 1989,

[19] Fraleigh, J.B.–Beauregard, R.A.: *Linear Algebra,* Addison–Wesley, 1990,

Some notions and theorems of elementary analysis (limit of functions, continuity, Lagrange's mean value theorem, implicit mapping theorem, etc.) are used without any reference; the following books are recommended to be consulted:

[20] Zamansky, M.: *Linear Algebra and Analysis,* Van Nostrand, 1969,

[21] Rudin, W.: *Principles of Mathematical Analysis,* McGraw Hill, 1976,

[22] Aliprantis, C.D.–Burkinshow, O.: *Principles of Real Analysis,* Arnold, 1981,

[23] Haggarty, R.: *Fundamentals of Mathematical Analysis,* Addison-Wesley, 1989,

[24] Adams, R.A.: *Calculus: a Complete Course,* Addison–Wesley, 1991.

From the theory of differential equations only the well-known existence and uniqueness theorem is used which can be found e.g. in

[25] Hyint-U Tyn: *Ordinary Differential Equations,* North-Holland, 1978,

[26] Birkhoff, G.–Rota, G.C.: *Ordinary Differential Equations,* Wiley, 1989.

The present book avoids the theory of smooth manifolds though it would be useful for the investigation of the space of general observers and necessary for the treatment of general relativistic spacetime models. The following books are recommended to the reader interested in this area:

[27] Boothby, W.M.: *An Introduction to Differentiable Manifolds and Riemannian Geometry,* Academic Press, 1975,

[28] Choquet-Bruhat, Y.–Dewitt-Morette, C.: *Analysis, Manifolds and Physics,* North-Holland, 1982,

[29] Abraham, R.–Marsden, J.E.–Ratiu, T.: *Manifolds, Tensor Analysis, and Applications,* Springer, 1988.

Nonrelativistic and special relativistic spacetime models involve some elementary facts about certain Lie groups. Those who want to get more knowledge on Lie groups can study, e.g. the following books:

[30] Warner, F.W.: *Foundations of Differentiable Manifolds and Lie Groups,* Springer, 1983,

[31] Sattinger, R.H.–Weawer, O.L.: *Lie groups and Lie algebras with Applications to Physics, Geometry and Mechanics,* Springer, 1986.

Since the first edition of this book, the following articles have been based on the present theory of spacetime to solve some problems:

[32] Matolcsi, T.– Gohér, A.: *Spacetime without reference frames and its application to the Thomas rotation,* Publications in Applied Analysis, **5** (1996), 1–11,

[33] Matolcsi, T.–Gruber, T.: *Spacetime without reference frames: An application to the kinetic theory,* International Journal of Theoretical Physics, **35** (1996), 1523–1539,

[34] Matolcsi, T.: *Spacetime without reference frames: An application to synchronizations on a rotating disk,* Foundations of Physics, **28** (1998), 1685–1701,

[35] Matolcsi, T.– Gohér, A.: *Spacetime without reference frames: An application to the velocity addition paradox,* Studies In History and Philosophy of Science Part B: Studies In History and Philosophy of Modern Physics, **32** (2001), 83–99,

[36] Farkas, Sz.–Kurucz, Z.–Weiner, M.: *Poincaré covariance of relativistic quantum position,* International Journal of Theoretical Physics, **41** (2002), 79–88,

[37] Matolcsi, T.–Matolcsi, M.–Tasnádi, T.: *On the relation of Thomas rotation and angular velocity of reference frames,* General Relativity and Gravitation, **39** (2007), 413–426,

[38] Matolcsi, T.–Ván, P.: *On the objectivity of time derivatives,* Atti dell'Accademia Peloritana dei Pericolanti Classe di Scienze Fisiche, Matematiche e Naturali, **86** (2008), Suppl. 1, C1S0801015,

[39] Ván, P.: *Kinetic equilibrium and relativistic thermodynamics,* EPJ Web of Conferences 13 (HCBM 2010 – International Workshop on Hot and Cold Baryonic Matter), 2011, 07004,

[40] Fülöp, T.–Ván, P.: *Kinematic quantities of finite elastic and plastic deformation,* Mathematical Methods in the Applied Sciences, **35** (2012), 1825–1841,

[41] Fülöp, T.–Ván, P.–Csatár, A.: *Elasticity, plasticity, rheology and thermal stress – an irreversible thermodynamical theory,* in: Pilotelli, M.–Beretta, G.P. (eds.): Proceedings of the 12th Joint European Thermodynamics Conference, JETC 2013, Cartolibreria Snoopy, 2013, 525–530,

[42] Fülöp, T.: *Objective thermomechanics,* arXiv:1510.08038 (2015),

[43] Asszonyi, Cs.–Csatár, A.–Fülöp, T.: *Elastic, thermal expansion, plastic and rheological processes – theory and experiment,* Periodica Polytechnica–Civil Engineering, **60** (2016), 591–601,

[44] Ván, P.: *Galilean relativistic fluid mechanics,* Continuum Mechanics and Thermodynamics, **29** (2017), 585–610,

[45] Ván, P.–Pavelka, M.–Grmela, M.: *Extra mass flux in fluid mechanics,* Journal of Non-Equilibrium Thermodynamics, **42** (2017), 133–151.

www.ingramcontent.com/pod-product-compliance
Lightning Source LLC
Chambersburg PA
CBHW020827160426
43192CB00007B/555
* 9 7 8 1 9 2 7 7 6 3 9 4 0 *